Techniques for the Seismic Rehabilitation of Existing Buildings

FEMA 547 – October 2006

Prepared by:
Rutherford & Chekene (R & C) Consulting Engineers (Subconsultant) under contract with National Institute of Standards and Technology (NIST).

Project funding was provided by the Federal Emergency Management Agency (FEMA) through an Interagency Agreement - EMW-2002-IA-0098 with the National Institute of Standards and Technology (NIST).

Additional funding was provided by the following agencies:

    General Services Administration (GSA)
    Air Force Civil Engineer Support Agency (AFCESA)
    Naval Facilities Engineering Command (NAVFAC)
    U.S. Bureau of Reclamation (USBR)

Cover Photo Courtesy of Kelly Peterson, University of Utah

Cover Photo: Unbonded Steel Braces are used as part of the seismic rehabilitation of the J. Willard Marriott Library on the University of Utah campus. FEMA provided partial funding of this project through the Pre-Disaster Mitigation Competitive (PDM-C) grant program.

This page is intentionally left blank.

Notice

Any opinions, findings, conclusions, or recommendations expressed in this publication do not necessarily reflect the views of the Federal Emergency Management Agency, the General Services Administration (GSA), the Air Force Civil Engineer Support Agency (AFCESA), the National Institute of Standards and Technology (NIST), the Naval Facilities Engineering Command (NAVFAC), and Rutherford & Chekene Consulting Engineers (R&C), and R&C's subconsultants. Additionally, neither FEMA, R&C nor its subconsultants, AFCESA, FEMA, GSA, NIST, NAVFAC, USBR, or other ICSSC member agencies, nor any of their employees, makes any warranty, expressed or implied, nor assumes any legal liability or responsibility for the accuracy, completeness, or usefulness of any information, product, or process included in this publication. Users of information from this publication assume all liability arising from such use.

Preface

This seismic rehabilitation techniques document is part of the National Earthquake Hazards Reduction Program (NEHRP) family of publications addressing seismic rehabilitation of existing buildings. It describes common seismic rehabilitation techniques used for buildings represented in the set of standard building types in FEMA seismic publications. This document supersedes *FEMA 172: NEHRP Handbook for Seismic Rehabilitation of Existing Buildings*, which was published in 1992 by the Federal Emergency Management Agency (FEMA). Since then, many rehabilitation techniques have been developed and used for repair and rehabilitation of earthquake damaged and seismically deficient buildings. Extensive research work has also been carried out in support of new rehabilitation techniques in the United States, Japan, New Zealand, and other countries. Available information on rehabilitation techniques and relevant research results for commonly used rehabilitation techniques are incorporated in this document.

The primary purpose of this document is to provide a selected compilation of seismic rehabilitation techniques that are practical and effective. The descriptions of techniques include detailing and constructability tips that might not be otherwise available to engineering offices or individual structural engineers who have limited experience in seismic rehabilitation of existing buildings. A secondary purpose is to provide guidance on which techniques are commonly used to mitigate specific seismic deficiencies in various model building types.

FEMA sincerely thanks all of the federal agencies that contributed funds toward completing this report as well as the members of the Interagency Committee for Seismic Safety in Construction (ICSSC) Subcommittee 1, the Technical Update Team, and all of the federal and private sector partners for their efforts in development, review and completion of this publication.

Acknowledgments

Private sector consultants working with the Subcommittee 1, *Standards for New and Existing Buildings*, of the Interagency Committee on Seismic Safety in Construction (ICSSC) produced this document. The following persons contributed to this document.

Contractors

Technical Update Team

>William T. Holmes, Team Leader, Rutherford & Chekene, San Francisco
>Bret Lizundia, Project Manager and Editor, Rutherford & Chekene, San Francisco
>
>James O. Malley, Degenkolb Engineers, San Francisco
>Kelly Cobeen, Cobeen & Associates Structural Engineering, Inc., Lafayette, California

Contractor Staff

>Rutherford & Chekene: Afshar Jalalian, Gyimah Kasali, Mark Moore, Rich Niewiarowski, and Karl Telleen
>Degenkolb Engineers: Jack Hsueh

Consultant on Foundation Rehabilitation Techniques

>Craig Comartin, CDComartin, Inc., Stockton, California

Other

>Figures shown in Table 4-1 and repeated later in the text were adapted from those originally rendered by Anthony Alexander of Palo Alto, California for a separate FEMA-funded project.

Project Review Panel

J. Daniel Dolan, Washington State University
Terry Dooley, Consultant
Kurt Gustafson, American Institute of Steel Construction
Robert D. Hanson, FEMA Consultant
Neil M. Hawkins, Consultant
James R. Harris, J. R. Harris and Company
Bela Palfalvi, General Services Administration (representing ICSSC)
Daniel Shapiro, SOH and Associates

FEMA Project Officer
Cathleen Carlisle, DHS-FEMA, Washington, DC

ICSSC Subcommittee 1

H. S. Lew, Chair	National Institute of Standards and Technology
John Baals	Department of Interior, U.S. Bureau of Reclamation
Krishna Banga	Department of Veteran Affairs
Rosana Barkawi	Department of Agriculture, U.S. Forest Service
James Binkley	U.S. Postal Service
Larry Black	Department of Army
Cathleen Carlisle	Federal Emergency Management Agency
James A. Caulder	Air Force Civil Engineering Support Agency
Cathy Chan	Department of Justice, Bureau of Prisons
Harish Chander	Department of Energy
Anjana Chudgar	U.S. Army Corps of Engineers
Ronald Crawford	Department of Housing and Urban Development
Joseph Corliss	Department of Health and Human Services
Nathaniel Foster	Tennessee Valley Authority
Freda Gerard	International Broadcasting Bureau
Colleen Geraghty	Department of Labor, Office of Safety and Health
Asadour Hadjian	Defense Nuclear Facility Safety Board
Owen Hewitt	Naval Facilities Engineering Command
Howard Kass	National Aeronautics and Space Administration
Regina Larrabee	Department of Commerce
Catherine Lee	General Services Administration
S. C. Liu	National Science Foundation
Rita Martin	Department of Transportation
Thomas Myers	Smithsonian Institution
Thomas Nelson	Lawrence Livermore National Laboratory
Dai Oh	Department of State, Office of Foreign Buildings
Bela Palfalvi	General Services Administration
Erdal Safak	Department of Interior, U.S. Geological Survey
Subir Sen	Department of Energy
C.A. Stillions	Architect of the Capital
Lawrence Swanhorst	Environmental Protection Agency
Steven Sweeney	Army Construction & Engineering Research Laboratory
Doris Turner	Federal Aviation Administration
Terry Wong	National Park Services

This page is intentionally left blank.

Table of Contents

PART 3 - REHABILITATION TECHNIQUES FOR DEFICIENCIES COMMON TO MULTIPLE BUILDING TYPES

Chapter 22 - Diaphragm Rehabilitation Techniques

List of Figures

List of Tables

Chapter 1 - Introduction

1.1   Overview

A considerable number of buildings in the existing building stock of the United States present a risk of poor performance in earthquakes because there was no seismic design code available or required when they were constructed, because the seismic design code used was immature and had flaws, or because original construction quality or environmental deterioration has compromised the original design.

The practice of improving the seismic performance of existing buildings—known variously as seismic rehabilitation, seismic retrofitting, or seismic strengthening—began in the U.S. in California in the 1940s following the Garrison Act in 1939. This Act required seismic evaluations for pre-1933 school buildings. Substandard buildings were required to be retrofit or abandoned by 1975. Many school buildings were improved by strengthening, particularly in the late 1960s and early 1970s as the deadline approached. Local efforts to mitigate the risks from unreinforced masonry buildings (URMs) also began in this time period. In 1984, the Federal Emergency Management Agency (FEMA) began its program to encourage the reduction of seismic hazards posed by existing older buildings throughout the country. This program has included development of many resources to assist engineers and other stakeholders to reduce this risk; guidance on evaluation, costs and priorities; and ultimately, a comprehensive, performance-based, rehabilitation design guideline, FEMA 273, *NEHRP Guidelines for the Seismic Rehabilitation of Buildings* (FEMA, 1997a)—which was converted to FEMA 356 (FEMA, 2000a) as an American Society of Civil Engineers (ASCE) prestandard. At this writing, ASCE is developing a standard entitled ASCE 41, *Seismic Rehabilitation of Existing Buildings*, using FEMA 356 as a basis.

Recognizing that building rehabilitation design is far more constrained than new building design and that special techniques are needed to insert new lateral elements, tie them to the existing structure, and generally develop complete seismic load paths, a document was published for this purpose in 1992. FEMA 172, *NEHRP Handbook of Techniques for the Seismic Rehabilitation of Existing Buildings* (FEMA, 1992b), was intended to identify and describe generally accepted rehabilitation techniques. The art and science of seismic rehabilitation has grown tremendously since that time with federal, state, and local government programs to upgrade public buildings, with local ordinances that mandate rehabilitation of certain building types, and with a growing concern among private owners about the seismic performance of their buildings. In addition, following the demand for better understanding of performance of older buildings and the need for more efficient and less disruptive methods to upgrade, laboratory research on the subject has exploded worldwide, particularly since the nonlinear methods proposed for FEMA 273 became developed.

The large volume of rehabilitation work and research now completed has resulted in considerable refinement of early techniques and development of many new techniques, some confined to the research lab and some widely used in industry. Like FEMA 172, this document describes the techniques currently judged to be most commonly used or potentially to be most useful. Furthermore, it has been formatted to take advantage of the ongoing use of typical building types in FEMA documents concerning existing buildings, and to facilitate the addition of techniques in the future.

1.2   Purpose and Goals

The primary purpose of this document is to provide a selected compilation of seismic rehabilitation techniques that are practical and effective. The descriptions of techniques include detailing and constructability tips that might not be otherwise available to engineering offices or individual structural engineers who have limited experience in seismic rehabilitation of existing buildings. A secondary purpose is to provide guidance on which techniques are commonly used to mitigate specific seismic deficiencies in various model building types.

The goals of the document are to:

> Describe rehabilitation techniques commonly used for various model building types
> Incorporate relevant research results
> Discuss associated details and construction issues
> Provide suggestions to engineers on the use of new products and techniques

1.3   Audience

This document was written primarily for engineers who are inexperienced in seismic rehabilitation, or who provide these services infrequently. Secondarily, the material will be useful for architects and project managers coordinating rehabilitation projects or programs to better appreciate the potential scope and construction needs of such work.

1.4   Scope

This document is intended to describe the most common seismic rehabilitation techniques used for each type of building represented in the set of standard building types often used in FEMA seismic publications (see Chapter 4). The basics of seismic building engineering are not included herein nor are methods and procedures to seismically evaluate buildings.

It is presumed that the user has a completed seismic evaluation of the building-of-interest, has concluded that some level of retrofit is appropriate, and has identified the seismic deficiencies to be corrected to achieve the desired performance objective.

In this document, *technique* is used to describe a local action consisting of insertion of a new lateral force-resisting component or enhancement of the seismic resistance of an in-situ component in an existing building. A complete seismic rehabilitation *scheme* may consist of the use of several techniques. Detailed guidance on the strategies to develop such overall schemes is not included in this document, although a general discussion of the topic is given in Chapter 3. The overall organization of the document is intended to lead the user toward selection of realistic, practical, and cost-effective techniques to mitigate a given deficiency.

The building types making up the FEMA set are described in Chapter 4. The building descriptions, performance characteristics, and potential mitigation techniques included are aimed at a broad, but not all-inclusive, range of buildings that fit into each category. The information may not apply to all buildings in the category, particularly those with configuration characteristics such as unusual story height or number of stories, or extreme irregularities. There are also buildings that do not fit neatly into one of the standard building types, but are combinations of standard types. Useful guidance can be obtained for such buildings by reviewing the recommendations for each type that is partially represented in Part 2.

Certain important rehabilitation techniques, such as seismic isolation or the addition of damping devices, are complex, far reaching, and on a different scale than the common techniques included here for each building type. Although these techniques are described briefly in Chapter 24, they are not described in the same level of detail as more standard techniques. Users are encouraged to consider such techniques and seek more complete guidance from text books, conference and seminar proceedings, or from specialty consultants.

A large number of research projects have been completed or are ongoing to develop new products or techniques for seismic rehabilitation in the United States and around the world. This document has included the most commonly used techniques at the time of this writing. For the rehabilitation of any specific building, products or techniques not included herein may be the most appropriate and economical.

Guidance for selection of the most appropriate technique or combination of techniques is covered in general in Chapter 3. Overlapping and sometimes conflicting characteristics of each rehabilitation project—such as performance objectives, cost, disruption to occupants, and aesthetics—most often control development of the structural rehabilitation scheme and cannot be differentiated by building type in the context of this document.

Seismic rehabilitation of nonstructural components is not included in this document. This broad category would include space-enclosing elements such as cladding, partitions, and ceilings; building service systems such as mechanical, electrical, and plumbing elements; and contents such as medical or laboratory equipment, storage shelves or racks, and furniture.

1.5   Other Resources

Technical design standards and analysis techniques can be obtained in documents such as:

> *Standard for the Seismic Evaluation of Buildings,* ASCE 31-03 (ASCE, 2003)
> *Prestandard and Commentary for the Seismic Rehabilitation of Buildings*, FEMA 356 (FEMA, 2000a)
> *NEHRP Commentary on the Guidelines for Seismic Rehabilitation of Buildings*, FEMA 274 (FEMA, 1997b)
> *Seismic Evaluation and Retrofit of Concrete Buildings,* ATC 40 (ATC, 1996)
> *Recommended Seismic Evaluation and Upgrade Criteria for Existing Welded Steel Moment-Frame Buildings*, FEMA 351 (FEMA, 2000b)
> *Improvement of Nonlinear Static Seismic Procedures*, FEMA 440 (FEMA, 2005)
> *Evaluation of Earthquake Damaged Concrete and Masonry Wall Buildings: Basic Procedures Manual,* FEMA 306, (FEMA, 1999)
> *International Existing Building Code,* 2003 Edition (ICC, 2003)
> *Uniform Code for Building Conservation,* 1997 Edition (ICBO, 1997)

The benefits to building owners of performance-based design, methods of managing seismic risk, and cost-benefit of seismic rehabilitation are discussed in:

>   *Primer for Design Professionals—Communicating with Owners and Managers of New Buildings on Earthquake Risk*, FEMA 389 (FEMA, 2004)
>   *Planning for Seismic Rehabilitation: Societal Issues*, FEMA 275 (FEMA, 1997c)
>   *Financial Management of Earthquake Risk*, (EERI, 2000)
>   *Typical Costs for Seismic Rehabilitation of Existing Buildings,* (FEMA, 1994 and 1995)

A series on incremental seismic strengthening of selected occupancy types includes the following documents:

>   *Incremental Seismic Rehabilitation of School Buildings (K-12)*, FEMA 395 (FEMA 2003a)
>   *Incremental Seismic Rehabilitation of Hospital Buildings*, FEMA 396 (FEMA, 2003b)
>   *Incremental Seismic Rehabilitation of Office Buildings*, FEMA 397 (FEMA, 2003c)
>   *Incremental Seismic Rehabilitation of of Multifamily Apartment Buildings*, FEMA 398 (FEMA, 2004a)
>   *Incremental Seismic Rehabilitation of Retail Buildings*, FEMA 399 (FEMA 2004b)

Many of these publications can be found on the FEMA-National Earthquake Hazard Reduction Program (NEHRP) website: http://www.fema.gov/plan/prevent/earthquake/nehrp.shtm.

1.6   Organization of the Document

As shown in Figure 1.6-1, the document is divided into three parts:

>   Part 1 (Chapters 1-3) provides background on seismic evaluation, categories of seismic deficiencies, classes of rehabilitation techniques, and general strategies to develop rehabilitation schemes.
>   Part 2 (Chapters 4-21) contains detailed descriptions of seismic deficiencies that are characteristic of each FEMA model building type and techniques commonly used to mitigate them.
>   Part 3 (Chapters 22-24) contains chapters on seismic rehabilitation techniques common to multiple building types such as those related to diaphragms and foundations. A chapter is also included in Part 3 describing significant global techniques that could be applied to any building, such as seismic isolation or the addition of damping.

An important aspect of the organization is to provide for flexible expansion of the material with future stand-alone printed documents, digital media, or with complete republication. Examples of such expansions include a chapter on nonstructural risk mitigation and descriptions of additional techniques not included in this edition or developed from future research results.

Part 1: Overview

1. Introduction
2. Seismic Vulnerability
3. Seismic Rehabilitation

Part 2: Rehabilitation Techniques Associated with Individual FEMA Model Building Types

4. FEMA Model Building Types
5. W1
6. W1A
7. W2
8. S1/S1A
9. S2/S2A
10. S4
11. S5/S5A
12. C1
13. C2b
14. C2f
15. C3/C3A
16. PC1
17. PC2
18. RM1t
19. RM1u
20. RM2
21. URM

Chapter Organization

12.1 Description of the Model Building Type
12.2 Seismic Response Characteristics
12.3 Common Seismic Deficiencies and Applicable Rehabilitation Techniques
12.4 Detailed Description of Techniques
    12.4.1 Add Steel Braced Frame
    12.4.2 Add Concrete or Masonry Shear Wall
    12.4.3 Provide a Collector
    12.4.4 FRP Overlay of a Concrete Column
    12.4.5 Concrete/Steel Overlay of Column
    12.4.6 Enhance Concrete Moment Frame
12.5 References

Part 3: Rehabilitation Techniques Common to Multiple Model Building Types

22. Diaphragm Rehabilitation Techniques
23. Foundation Rehabilitation Techniques
24. Reducing Seismic Demand

Figure 1.6-1: Organization of Chapters and Parts

1.6.1 Part 1 – Overview

Chapter 2 gives a brief overview of evaluation methods and how seismic deficiencies, in general, can be placed into categories. A set of categories of seismic deficiencies is defined, both because such categories are useful to describe appropriate retrofit measures and also because the categories are useful as an organization of the chapters covering building types.

Chapter 3 briefly summarizes various codes, standards, and guidelines that are normally used to define design procedures for seismic rehabilitation. These documents provide the numerical parameters for design but seldom describe the techniques for strengthening existing components or for adding new lateral force-resisting elements to an existing building.

To relate various seismic rehabilitation techniques to seismic deficiencies, classes of techniques are established and described. Similar to the categories of seismic deficiencies defined in Chapter 2, these classes of techniques provide a consistent organization for the chapters covering building types.

Finally, Chapter 3 includes a description of socio-economic characteristics that are common to most seismic rehabilitation projects and often control the selection of the rehabilitation scheme.

1.6.2 Part 2 – Rehabilitation Techniques for FEMA Model Buildings

This document is primarily organized around the FEMA model building types, first categorized in ATC 14 (ATC, 1987) in the late 1980s and then carried forward into FEMA 178 (FEMA, 1992a) and almost all succeeding FEMA publications on existing buildings. It is expected that most users of this document will be interested in finding information on a particular building or building type, which suggested this organization. Each building type is therefore assigned a chapter. Common seismic deficiencies for each building type are identified and mitigation techniques suggested, although it is recognized that most buildings will have multiple deficiencies and may require a combination of mitigating actions. The rehabilitation techniques commonly used for each building type are identified in each chapter and, if closely associated with the building type, described in detail in that chapter. References are given to other chapters for other applicable techniques.

To direct the user to appropriate chapters, the model buildings are briefly described in Chapter 4 at the beginning of Part 2.

1.6.3 Part 3 – Rehabilitation Techniques for Deficiencies Common to Multiple Building Types

Although certain diaphragm and foundation deficiencies will be found more often in one building type than another, the issues and mitigation techniques are cross-cutting and therefore grouped together in Part 3 in Chapters 22 and 23.

Two important rehabilitation techniques, seismic isolation and added damping, can be applied to any building type, are global in nature, and cannot be described as a local technique in the context of Part 2. These techniques are therefore described independently in Chapter 24.

1.7   Disclaimers

The seismic rehabilitation techniques and details in this document are intended to provide guidance to qualified design professionals. Development of schemes that employ one or more techniques in this document is the sole responsibility of the engineer of record for the project. The details are not to be used in an actual rehabilitation project without review for technical and geometric applicability. In all cases, the details must be completed with additional project specific information.

Some techniques included in the document have been developed using laboratory research. Conclusions from selected research and resulting product characteristics have been included in the document as a starting point for the design engineer. The adequacy of research methods and conclusions has not been verified as part of the development of this document. The search for applicable research and evaluation of results was not exhaustive, particularly for research outside the United States. Inclusion of research or products does not represent endorsement, and exclusion does not necessarily represent lack of confidence.

1.8   References

ASCE, 2003, *Standard for the Seismic Evaluation of Buildings*, ASCE 31-03, Structural Engineering Institute of the American Society of Structural Engineers, Reston, VA.

ATC, 1987, *Evaluating the Seismic Resistance of Existing Buildings*, ATC-14, Applied Technology Council, Redwood City, CA.

ATC, 1996, *The Seismic Evaluation and Retrofit of Concrete Buildings*, ATC-40, Applied Technology Council, Redwood City, CA.

EERI, 2000, *Financial Management of Earthquake Risk*, Earthquake Engineering Research Institute, Oakland, CA.

FEMA, 1992a, *NEHRP Handbook for the Seismic Evaluation of Existing Buildings*, FEMA 178, Federal Emergency Management Agency, Washington, D.C.

FEMA, 1992b, *NEHRP Handbook for the Seismic Rehabilitation of Existing Buildings*, FEMA 172, Federal Emergency Management Agency, Washington, D.C.

FEMA, 1994, *Typical Costs for the Seismic Rehabilitation of Existing Buildings, Volume 1, Summary*, FEMA 156, Second Edition, Federal Emergency Management Agency, Washington, D.C.

FEMA, 1995, *Typical Costs for the Seismic Rehabilitation of Existing Buildings, Volume 2, Supporting Documentation*, FEMA 157, Second Edition, Federal Emergency Management Agency, Washington, D.C.

FEMA, 1997a, *NEHRP Guidelines for the Seismic Rehabilitation of Buildings*, FEMA 273. Federal Emergency Management Agency, Washington, D.C.

FEMA, 1997b, *NEHRP Commentary on the Guidelines for Seismic Rehabilitation of Buildings*, FEMA 274, Federal Emergency Management Agency, Washington, D.C.

FEMA, 1997c, *Planning for Seismic Rehabilitation: Societal Issues*, FEMA 275, Federal Emergency Management Agency, Washington, D.C.

FEMA, 1999, *Evaluation of Earthquake Damaged Concrete and Masonry Wall Buildings: Basic Procedures Manual*, FEMA 306, Federal Emergency Management Agency, Washington, D.C.

FEMA, 2000a, *Prestandard and Commentary for the Seismic Rehabilitation of Buildings*, FEMA 356, Federal Emergency Management Agency, Washington, D.C.

FEMA, 2000b, *Recommended Seismic Evaluation and Upgrade Criteria for Existing Welded Steel Moment-Frame Buildings*, Federal Emergency Management Agency, Washington, D.C.

FEMA 2003a, *Incremental Seismic Rehabilitation of School Buildings (K-12)*, FEMA 395, Federal Emergency Management Agency, Washington, D.C.

FEMA 2003b, *Incremental Seismic Rehabilitation of Hospital Buildings*, FEMA 396, Federal Emergency Management Agency, Washington, D.C.

FEMA 2003c, *Incremental Seismic Rehabilitation of Office Buildings*, FEMA 397, Federal Emergency Management Agency, Washington, D.C.

FEMA 2004a, *Incremental Seismic Rehabilitation of of Multifamily Apartment Buildings*, FEMA 398, Federal Emergency Management Agency, Washington, D.C.

FEMA 2004b, *Incremental Seismic Rehabilitation of Retail Buildings*, FEMA 399, Federal Emergency Management Agency, Washington, D.C.

FEMA, 2004, *Primer for Design Professionals—Communicating with Owners and Managers of New Buildings on Earthquake Risk*, FEMA 389, Federal Emergency Management Agency, Washington, D.C.

FEMA, 2005, *Improvement of Nonlinear Static Seismic Procedures*, FEMA 440, Federal Emergency Management Agency, Washington, D.C.

ICBO, 1997, *Uniform Code for Building Conservation*, 1997 Edition, International Conference of Building Officials, Whittier, California.

ICC, 2003, *International Existing Building Code*, 2003 Edition, International Conference of Building Officials, Country Club Hills, IL.

Chapter 2 - Seismic Vulnerability

2.1   Introduction

In this document, a seismic deficiency is defined as a condition that will prevent a building from meeting the designated seismic performance objective. The performance objective for a building may be established by the choice of a prescriptive evaluation standard, or when using performance-based standards or guidelines, may be selected from a range of defined performance levels. A building evaluated against standards intended to minimize damage and to allow occupancy soon after the event may have significantly more deficiencies than the same building evaluated only to prevent collapse. Typically, techniques useful to mitigate a particular type of deficiency remain the same regardless of the performance objective, but the extent of the mitigating measure required may differ.

The seismic protection systems for nonstructural components in a building have a profound effect on building seismic performance, particularly for higher performance levels and particularly in the weeks immediately following an event. However, the techniques for seismic retrofit of nonstructural components are relatively straightforward, and this document is devoted to structural issues.

The most important issue when beginning to evaluate the seismic capabilities of an existing building is the availability and reliability of structural drawings. Detailed evaluation is impossible without framing and foundation plans, layouts of primary lateral force elements, reinforcing for concrete structures, and connection detailing for steel structures. Developing as-builts from field information is extremely difficult, particularly for reinforced concrete, reinforced masonry, or structural steel buildings. In most cases, such structures must be seismically rehabilitated by placement of a new lateral force-resisting system, with enough physical testing performed to determine overall deformation capacity of the existing structure. This chapter and this entire document assume that sufficient information is available to perform a seismic evaluation that will identify all significant deficiencies.

There are many different procedures and standards for seismic evaluation available to engineers, ranging from highly prescriptive sets of rules developed for a single building type to determination of probable performance considering nonlinear cyclic response to earthquake time histories. These methods are not delineated or described in detail here, nor are the basic principles of building seismic design. Instead, it is assumed that the user has already appropriately completed a seismic evaluation of some sort and has thus identified seismic deficiencies targeted for mitigation.

This chapter describes the evaluation process in general terms and introduces categories of seismic deficiencies used throughout the document.

2.2   Seismic Evaluation

Seismic evaluation of older buildings may be commissioned as part of a municipal, regional, state, or federal risk reduction program that includes mandatory evaluation and rehabilitation of certain buildings. In these cases, the buildings may be identified by type of structural framing system, by age, by location, or by a combination of these risk factors. Seismic evaluations may also be required 1) by local building officials when alterations are made to a building such as a change in occupancy, addition, or revision to the structural system; or 2) as part of an owner's voluntary seismic risk analysis. Lastly, building owners simply may be concerned about their economic investment or about post-earthquake use of the buildings. Evaluations that are mandated by the governing jurisdiction normally specify a minimum standard to be met. Evaluations performed voluntarily by owners are often performance-based—the seismic performance of the building is estimated by the engineer, rather than the building characteristics being compared to a set of prescriptive rules.

Some types of evaluation techniques are briefly described below.

2.2.1   Comparison with Requirements for New Buildings

Until FEMA began an initiative to reduce the seismic risk from existing buildings in the mid-1980s, there were very few standards or guidelines applicable to existing buildings. California engineers had developed rules for evaluation and retrofit of unreinforced masonry bearing walls buildings, but there was little else. Therefore, seismic adequacy was often determined by comparing the older building to the requirements for new buildings. This comparison is often difficult or impossible because the older building may include structural materials or systems prohibited in the code for new buildings, and it is often impractical to completely remove materials or change structural systems. Commonly, a completely new complying seismic system was introduced, often at great disruption and cost. Some jurisdictions still use this standard, particularly in cases of complete building renovation, but some form of performance-based equivalent is preferred.

2.2.2   Prescriptive Standards

The most notable document available for seismic evaluation of existing buildings is *ASCE 31-03: Seismic Evaluation of Existing Buildings* (ASCE, 2003), originally developed by FEMA as *FEMA 310: Handbook for the Seismic Evaluation of Existing Buildings – A Prestandard* (FEMA, 1998). FEMA 310 was converted to ASCE 31 as part of the Amercian Society of Civil Engineers standardization process. ASCE 31-03 is intended for use on older building and recognizes that older and out-moded structural systems may be incorporated in these buildings. The seismic life safety provided by a building is judged adequate if the requirements are met and many jurisdictions accept this level of performance for their community.

The federal government has also developed *Standards of Seismic Safety for Existing Federally Owned and Leased Buildings* (NIST, 2002) that includes policy in addition to evaluation standards.

Other prescriptive standards have also been developed, primarily for specific building types, such as unreinforced masonry bearing walls, timber residential construction, and tilt-up concrete buildings (ICBO, 1997; ICC, 2003). Local jurisdictions also may have a particular interest in a narrowly described building type within their region that is common and/or hazardous and may develop an appropriate minimum standard.

2.2.3   Performance-Based Evaluation Using Expected Nonlinear Response

The most sophisticated and complex seismic evaluation is performed using analytical techniques that explicitly consider the expected nonlinear response of the structure in strong shaking. Such analysis can be performed for selected past ground motions or using slightly simplified techniques such a pushover analysis, as described in ATC 40 (ATC, 1996), FEMA 356 (FEMA, 2000) , and FEMA 440 (FEMA, 2005). The results of such an analysis must be compared to responses associated with certain performance levels such as Immediate Occupancy, Life Safety, or Collapse Prevention. In order to use these techniques for evaluation, the governing jurisdiction or the owner must select the minimum acceptable performance for the building.

2.3   Categories of Seismic Deficiencies

Regardless of the evaluation method used, failure to meet the stipulated criteria will identify certain seismic deficiencies. It is convenient for the purposes of discussion and for developing strategies for seismic rehabilitation to place these deficiencies into *categories*. It is recognized that many building characteristics identified as a deficiency by a seismic evaluation could be identified in more than one category. For example, a shear wall structure with inadequate length of walls will probably have a deficiency in both global strength and stiffness. Similarly, a one-story tilt-up building with an inadequate diaphragm could be listed with inadequate global strength, inadequate global stiffness, or a diaphragm deficiency. Fortunately, these distinctions are not of great importance, because the options of mitigation techniques for a given deficient building characteristic are generally the same regardless of the category in which it is placed. As indicated above, the categories of seismic deficiencies, coupled with somewhat parallel classes of rehabilitation techniques described in Chapter 3, are incorporated to provide a convenient organizational format for Part 2.

The categories of deficiencies used in this document are described below. In Part 2, the categories of deficiencies present in an individual building will lead a user to consider certain techniques for rehabilitation. Therefore, efficient use of this document is dependent on the user understanding the nature of the seismic deficiencies of the building targeted for rehabilitation. More building-specific seismic deficiencies that may be characteristic of each building type are described in each chapter of Part 2.

2.3.1   Global Strength

A deficiency in global strength is common in older buildings either due to a complete lack of seismic design or a design to an early code with inadequate strength requirements. However, it is seldom the only deficiency and the results of the evaluation must be studied to identify deficiencies that may not be mitigated solely by adding strength.

Global strength typically refers to the lateral strength of the vertically oriented lateral force-resisting system at the effective global yield point, (as defined in documents that use simplified nonlinear static procedures based on "pushover" curves), but these concepts will not be described in detail here. Refer to FEMA 356 (FEMA, 2000) for details. For degrading structural systems characterized by a negative post-yield slope on the pushover curve, a minimum strength requirement may also apply as indicated in FEMA 440 (FEMA, 2005). In certain cases, the strength will also affect the total expected inelastic displacement and added strength may reduce nonlinear demands into acceptable ranges.

If prescriptive equivalent lateral force methods or linear static procedures have been used for evaluation or preliminary rehabilitation analysis, inadequate strength will directly relate to unacceptable demand-to-capacity ratios within elements of the lateral force-resisting system.

2.3.2   Global Stiffness

Although strength and stiffness are often controlled by the same existing elements or the same retrofit techniques, the two deficiencies are typically considered separately. Failure to meet evaluation standards is often the result of a building placing excessive drift demands on existing poorly detailed components.

Global stiffness refers to the stiffness of the entire lateral force-resisting system although the lack of stiffness may not be critical at all levels. For example, in buildings with narrow walls, critical drift levels occur in the upper floors. Conversely, critical drifts most often occur in the lowest levels in frame buildings. Stiffness must be added in such a way that drifts are efficiently reduced in the critical levels.

Given an adequate minimum strength level, global nonlinear displacements and thus demands on most components in the building are more effectively reduced by increased initial stiffness than by increased global strength.

2.3.3   Configuration

This deficiency category covers configuration irregularities that adversely affect performance. In codes for new buildings, these configuration features are often divided into plan irregularities and vertical irregularities. Plan irregularities are features that may place extraordinary demands on elements due to torsional response or the shape of the diaphragm. Vertical irregularities are created by uneven vertical distribution of mass or stiffness between floors that may result in concentration of force or displacement at certain levels. In older existing buildings, such irregularities were seldom taken into consideration in the original design and therefore normally require retrofit measures to mitigate.

In prescriptive evaluation methods, features that qualify as irregularities are defined by rules, similar to the rules used for new buildings. Evaluation methods that explicitly consider nonlinear behavior will normally identify concentrations of force or displacement due to configuration and the components affected by these concentrations will be shown to have inadequate capacity.

2.3.4   Load Path

Although all of the deficiencies described have significant effects on seismic performance, a break in the load path, or inadequate strength in the load path, may be considered overarching because this deficiency will prevent the positive attributes of the seismic system from being effective. The load path is typically considered to extend from each mass in the building to the supporting soil. For example, for a panel of cladding, this path would include its connection to the supporting floor or floors, the diaphragm and collectors that deliver the load to components of the primary lateral force-resisting system (walls, braces, frames, etc.), continuity of these components to the foundation, and finally the transfer of loads between foundation and soil. If the connection of the cladding panel or exterior wall fails and the element falls away from the building, the adequacy of the balance of the load path is moot. Similarly, if a new shear wall element is added to the exterior of a building as a retrofit measure, its strength and stiffness will have no effect if it is not connected adequately to the floor diaphragms.

Many load path deficiencies are difficult to categorize because the strength deficiency may be considered to be part of another element. For example, an inadequate construction joint in a shear wall could be considered a load path deficiency or a shear wall deficiency in the category of global strength. As previously mentioned, the categorization does not make too much difference as long as the deficiency is recognized and mitigated. In this document, local connections of panels and walls to the diaphragm, and collectors or other connections to the lateral force-resisting elements are considered load path issues. Inadequacies within a lateral element such as a shear wall, braced frame, or moment frame are generally associated with the element and not considered a load path issue. Inadequacies at the foundation level are generally considered foundation deficiencies.

2.3.5   Component Detailing

Detailing, in this context, refers to design decisions that affect a component's or system's behavior beyond the strength determined by nominal demand, often in the nonlinear range. Perhaps the most common example of a detailing deficiency is poor confinement in concrete gravity columns. Often in older concrete buildings, the expected drifts from the design event will exceed the deformation capacity of such columns, potentially leading to degradation and collapse. Although the primary gravity load design is adequate, the post-elastic behavior is not, most often due to inadequate configuration and spacing of ties.

Another common example is a shear wall that has adequate length and thickness to resist the design shear and moment, but that has been reinforced such that its primary post-elastic behavior will be degrading shear failure rather than more ductile flexural yielding. Examples in structural steel include braced frames with brittle and weak connections that are unable to develop the diagonal brace, or brittle beam-column connections in moment frames that are unable to develop the capacity of the frame elements.

Identification of detailing deficiencies is significant in selection of mitigation strategies because acceptable performance often may be achieved by local adjustment of detailing rather than by adding new lateral force-resisting elements. In the case of gravity concrete columns, acceptable performance often can be more efficiently achieved by enhancing deformation capacity (e.g. by adding confinement) than by reducing global deformation demand (e.g. by adding lateral force-resisting elements).

2.3.6   Diaphragms

The primary purpose of diaphragms in the overall seismic system is to act as a horizontal beam spanning between lateral force-resisting elements. In this document, deficiencies affecting this primary purpose, such as inadequate shear or bending strength, stiffness, or reinforcing around openings or re-entrant corners, are placed in this category. Inadequate local shear transfer to lateral force-resisting elements or missing or inadequate collectors are categorized as load path deficiencies.

Since the purpose, configuration, typical deficiencies, and retrofit of diaphragms are essentially independent of specific building types, techniques for rehabilitation are in Part 3, Chapter 22.

2.3.7   Foundations

Foundation deficiencies can occur within the foundation element itself, or due to inadequate transfer mechanisms between foundation and soil. Element deficiencies include inadequate bending or shear strength of spread foundations and grade beams; inadequate axial capacity or detailing of piles and piers; and weak and degrading connections between piles, piers, and caps. Transfer deficiencies include excessive settlement or bearing failure, excessive rotation, inadequate tension capacity of deep foundations, or loss of bearing capacity due to liquefaction.

Analysis and identification of transfer deficiencies is problematic due to recognition that structural movement within the soil may be beneficial, or at least not detrimental, depending on the performance objective. Mitigation of apparent transfer deficiencies is often expensive and disruptive, adding incentive to more carefully consider their effects. Explicit modeling of soil resistance to foundation movement therefore is becoming more common and can affect the overall dynamic characteristics of the structure as well as base fixity of rigid elements.

Similarly, the potential for liquefaction at the site is only a deficiency if the projected surface settlement is expected to compromise the performance objective for the building.

This document assumes that apparent deficiencies in structure-soil transfer mechanisms have been confirmed by analysis to warrant mitigation.

Similar to diaphragms, the issues surrounding foundation retrofit are generally independent of specific building types, and have been placed in Part 3, Chapter 23.

2.3.8   Other Deficiencies

Deficiencies that do not fit into one of the categories described above can be identified but are highly variable and unique. In some cases, such as certain geologic hazards or interaction with adjacent buildings, the hazard is created off the building site and may be out of the control of the building owner. Standard mitigation techniques cannot be identified for such conditions and are not included in this document. The significance of these deficiencies with regard to the designated performance objective must be discussed with the owner and if appropriate and feasible, mitigation actions developed. In rare cases, replacement of the building, abandonment of the site, or creation of a redundant facility may be indicated.

Some of these potential deficiencies are briefly discussed below.

Geologic Hazards

On-site liquefaction can be categorized as a foundation deficiency and mitigated if deemed necessary. However, the liquefaction and/or lateral spread of adjacent off-site soils can disrupt utility service to the site or even cause lateral movement of the building.

Up-slope, offsite landslides or upstream dam failure and flooding can also be identified in geologic hazard studies. Similarly, potential slide planes may pass under the site but extend beyond the site in such a way that mitigation within the site is impractical.

Although a rare condition, active fault traces can pass through the site or through the building footprint.

Most of these hazards will not be identified unless a detailed geological hazard study is performed, which may not be justified unless exceptionally high performance is needed, or if required by the local jurisdiction. If identified, the risk of receiving unacceptable damage must be weighed against the cost of local mitigation or alternate means of meeting the owner's requirements.

The potential effects of these hazards on building foundations and possible mitigating actions are discussed in Section 23.10.

Adjacent Buildings

When the gap between buildings is insufficient to accommodate the combined seismic deformations of the buildings, both may be vulnerable to structural damage from the "pounding" action that results when the two collide. This condition is particularly severe when the floor levels of the two buildings do not match and the stiff floor framing of one building impacts on the more fragile walls or columns of the adjacent building.

For conditions created by expansion joints that are commonly found in buildings, the slabs usually align, and the pounding damage is normally assumed to be a local problem. However, if the lateral systems on either side of the joint are of considerably different stiffness or strength, an independent analysis of both portions may be inappropriate as loads can be transferred from one portion to the other.

For conditions along property lines or involving party walls, the two buildings likely have different ownership, and practical and legal issues may be more significant that technical ones. Without a high level of cooperation, performance to the satisfaction of both owners may not be possible.

When one owner owns both adjacent buildings, these legal issues no longer apply, and tying the buildings together can focus on the technical issues. Like expansion joints in large buildings, if expansion and contraction movements between the structures are expected to be minimal, these joints can be structurally closed, eliminating the pounding problem and often increasing the options for the location of new seismic elements.

Deterioration of Structural Materials

Structural materials that are damaged or seriously deteriorated may have an adverse effect on the seismic performance of an existing building during a severe earthquake. Methods and techniques for repair of poor workmanship, deterioration, fire, settlement, or earthquake damage are not covered in this manual. If significant damage is suspected in a building, a condition assessment should be developed and carried out prior to development of a final seismic strengthening scheme. The significance of the damage or deterioration must be evaluated with respect to both the existing condition and the proposed seismic strengthening of the building. Structural condition assessment is not covered in this document, but appropriate procedures and measures are well documented (Ratay, 2005).

Timber: Common problems with timber members that require rehabilitation include termite attack, fungus ("dry rot" or "damp rot"), warping, splitting, checking due to shrinkage, strength degradation of fire-retardant wood structural panel in areas where high temperatures exist, or other causes.

Unreinforced masonry: The weakest element in older masonry usually is the mortar joint, particularly if significant amounts of lime were used in the mortar and the lime was subsequently leached out by exposure to the weather. Thus, cracks in masonry walls caused by differential settlement of the foundations or other causes generally will occur in the joints; however, well-bonded masonry occasionally will crack through the masonry unit.

Unreinforced concrete: Unreinforced concrete may be subject to cracking, spalling, and disintegration. Cracking may be due to excessive drying shrinkage during the curing of the concrete or differential settlement of the foundations. Spalling can be caused by exposure to extreme temperatures or the reactive aggregates used in some western states. Disintegration or raveling of the concrete is usually caused by dirty or contaminated aggregates, old or defective cement, or contaminated water (e.g., water with a high salt or mineral content).

Reinforced concrete or masonry: Reinforced concrete and masonry are subject to the same types of deterioration and damage as unreinforced concrete and masonry. In addition, poor or cracked concrete or masonry may allow moisture and oxygen to penetrate to the steel reinforcement and initiate corrosion. The expansive nature of the corrosion byproducts can fracture the concrete or masonry and extend and accelerate the corrosion process.

Structural steel: Poorly configured structural steel members may trap moisture from rainfall or condensation under conditions that promote corrosion and subsequent loss of section for the steel member. Even well-configured steel members exposed to a moist environment require periodic maintenance (i.e., painting or other corrosion protection) to maintain their effective load-bearing capacity. Older structural steel buildings often have little or no vapor barrier, particularly at the perimeter where failures in the weatherproofing of the cladding can lead directly to exposure to moisture. Light structural steel members (e.g., small columns or bracing members) in some installations may be subject to damage from heavy equipment or vehicles. While such damage may have no apparent detrimental effect on the vertical-load-resisting capacity of the steel member, its reserve capacity for resisting seismic forces may be seriously impaired.

2.4   References

ASCE, 2003, *Standard for the Seismic Evaluation of Buildings*, ASCE 31-03, Structural Engineering Institute of the American Society of Structural Engineers, Reston, VA.

ATC, 1996, *The Seismic Evaluation and Retrofit of Concrete Buildings*, ATC-40, Applied Technology Council, Redwood City, CA.

FEMA, 1998, *Handbook for the Seismic Evaluation of Buildings – A Prestandard*, FEMA 310, Federal Emergency Management Agency, Washington, D.C.

FEMA, 2000, *Prestandard and Commentary for the Seismic Rehabilitation of Buildings*, FEMA 356, Federal Emergency Management Agency, Washington, D.C.

FEMA, 2005, *Improvement of Nonlinear Static Seismic Procedures*, FEMA 440, Federal Emergency Management Agency, Washington, D.C.

ICBO, 1997, *Uniform Code for Building Conservation*, 1997 Edition, International Conference of Building Officials, Whittier, CA.

ICC, 2003, *International Existing Building Code*, 2003 Edition, International Conference of Building Officials, Country Club Hills, IL.

NIST, 2002, *Standards of Seismic Safety for Existing Federally Owned and Leased Buildings*, ICSSC RP 6, National Institute of Standards and Technology, Gaithersburg, MD.

Ratay, Robert T., 2005, *Structural Condition Assessment,* John Wiley & Sons, Inc, Hoboken, NJ.

Chapter 3 - Seismic Rehabilitation

3.1   Introduction

This document is primarily intended to provide descriptions of individual construction
techniques used in seismic rehabilitation rather than to give complete guidance on the far more
subtle process of developing and designing complete rehabilitation schemes. Although the latter
may be useful to engineers inexperienced in seismic retrofit or seismic design in general, the
schematic design process for seismic rehabilitation is complex and, not unlike other civil
engineering design, often involves more art than science.

Classes of rehabilitation methods are given in this chapter that address one or more of the
potential categories of deficiencies described in Chapter 2. As previously mentioned, these
categories and classes are somewhat arbitrary and sometimes overlap. However, they are
intended to form a framework and logic for development of alternate overall rehabilitation
schemes. This chapter describes the classes of rehabilitation methods and issues that commonly
must be considered when developing overall schemes.

3.2   Rehabilitation Standards

Seismic rehabilitation guidelines and standards have developed parallel with, but somewhat
behind, seismic evaluation documents. Often, however, they are the same. For example,
minimum standards for URM buildings, developed in California, specified sets of configuration,
maximum stress, and minimum inter-tie rules that were required. When used in an evaluation
mode, the evaluator noted what was missing or deficient. When used in the rehabilitation mode,
the engineer provided what was missing or added strength to eliminate deficiencies.

However, it may not always be true that the evaluation standard and the rehabilitation standard
are the same. Some engineers and policy-makers believe that the evaluation threshold should be
set at a very minimum acceptable level because of the cost and disruption of rehabilitation, but
that once rehabilitation is required, a higher, more reliable standard should be used. This is
currently the case with the most commonly used documents, ASCE 31-03 (ASCE, 2003) for
evaluation and FEMA 356 (FEMA, 2000) for rehabilitation. Slightly different methods are used
which can lead to slightly different levels of deficiency and the general level of expected
performance has also been set lower in ASCE 31-03.

The types of common standards and guidelines used to seismically rehabilitate buildings are
described below.

3.2.1   Mitigation of Evaluation Deficiencies

Most commonly, the scope of rehabilitation is determined by directly addressing the deficiencies
determined by evaluation. This is certainly the case when using building type-specific codes
such as the IEBC (ICC, 2003) and local ordinances, because the evaluation and retrofit standards
are one and the same. Similarly, the Simplified Rehabilitation method contained in FEMA 356
is based on use of the evaluation standard, ASCE 31-03 as a basis for design of rehabilitation
measures. In the rare case where the code for new buildings is used as a standard for existing

buildings, the rehabilitation would also be determined by directly addressing deficiencies from an evaluation.

3.2.2   Rehabilitation Design Based on Nonlinear Response

Few, if any, evaluation methods fully consider nonlinear response (unless FEMA 356 itself is used to evaluate), so if rehabilitation designs are determined in this manner, the extent of retrofit, and in some cases, the entire strategy of retrofit, may differ from merely eliminating the evaluation deficiencies. Nonlinear techniques are intended to more reliably predict performance, so when this is desirable—rather than meeting an arbitrary standard—these methods of analysis and design of rehabilitation measures are indicated.

3.3   Classes of Rehabilitation Measures

In most cases, the primary focus for determining a viable retrofit scheme is on vertically oriented components (e.g. column, walls, braces, etc.) because of their significance in providing either lateral stability or gravity load resistance. Deficiencies in vertical elements are caused by excessive inter-story deformations that either create unacceptable force or deformation demands. However, depending on the building type, the walls and columns may be adequate for seismic and gravity loads, while the building is inadequately tied together, forming a threat for partial or complete collapse in an earthquake. In order to design an efficient retrofit scheme, it is imperative to have a thorough understanding of the expected seismic response of the existing building and all of its deficiencies.

In the traditional sense of improving the performance of the existing structure, there are three basic *classes* of measures taken to retrofit a building:

> Add elements, usually to increase strength or stiffness
> Enhance performance of existing elements, increasing strength or deformation capacity
> Improve connections between components, assuring that individual elements do not become detached and fall, a complete load path exists, and that the force distributions assumed by the designer can occur

The types of retrofit measures often balance one another in that employing more of one will mean less of another is needed. It is obvious that providing added global stiffness will require less deformation capacity for local elements (e.g. individual columns), but it is often less obvious that careful placement of new lateral elements may minimize a connectivity issue such as a diaphragm deficiency. Important connectivity issues such as wall-to-floor ties, however, are often independent and must be adequately supplied.

In addition to improving the strength or ductility of the existing structural elements, there are less traditional methods of improving the performance of the overall structure. These methods can be categorized as follows:

> Seismic demand can be reduced by removing upper floors or other mass from the structure, adding damping devices to reduce displacement, or seismically isolating all or part of the structure.

Selected elements can be removed or weakened to prevent damaging interaction between different systems, to eliminate damage to the element or to minimize a vertical or horizontal irregularity.

This document uses these five *classes* of retrofit measures, in conjunction with the *categories* of seismic deficiencies described in Section 2.3 as a framework to present specific retrofit techniques. The matrices in each chapter of Part 2 list rehabilitation techniques according to these classes of retrofit measures and the deficiency that they mitigate. Retrofit methods that are relatively independent of the model building being considered are described in Part 3.

The classes of retrofit measures are discussed in more detail below.

3.3.1   Add Elements

This is the most obvious and most general class of retrofit measures. In many cases, new shear walls, braced frames, or moment frames are added to an existing building to mitigate deficiencies in global strength, global stiffness, configuration, or to reduce the span of diaphragms as described in Sections 2.3.1, 2.3.2, 2.3.3, or 2.3.6 respectively. New elements can also be added as collectors to mitigate deficiencies in load path as described in Section 2.3.4.

Retrofit schemes are developed with a balance of additional elements and enhanced existing elements (see Section 3.3.2) that best fit the socio-economic demands described in Section 3.4.2. Either adding new elements or enhancing the strength of existing elements could create a load path issue. The designer must assure that the new loads attracted to these elements can be delivered by other existing components. Therefore, eliminating a deficiency in *Global Strength* or *Global Stiffness* may create a deficiency in *Load Path* that did not exist initially.

3.3.2   Enhance Performance of Existing Elements

Rather than providing retrofit measures that affect the entire structure, deficiencies can also be eliminated at the local, component level. This can be done by enhancing the existing shear or moment strength of an element, or simply by altering the element in a way that allows additional deformation without compromising vertical load-carrying capacity.

Given that certain components of the structure will yield when subjected to strong ground motion, it is important to recognize that some yielding sequences are almost always preferred: beams yielding before columns, bracing members yielding before connections, flexural yielding before shear failure in columns and walls. These relationships can be determined by analysis and controlled by local retrofit in a variety of ways. For example:

> Columns in frames and connections in braces can be strengthened, and the shear capacity of columns and walls can be enhanced to be stronger than the shear that can be delivered by the flexural strength.
> Concrete columns can be wrapped with steel, concrete, or other materials to provide confinement and shear strength. Composites of glass or carbon fibers and epoxy are becoming popular to enhance shear strength and confinement in columns.
> Concrete and masonry walls can be layered with reinforced concrete, plate steel, and other materials such as fiber composites.

An indirect method of mitigating an unreasonably small drift capacity of a gravity element or system is to provide a supplemental gravity support system. In some situations, the cost of adding sufficient new global strength and stiffness or of increasing deformation capacity of certain gravity elements is excessive. For seismic performance primarily aimed at life safety, adding supplemental gravity supports might provide efficient mitigation. A common example of this practice is the supplemental support required for concentrated wall-supported loads in unreinforced masonry bearing wall buildings contained in most standards for retrofit. Supplemental support techniques have also been used in several cases for parts or all of concrete gravity systems.

Although enhancement of performance of existing elements can provide strength and stiffness for deficiencies similar to adding elements, these measures are most commonly used to mitigate inadequate component detailing as described in Section 2.3.5.

3.3.3   Improve Connections Between Components

The class of rehabilitation technique is almost exclusively targeted at mitigation of load path deficiencies as described in Section 2.3.4. With the exception of collectors, a deficiency in the load path is most often created by a weak connection, rather than by a completely missing link. However, some poor connections, particularly between beam and supporting column, are not directly in the primary seismic load path but still require strengthening to assure reliable gravity load support during strong shaking.

3.3.4   Reduce Demand

For buildings that contain a complete but relatively weak lateral system and that also have excess space or a site where supplementary space can be constructed, removal of several top floors may prove to be an economical and practical method of providing acceptable performance. However, like schemes that require strengthening, the noise and disruption or removing floors must be considered, particularly if the remaining floors are to remain occupied. In many cases, little or no retrofit work may be required on the lower floors, although due to a shortened period, the acceleration response of the base may be increased. This issue is discussed further in Chapter 24.

Techniques to reduce demand on the seismic system by modification of dynamic response of a structure are also included in this class. Perhaps the most notable example is seismic isolation, although this procedure is relatively expensive compared to alternate techniques and is normally employed in existing buildings for historic preservation or for occupancies that cannot be disturbed. A technique to modify response that is often economically competitive with traditional rehabilitation is the addition of damping in a structure. The added damping may reduce deformations sufficiently to prevent unacceptable damage in the existing system. Systems that actively control dynamic response have also been the subject of research, but have not made their way into common use. Further descriptions of response modification techniques are given in Part 3.

3.3.5   Remove Selected Components

Lastly, deformation capacity can be enhanced locally by uncoupling brittle elements from the deforming structure, or by removing them completely. Examples of this procedure include placement of vertical sawcuts in unreinforced masonry walls to change their behavior from shear failure to a more acceptable rocking mode and to create slots between spandrel beams and columns to prevent the column from being a "short column" prone to shear failure.

3.4   Strategies to Develop Rehabilitation Schemes

3.4.1   Technical Considerations

The first overview by a retrofit designer should be studying the deficiencies identified by the evaluation. Typical deficiencies are categorized by model building type in Part 2 and a table for each is give that relates the deficiencies to common mitigation techniques.

Some common seismic deficiencies are very localized and can be efficiently mitigated by narrowly targeting the retrofit activity. For example, for some one-story and two-story masonry or concrete wall buildings, the only deficiency may be out-of-plane wall ties to the diaphragm. Similarly, adequate resistance to overturning for a discontinuous shear wall may be made available by no more than providing confinement to the supporting column. Load path issues should be completely identified because there are often few choices for mitigation.

Next, the appropriate deficiency table in Part 2 should be studied to identify if a potential mitigation technique is effective for more than one deficiency present in the building. Adding strength or stiffness is very common, and a few new elements may solve strength, drift, and configuration problems.

When adding new lateral force-resisting elements such as shear walls, moment frames, or braced frames, several issues should be considered: Is the deformation compatible with the existing lateral force-resistsing or gravity load-carrying system? Will the new system sufficiently relieve the existing structure of load or deformation at all levels? Is the new system adding significant mass to the structure? Will this mass invalidate the previous evaluation? Will extensive new foundations be needed for the new system?

For any early trial scheme, review that the altered structure will:

> Have a complete load path
> Have sufficient strength and stiffness to meet the design standard
> Be compatible with and will adequately protect the existing lateral and gravity system
> Have an adequate foundation to assume a fixed base building, or have appropriately considered foundation flexibility in the design

3.4.2   Nontechnical Considerations

The solution chosen for retrofit is almost always dictated by building-user oriented issues rather than by merely satisfying technical demands. There are five basic issues that are of concern to building owners or users:

> Construction cost
> Seismic performance
> Short-term disruption of occupants

Long-term functionality of building

Aesthetics, including consideration of historic preservation

All of these characteristics are always considered, but an importance will eventually be put on each of them, either consciously or subconsciously, and a combination of weighting factors will determine the scheme chosen.

3.4.3   Cost

Construction cost is always important and is balanced against one or more other considerations deemed significant. However, sometimes other economic considerations, such as the cost of disruption to building users or the value of contents to be seismically protected, can be orders-of-magnitude larger than construction costs, thus lessening its importance.

3.4.4   Seismic Performance

If the governing jurisdiction is requiring seismic strengthening, either due to extensive remodeling or structural alteration, a design standard and resulting seismic performance expectation will normally be specified. When seismic rehabilitation is voluntary, the benefit-cost relationship of various performance levels may be considered explicitly, but in any case, the seismic performance factor will become important in the development of the scheme.

Typically, in either situation, perceived qualitative differences between the probable performance of different schemes were often used to assist in choosing a scheme. Now that performance-based design is integral to most rehabilitation, specific performance objectives are often set prior to beginning development of schemes. Objectives that require a limited amount of damage or "continued occupancy" will severely limit the retrofit methods that can be used and may control the other four issues.

3.4.5   Short-term Disruption of Occupants

Often retrofits are done at the time of major building remodels and this issue is minimized. However, in cases where the building is partially or completely occupied, this parameter commonly becomes dominant and controls the design.

To minimize disruption, schemes are often explored that place strengthening elements outside the building the building envelope. Concrete shear walls, pier-spandrel frames, and steel braced frames placed adjacent to or within the plane of exterior walls have been used in this way. Shear connection of the diaphragms to these new elements must be carefully considered. External elements that can also provide new strength and stiffness perpendicular to the exterior wall have also been used. In this case, a collector normally must be run into the building to connect the new element to the floor diaphragms. Installation of this collector may disrupt the internal systems, finishes, and occupants of the building to the extent that nullifies the exterior location of the new lateral element. Although there are many examples of exterior solutions that have been installed with continuous occupancy of the building, acceptability of the noise, dust, and vibration associated with the construction, as well as the potential disruption of access and egress, must be carefully considered during planning and design.

3.4.6   Long-term Functionality of Building

The addition of shear walls or braced frame in the interior of a building will always change the functional use plan. If the seismic work is being done as part of a general renovation, new functional spaces can often be planned around the new elements. However, such permanent structural elements will always reduce the flexibility of future replanning of the space. This characteristic is often judged less important than the other four and is therefore sacrificed to satisfy other goals. Often the planning flexibility is only subtlety changed. However, it can be significant in building occupancies that need open spaces such as retail spaces and parking garages.

3.4.7   Aesthetics

In historic buildings, considerations for preservation of historic fabric usually control the design. In many cases, even performance objectives are controlled by limitations imposed by preservation. In non-historic buildings, aesthetics is commonly stated as a criterion, but, in the end, is often sacrificed, particularly in favor of minimizing cost and disruption to tenants.

3.5   Other Common Issues Associated with Seismic Rehabilitation

3.5.1   Constructability

The options to obtain adequate access to the location of construction within the building as well as a sufficient local construction space are far more limited in a seismic rehabilitation project than in new construction. In addition, there may be issues related to undercutting existing footings, providing temporary shoring of gravity elements, or providing temporary lateral support for certain elements of the structure, certain floors, or even for the whole building. The design engineer must consider these issues when conceiving a rehabilitation scheme; the reality of field conditions may render a scheme physically or economically infeasible.

To control their liability for site construction safety, engineers have generally avoided specification of "means and methods" of construction as part of the construction documents. This concern is no less true for rehabilitation projects, but, in cases where significant structural alteration is required, it is often difficult to develop a realistic scheme without a thorough understanding of probable construction methods.

3.5.2   Materials Testing

Destructive testing of existing material can be disruptive and expensive. Care should be taken in designing a program that suits the building-specific conditions. If basic information is available on structural materials, it is often prudent to delay testing until preliminary evaluation is completed to identify critical existing components, or on the other hand, to determine that the material strengths are relatively unimportant and material testing can be minimized.

3.5.3   Disruption to Building Systems and Replacement of Finishes

The significance of conflicts with mechanical, electrical, or plumbing distribution systems or equipment should be considered during development of rehabilitation schemes. Temporary disruptions of services may be acceptable if the building is not to be occupied during construction, but may need to be limited if the building is occupied. High costs may be associated with permanent changes in routing or relocation of equipment due to the seismic work.

Similarly, the cost and disruption of removal and replacement of finishes or cladding to gain access to the structure must be considered. In addition to certain finishes being unique and expensive or historic, the construction associated with gaining this access normally requires evacuation and closing off of the local area.

3.5.4   Concealed Conditions

Even when original construction drawings are available and certain material tests have been performed to gain confidence in the knowledge of existing conditions, different conditions may be exposed during construction. In addition to attempting to minimize the importance of such possibilities by field exposures and design, the design professional of record should also be engaged during construction, in order to properly assess such discoveries and to enable design of mitigating measures consistent with the overall scheme.

3.5.5   Quality Assurance

Quality assurance programs are probably more important in rehabilitation projects than with new construction. Given no control of existing conditions, the margin for error is often small. In addition, as indicated in Section 3.5.4, conditions in the field are often different than assumed and effective revisions often need to be developed.

3.5.6   Detailing for New Elements

In almost all codes, new elements installed into existing buildings as part of a seismic rehabilitation must meet the detailing requirements for new construction. For example, minimum reinforcing of concrete walls or columns, slenderness ratios of braces and connection details must be in accordance with new code requirements. With designs that utilized nonlinear analysis, deformation capacities will have been set using an assumed detailing pattern from the code from new buildings, and that level of detailing must then be provided.

3.5.7   Vulnerability During Construction

Installation of new seismic elements within an existing building often requires demolition of parts of the gravity load system as well as the effective lateral load system. Although safety during construction is the contractor's responsibility under the "means and methods" principle, the engineer designing the seismic rehabilitation may be in a position to identify global weaknesses in the gravity or lateral load system that could develop during construction. Such conditions should be pointed out in the contract documents, although temporary strengthening measures that might be needed during construction should be designed by the contractor's engineer.

3.5.8   Determination of Component Capacity by Testing

There are many unique components in existing buildings for which no data are available to define strength and/or deformation capacity. If certain components or connections occur in multiple locations and will potentially require extensive and costly retrofit, in-situ or laboratory testing may be justified. The cost of such testing and the possibility of acceptable performance

must be judged against potential savings in the cost of rehabilitation. Experts in material
behavior and testing should be consulted to assist with such evaluations.

3.5.9   Incremental Rehabilitation

Disruption to occupants can be minimized if seismic rehabilitation is combined with other
maintenance or renovation work. This may lead to phased or incremented construction. The
potential to implement this type of seismic improvements is documented in a series of FEMA
documents, FEMA 395 to FEMA 400 and FEMA 420.

3.6   Issues with New Techniques or Products

Part 2 discusses many commonly employed seismic rehabilitation techniques. However, it not
possible to include all currently available techniques in the document; and there will always be
new techniques, products, research, and approaches developed in the future. There is no
substitute for engineering judgment. When considering a rehabilitation technique or product, the
design engineer should consider the following issues.

Prior Use

    Has the approach been used successfully before?

    How long have previous installations been in place?

    Have the installed rehabilitation measures been through actual seismic events?

Testing

    General quality of testing.

    General quality of documentation.

    Was the testing performed by the manufacturer or by an independent entity?

    Relevance of test to actual elements.

    Type of testing: monotonic, cyclic quasistatic, dynamic.

    Number of specimens.

    How far into the nonlinear range did the testing go?

    Why was the test stopped?

    Were test results placed in performance-based design limit states?

Construction

    Is the technique limited to only certain specialized subcontractors?

    Can the technique be documented sufficiently to be bid?

    Does installation involve noise, dust, vibration, harmful vapors, and/or danger?

    Are special tools or set-ups needed?

Long-Term Stability of Mitigation Materials

    Do the materials creep, crack, shrink, lose strength, debond, rust/corrode, etc. over time?

    Can they be placed in exterior environments?

    Are there fire safety requirements or concerns?

    Are there temperature range limitations?

    Are coefficients of thermal expansion compatible with adjacent materials?

Do they react with other materials, such as galvanic corrosion from dissimilar metals, efflorescence in masonry, or breakdowns from ultraviolet light?

Are moisture issues appropriately addressed or can they be mitigated sufficiently when the technique is used?

Aesthetic and Historic Preservation

Is the technique suitable for sensitive structures?

Is it reversible?

Code Considerations

Are building code procedures and design methodologies available or applicable?

Does the product have approvals?

Quality Assurance

Can an adequate field quality assurance program be developed to verify that in-situ properties meet design assumptions?

Can a typical testing lab perform the inspection or testing or is special expertise needed?

Cost

Is adequate information available on pricing to make decisions during design?

Is the work or product best procured lump-sum or by unit price?

Is the cost worth the benefit?

3.7   References

ASCE, 2003, *Standard for the Seismic Evaluation of Buildings*, ASCE 31-03, Structural Engineering Institute of the American Society of Structural Engineers, Reston, VA.

FEMA, 2000, *Prestandard and Commentary for the Seismic Rehabilitation of Buildings*, FEMA 356, Federal Emergency Management Agency, Washington, D.C.

ICC, 2003, *International Existing Building Code*, 2003 Edition, International Conference of Building Officials, Country Club Hills, IL.

Chapter 4 - FEMA Model Building Types

4.1   Introduction

This document is primarily organized around the FEMA model building types. It is expected that most users of this document will be interested in finding information on a particular building or building type, which suggested this organization. Each building type is therefore assigned a chapter. Common seismic deficiencies for each building type are identified and mitigation techniques suggested, although it is recognized that most buildings will have multiple deficiencies and may require a combination of mitigating actions. The rehabilitation techniques commonly used for each building type are identified in each chapter and, if closely associated with the building type, described in detail in that chapter. References are given to other chapters for other applicable techniques.

4.2   History of Development

Several sets of standard structural types have been created to describe the building inventory of the U.S. Initially, these model building types were developed for the purposes of assigning fragility relationships to inventories of buildings for loss estimation in ATC 13, (ATC, 1985). Studies of buildings for development of ATC 14, *Evaluating the Seismic Resistance of Existing Buildings* (ATC, 1987), indicated a large number of types in existence, but identified 15 primary types around which evaluation considerations could be grouped.

ATC 14 was later adapted for use in the FEMA series as FEMA 178, *NEHRP Handbook for Seismic Evaluation of Buildings* (FEMA, 1992a). This set of building types has subsequently been used extensively in other FEMA documents related to existing buildings, including FEMA 154 (FEMA, 1988), FEMA 227 (FEMA, 1992b), and FEMA 156 (FEMA, 1995).

When FEMA 178 was converted to a prestandard for input to the ASCE standards adoption process (FEMA 310 [FEMA, 1998] and ASCE 31-03 [ASCE, 2003]), the distinction between similar building types with flexible and rigid diaphragms was included by adding the suffix "A" to the alpha-numeric designation. For example, the definition of Building Type **S1**, Steel Moment Frames, was refined to designate steel moment frames with rigid diaphragms, and Building Type **S1A** was designated as steel moment frames with flexible diaphragms.

However, this new designation was not assigned consistently. For example, **W1A** was defined to represent a **W1** of larger size, rather than one with a flexible diaphragm; the designations **RM1** and **RM2** were used to differentiate flexible and rigid diaphragms in reinforced masonry buildings; finally, for the **URM** building type, the suffix A indicates a rigid diaphragm rather than a flexible diaphragm.

4.3   Model Building Type Refinements in this Document

Rather than causing additional inconsistency between documents, this document uses the pre-established model buildings types and designations described above. However, for the purposes of relating retrofit techniques to building types, additional minor refinements to the building type designations are convenient and clarifying. Specifically, concrete shear wall buildings (Building

Type **C2**) have been split into two groups, those with essentially complete gravity frames (Building Type **C2f**) and those primarily using bearing walls (Building Type **C2b**). Similarly, reinforced masonry buildings (Building Type **RM1**) have been split into two groups, those that are very similar to concrete tilts ups (Building Type **RM1t**) and those that are very similar to older, unreinforced masonry buildings (Building Type **RM1u**). Using these refinements, building performance characteristics, common seismic deficiencies, and applicable mitigation techniques can be more clearly described.

Finally, building types that are less common or that seldom require retrofit have not been included or have been de-emphasized in this document, although techniques suggested for a similar building will generally be applicable. The excluded building types include Building Type S3, Steel Light Frames, and the following sub-types designated by the "A" suffix: C2A, PC1A, PC2A, and URMA.

4.4   Description

The model building types are summarized below. Detailed descriptions can be found in each dedicated chapter. Many real buildings have characteristics from more than one model building type. Useful information can still be obtained by referring to chapters for similar building types.

Table 4-1: Model Building Types	
 W1: Wood Light Frames	Building Type **W1** consists of one- and two-family detached dwellings of one or more stories. Floor and roof framing are most commonly woodframe joists and rafters supported on wood stud walls. The first floor may be slab-on-grade or framed. Lateral forces in **W1** buildings are resisted by woodframe diaphragms and shear walls.
 W1A: Multistory, Multi-Unit Residential Woodframes	Building Type **W1A** is similar to Building Type **W1** in use of light-frame wall, floor and roof construction, but includes large multi-family, multistory buildings. In **W1A** buildings, second and higher stories are almost exclusively residential use, while the first story can include any combination of parking, common areas, storage, and residential units. Post and beam framing often replaces bearing walls in non-residential areas. Multi-family residential buildings with commercial space at the first story are included in Building Type **W1A** due to similar building characteristics. Lateral forces in **W1A** buildings are primarily resisted by woodframe diaphragms and shear walls.

Table 4-1: Model Building Types (continued)

 W2: Woodframes, Commercial and Industrial	Building Type **W2** consists of commercial, institutional, and smaller industrial buildings constructed primarily of wood framing. The first floor is most commonly slab-on-grade, but may be framed. Floor and roof framing may include wood joists, wood or steel trusses, and glulam or steel beams, with wood posts or steel columns. Post and beam framing is common at storefronts or garage openings. Lateral forces in **W2** buildings are primarily resisted by woodframe diaphragms and shear walls, sometimes in combination with isolated concrete or masonry shear walls, steel braced frames, or steel moment frames. Diaphragm spans may be significantly larger than in **W1** and **W1A** buildings.
 S1/S1A: Steel Moment Frames	Building Type **S1** consists of an essentially complete frame assembly of steel beams and columns. Lateral forces are resisted by moment frames that develop stiffness through rigid connections of the beam and column created by angles, plates, and bolts, and/or by welding. Floors are cast-in-place concrete slabs or metal decks infilled with concrete. Building Type **S1A** is similar but has floors and roofs that act as flexible diaphragms such as wood or untopped metal deck.
 S2/S2A: Steel Braced Frames	Building Type **S2** consists of a frame assembly of steel beams and columns. Lateral forces are resisted by diagonal steel members placed in selected bays. Floors are cast-in-place concrete slabs or metal decks infilled with concrete. Building Type **S2A** is similar but has floors and roofs that act as flexible diaphragms such as wood or untopped metal deck.

Table 4-1: Model Building Types (continued)	
 'Punched' concrete exterior walls are an alternate shear wall configuration *Vertical shafts often constructed of concrete* *Concrete slab or concrete over metal deck floor* *Steel beams and columns* *Concrete walls placed in selected interior and exterior bays in each direction* **S4: Steel Frames with Concrete Shear Walls**	Building Type **S4** consists of an essentially complete frame assembly of steel beams and columns. The floors are concrete slabs or concrete fill over metal deck. These buildings feature a significant number of concrete walls effectively acting as shear walls, either as vertical transportation cores, isolated in selected bays, or as a perimeter wall system. The steel column and beam system may act only to carry gravity loads or may have rigid connections to act as a moment frame to form a dual system.
 Interior partitions or shaft walls often built with clay tile *Steel beams and columns* *Multi-wythed brick masonry exterior with one or more wythes built within the column/beam envelope as 'infill'* *Floors usually formed concrete* **S5/S5A: Steel Frames** **with Infill Masonry Shear Walls**	Building Type **S5** is normally an older building that consists of an essentially complete gravity frame assembly of steel floor beams or trusses and steel columns. The floor consists of masonry flat arches, concrete slabs or metal deck and concrete fill. Exterior walls, and possibly some interior walls, are constructed of unreinforced masonry, tightly infilling the space between columns and between beams and the floor such that the infill interacts with the frame to form a lateral force-resisting element.
 Vertical shafts of nonstructural materials *Concrete beams and columns* *Nonstructural exterior cladding is often window wall or panelized construction* *Selected bays in each direction constructed as moment frames* *Floors: most often formed or precast concrete* **C1: Concrete Moment Frames**	Type **C1** buildings consist of concrete framing, either a complete system of beams and columns or columns supporting slabs without gravity beams. Lateral forces are resisted by cast-in-place moment frames that develop stiffness through rigid connections of the column and beams.

Table 4-1: Model Building Types (continued)

C2b: Concrete Shear Walls (Bearing Wall Systems)	Building Type **C2** covers buildings with concrete walls. For this document, the type is split into **C2b** and **C2f**. Building Type **C2b** is usually all concrete with flat slab or precast plank floors and concrete bearing walls. Little, if any, of the gravity loads are resisted by beams and columns.
C2f: Concrete Shear Walls (Gravity Frame Systems)	Building Type **C2f** has a column and beam or column and slab system that essentially carries all gravity load. Lateral loads are resisted by concrete shear walls surrounding shafts, at the building perimeter, or isolated walls placed specifically for lateral resistance.
C3/C3A: Concrete Frames with Infill Masonry Shear Walls	Building Type **C3** is normally an older building with an essentially complete gravity frame assembly of concrete columns and floor systems. The floors can consist of a variety of concrete systems including flat plates, two-way slabs, and beam and slab. Exterior walls, and possibly some interior walls, are constructed of unreinforced masonry, tightly infilling the space between columns horizontally and between floor structural elements vertically, such that the infill interacts with the frame to form a lateral force-resisting element.

Table 4-1: Model Building Types (continued)

PC1: Tilt-Up Concrete Shear Walls | Building Type **PC1** is constructed with concrete walls, cast on site and tilted up to form the exterior of the building. **PC1** buildings are used for many occupancy types including warehouse, light industrial, wholesale and retail stores, and office. The majority of these buildings are one story; however, tilt-up buildings of up to three and four stories are common, and a limited number with more stories exist. For many years, tilt-up buildings have been primarily large box-type buildings with the tilt-up walls at the building perimeter; however, in recent years, tilt-up construction has been used in more complex and varied commercial building configurations. Lateral forces in **PC1** buildings are resisted by flexible wood or steel roof diaphragms and tilt-up concrete shear walls. Floor diaphragms are most commonly composite steel decking. |
|

PC2: Precast Concrete Frames with Shear Walls | Buildings designated as **PC2** include wide ranging combinations of precast and cast-in-place concrete elements. Precast members may be limited to a floor system of hollow core or T-beam construction, or may include all elements of the gravity and lateral load systems. For this document, Building Type **PC2** includes concrete wall or frame buildings in which any of the horizontal or vertical elements of the lateral load system are of precast concrete, except for flexible diaphragm buildings which are addressed as Building Type **PC1**. |

Table 4-1: Model Building Types (continued)

**RM1t: Reinforced Masonry Bearing Walls
(Similar to Tilt-Up Concrete Shear Walls)**

**RM1u: Reinforced Masonry Bearing Walls
(Similar to Unreinforced Masonry Bearing Walls)**

Building Type **RM1** is constructed with reinforced masonry (brick cavity wall or concrete masonry unit) perimeter walls with a wood or metal deck flexible diaphragm.

For this document, Building Type **RM1** is separated into two categories, **RM1u**, which is multistory, and typically has interior CMU walls and shorter diaphragm spans, and **RM1t**, a large, typically one-story buildings similar to concrete tilt-ups.

**RM2: Reinforced Masonry Bearing Walls
(Similar to Concrete Shear Walls w/ Bearing Walls)**

Building Type **RM2** consists of reinforced masonry walls and concrete slab floors that may be either cast-in-place or precast. This building type is often used for hotel and motels and is similar to the concrete bearing wall type **C2.**

Table 4-1: Model Building Types (continued)	
2-4 wythe brick masonry exterior bearing walls Wood joists or trusses with wood sheathing Wood stud bearing walls or post and beam construction on interior Wood joists bearing on masonry wall **URM: Unreinforced Masonry Bearing Walls**	Building Type **URM** consists of unreinforced masonry bearing walls, usually at the perimeter and usually brick masonry. The floors are typically of wood joists and wood sheathing supported on the walls and on interior post and beam construction.

4.5   References

ASCE, 2003, *Seismic Evaluation of Buildings*, ASCE 31-03, Structural Engineering Institute of the American Society of Civil Engineers, Reston, VA.

ATC, 1985, *Earthquake Damage Evaluation Data for California*, ATC 13, Applied Technology Council, Redwood City, CA.

ATC, 1987, *Evaluating the Seismic Resistance of Existing Buildings*, ATC 14, Applied Technology Council, Redwood City, CA

FEMA, 1988, *Rapid Visual Screening of Buildings for Potential Seismic Hazards*, ATC 154, Federal Emergency Management Agency, Washington, D.C.

FEMA, 1992a, *NEHRP Handbook for the Seismic Rehabilitation of Existing Buildings*, FEMA 172, Federal Emergency Management Agency, Washington, D.C.

FEMA, 1992b, *A Benefit-Cost Model for the Seismic Rehabilitation of Buildings*, FEMA 227, Federal Emergency Management Agency, Washington, D.C.

FEMA,1995, *Typical Costs for Seismic Rehabilitation of Existing Buildings*, FEMA 156, Second Edition, Federal Emergency Management Agency, Washington, D.C.

FEMA, 1998, *Handbook for the Seismic Evaluation of Buildings—A Prestandard*, FEMA 310, Federal Emergency Management Agency, Washington, D.C.

Chapter 5 - Building Type W1: Wood Light Frames

5.1   Description of the Model Building Type

Building Type **W1** consists of one- and two-family detached dwellings of one or more stories.
Floor and roof framing are most commonly wood joists and rafters supported on wood stud walls
(called woodframe or wood light-frame). The first floor may be slab-on-grade or a raised framed
floor. Lateral forces in **W1** buildings are resisted by woodframe diaphragms and shear walls.
Chimneys, where present, consist of solid brick masonry, masonry veneer, or woodframe with
internal metal flues. Although materials for detached one- and two-family dwellings vary beyond
woodframe, this chapter will focus on this most common type of construction. Figure 5.1-1
provides one illustration of this building type.

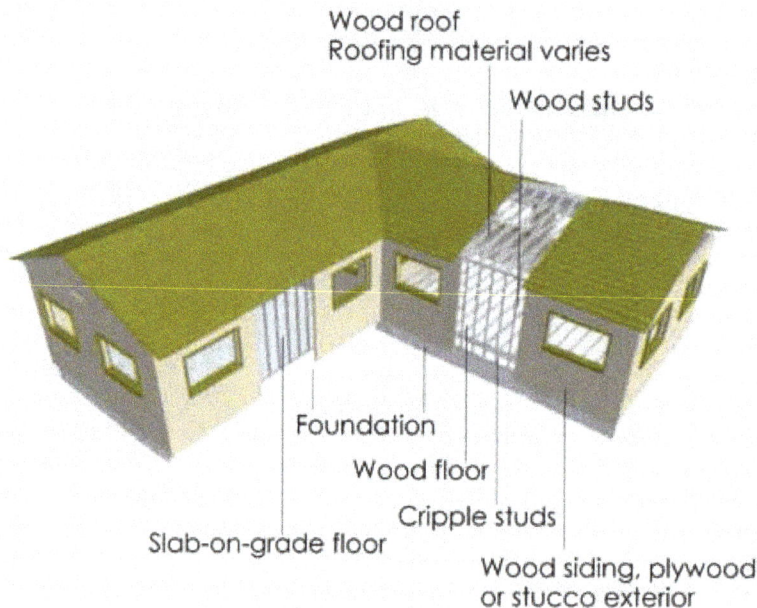

Figure 5.1-1: Building Type W1: Wood Light Frames
One- and Two-Family Detached Dwelling

Design Practice

W1 buildings recently constructed near population centers may have a partial or complete
engineered design; however, most **W1** buildings will have been designed using prescriptive
provisions (conventional construction). Where prescriptive design has been used, it can generally
be expected that no numerical check of sheathing, fastening, wall overturning, or other load-path
connections has been performed, and that no fastening or connections beyond basic fastening
schedules have been used. In engineered design, the extent of analysis and detailing can vary

from a check of in-plane shear capacity of shear walls, to exhaustive design and detailing. Minimum fastening and connection needs to be assumed unless more is known to exist.

Walls

Wall bracing materials and detailing vary depending on dwelling age and location. Except for recently constructed or rehabilitated **W1** buildings, it is most common for the finish material to also serve as the shear wall bracing material. Common interior finish and bracing materials include plaster over wood lath, plaster over gypsum lath (button board) and gypsum wallboard. Common exterior finish and bracing materials include board siding, shingles, panel siding, and stucco. Finish materials such as vinyl siding and EIFS are not included with these bracing materials due to low stiffness and negligible bracing capacity. Wall sheathing is sometimes present in addition to finish materials. In older **W1** buildings, lumber sheathing--applied horizontally, vertically or diagonally--was often used. In newer buildings, wood structural panel (plywood or oriented strand board) sheathing is most often used. Because interior and exterior finish materials often also serve as bracing materials in **W1** buildings, it is difficult to differentiate between structural and nonstructural materials.

Early **W1** building construction used post and beam wall framing systems in lieu of closely spaced studs. Most construction shifted to stud systems between the mid 1800s and early 1900s; however, some post and beam construction is still built. Except where braced frames or knee-braces provide alternate lateral force-resisting systems, post and beam wall systems still rely on wall finishes or sheathing to resist in-plane lateral loads. Stud systems were first constructed using balloon-framed walls, in which individual stud members extended from the foundation to the very top of the framed wall. This height often included cripple walls plus two stories. When walls are balloon framed, floor framing is hung off of the interior face of the studs. In the early 1900s, most framing changed from balloon framed to platform framed, in which the wall framing stops at the underside of each floor, and the floor framing sits on top of wall framing rather than hanging off the face. These two wall framing systems have important differences for detailing load transfer, chords, and collectors for shear walls and diaphragms.

Floor and Roof Diaphragms

Floor and roof diaphragm materials and detailing vary depending on building age and location. In older buildings, solid lumber sheathing is most often applied straight or diagonally under built-up and membrane roofs, and spaced lumber sheathing is found under shingle and tile roofs. In older buildings, floor sheathing is often solid lumber sheathing applied horizontally or diagonally. In some cases, hardwood floors form both the sheathing and the floor finish. In newer or rehabilitated buildings, wood structural panel (plywood or oriented strand board) floor and roof sheathing is most common. The strength of wood structural panel diaphragms varies depending on whether they are blocked (interior sheathing panel edges supported on and edge nailed to blocking) or unblocked (interior edges not supported or nailed), sheathing panel layout, and sheathing nail size and spacing. The presence or absence of diaphragm chords and collectors also affects the diaphragm strength and stiffness. As is true with shear walls, the level of design detailing for diaphragms can vary significantly.

Plank and beam framing became popular in the mid-1900s and is still in use today. This system uses 2x or thicker straight lumber plank sheathing for floors and roofs, supported on beams spaced between four and eight feet on center. The planking is often left exposed on the underside for the ceiling below. Publications such as *Plank and Beam Framing for Residential Buildings* (AF&PA, 2003) describe this construction type. In the western U.S., wood structural panel overlays are often applied over the lumber sheathing to provide diaphragms for engineered designs. Shear walls are used as vertical elements to resist lateral loads.

Distribution of seismic forces to the vertical elements of the lateral force-resisting system is influenced in part by the diaphragm stiffness. The selection of a flexible or rigid diaphragm model for purposes of force distribution is controversial at the time of this update, and details of building analysis are beyond the scope of this document. It is recommended that the reader refer to applicable building codes, local jurisdiction requirements, and the local standard of practice.

System Between Lowest Framed Floor and Grade

Where the lowest occupied floor in a **W1** building is woodframed, there are a large number of structural systems that can occur between the framed floor and grade. Figure 5.1-2 illustrates some of the common systems for level building sites. The type of system can vary based on region, building age, soils, type of site, exposure to environmental hazard such as flood, etc. These may be foundation systems, or may include superstructure sitting on top of the foundation. Common weaknesses in these systems include 1) lack of a load path for lateral loads (Figure 5.1-2A), 2) limited lateral load resistance, and 3) lack of adequate connection to transfer lateral loads to the foundation.

Cripple walls (Figure 5.1-2C) are one common system between the framed floor and grade. Cripple walls are wood stud framed walls that extend from the top of a foundation to the underside of the first framed floor. Cripple walls often enclose an uninhabited crawl space, but may also sit on top of partial height concrete or masonry walls in a basement. In past earthquakes, dwelling drift and damage has often been concentrated in cripple walls.

W1 buildings are often supported on continuous perimeter foundations or foundation walls (Figure 5.1-2D) in combination with continuous or isolated interior pier footings. Alternate foundation types may be used locally or regionally. Materials for continuous perimeter foundations or foundation walls vary depending on age and location. Many older dwelling foundations use unreinforced concrete, brick masonry, or stone masonry. Today, use of lightly reinforced concrete continuous foundations is most common in the western states, and unreinforced concrete masonry and brick masonry (pier and curtain wall) are common in other regions. In the 1970s and 1980s, use of post-tensioned slab-on-grade foundations became common in some areas with highly expansive soils; these present additional issues for anchorage that are discussed in Section 5.4.4.

5.2   Seismic Response Characteristics

The dynamic response of **W1** buildings is very short period due to the stiffness of wall bracing and finish materials. Deflection and inelastic behavior occur primarily in the walls, while the floor and roof diaphragms remain close to elastic. Likewise, damage is mostly seen in the walls rather than the floor or roof systems.

FRAMED FIRST
FLOOR

FRAMED FIRST
FLOOR

INTERMITTENT
POST AT
BUILDING
PERIMETER

INTERMITTENT
CONCRETE
OR MASONRY
PIER AT
BUILDING
PERIMETER

A

B

FRAMED FIRST
FLOOR

FRAMED FIRST
FLOOR

FRAMED
CRIPPLE
WALL WITH
FINISH AT
BUILDING
PERIMETER

CONTINUOUS
CONCRETE
OR MASONRY
FOUNDATION
WALL AT
BUILDING
PERIMETER

C

D

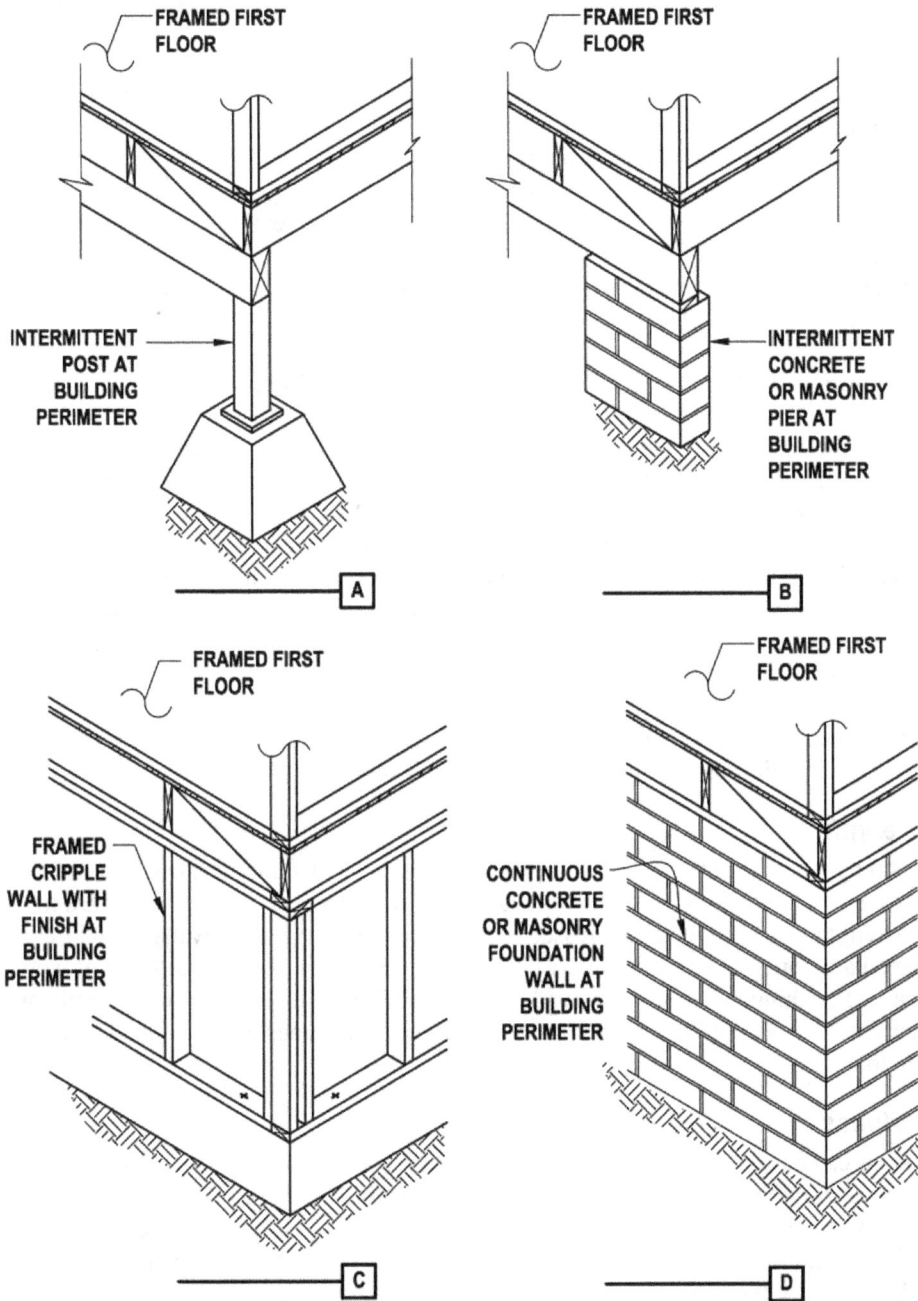

Figure 5.1-2: Systems Between First Framed Floor and Grade – Level Lot Sites

5.3   Common Seismic Deficiencies and Applicable Rehabilitation Techniques

Life-safety performance of **W1** buildings has generally been very good. A limited number of vulnerable configurations, however, have repeatedly resulted in significant damage, and in a few instances loss of life. In **W1** buildings, damage to wall finish materials has contributed notably to repair costs. Wood chapters in two recently published earthquake engineering handbooks provide overviews of earthquake performance for woodframe buildings and extensive lists of references describing extent and details of damage (Dolan, 2003; Cobeen, 2004). Of the many discussions of performance, of particular note due to extent and detail is the Northridge earthquake reconnaissance report (EERI, 1996).

It is not the objective of this document to address rehabilitation of buildings for wind loads; however, many of the rehabilitation measures that increase the strength and stiffness of the primary lateral-force-resisting system for seismic loads will also provide increased resistance to wind loads. Included is the addition of strength and stiffness in diaphragms, shear walls, and their connections. For load path connections, locations of greatest vulnerability and therefore priority items for seismic rehabilitation tend to be located at the base of the structure where seismic demand is greatest, such as anchorage to the foundation. In contrast, for wind rehabilitation, load path connections of greatest vulnerability and highest priority tend to be at the top of the structure, including roofing attachment to roof sheathing, roof sheathing attachment to rafters, rafter attachment to walls, etc.

See below for general discussion and Table 5.3-1 for a detailed compilation of common seismic deficiencies and rehabilitation techniques for Building Type **W1**.

Global Strength

Inadequate strength, particularly in lower stories of multistory **W1** buildings, has caused extensive damage to bracing and finish materials but has not generally resulted in hazard to life. Inadequate strength is most directly addressed by enhancing existing shear walls or adding new vertical elements. In one- and two-family dwellings this most often involves addition of wood structural panel sheathing and associated load path connections to an existing framed wall. While not commonly used in **W1** buildings, steel moment and braced frames may be added to address global strength.

Global Stiffness

Global stiffness can occasionally be an issue in **W1** buildings, particularly where archaic materials such as horizontal or vertical straight lumber sheathing are used for bracing and finish materials. In dwellings this is most likely to occur in unfinished garage, crawlspace or basement areas. This is a common condition for garage side walls in dense urban areas such as San Francisco. Where these types of sheathing are used, strength is usually an issue as well as stiffness. As with global strength, typical rehabilitation measures include enhancing existing shear walls or adding new vertical elements. Applicable rehabilitation measures are discussed in the *Global Strength* section.

Table 5.3-1: Seismic Deficiencies and Potential Rehabilitation Techniques for W1 Buildings

Deficiency		Rehabilitation Technique				
Category	Deficiency	Add New Elements	Enhance Existing Elements	Improve Connections Between Elements	Reduce Demand	Remove Selected Components
Global Strength	Insufficient in-plane wall strength	Wood structural panel shear wall [5.4.1] Steel moment frame [6.4.1] Steel braced frame [7.4.1]	Enhance woodframe shear wall [5.4.1]	Shear wall uplift anchorage and compression posts [6.4.4]	Replace heavy roof finish with light finish	
Global Stiffness	Insufficient in-plane wall stiffness	Wood structural panel shear wall [5.4.1] Steel moment frame [6.4.1] Steel braced frame [7.4.1]	Enhance woodframe shear wall [5.4.1]	Shear wall uplift anchorage and compression posts [6.4.4]		
Configuration	Missing or inadequate cripple wall bracing	Add woodframe cripple wall Add continuous foundation and foundation wall	Enhance woodframe cripple wall [5.4.4]			
	Open front	Wood structural panel shear wall [5.4.1] Collector [5.4.2] Moment frame [6.4.1]	Enhance woodframe walls perpendicular to open front [5.4.1] Detailing of narrow woodframe shear wall piers			
	Hillside	Wood structural panel shear wall [5.4.5]	Enhance woodframe shear wall [5.4.5]	Anchor base level diaphragm to uphill foundation [5.4.5]		
Load Path	Inadequate shear anchorage to foundation			Anchorage to foundation [5.4.3]		

Table 5.3-1: Seismic Deficiencies and Potential Rehabilitation Techniques for W1 Buildings					
Deficiency		Rehabilitation Technique			
Category	Deficiency	Add New Elements	Enhance Existing Elements	Improve Connections Between Elements	Reduce Demand	Remove Selected Components
Load Path (continued)	Inadequate shear wall overturning load path		Supplement framing supporting woodframe shear wall [6.4.3]	Shear wall uplift anchorage and compression posts [6.4.4]		
	Inadequate shear transfer in wood framing			Enhance load path for shear [5.4.1]		
	Inadequate collectors to shear walls		Enhanced existing collector	Add collector [6.4.5], [7.4.2]		
Component Detailing	Unreinforced & unbraced chimney		Infill chimney [5.4.6] Brace chimney [5.4.6]		Reduce unsupported chimney height [5.4.6]	Remove chimney [5.4.6]
Diaphragms	Inadequate in-plane strength and/or stiffness		Enhance diaphragm [22.2.1] Diaphragm overlay [22.2.1]		Replace heavy roof finish with lighter finish	
	Inadequate chord capacity		Enhance chord members and connections [22.2.2]			
	Excessive stresses at openings and irregularities		Enhance diaphragm detailing			
	Re-entrant corner		Enhance diaphragm detailing			
Foundations	See Chapter 23					
[ ] Numbers noted in brackets refer to sections containing detailed descriptions of rehabilitation techniques.						

Configuration

Several **W1** building configurations have been observed to be vulnerable to damage, in some cases resulting in full or partial collapse. Vulnerable configurations include buildings with inadequate bracing systems between the lowest framed floor and grade, open front building portions, and split-level buildings. Primary rehabilitation measures are specific to each of these configurations.

For level site buildings, inadequate bracing systems between the lowest framed floor and grade commonly include inadequately braced cripple walls and perimeter post and pier systems that do not provide a path for seismic forces. Inadequately braced cripple walls are commonly enhanced with sheathing and anchorage to the foundation. Where post and pier systems occur at the building perimeter, it is generally necessary to add a continuous footing and either a foundation stem wall or braced cripple wall. See related discussion of anchorage to the foundation in the Load Path section.

Hillside buildings can be vulnerable when large variations occur in the stiffness of the system between the lowest framed floor and grade. Generally, the bracing for lateral loads at the uphill side will be significantly stiffer that the downhill side, attracting a much higher force. Flexible downhill systems permit significant deflection and diaphragm rotation. Hillside buildings can be improved by anchoring floor diaphragms to the uphill foundation, and by enhancing strength and stiffness of downhill bracing systems.

Open front building portions occur when an exterior wall contains little or no bracing at any story level; common occurrences include garage fronts and window walls. Open front building portions can be rehabilitated by the addition or enhancement of shear walls or the addition of collectors, transferring seismic loads to portions of the building that have adequate shear walls.

Split-level buildings have vertical offsets in the top of floor framing in adjacent portions of the building (i.e. sunken living room). Where floor framing with varying top elevations frames into a common wall, earthquake loading may cause one level of framing to separate from the wall, potentially causing local floor collapse. This behavior was seen in the San Fernando earthquake (ATC, 1976). Vulnerable split-level buildings are commonly rehabilitated by improving connections between framing on either side of the floor offset.

Load Path

The highest priority and most cost effective rehabilitation measure for **W1** buildings is ensuring that the home is adequately anchored to the supporting foundation. Anchorage may use anchor bolts or proprietary retrofit anchors, and it may be done alone or in combination with cripple wall enhancement. In addition, a systematic evaluation of the seismic force-resisting system will often result in the need to rehabilitate load path connections. Load path improvements include shear anchorage to the foundation, uplift anchorage to the foundation, shear transfer load path in the wood framing, uplift load path in the wood framing, and collectors to shear walls. Rehabilitation measures primarily involve the addition of fasteners and connector hardware.

Component Detailing

Many **W1** buildings contain unreinforced, unbraced masonry chimneys, for which rehabilitation measures include removal, partial removal, infill, and bracing. Appendages such as exit stairs, porches and decks, and their roofs are commonly rehabilitated by improving seismic attachment to the main building structure. Inadequately anchored stone or masonry veneer in **W1** buildings, if addressed, is most commonly removed, or removed and replaced with properly anchored veneer.

Diaphragm Deficiencies

A systematic evaluation may identify deficiencies in the diaphragm systems, including inadequate diaphragm strength and/or stiffness, inadequate shear transfer to walls, and inadequate detailing at large diaphragm openings and re-entrant corners. Diaphragm deficiencies have not stood out as a source of damage to **W1** buildings. The removal and replacement of existing roofing, as part of regular building upkeep, often provides an opportunity for existing straight lumber sheathed diaphragms or spaced sheathing to be overlain with wood structural panel sheathing. Even though the roof diaphragm is seldom a top priority for **W1** building rehabilitation, this can provide an opportunity to tie the roof together and achieve more monolithic behavior at a nominal cost. Rehabilitation measures can be found in Chapter 22.

Foundation Deficiencies

Common seismic deficiencies in foundations undergoing systematic rehabilitation include inadequate strength for shear wall overturning forces. Rehabilitation measures for foundation deficiencies are discussed in Chapter 23. Other deficiencies such as deteriorated foundations, sliding on unreinforced cold joints and settlement in cut and fill sites are not addressed by this document.

5.4   Detailed Description of Techniques Primarily Associated with This Building Type

5.4.1   Add New or Enhance Existing Shear Wall

Deficiency Addressed by Rehabilitation Technique

This rehabilitation technique addresses insufficient global strength and/or stiffness, as well as local areas of insufficient strength and/or stiffness such as at open front conditions. Discussion is applicable to **W1, W1A,** and **W2** buildings. In **W1** buildings, it is most common for insufficient strength or stiffness to be local rather than global.

Description of the Rehabilitation Technique

This rehabilitation technique involves adding a new shear wall or enhancing an existing shear wall. The primary focus of the discussion is addition of wood structural panel (plywood or OSB) sheathing, fastening and connections to an already existing framed wall, as this is most common in **W1** buildings. Additions or alterations may lend themselves to adding a new shear wall. Addition of a completely new shear wall and other options for enhancement of existing walls are discussed in Section 6.4.2.

As a fundamental element of shear wall addition or enhancement, this section includes discussion of load path for transfer of forces into and out of the shear wall. This discussion is applicable to building types **W1**, **W1A**, and **W2**, as well as other buildings types with wood floor and roof diaphragms.

This section also discusses a few general topics relating to rehabilitation of existing woodframe buildings, including wood shrinkage, pre-drilling for fasteners, and wood species. The issues are applicable to building types **W1**, **W1A**, and **W2**, as well as other buildings types with wood floor and roof diaphragms.

Design Considerations

Research basis: Research that specifically discusses addition of shear walls in one- and two-family detached dwellings has not been identified; however, a significant amount of testing and analysis on new shear walls and shear wall buildings can be considered applicable to this use. Primary references for shear wall testing are APA (1999a, 1999b), City of Los Angeles & SEAOSC (1996), Salenikovich (2000), Gatto and Uang (2002), and Pardoen et al. (2003). Testing of slender walls can be found in ATC R-1 (ATC, 1995). Testing of perforated shear walls can be found in Heine (1997). Testing of walls designed for continuity around openings can be found in Kolba (2000). Research addressing specific connections within the shear wall load path is referenced in the following discussions.

Shear wall design method: New shear walls are primarily designed in accordance with provisions of the IBC (ICC, 2003a) or the *Wind and Seismic Supplement* (AF&PA, 2005b). Requirements for new shear walls should be used for design of new shear walls in existing residences in addition to the considerations addressed in this section. The IBC and the *Wind and Seismic Supplement* recognize three methods of analyzing wood structural panel shear walls: segmented, designed for continuity around openings, and perforated shear walls (Figure 5.4.1-1). All of these methods are acceptable. Design for continuity around openings will allow for use of slender wall piers, where necessary. The third method – perforated shear wall--was developed particularly for residential construction in order to minimize the required overturning restraint hardware.

Shear wall location: Analytical studies have shown that one- and two-family dwellings will tend to have a concentration of deformation demand in the first story (lowest framed story) (Isoda, Folz, and Filiatrault, 2002) and (Cobeen, Russell, and Dolan, 2004). Therefore, under most circumstances, shear walls added in the lowest story are likely to have a larger impact on building performance than those added in upper stories, and lower stories should generally be given higher priority.

In order for shear walls to function as part of the structural system, it is necessary to design for transfer of in-plane load from the diaphragm being supported into the wall top and transfer of in-plane and overturning loads out at the wall base. In addition, the size and aspect ratio need to be adequate to meet demand, and significant disruptions over the height of the wall should be minimized. These considerations guide preferable locations for shear walls.

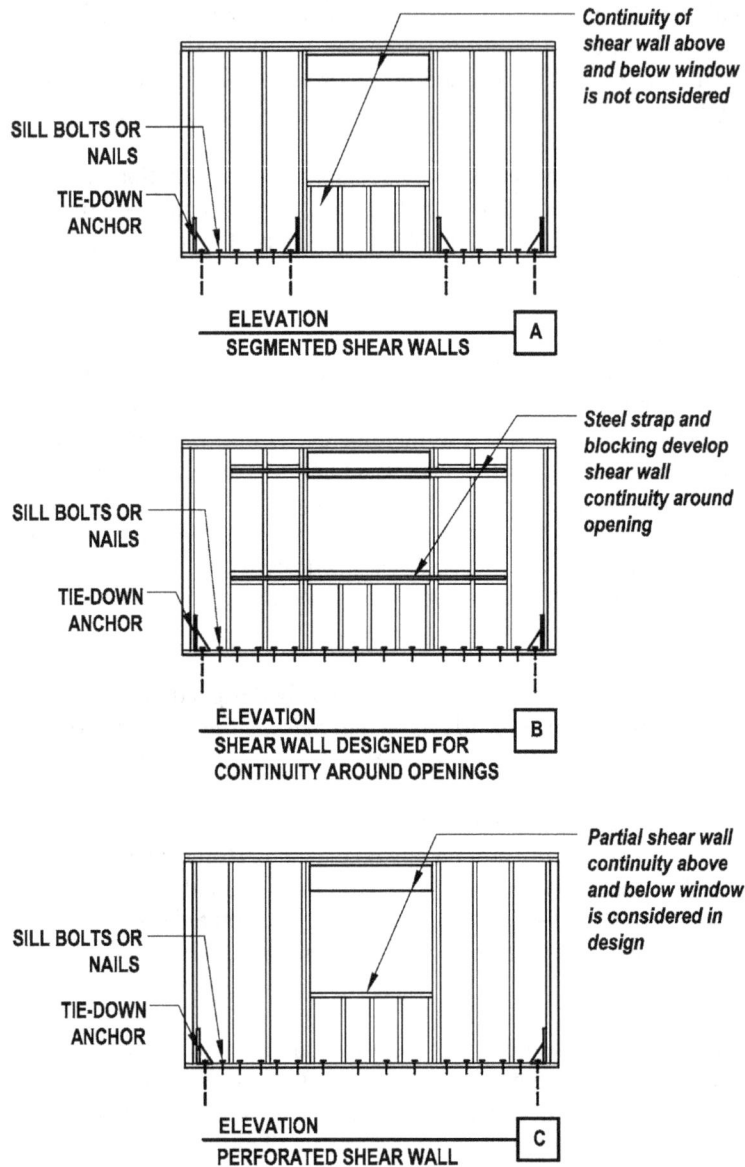

Figure 5.4.1-1: Shear Wall Design Methods

Preferred shear wall locations:

Exterior walls generally have inherent continuity of load path framing at the wall top and to a bearing foundation at the base (Figure 5.4.1-2). Conditions that can make exterior walls less effective include wall locations that are detached from the floor diaphragm (along stair opening or back of light-frame fireplace), walls that are balloon framed (wall studs continuous past floor framing), and walls that have interruptions over their height (low roof framed into side of wall). Most of these conditions can be addressed with additional load path detailing, however, at greater cost and disruption.

Interior bearing walls, like exterior walls, generally have inherent continuity of load path at wall top and bottom.

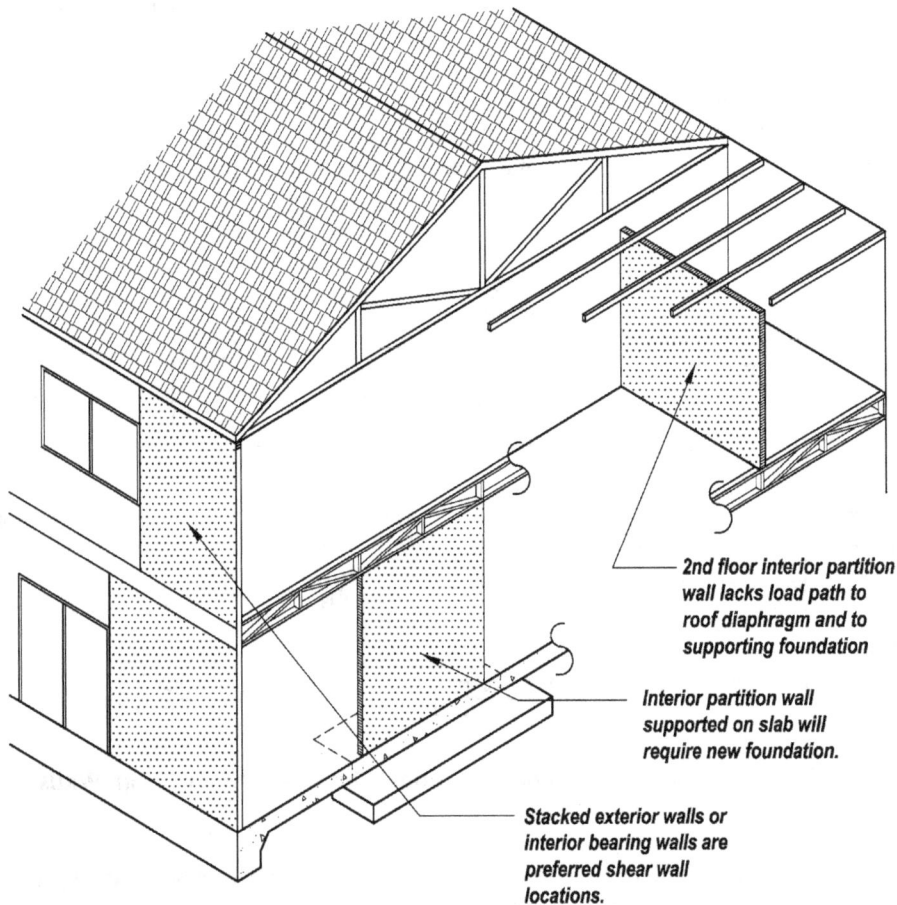

2nd floor interior partition wall lacks load path to roof diaphragm and to supporting foundation

Interior partition wall supported on slab will require new foundation.

Stacked exterior walls or interior bearing walls are preferred shear wall locations.

Figure 5.4.1-2: Preferred and Less Preferred Shear Wall Locations

Less preferable shear wall locations:

Interior partition walls can be problematic due to lack of load path continuity at both the top and bottom of the wall. Inadequate support for overturning forces is generally the most difficult problem to solve. It is easiest to use second floor walls that are continuously supported on framed first story walls (Figure 5.4.1-3A) allowing transmission of uplift and downward loads to the foundation. As a second choice, it may be possible to use a section of an upper story wall that can be vertically supported by posts at each end, again allowing transmission of uplift and downward loads to the foundation (Figure 5.4.1-3B). In both these support cases, the overturning stiffness of the shear wall should not be significantly different than shear walls located at the building exterior. As a last choice, floor framing systems can be enhanced to support interior shear walls (Figure 5.4.1-3C).

Interior partition walls in residences with truss roof and/or floor systems (Figure 5.4.1-2) require special attention to wall location, and analysis and detailing in order to avoid damaging the trusses.

Bathroom and kitchen plumbing walls can be problematic for use as bracing walls because of penetrations through the wall sheathing and because piping often results in breaks in the top and bottom plates serving as chords and collectors.

Walls oriented at an angle to the primary framing direction can pose particularly difficult detailing issues.

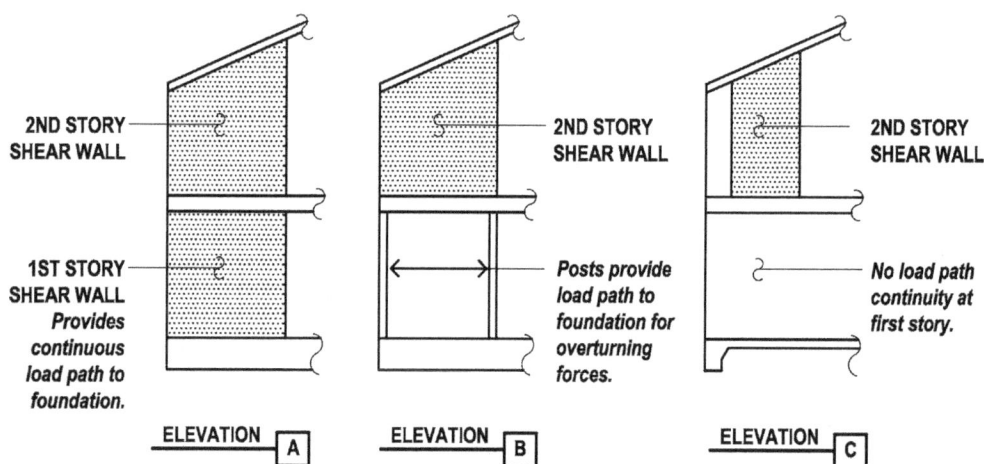

Figure 5.4.1-3: Overturning Support Conditions for Upper Story Shear Walls

The addition of shear walls is often most needed in portions of residences where existing walls are too slender to provide effective bracing. Use of properly detailed wood structural panel shear walls assists in making slender shear walls effective in providing resistance. It may become necessary, however, to reduce window openings in order to provide adequate lengths of shear wall.

All of the above limitations are only in response to the physical configuration of the residence. Other considerations in choice of wall locations include the level of disruption that is acceptable to the occupant and other planned work that may provide access for rehabilitation.

Adequacy of foundation: Addition or modification to existing foundations can often be the most expensive portion of adding shear walls in existing residences. Shear walls produce concentrated uplift and downward loads at each end. Engineered shear walls are seldom added without addition of uplift anchorage. Where the shear wall is long enough and the overturning forces low, the forces on the foundation can be modeling as two separate vertical forces, one up and one down. The downward load must be transferred to the supporting soils. Where the required bearing length does not exceed twice the depth of the foundation, the foundation capacity is not critical to footing resistance. Where a greater length is required, foundation shear and flexure capacity come into play. For the uplift anchorage it is necessary to have the foundation span far enough to mobilize dead load to resist uplift. Where slender walls are used, concentrated moments are introduced into the foundation by the closely spaced uplift and downward forces. This is particularly true of slender proprietary walls.

Construction and capacity of the foundation will significantly impact the ability to withstand these concentrated forces. Continuous concrete foundations or foundation walls with reinforcing are preferred. Anchorage, shear capacity and flexure capacity can be particularly problematic with existing unreinforced brick masonry foundations, unreinforced concrete masonry foundations, partially grouted concrete masonry foundations, and isolated foundations of any material. Addition of new foundations is often required. New foundations cast along side and tied into existing foundations can have the advantage of mobilizing the resisting weight of the existing foundation, as can new foundations that run between and dowel into existing foundations.

Figure 5.4.1-4 shows a new continuous footing cast alongside an existing footing. Adhesive anchors are drilled into the existing footing at a regular spacing so that if the new footing uplifts, it will also pick up the existing footing. The adhesive anchor can be a bolt, as shown, or reinforcing steel designed for shear friction. The bolt or reinforcing is designed to transfer the required vertical resisting load. Design must consider concrete anchor capacity including edge distance effects. Reinforcing steel should be anchored on both sides of the interface to develop the bar yield. Preparation of the existing concrete surface would normally involve cleaning only; intentional roughening is generally not practical.

Detailing Considerations

General: A few topics deserve general consideration before getting into the specifics of shear wall detailing, including shrinkage, predrilling, wood species, corrosion issues, and condition assessment of existing buildings. Shrinkage of wood framing members is an issue that must be considered in design of both new and existing wood buildings. Shrinkage of wood framing as it drops to equilibrium moisture content is accommodated in new construction every day. Whether in new construction or rehabilitation, the primary concern is differential shrinkage where members subject to shrinkage might act in a system with members subject to lesser shrinkage, no shrinkage, or possibly even slight swelling. In rehabilitation, new framing members subject to shrinkage may need to be added in parallel to members that are already at equilibrium moisture content. Shrinkage in the length of framing members is negligible. The primary shrinkage of concern is in the width of members. With a combination of radial and tangential directions, shrinkage on the order of 6% or ¾" in 12 inches is reasonably possible. This could mean a gap of 3/4 inches developing between blocking and the diaphragm above in a shear transfer or similar connection, greatly reducing the resistance provided.

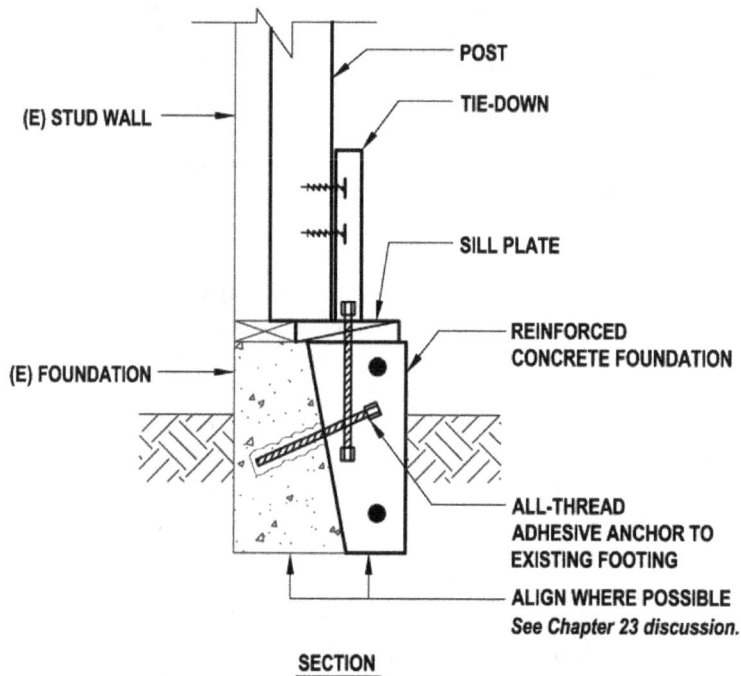

**Figure 5.4.1-4: New Continuous Foundation Cast Along Side Existing
to Provide Capacity for Tie-Down Anchor**

Effects of shrinkage are best mitigated by use of dry framing members and detailing to minimize reliance on configurations susceptible to shrinkage problems. Equilibrium moisture content for enclosed buildings is most often in the range of 7 to 12 percent. The closer new framing is to this range at time of installation, the less the potential shrinkage problems. This can be accomplished by setting aside framing (purchased green, at MC19, or at MC15) in a protected location to dry. In a dry season, the moisture content can drop significantly in the range of several weeks to several months. Another approach is to use engineered wood members such as glulams, which are manufactured at low moisture content. Laminated veneer lumber (LVL) and similar engineered wood products can also be used; however, the manufacturers restrict the size and spacing of nails into the top and bottom faces of these members due to concerns of splitting along lamination lines; this limits these members to use for low to moderate shear transfer loads.

Splitting of wood framing due to new fastening during rehabilitation is of significant concern. Nails that can easily be driven into new framing can be very difficult in existing framing, and splitting can occur. The current building code approach to splitting of members is primarily a performance approach. If members are split, the fasteners are not considered to provide capacity. This approach is of little help once splitting of critical structural members has already occurred. Repair and replacement of existing members can be very difficult. Predrilling for nails and other fasteners prior to installation will substantially reduce the risk of splitting framing members. Details of predrilling requirements are given in the NDS (AF&PA, 2005a).

Wood species is another item of general concern for detailing. The design values of wood fasteners and shear walls are a function of the framing density and, therefore, the wood species being fastened. The species of framing used may have varied over time. Older buildings may be framed with species that are no longer commonly used. Fastener, shear wall and diaphragm values need to be adjusted for the framing used. In very occasional cases, it might be desirable to determine the density of existing framing in order to identify the best choice of fastener values.

Corrosion of fasteners and connectors due to pressure preservative treatments is currently a concern for new construction due to recent changes in treatment formulation. This concern and related cautions regarding use of corrosion resistant fasteners and connectors is equally applicable where preservative treated wood is added in rehabilitation.

In woodframe buildings, deterioration of the structure can particularly impact seismic performance and the ability to implement seismic rehabilitation measures. For this reason it is important that condition assessment of critical elements of the existing woodframe structures be considered. See Section 2.3.8 of this document for additional discussion.

Sheathing and fastening: Added sheathing will generally be wood structural panel sheathing (plywood or OSB). In very unusual circumstances, addition of diagonal lumber sheathing might occur. The choice of extent and unit shear for sheathing and fastening is a balance between cost and performance. In general, providing more sheathing at a lower shear capacity results in less building deformation and better building performance. As with any system, well-distributed resistance is always better than heavy concentrations of resistance in local areas. In addition, when sheathing fastening is being added to existing dry wood members, close fastener spacing increases the possibility of member splitting. This is particularly true in members on which sheathing panel edges abut. Under the IBC and *Wind and Seismic Supplement*, use of close nail spacing on shear walls will trigger a requirement for minimum 3x studs at adjoining panel edges. Since 3x framing will seldom already occur in an existing wall, two options generally result. First 3x or 4x members can be added, and wood structural panel sheets lain out to fall on these members, or a new 2x stud can be added along side an existing stud, and the two "stitch-nailed" to provide adequate interconnection. Shear walls with stitch-nailed 2x's at abutting panel edges were tested recently by APA and found to provide acceptable behavior (APA, 2003). A provision permitting "stitch-nailing" has been incorporated into the 2004 supplement to the IBC (ICC, 2003a).

The IBC requires the use of 3x foundation sill plates for shear wall unit shears over 350 plf, while the NEHRP Provisions (FEMA, 2003) permit 2x plates in combination with steel plate washers on anchor bolts. In rehabilitation work, it is seldom practical to replace or modify the existing foundation sill, so practice is to retain the existing sill. The IBC and predecessor UBC (ICBO, 1997a) requirements for 3x sills primarily address cross-grain splitting of foundation sill plates, observed in the Northridge earthquake and laboratory testing (SEAOC, 1999). In recent testing of shear wall anchorage to foundations (Mahaney & Kehoe, 2002), as discussed in Cobeen, Russell, and Dolan (2004), the best performance of foundation anchorage was seen with 3x foundation sill plates; however, significant numbers of loading cycles were resisted by 2x plates with steel plate washers, supporting continued use of 2x plates in rehabilitation. Where a performance objective more stringent than one such as the FEMA 356 Basic Safety Objective is being used, however, replacement with 3x sills should be considered.

It is recommended that existing finishes be removed, allowing new structural sheathing to be installed directly over framing whenever possible. This permits an opportunity to observe the condition and fastening of existing framing, to install shear and overturning connections, and to add boundary member framing if required. The IBC permits wood structural panel sheathing to be installed over gypsum wallboard for fire-rating purposes; increased nail size is required. Increasing the distance from the center of sheathing nails to the edge of sheathing panels from 3/8-inch to ¾-inch has been seen to reduce fastener failure due to tear-out at the panel edge and greatly toughen the shear wall (Cobeen, Russell and Dolan, 2004). This is easily accomplished at top plates, bottom plates and end posts where only one row of edge nailing needs to be provided. It requires use of wider framing at interior wood structural panel joints where two panels abut and are edge nailed to a single framing member.

Buildings that have exterior wood structural panel siding present a unique opportunity to improve sheathing fastening without opening up finishes. In many cases only one of the two edges at abutting panels will be properly nailed. Providing full edge nailing on both panels can improve shear capacity. Nailing may be exposed on the siding exterior, or may be under trim boards which can be removed and replaced. Corrosion resistant fasteners are needed for siding nailing.

Sheathing to framing fastening with staples and use of wood structural panel overlays are discussed in Section 6.4.2.

Shear transfer criteria (when using FEMA 356): FEMA 356 (FEMA, 2000) identifies fasteners used to transfer forces from wood to wood or wood to metal as being deformation-controlled actions. When coupled with several relatively high *m*-factors for static procedure acceptance criteria, this can result in less fastening being required by FEMA 356 than the current building codes. At the same time, the shear wall sheathing fastening is identified as the desired location of inelastic behavior, which suggests that fastening for shear transfer into the shear wall should be force-controlled and more fastening provided. Because shear transfer nailing has only rarely been seen as a critical weak link in earthquake performance to date, it is recommended that current building code requirements be used for a basic safety objective. For a higher performance objective where inelastic behavior of the shear wall is anticipated, the proportioning of fastening relative to anticipated shear wall demand should be considered.

Shear transfer into top of wall: The addition of sheathing and fastening is not of value unless shear forces can be transferred into the top of the wall. Where sheathing is added to an existing wall line, the wall top plates will most often serve as the collector. Where top plates are not present, or are not continuous for a reasonable distance, a supplemental collector should be provided.

Figure 5.4.1-5 shows a series of top of shear wall details where the shear is being transferred from a roof diaphragm into the top of the wall. Since most diaphragms in residential construction are not blocked, the unit capacity of the new shear wall is likely to be higher than the unit

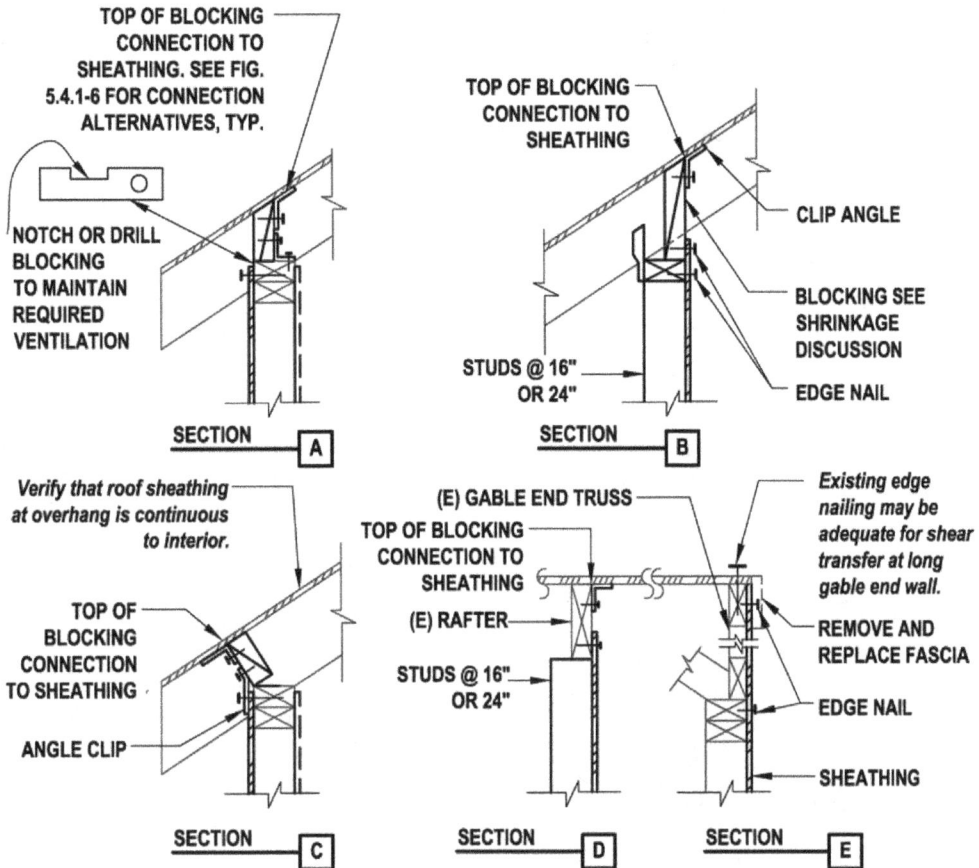

Figure 5.4.1-5: Load Path from Roof Diaphragm to Top of Shear Wall

capacity of the diaphragm above. For lightly loaded shear walls, the minimum length of the diaphragm to be connected into the shear wall can be calculated, and a collector provided to tie the diaphragm into the top of wall. For highly loaded vertical elements, it is recommended that the collector extend for the full diaphragm length, as discussed in Section 7.4.2.

In new construction, attachment of floor or roof sheathing to shear walls below typically requires nailing through the sheathing into framing below, as shown in Figure 5.4.1-6A. While this attachment remains the preferred approach, installation of nailing is not possible where roof or floor finishes cannot be removed. Figures 5.4.1-6B, 5.4.1-6C and 5.4.1-6D show alternative attachments of roof or floor sheathing. Significant cautions are applicable when using either of the alternative approaches, as detailed below.

Limited testing of the connections shown in Figures 5.4.1-6B and 5.4.1-6D occurred in the CUREE-Caltech Woodframe Project (Mosalam et al., 2002). The purpose of the testing was to find the best method of attaching new steel moment frames to existing wood buildings. The specimens used 12-inch deep joists and blocks in two 16-inch bays and tested angle clip connections monotonically and cyclically and adhesive connections monotonically. Due to the geometry of the test specimen, overturning behavior was significant. Both methods of attachment increased the load capacity beyond that for minimum framing nailing. The attachment of the blocking to the sheathing was not a controlling factor in any of the tests.

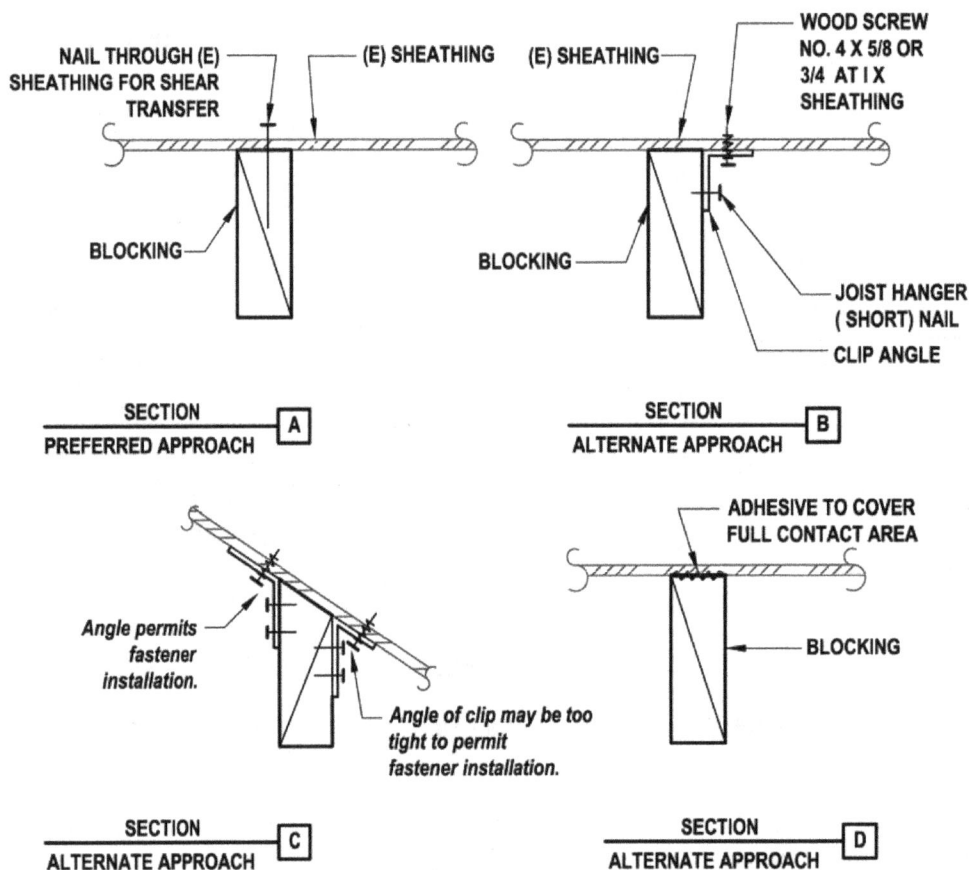

Note: Wood screw length is exaggerated for clarity. A properly sized wood screw will not penetrate the top face of the sheathing.

Figure 5.4.1-6: Attachment of Blocking to Existing Sheathing

Where unit shears are low and a nailed sheathing to framing connection is not possible, connection of sheathing to framing using steel clip angles provides a possible alternative (Figures 5.4.1-6B and 5.4.1-6C). The clip angle is generally attached to the framing with nails and to the sheathing with wood screws. NDS requires a minimum penetration of six times the wood screw diameter into the sheathing (note that the length of the screw point is included when calculating the 6 diameter penetration). This minimum penetration requirement results in use of very small screws, with very small capacities, making this connection type practical only when unit shears are low. If Number 4 screws are used (the smallest size generally available) the penetration into 1x sheathing with an actual dimension of 5/8 inches will be just short of meeting this penetration requirement. The Number 6 screws used in ¾-inch plywood in testing also fell just short. Use of increased penetration is encouraged whenever possible.

Along with caution due to the low capacity of the screws, two other significant cautions should be considered. First, during installation of the wood screws into the sheathing it is very easy to overdrive the screw, stripping out attachment to the sheathing. This is particularly easy when installation is with a screw gun, and it is even more so when the wood screw is connecting a steel clip, because the drawing of the screw head against the sheathing is not visible to the installer. Second, if the screw used is too long, it will penetrate the top surface of the sheathing. Care must be taken to not penetrate where the top surface is roofing or a sensitive finish. The thickness of the clip angle and protrusion of the fastener head generally reduce the screw penetration by 1/16 to 1/8 inch, which should be considered in specifying screw length. Considerable attention to quality control and quality assurance is recommended if this detail is to be used.

Where use of a nailed connection is not possible, adhesive connection from sheathing to framing provides a second alternative (Figure 5.4.1-6C). Adhesive attachment of sheathing to framing is discouraged in diaphragm assemblies in which inelastic behavior is anticipated, such as long-span and high load diaphragm systems. This is because adhesive connections do not allow slip between the sheathing and framing and do not permit energy dissipation, which generally occurs through nail bending. As a result, a glued diaphragm would be anticipated to behave nearly elastically up to a failure load and then fail in a brittle manner. In addition, adhesive sheathing to framing connections will be significantly stiffer than nailed connections, attracting higher loads to the adhesive where both types of attachment are used in combination. For these same reasons, use of adhesive in shear walls resisting seismic forces is not recommended, although the *NEHRP Provisions* do permit use in Seismic Design Category A, B or C, using and *R*-factor of 1.5.

In most **W1** and **W1A** buildings, however, it is anticipated that inelastic behavior will be concentrated in shear walls and other vertical elements, making use of adhesives in diaphragm connections an alternative. It is recommended that, when used, adhesive connections be designed for maximum expected forces (either overstrength forces, or using a very small *R*-factor or *m*-factor).

Adhesives used in recent testing have included cartridge types, applied using caulking guns, and spray-on self-expanding foam adhesives. Foam adhesives are also being used for attachment of roof sheathing to framing in high-wind regions. In this case the adhesive improves both wind and seismic resistance. Cautions when using adhesive sheathing to framing connections include the following: first, great care must be taken in ensuring that adhesives do not harden before blocking placement, as adhesives can have limited pot lives. Second, adhesives should be used in connections that minimize overturning rotation (continuous joists or shallow blocks) so that tension on the glue joint is minimized. Again, significant attention to quality control and quality assurance are recommended when using this connection alternative.

When attaching to the roof, required roof cross-ventilation needs to be maintained. This can influence details both at the roof perimeter and interior, as shown in Figure 5.4.1-5.

Load transfer from a roof diaphragm through a roof truss system into the top of a shear wall can be very complicated at both bearing and nonbearing partition walls. The complication comes from two sources. First, the shear wall must be extended through the roof truss system. Where the shear wall is parallel to truss members, this may simply mean placing the shear wall off the roof truss line and extending it to the roof sheathing. Where the shear wall is perpendicular to the roof framing, infill panels between the roof trusses are added to act as shear wall extensions (similar to Figure 5.4.2-2). Second, because existing nonbearing walls will often be attached with clips that permit vertical movement of the truss, the addition of a shear wall can create an unintended reaction, changing truss forces, and if between truss panel points, potentially leading to fracture of a truss chord. Connections are best made to existing trusses at truss panel points and should never be made without evaluating the potential change in truss forces.

Figures 5.4.1-7 and 5.4.1-8 show a series of details where shear is transferred into the top of a shear wall at a framed floor level. Note that Figure 5.4.1-8A shows an existing balloon framed condition prior to rehabilitation. Figures 5.4.1-8B through 5.4.1-8D show rehabilitation alternatives.

Shear transfer out of wall: Second story or higher shear walls will generally be supported on wood floor framing. Figure 5.4.1-9 illustrates common details for shear transfer at the wall base. See also the following discussion of overturning forces.

First story shear walls may be supported directly on foundations, or on framed floor systems supported on foundations or foundation walls. A detailed discussion of shear transfer anchorage to existing foundations can be found in Section 5.4.3. See also the following discussion of overturning forces.

Disruption over height of wall: Where shear wall sheathing cannot be placed in a continuous plane over the full height of a shear wall, additional detailing for continuity is needed. Disruption of the shear wall sheathing occurs most often where a floor or roof frames into the wall between floor levels, such as at a stair side or landing, a one-story roof hitting the side of a two-story section, a split-level floor, or a deck ledger. Testing done in the CUREE-Caltech project showed that shear wall studs that lose support from the sheathing can fail in weak axis bending. The same vulnerability could potentially occur where shear wall sheathing stops below an obstruction and then starts again above. Figure 5.4.1-10 shows methods for maintaining shear wall continuity across this type of disruption.

Overturning at wall base: Figure 5.4.1-3 illustrates a series of overturning support conditions that may occur at the base of second story shear walls. Continuity for the uplift and downward loads are required at each end of the upper story wall. Figure 5.4.1-11 shows common detailing for the overturning load path. Plumbing, electrical and mechanical utilities often run through the floor framing, greatly complicating addition of new floor framing members under second story shear walls.

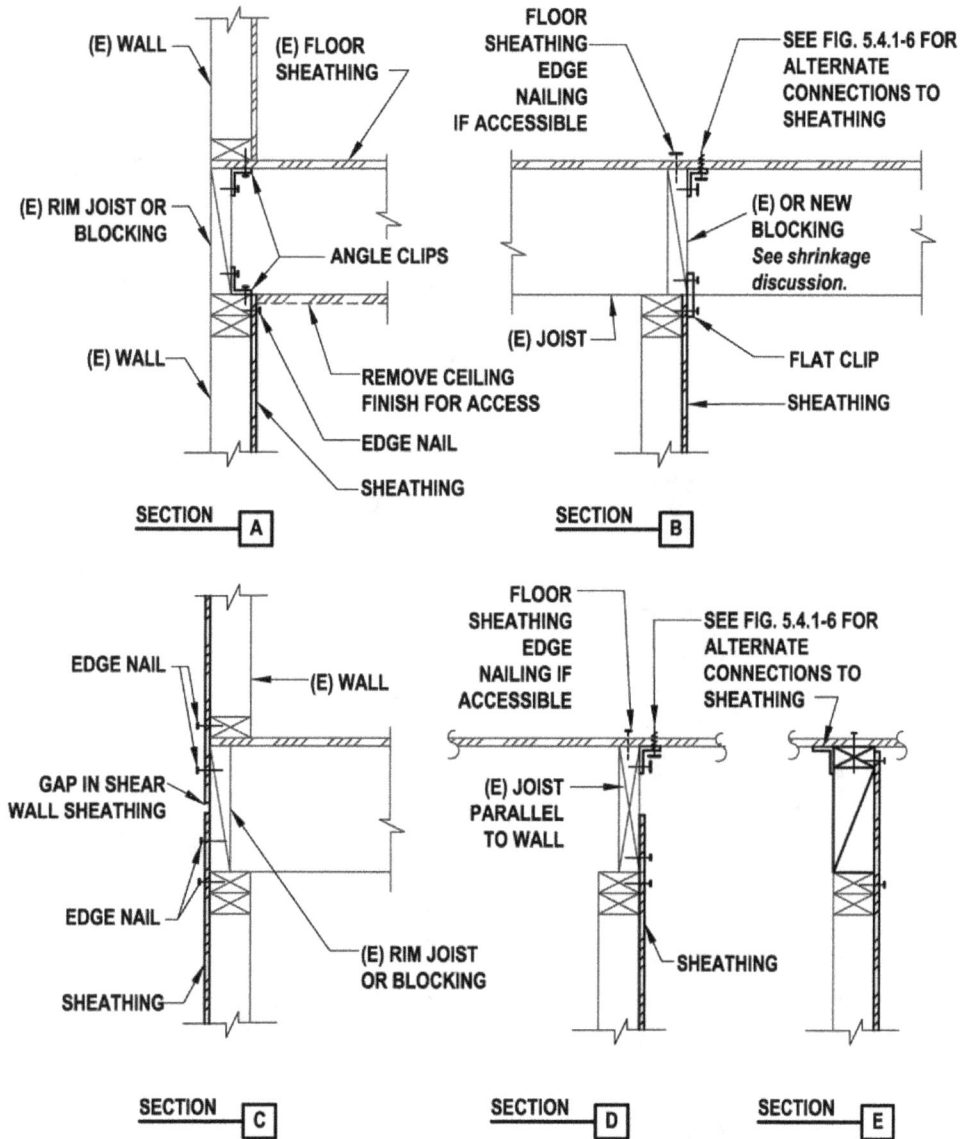

Figure 5.4.1-7: Load Path From Floor Diaphragm to Lower Story Shear Wall

Figure 5.4.1-12 illustrates common overturning support conditions at the base of first story shear walls. See also the earlier discussion of foundation design issues. See Section 6.4.4 for discussion of anchorage to concrete issues under recent ACI 318 (ACI, 2005) provisions.

SECTION A

(E) BALLOON FRAMED WALL

(E) FLOOR SHEATHING

(E) NAILING AT JOIST AND STUD LAP

(E) WALL

(E) JOIST

(E) CONTINUOUS RIBBON

(EXISTING CONSTRUCTION ONLY)

SECTION A

SECTION B

(E) BALLOON FRAMED WALL

(E) FLOOR SHEATHING

3X OR 4X BLOCKING ALSO SERVES AS FIRE STOP

SEE FIG. 5.4.1-6 FOR SHEATHING TO FRAMING CONNECTION

BLOCKING *See shrinkage discussion.*

(E) JOIST

REMOVE (E) RIBBON

SECTION B

SECTION C

SEE FIG. 5.4.1-6 FOR ALTERNATE CONNECTIONS TO SHEATHING

(E) FLOOR SHEATHING

BLOCKING *See shrinkage discussion.*

(E) JOIST

FLAT CLIP

2X NAIL PRIOR TO PLACING WALL SHEATHING

SECTION C

SECTION D

CLIP ANGLE WITH WOOD SCREWS TO FLOOR

2X BLOCKING ALSO SERVES AS FIRE STOP

SECTION D

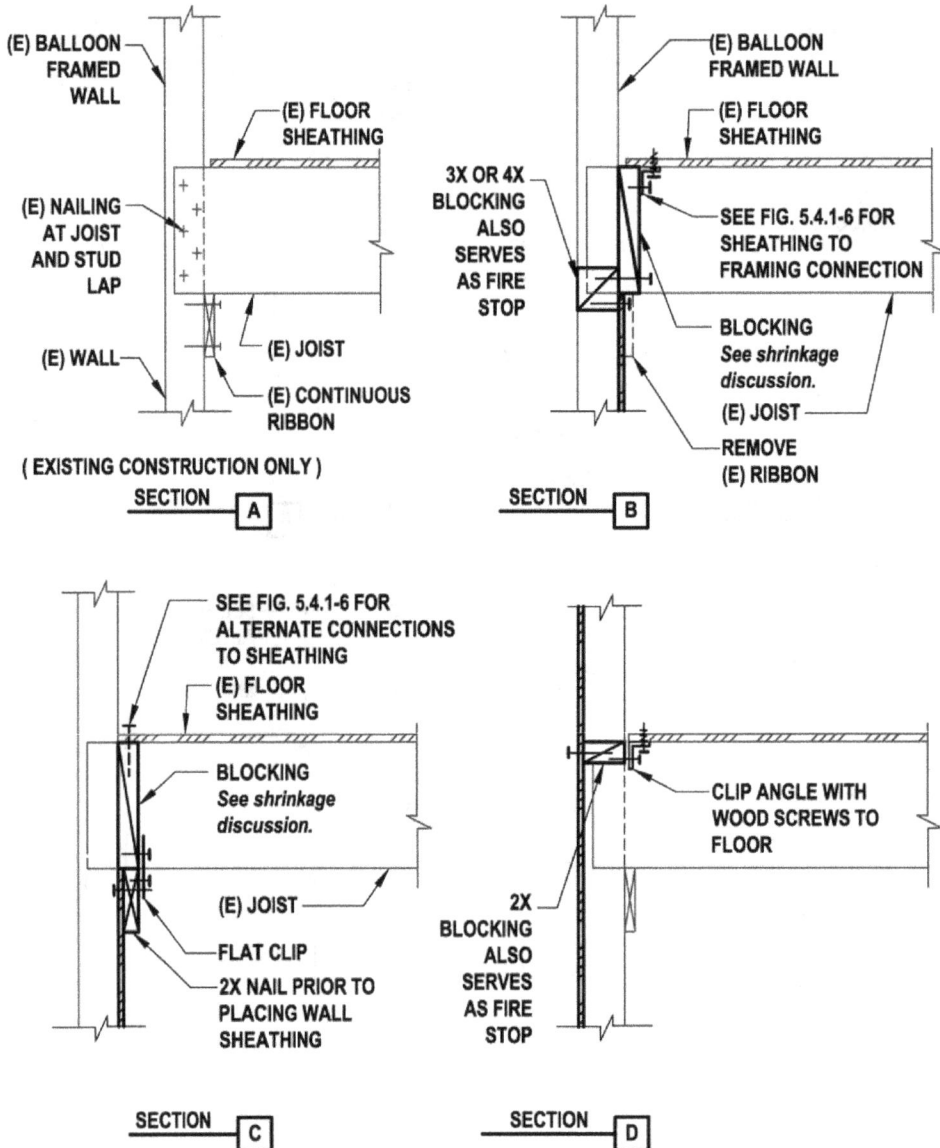

Figure 5.4.1-8: Load Path From Floor Diaphragm to Lower Story Shear Wall – Balloon Framing

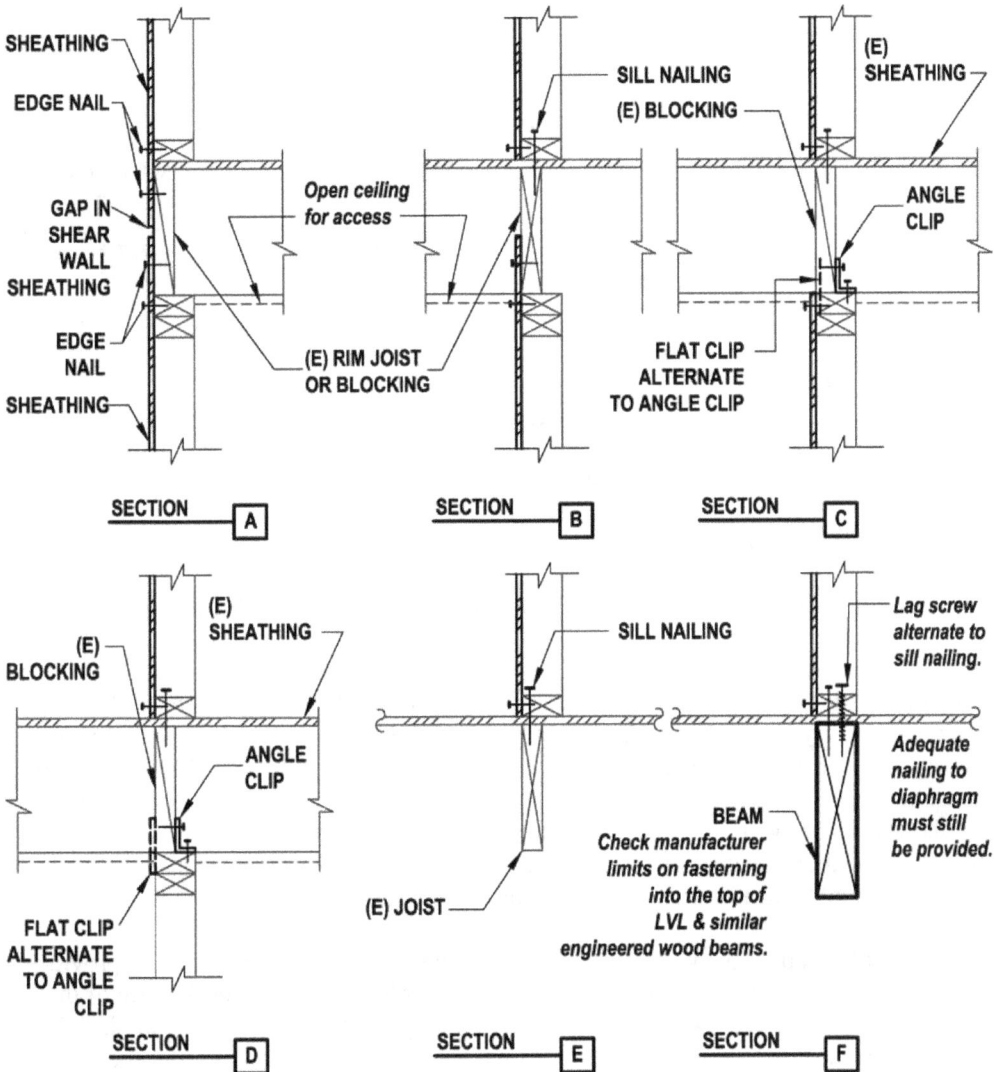

Figure 5.4.1-9: Load Path from Upper Story Wall-To-Floor Diaphragm

NOTES:

1. In A, B & C in-plane shear is transferred from the wall above to the wall below.
2. In D in-plane shear is transferred from the wall above to the top plates below. The top plate serves as a collector. Splice detail should be provided.
3. In E and F in-plane shear is transferred from the wall above to the floor diaphragm. Nailing adequate to transfer forces into the diaphragm is required.

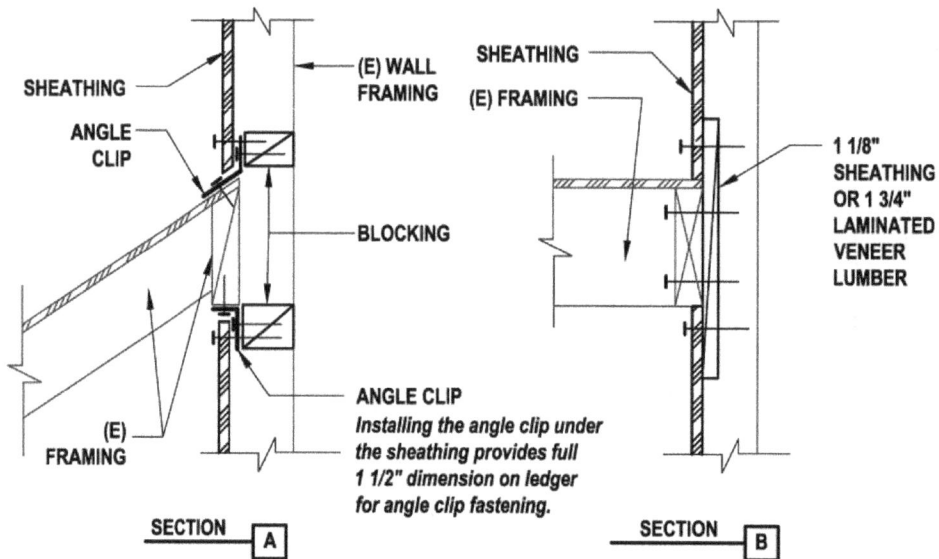

Figure 5.4.1-10: Load Path at Disruption in Shear Wall Sheathing

Reduction of slender shear wall height: Shear walls at the sides of garage doors and other large openings are often very slender and, therefore, develop significant overturning forces. One easy and relatively inexpensive approach of modestly reducing overturning forces and increasing wall stiffness is shown in Figure 5.4.1-13. A steel collector strap is run across the full length of the wall near the bottom of the door header. This strap will effectively reduce the shear wall height to the height of the door opening; in addition, limited moment fixity may develop at the wall top. The strap is nailed to the header and to blocking added in line with the bottom of the header.

The strap is best placed over the wood structural panel sheathing, so that strap nailing provides shear transfer to the sheathing. Alternately the strap can be placed on the opposite face of the framing, however fastening of the sheathing to the blocking and header is also required. This approach can be used alone, or in combination with rehabilitation of anchorage and sheathing.

Cost, Disruption, and Construction Considerations

Addition of sheathing and fastening to woodframe shear walls can often occur while the dwelling is occupied. Work will generally progress faster, however, without occupants. Where feasible, work on the outside face of exterior walls often provides not only the best access, but also the best load-transfer detailing options. Added sheathing that increases the thickness of a shear wall will require adjustment of trim at openings and reworking of water barrier detailing at windows and doors. Completely sheathing an exterior wall, including areas above and below windows and doors gives not only improved structural performance, but also the best surface for correctly installing windows and water barriers.

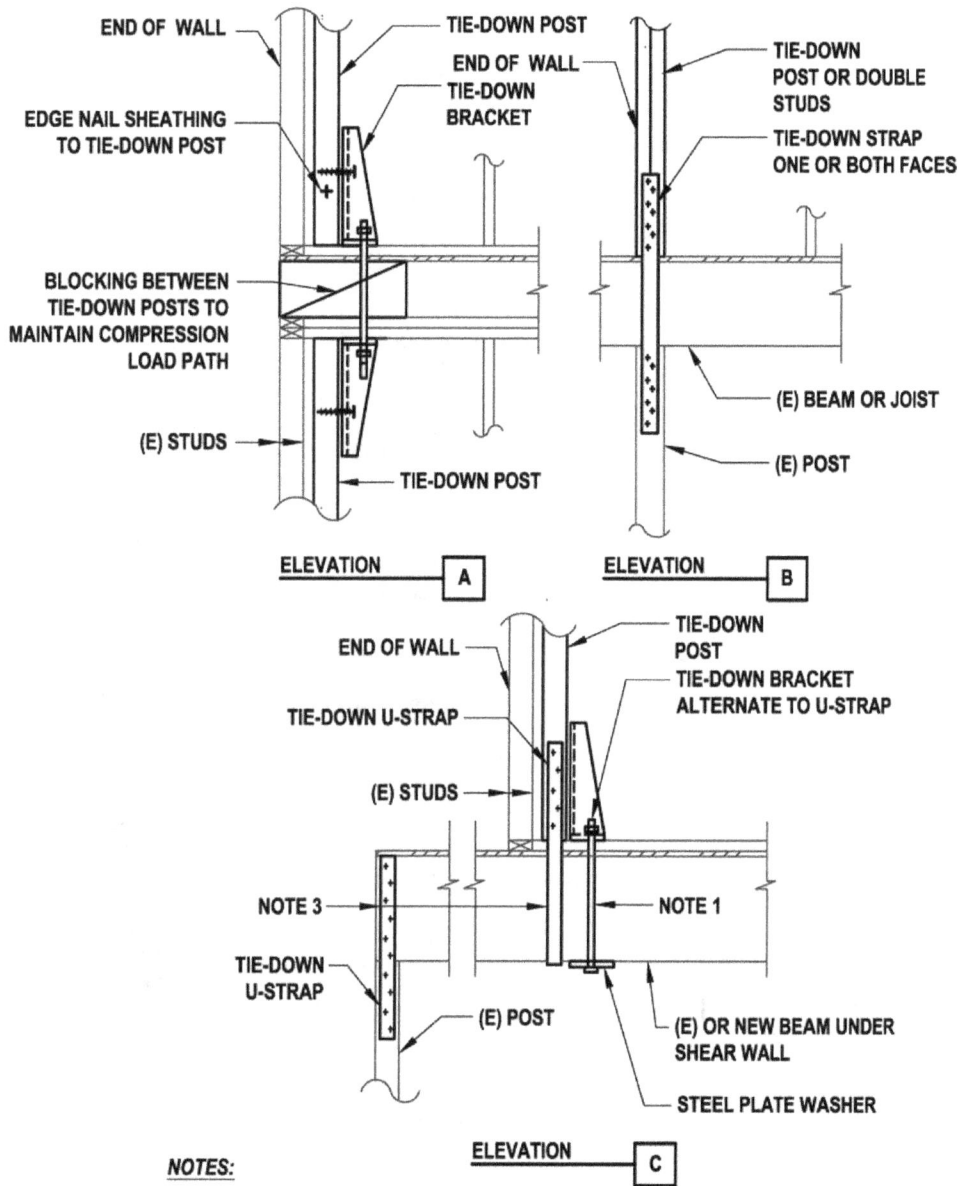

NOTES:

1. *Where beam is engineered lumber (glulam, LVL, etc.) consult with manufacturer regarding acceptability and design effect of bolt hole drilled through beam depth.*
2. *Shear wall sheathing omitted for clarity.*
3. *Provide shims where tie-down and beam are different widths.*

Figure 5.4.1-11: Load Path for Overturning (Tension and Compression) at Upper Story Shear Wall

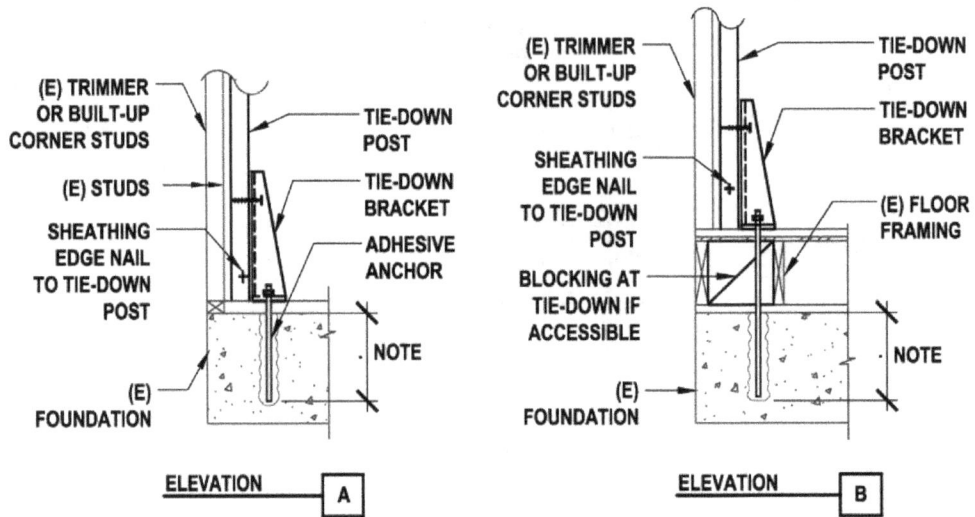

Note: Adhesive anchor embedment length into existing foundation as
required by ICCES evaluation report.

Figure 5.4.1-12: Load Path for Overturning (Tension and Compression) at Foundation

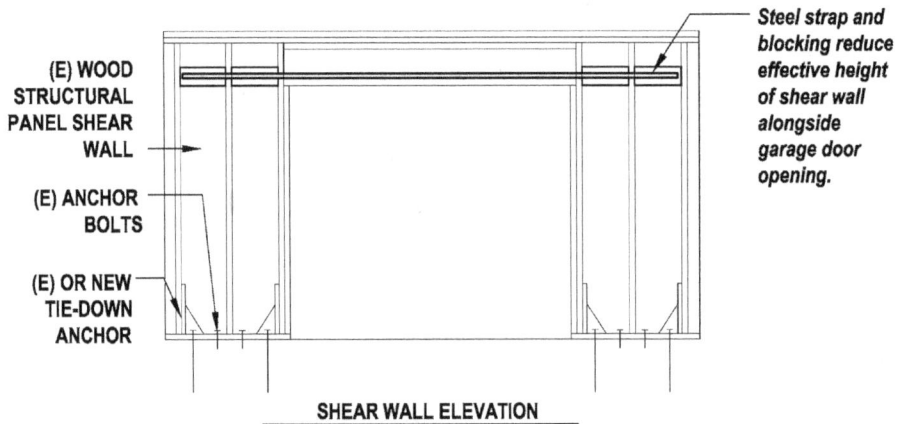

Figure 5.4.1-13: Reduction of Slender Shear Wall Effective Height

As in new construction, it can be a challenge to assure that rehabilitation measures are constructed with the fastener (nail, staple, screw, etc.) type and size that has been assumed in design and construction documents. Use of improper type and size often results in reduced rehabilitation measure capacity. Most nails are placed with nail guns. Most gun nails are ordered by diameter and length. Indications of type and pennyweight continue to be misleading. The only way to verify that required fasteners are being used is to measure them with calipers or a similar device. Fasteners connecting sheathing to framing should not be overdriven (not break the face of the sheathing). Where overdriving occurs, fastener capacity may be reduced up to 40%.

Proprietary Concerns

There are no proprietary concerns with this rehabilitation technique other than the use of proprietary connectors and adhesives as part of the assemblage.

5.4.2   Add Collector at Open Front

Deficiency Addressed by Rehabilitation Technique

This rehabilitation technique addresses configuration deficiencies created by an open front condition such as at a garage or window wall.

Description of the Rehabilitation Technique

Often in **W1** buildings, an open front will occur at a portion or wing of the building, while adequate shear walls are provided in an adjacent portion. A common example of this is a lack of shear wall at the front of a garage, while sufficient bracing exists in the adjacent portion, as illustrated in Figures 5.4.2-1 and 5.4.2-2. Where this type of condition occurs, a collector can be used to transfer seismic forces generated in the open front portion to the adjacent portion with adequate shear walls. In woodframe construction, most shear walls are capped by double top plates that can be used as collectors. Figure 5.4.2-1 shows the collector connecting from the roof diaphragm at the garage to top plate collectors at the front of the house. Figure 5.4.2-2 shows the collector connecting from the second floor diaphragm above the garage to double top plates at the front of the house.

Design and Detailing Considerations

Research basis: No research applicable to this rehabilitation measure has been identified.

General design: Collector connections like the ones illustrated in Figures 5.4.2-1 and 5.4.2-2 are often complex, and they can include both vertical and horizontal offsets between bracing lines. Rehabilitation is seldom inexpensive, and alternatives such as added shear walls should always be considered. If used, however, collectors should help to mitigate differential movement between the one-story and two-story portions of the building and to reduce resulting damage. The collector will most often but not always need to resist both tension and compression.

Figure 5.4.2-1 shows one of several possible methods of providing a collector. In the illustrated approach, a steel strap ties the top plates from the garage open-front to wood structural panel sheathed infill panels between the roof trusses in the one-story portion of the building. The infill panels transfer the load from the truss bottom chord up to the roof diaphragm, where loads can be carried to the shear walls. In Figure 5.4.2-1, a vertical eccentricity exists between the collector

WOOD STRUCTURAL
PANEL SHEATHING

ROOF DIAPHRAGM
SHEATHING
REQUIRED. NOT
SHOWN FOR
CLARITY

DOUBLE TOP
PLATE
COLLECTOR

SHEAR WALL

STEEL STRAP
COLLECTOR

OPEN FRONT AT FIRST STORY

ONE STORY ⟷ 2 STORY

B

SHEATHING
EDGE
NAILING

WOOD
STRUCTURAL
PANEL
SHEATHING
BETWEEN
TRUSSES
(TYPICAL)

2X4
EACH FACE

STEEL STRAP
COLLECTOR
CONTINUOUS

SPECIFY
LENGTH OF
LAP, TYPE
AND SPACING
OF NAILS

SHEATHING

CONNECTION
TO ROOF
DIAPHRAGM

2X4

NAIL 2X4
TO (E)
2X4 TO
DEVELOP
SHEATHING
CAPACITY

STRAP

(E) DOUBLE TOP
PLATE COLLECTOR
TO SHEAR WALL

ELEVATION   A

SECTION   B

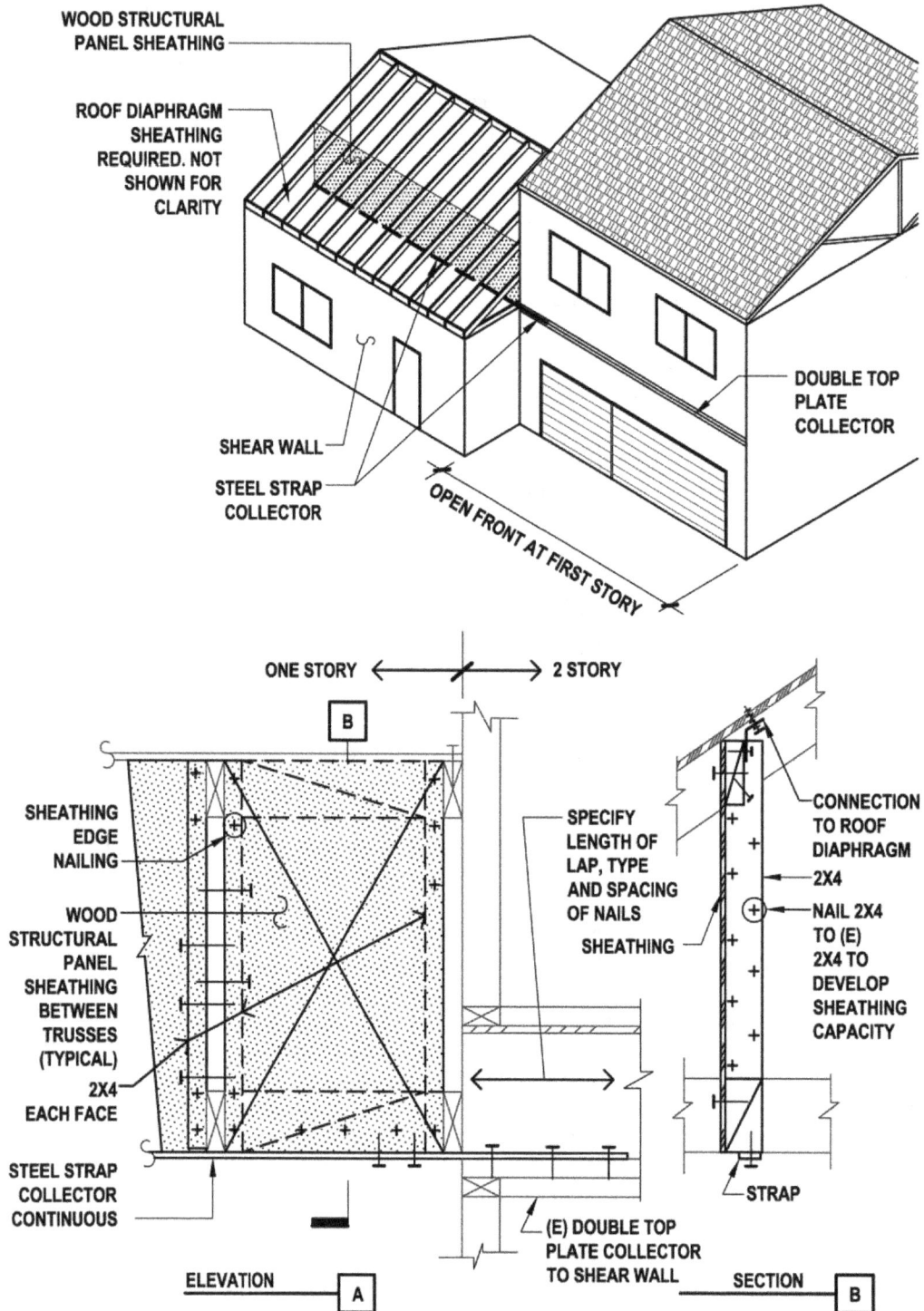

Figure 5.4.2-1: Collector from Garage Open Front to Adjacent Dwelling

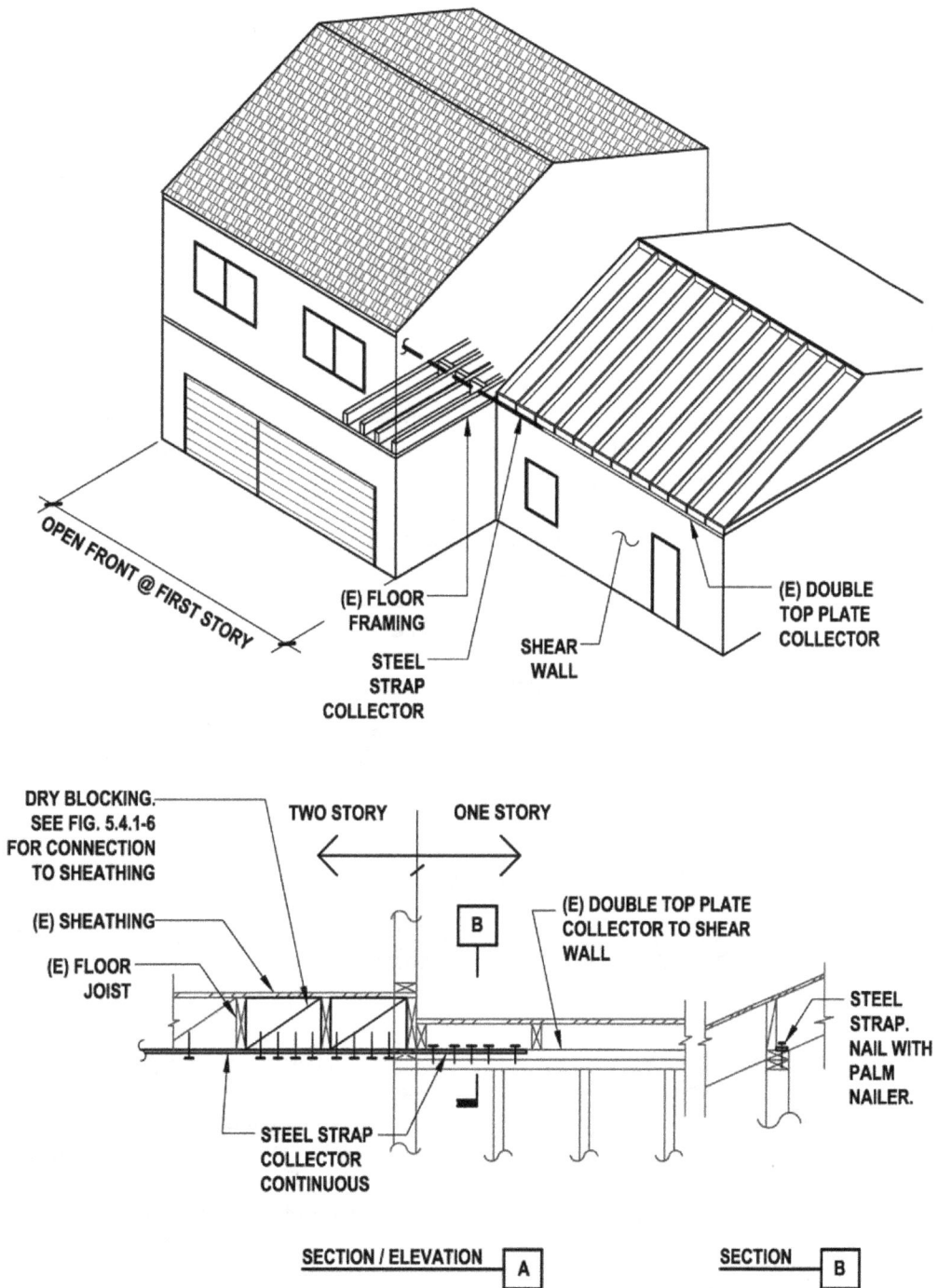

OPEN FRONT @ FIRST STORY

(E) FLOOR
FRAMING

STEEL
STRAP
COLLECTOR

SHEAR
WALL

(E) DOUBLE
TOP PLATE
COLLECTOR

DRY BLOCKING.
SEE FIG. 5.4.1-6
FOR CONNECTION
TO SHEATHING

(E) SHEATHING

(E) FLOOR
JOIST

TWO STORY   ONE STORY

B

(E) DOUBLE TOP PLATE
COLLECTOR TO SHEAR
WALL

STEEL
STRAP.
NAIL WITH
PALM
NAILER.

STEEL STRAP
COLLECTOR
CONTINUOUS

SECTION / ELEVATION A

SECTION B

Figure 5.4.2-2: Collector from Garage Open Front to Adjacent Dwelling

level and the roof diaphragm in the one-story portion. This eccentricity is resolved by continuing the sheathing infill panels for the full width of the one-story building so that the vertical reaction can be resisted at the exterior walls. An alternate approach would be to install a wood structural panel ceiling diaphragm on the underside of the roof trusses, in which case no vertical eccentricity would exist.

Figure 5.4.2-2 shows a steel strap from the underside of floor joist blocking above the garage to top plates in the adjacent framed wall. The floor blocking transfers load from the strap to the floor diaphragm above. The depth of the floor blocking creates a small vertical eccentricity, causing the blocks to overturn. End nails or toenails at each end of the blocking generally resist this overturning. See Figure 5.4.1-6 and related discussion for connection to the floor and roof diaphragm sheathing.

Deformation of collector: The collector will only be able to protect the open front against excessive drift if the deformation in the collector system is kept to a minimum. Elongation of the steel collector strap and nail slip are likely to be the primary contributors to deformation. Loads in the strap and nails should be kept moderate.

Other parts of the load path: When the double top plate serves as a portion of the collector, breaks in the double top plates may require steel straps in order to provide adequate capacity. In order to complete the load path, diaphragm capacity, roof diaphragm connections to the top plates, and splices in the top plates should all be checked.

Cost, Disruption and Construction Considerations

Installing the collector connection shown involves opening up ceiling finishes in both portions of the building and extensive work infilling between the roof trusses. Other solutions to bracing of the open front should be explored.

Proprietary Concerns

There are no proprietary concerns with this rehabilitation technique, other than the use of proprietary connectors as part of the assemblage.

5.4.3   Add or Enhance Anchorage to Foundation

Deficiency Addressed by Rehabilitation Technique

This rehabilitation technique addresses insufficient shear connection between woodframe dwellings and their foundations. The highest priority and most cost effective rehabilitation measure for **W1** buildings is ensuring that the home is adequately anchored to the supporting foundation. This technique is equally applicable to **W1A** and **W2** buildings. Enhanced anchorage may be provided from the foundation to first story walls, to floor framing, or to cripple walls. Enhanced anchorage is often used in combination with cripple wall enhancement as discussed in Section 5.4.4.

Description of the Rehabilitation Technique

Foundation anchorage can often simply involve anchor bolts connecting a foundation sill plate to the supporting continuous foundation or foundation wall. The intent is to transfer the earthquake horizontal base shear from the foundation sill plate into the foundation; nominal uplift capacity is often also provided by the anchorage. The primary objective is to keep the foundation sill and framed building above from sliding relative to the foundation under earthquake loading. Shear transfer to isolated footings or short foundation piers is not recommended without evaluation of the footing and transfer to the supporting soils. Where configuration and access prohibit installation of anchor bolts, proprietary anchors are used to transfer horizontal shear to the foundation.

Figure 5.4.3-1 illustrates common anchorage details using anchor bolts to existing concrete foundations. Figure 5.4.3-2 illustrates an anchor bolt connection to an existing masonry foundation. Where possible, anchor bolts remain the preferred method of anchorage to foundations. Where the existing foundation is concrete masonry (Figure 5.4.3-2), grout may not exist in all masonry cells. The existence of grout at the added anchor should be confirmed. Where anchorage into grouted cells is not possible, cutting out face shells and pouring grout around an added anchor bolt is a preferred alternative. As a second alternative, adhesive anchors intended for connection to hollow bases can be used; however, capacities are very low. A combination of anchorage to grouted and ungrouted cells is not recommended.

Steel plate washers need to be provided at each added anchor bolt between the foundation sill plate and the nut. Current codes require that the steel plate washer be a minimum of ¼"x3x3", and they allow a slotted hole to accommodate bolt location tolerances. Shear wall anchorage to foundations has been tested by Mahaney and Kehoe (2002).

Installation of anchor bolts in first story shear walls involves the removal of finish materials. Where shear wall rehabilitation per Section 5.4.1 is already being provided, finishes will generally be removed in order to access framing. Where finishes or structural sheathing are not otherwise going to be removed, it is possible to create access for anchor bolt installation by removing finishes over the bottom two to three feet of the wall (Figure 5.4.3-1A). Where structural sheathing is removed for access, blocking needs to be provided at all sheathing panel edges so that edge nailing can be provided when the sheathing is replaced.

In the configuration shown in Detail 5.4.3-1D, the existing foundation sill is wider than the existing studs. 2x4 blocking is added between the studs and nailed down to the foundation sill plate. In prescriptive provisions this is most often with four 10d common nails. Cripple wall retrofits using this base detail were tested by Chai, Hutchinson and Vukazich (2002) and performed well in testing. House inspectors, however, have reported seeing splitting of the 2x4 block in homes that have been retrofitted using this approach. Alternative fastening approaches include using nails with pre-drilling, using staples, and using wood screws between the block and the foundation sill plate. Another approach is to cut the foundation sill plate flush with the studs above so that blocking is not required. No testing is available to judge the relative performance of these approaches.

Addition of anchor bolts is often not possible with a crawlspace configuration due to inadequate vertical clearance for a rotary-hammer to drill down into the top of the foundation. Figure 5.4.3-3 illustrates some of the alternate proprietary anchors that can be used for these configurations. Although shown with stud walls above the foundation sill plate, these connections work equally

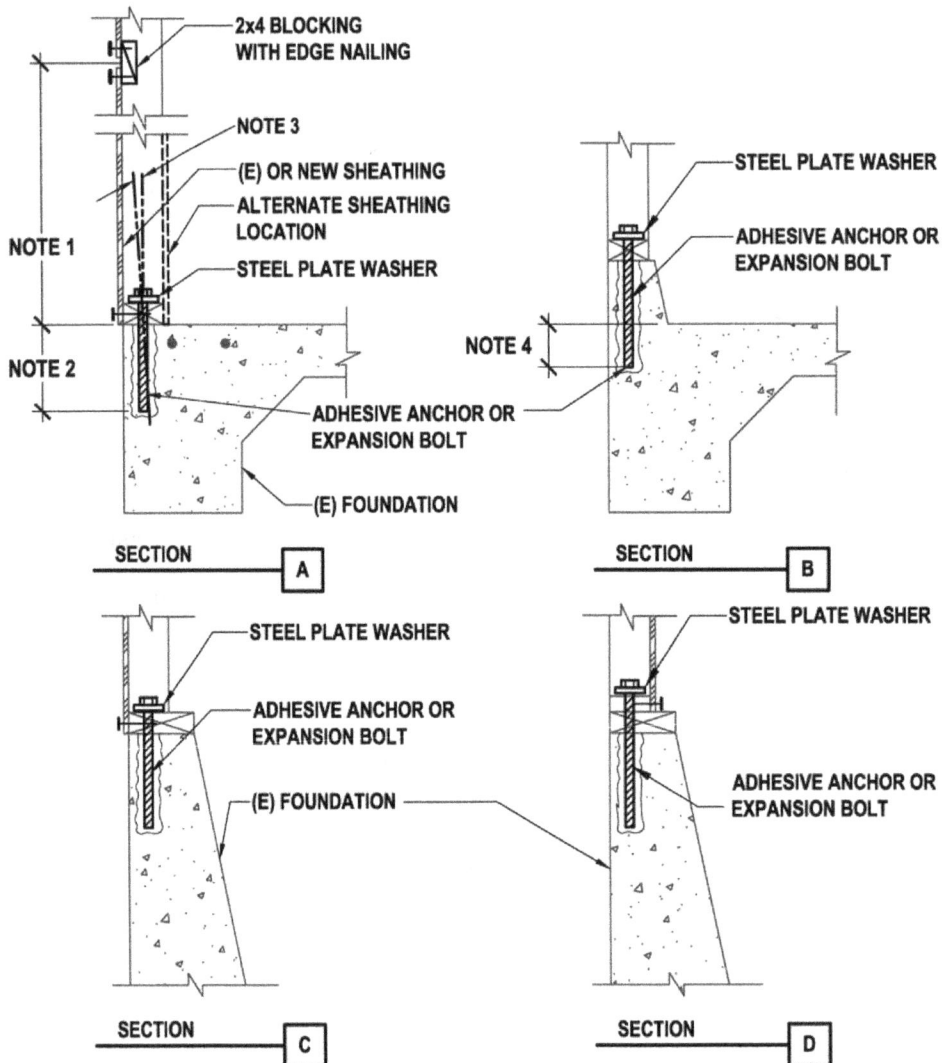

Figure 5.4.3-1: Added Anchor Bolt at Existing Concrete Foundation

NOTES:

1. A height of 2 or 3 feet is required for rotohammer access to drill anchor bolt holes. Existing sheathing
may need to be opened up to provide this access. Block and edge nail all panel edges when replacing sheathing.
2. Specify depth for adhesive anchor embedment into existing concrete.
3. It is acceptable to install anchor bolts at a small angle to vertical provided that concrete cover over the bolt is
maintained and that full bearing between the steel plate washers and foundation sill plate is maintained.
4. Where concrete curb length is short, extend anchor bolt below top of slab.

well when floor framing sits directly on the foundation plate. The steel angle connection in Figure 5.4.3-3B is generally not recommended as an alternate to anchor bolts for in-plane shear due to flexibility and potentially causing cross-grain splitting of the joists; other depicted anchor types resist in-plane shear much more effectively. ICC Evaluation Service reports should be consulted for anchors to the foundation and alternate proprietary anchors.

Pier-and-curtain wall foundations (Figure 5.4.3-4) are used in some areas of the southern United States. As-built anchorage for shear transfer between the wood framing and foundation is generally minimal to non-existent. Rehabilitation of anchorage to this type of foundation is not known to have been undertaken to date. One possible approach is a continuous steel angle from the underside of the floor framing to the inside face of the single-wythe curtain wall, anchored to the curtain wall with veneer anchors and to the wood with nails or screws. Care would need to be taken in drilling for veneer anchors. An alternate approach would be new concrete or masonry foundations from pier to pier, allowing use of cast-in anchor bolts to the foundation and nailed or screwed connections to the wood framing.

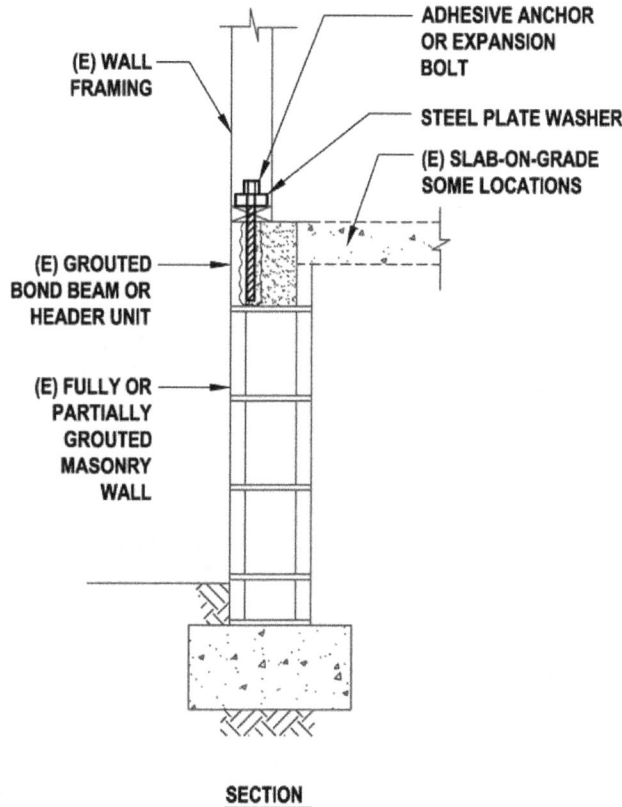

SECTION

**Figure 5.4.3-2: Added Anchor Bolt at Existing Partially
Grouted Concrete Masonry Foundation**

SECTION [A]

(E) FRAMING
PROPRIETARY CONNECTOR
ADHESIVE ANCHOR OR EXPANSION BOLT
(E) FOUNDATION
SPECIFY DEPTH

SECTION [B]

(E) FRAMING
PROPRIETARY CONNECTOR
(E) FOUNDATION
ADHESIVE ANCHOR OR EXPANSION BOLT
SPECIFY DEPTH

SECTION [C]

(E) FRAMING
PROPRIETARY CONNECTOR
(E) FOUNDATION
ADHESIVE ANCHOR OR EXPANSION BOLT
SPECIFY DEPTH

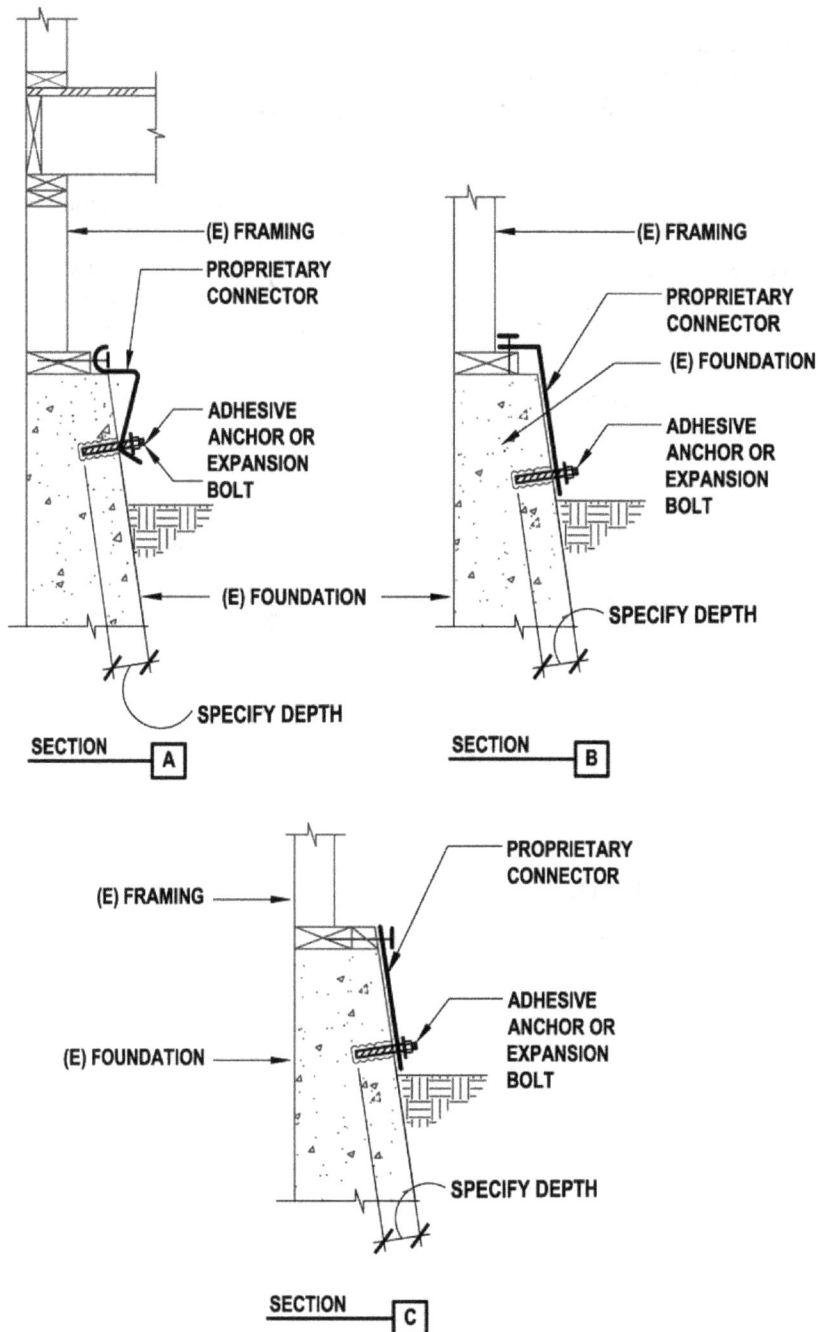

Figure 5.4.3-3: Anchorage to Existing Foundation Using Proprietary Connectors

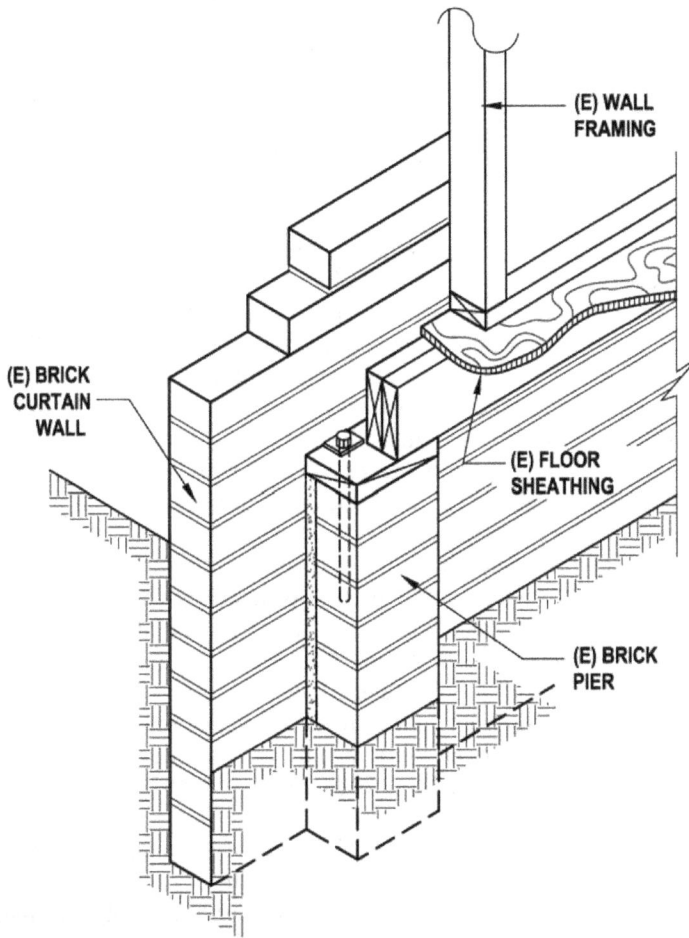

**Figure 5.4.3-4: Pier and Curtain Wall Foundation System with Inadequate
Load Path Between Shear Walls and Foundation**

Design and Detailing Considerations

Research basis: Testing of shear wall to foundation anchorage has been conducted by Mahaney
& Kehoe (2002). Testing of prescriptive cripple walls anchored to foundations has been
conducted by Chai, Hutchinson & Vukazich (2002).

Anchor type and installation: A variety of proprietary anchors are available for anchorage to
existing concrete and masonry foundations. Both manufacturer literature and ICC Evaluation
Service reports should be consulted for information on conditions of use, allowable loads, and
installation and inspection requirements. It is important to make sure that the anchor type is
appropriate for the material being connected to, is approved for seismic loads, and is appropriate
for weather and temperature exposure. Either adhesive or expansion anchors to the existing
foundation are commonly used; however, because expansion anchors create splitting tensile

forces, the proximity to the foundation edge and strength of existing foundation material may make use of adhesive anchors a better choice. In addition, some concerns have been raised regarding potential relaxation of expansion anchors under seismic loading. Use of powder-driven fasteners for anchorage to concrete or masonry is not recommended due to concerns regarding performance under cyclic loading (Mahaney & Kehoe, 2002). The diameter of drilled holes is specified in installation requirements for each anchor type; variation from this size often leads to inadequate anchor capacity.

Most manufacturers have caulking gun-like devices that make field placement of adhesives fairly simple and automatically mix two-part adhesives. Generally, these types of adhesives provide more than adequate strength, and there is no need to use more complicated high-strength adhesive types. The cleaning of holes prior to placing adhesive anchors is paramount for anchor capacity. When not well cleaned, the anchors can pull out at a small fraction of the design load. It is common to pull-test a portion of the adhesive anchors to verify adequate installation. The pull test load is usually in the range of one to two times the tabulated allowable stress design tension load. The bridge used for testing generally makes a concrete pull-out failure unlikely. The test load should not be near yield load for bolts or adhesive pull-out (bond) failure loads.

Use of nonshrink grout in lieu of adhesives for anchor bolt attachment is another possible installation alternative. This approach was commonly used prior to adhesives being readily available. If used, literature from the grout manufacturer should be consulted for installation requirements and anchorage design procedures. The hole drilled for anchor placement is often required to be 1/4–inch (or more) larger that the diameter of the anchor being placed. This size of hole may not be practical near the edge of a foundation and in weaker foundation materials. When using this approach, it is important that the anchorage design consider the implications of full expected seismic loads, rather than just code level loads.

Anchors will very often need to be installed near the exterior edge of a foundation. Typical anchor bolt placement in nominal 4-inch walls results in a distance from center of bolt to edge of concrete of 1-3/4 inches. Due to this edge distance, reductions in anchor capacity will likely apply. In addition, it is recommended that a minimum clear cover distance be maintained between the face of the anchor and the exterior face of the foundation. Where the exterior face of the foundation in the vicinity of the anchor bolt has been formed, ACI 318 Appendix D would require a clear cover of 1-1/2 inches in new installation. This provides reasonable guidance for rehabilitation also. In addition the placement of the anchors will be limited somewhat by the dimensions of the steel plate washers. Where possible, moving an anchor away from the edge of the foundation will result in a stronger foundation anchorage, but may not affect the wood to steel capacity. When the anchorage is at the base of a sheathed shear wall or cripple wall, it is best to keep the anchor as close as practical to the sheathed face of the studs in order to minimize risk of sill plate cross-grain splitting.

Configuration implications: Where foundation anchors are being installed in a crawl space, the design of anchorage to the existing foundation will be driven almost entirely by the configuration of the existing foundation, sill plate and framing configuration. A good look at existing conditions is needed before design is started. Limitations on access for materials and equipment will often limit anchorage methods.

Prescriptive and engineered anchorage: Prescriptive provisions for anchorage of foundation sill plates and cripple walls can be found in the International Existing Building Code – IEBC -- (ICC, 2003b). These were developed from similar or identical provisions in the GSREB (ICBO, 2001) and the UCBC (ICBO, 1997b). An extensive commentary to the GSREB Chapter 3 provisions has been developed by SEAOC Existing Buildings Committee (ICC, 2005). Some organizations have developed local adaptations of these provisions. The objective of these prescriptive provisions is reduction of earthquake hazard; they are intended to provide a reasonable level of improvement for the majority of buildings within their scoping limitations. **W1** buildings with unusual configurations, site slopes greater than one vertical in ten horizontal, or higher performance objectives should be addressed with an engineered design. An engineered design is recommended for all **W1A** and **W2** buildings because of higher loads and potential configuration issues.

Engineered design for anchorage without specifically identified superstructure shear walls: In cases where the prescriptive provisions are not applicable, it may be desirable to provide an engineered design for foundation anchorage, with or without cripple wall bracing. An engineered design allows load distribution to the cripple walls to be addressed for the specific building configuration and allows specific design for non-standard framing and foundation conditions. Where rehabilitation will be limited to anchorage to the foundation, it is common to make simplifying assumptions regarding force distribution. For small buildings, forces generated at and above the lowest framed floor are distributed by tributary area to the perimeter foundations. For larger buildings, force may also be distributed to interior foundations based on tributary area. In addition to providing foundation anchorage at engineered cripple walls, it is desirable to provide a minimum level of anchorage for all foundation sill plates to avoid loss of vertical support should building movement occur.

Engineered design for anchorage with specifically identified superstructure shear walls (see also Section 5.4.1): Where shear walls are being added or enhanced in the story above the crawlspace, the foundation anchorage design will need to specifically provide a load path for the shear wall reactions.

Adequacy of foundation: Shear anchorage of a woodframe building to a foundation generally puts modest demands on the foundation. In order to perform adequately, the foundation needs to resist local demands from the anchor installation (such as drilling as splitting tensile stresses if installing expansion anchors), and it needs to have enough continuity to distribute the seismic shear forces without local failure. Installation of shear anchorage into existing reinforced concrete or masonry footings or foundation walls is commonly done without any specific evaluation of the foundation capacity. Likewise, shear anchorage to an unreinforced concrete foundation in good condition is commonly done without specific evaluation. Evaluation is needed when any foundation shows signs of deterioration due to differential movement, moisture, or other causes. Foundations that are moving differentially should be stabilized prior to installation of anchorage. If not stabilized, further movement of the foundation can telegraph into deformation and damage in the building above.

Views on addition of shear anchorage between woodframe dwellings and unreinforced masonry foundations vary widely. In some regions, there is considerable concern that unreinforced brick foundations are fragile due to moisture driven deterioration and lack of confining overburden. Approaches to shear connections taken in these regions include casting new foundations alongside existing foundations and cutting out blocks of existing foundations in order to place a concrete key around added anchor bolts. In other regions, it is more common to recommend bolting woodframe dwellings directly to unreinforced brick masonry foundations that are in good condition. IEBC Chapter A3 requires an engineering evaluation of unreinforced masonry foundations, but does not provide details of the required evaluation. This allows some flexibility for anchorage practice to be determined locally based on local concerns, experience, foundation materials and construction practice. Load testing of anchorages should be considered as a quality assurance measure, particularly when new combinations of foundation materials and anchorage methods are being used. Addition of overturning anchors or concentrated loads requires specific evaluation of foundation capacity.

Special attention is needed where a masonry foundation is constructed of large cut stones because use of typical connections is impractical.

Prestressed foundations: Where foundations contain prestressing tendons, it is important to locate tendons prior to drilling for foundation anchorage. Tendons cut during drilling for anchorage may fail explosively, either along the length of the tendon or at the tendon anchorage, potentially causing injury and damage. Original design drawings identifying tendon locations and profiles are of great value in understanding placement. Alternately, post-tensioning experts can field locate tendon anchorages and profiles.

Alternate anchorage configurations: In California, encouragement of anchor bolting at the state, county and local government level has led to a noticeable amount of retrofit for anchorage to foundations. The lack of mandatory standards has led to a great variety of anchorage types being used, some appropriate for shear force transfer between the foundation and framing and some not. Where anchorage details used for prescriptive designs are not coming from national standards such as IEBC (ICC, 2003a), or guidance developed by local authorities, it is necessary to ascertain whether 1) the connection appropriately addresses the primary objective of preventing movement between the foundation sill plate and foundation, and 2) the capacity is comparable to the capacity that would have been provided by a prescriptive connection. In making this evaluation, consideration should be given to earthquake loading in both horizontal directions and a complete load path, additionally the occurrence of cross-grain tension should not be allowed.

Construction Considerations

Addition of foundation anchorage in a crawl space with minimum required code vertical clearance is difficult due to very cramped conditions; work areas are often hard to get to, let alone getting tools and supplies and executing work. New temporary access openings and disconnection of HVAC ducting may occasionally be needed to provide access to work.

Proprietary Concerns

There are no proprietary concerns with this rehabilitation technique, other than the use of proprietary connectors as part of the assemblage.

5.4.4   Enhance Cripple Wall

Deficiency Addressed by Rehabilitation Technique

This rehabilitation technique addresses enhancement of existing cripple walls. After addition of anchor bolts, as discussed in Section 5.4.3, enhancement of cripple walls is the most effective rehabilitation measure for older one- and two-family detached dwellings. Past earthquakes have repeatedly shown cripple walls to be a significant weak link in the performance of **W1** buildings. **W1A** and **W2** buildings with this configuration are equally susceptible. This rehabilitation measure is almost always done in conjunction with providing anchorage to the existing foundation (Section 5.4.3).

Description of the Rehabilitation Technique

This rehabilitation measure involves addition of wood structural panel shear wall sheathing to existing cripple walls and development of a load path into and out of the walls. The objective is to eliminate in-plane shear failure of the cripple walls, often resulting in the building falling off of the cripple walls and foundation.

Prescriptive provisions for rehabilitation of cripple walls can be found in the International Existing Building Code – IEBC -- (ICC, 2003b). These were developed from similar or identical provisions in the GSREB (ICBO, 2001) and the UCBC (ICBO, 1997b). An extensive commentary to the GSREB Chapter 3 provisions has been developed by SEAOC Existing Buildings Committee (ICC, 2005). Some organizations have developed local adaptations of these provisions (ABAG, 2005). The objective of these prescriptive provisions is reduction of earthquake hazard; they are intended to provide a reasonable level of improvement for the majority of buildings within their scoping limitations. **W1** buildings with unusual configurations, site slopes greater than one vertical in ten horizontal, cripple walls taller than 4 feet, or higher performance objectives should be addressed with an engineered design. An engineered design is recommended for all **W1A** and **W2** buildings because of higher loads and potential configuration issues.

The prescriptive provisions address:

> Shear transfer between floor framing and the cripple wall top plate
> Shear wall sheathing and fastening
> Anchorage of the foundation sill plate to the foundation (Section 5.4.3)

Figure 5.4.4-1 illustrates common cripple wall enhancement. The top of wall detail shows angle clips to a continuous rim joist or blocking. It is assumed that both the floor sheathing and sole plate above are nailed to the rim joist or blocking. If not, shear transfer per Figure 5.4.1-7 should be provided.

Where cripple walls are 14 inches tall or less, wood structural panel sheathing may no longer provide reliable bracing of the studs, and splitting of the studs becomes a more significant concern. For this configuration, use of solid blocking between studs is recommended in lieu of sheathing.

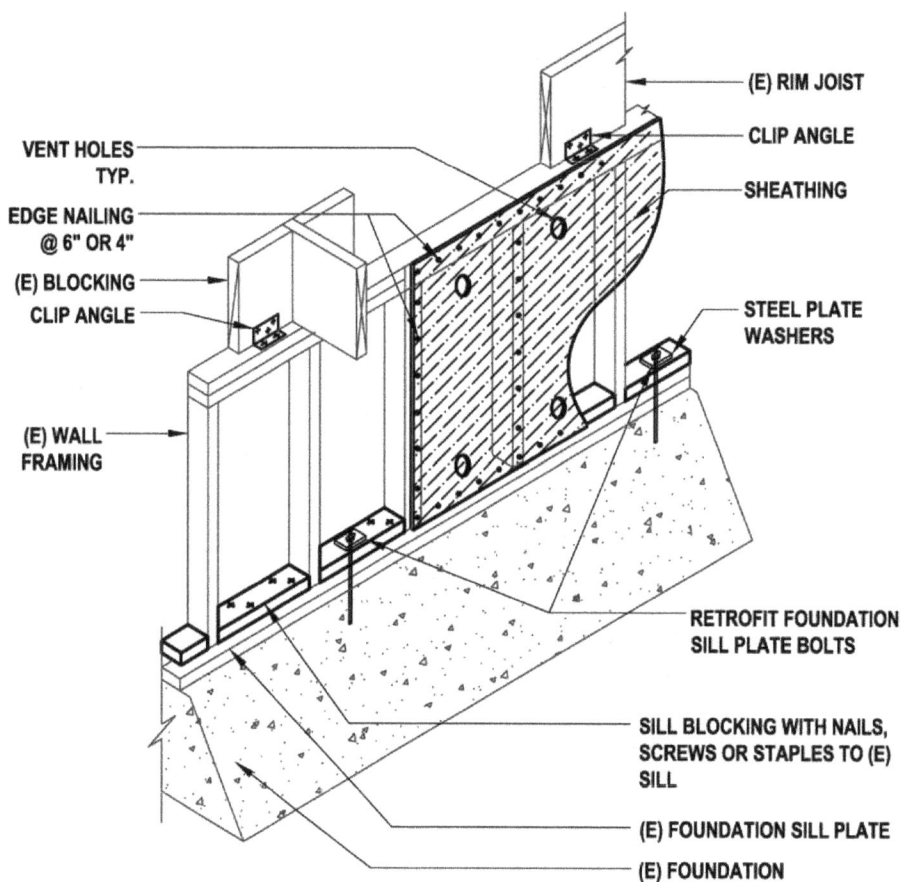

Figure 5.4.4-1: Cripple Wall Enhancement

An engineered design of cripple wall bracing would be anticipated to use very similar detailing, although additional fastening to further complete the load path may be desirable.

Design and Detailing Considerations

Research basis: Research into prescriptive methods for strengthening of cripple walls was conducted by Chai, Hutchinson & Vukazich (2002).

Bracing material vulnerability: Cripple walls have been seen in analytical studies and past earthquakes to often be subjected to much higher drifts than the occupied stories above. Wood structural panel sheathing is the preferred bracing material for cripple walls in order to accommodate required drifts without significant loss of capacity. Although still permitted for shear walls in new construction, stucco has not consistently provided adequate bracing of cripple walls. Often fasteners between the stucco and framing have withdrawn, resulting in damage and collapse. As a result, rehabilitation is encouraged for cripple walls not braced by either wood structural panel or diagonal lumber sheathing.

Horizontal force distribution: Where rehabilitation will be limited to the cripple walls and anchorage to the foundation, it is common to make simplifying assumptions regarding force distribution to the cripple walls. For small buildings, forces generated at and above the lowest framed floor are distributed by tributary area to the perimeter foundations. For larger buildings, force may also be distributed to interior cripple walls based on tributary area. Where buildings have had additions, cripple wall bracing may be needed on the foundation separating original and addition construction.

Where crawl spaces extend under framed decks and porches, it is necessary to provide cripple wall bracing at the perimeter of the enclosed building, as well as at the perimeter of the framed deck or porch. With this configuration it is sometimes necessary to alter the bracing approach to allow continued under-deck access. Other bracing approaches should have load-deflection behavior similar to the rest of the cripple walls, or the system should be evaluated considering the differences is behavior. Occasionally perimeter foundations are not complete between the enclosed dwelling and the deck or porch. The simplest solution is often to complete the foundation and add braced cripple walls.

Overturning anchorage: Tie-down anchors are not required by the IEBC provisions. This is primarily because the low unit shears in the sheathing (controlled by 15/32 sheathing and 8d common at 4" nailing) and a maximum wall height of four feet limit the overturning forces that are generated. Testing by Chai, Hutchinson & Vukazich (2002) indicates that good cripple wall behavior (strength, stiffness and energy dissipation) can occur with this construction. If the bracing unit shear capacity is increased or if the height of the cripple walls are increased, overturning anchorage may be required. See Sections 5.4.1 and 6.4.4 for discussion of overturning anchorage.

Ventilation and access: Existing access openings, ventilation openings and flood vents should not be reduced and, if possible, should be increased to meet code requirements during cripple wall bracing.

Construction Considerations

Moisture exposure: Elevated moisture can sometimes occur at cripple wall construction. Possible moisture sources include seasonal rain coming through cracks in the wall finish and high relative humidity at the building location. Decay in the existing cripple wall framing is a good indication that the rehabilitation work may also have a potential for decay. Where decay exists in existing framing, it should be repaired. Where no specific source of water can be identified and stopped, it is recommended that both replacement framing and new construction use preservative treated wood products and corrosion resistant fasteners and connectors. See Section 5.4.1 for further discussion.

Ventilation of stud bays: Where cripple wall studs are being sheathed on the interior face, it is recommended that ventilation holes be provided near the top and bottom of each stud bay to allow air circulation. Ventilation holes of 1-1/2 to 2 inches in diameter with centerline no closer than three inches to the panel edge will generally not reduce the effectiveness of the cripple wall bracing.

Variations in existing framing details: It is common to find variations in the framing details at the top of the cripple walls. The variations come from initial construction, repairs, and additions. Modification to typical details is often needed to address these conditions. Care should be taken that these modifications address the basic objective of transferring in-plane forces into the top of the cripple wall and providing capacity approximately equal to the detail being replaced.

Access: Access openings and under-floor clearance are likely to control the size of wood structural panel sheet that can practically be placed.

Proprietary Concerns

There are no proprietary concerns with this rehabilitation technique, other than the use of proprietary connectors as part of the assemblage.

5.4.5   Rehabilitate Hillside Home

Deficiency Addressed by Rehabilitation Technique

This rehabilitation technique addresses seismic vulnerabilities associated with hillside buildings. Buildings constructed on sites sloping downward from street level will often have cripple walls or skirt walls of widely varying heights around the building perimeter between grade and the lowest framed floor. The variation in height leads to widely varying shear wall stiffness. Seismic forces away from the hill can lead to the floor diaphragm pulling away from the uphill foundation (Figure 5.4.5-1A). Seismic forces across the hill can result in torsion due to stiff support on the uphill side and flexible support on the downhill side, also pulling the floor away from the uphill foundation and damaging stepped or sloped side cripple walls (Figure 5.4.5-1B). Similar behavior can result when steel rod bracing rather than cripple walls provide bracing between floor and grade. Collapse of hillside homes in the Northridge earthquake was attributed to this behavior. Information on damage from the Northridge earthquake and hillside building behavior can be found in City of Los Angeles & SEAOSC (1996), EERI (1996), von Winterfeldt et al. (2000) and Cobeen, Russell and Dolan (2004).

Description of the Rehabilitation Technique

The primary objective of this rehabilitation technique is to address hillside buildings that are vulnerable due to inadequate or missing bracing between the lowest framed floor and grade. A primary resource for this technique is voluntary rehabilitation provisions developed by the City of Los Angeles and included in the City of Los Angeles Building Code (City of Los Angeles, 2002). The objective of these provisions is to reduce the risk of death or injury. The provisions are indicated to be applicable to buildings constructed on a hillside slope in excess of one vertical to three horizontal. The rehabilitation measures described, however, may not be applicable to all **W1** buildings constructed on this slope.

The basic elements of the City of Los Angeles voluntary provisions include:

FIRST FLOOR

Stiff support at
uphill
foundation.
Load will go
here first.
Connection
needs to be
strengthened.

Flexible
support at
stepped
cripple wall.

STEPPED
FOUNDATION

SEISMIC FORCE

Diaphragm
deflection if
spanning to
stepped
cripple walls.

CRIPPLE WALL

SEISMIC FORCE AWAY FROM THE HILL   A

FIRST FLOOR

Diaphragm
rotation.

Stiff support
at uphill
foundation.

SEISMIC FORCE

STEPPED
FOUNDATION

Flexible support at
down hill cripple wall.

SEISMIC FORCE ACROSS THE HILL   B

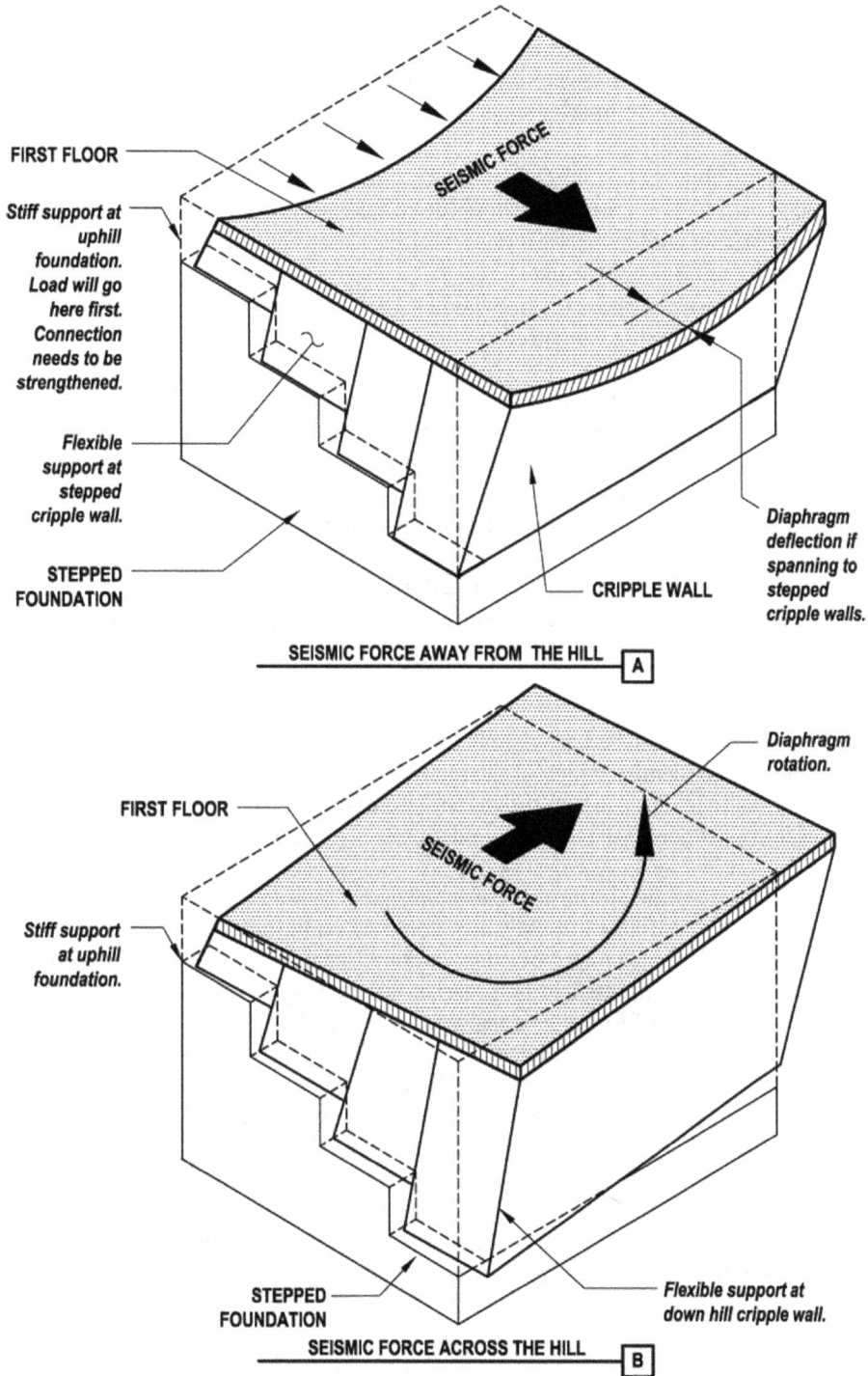

Figure 5.4.5-1: Hillside Home Response to Seismic Forces
Adapted from Von Winterfeldt, Roselund and Kitsuse (2000)

"Primary anchors" (designed for tributary seismic load) tying the floor diaphragm to the uphill foundation in line with each foundation extending in the downhill direction

"Primary anchors" where interior shear walls occur in contact with the base level diaphragm

"Secondary anchors" to the uphill foundation at a spacing not exceeding four feet

Foundation load path at primary anchors (or addition of tie-beam extending downhill from anchorage location)

Drift limits for tall downhill walls

Alternates to primary anchors include wood shear walls, steel braced frames, and rod bracing, all within specific limitations

The primary focus of this rehabilitation technique is providing direct tension anchorage from floor diaphragms to uphill foundations or foundation walls, as shown in Figure 5.4.5-2. This anchorage prohibits separation of the floor diaphragm from the uphill foundation or foundation wall, whether from direct tension or rotation. In doing so, the lateral and vertical load paths at the uphill foundation are maintained. The provisions require engineering evaluation and design.

Figure 5.4.5-2 illustrates a primary anchor at the exterior wall, in line with the stepped foundation, a primary anchor interior with a concrete tie-beam added in line, and a secondary anchor between the two, with no requirements for load path beyond anchorage to the uphill foundation.

To date, these are the only published provisions for addressing vulnerable hillside buildings. Further work is needed to identify which of the many possible hillside building configurations are vulnerable. At this time, there are no provisions addressing hillside buildings on pole or pier foundations where connection to the uphill foundation is not possible.

Damage observed following the Northridge earthquake also raised questions about the performance of stepped woodframe cripple walls, common on the sides of hillside buildings. It was suggested that seismic forces might be concentrating in the shortest uphill step of the woodframe walls, causing overstress and progressive failure. City of Los Angeles provisions require that the concentration of forces be considered in stepped cripple wall analysis. Testing of stepped cripple walls by the CUREE-Caltech Woodframe Project (Chai, Hutchinson and Vukazich, 2002) did not observe concentrations of seismic force, but instead saw well distributed forces and good performance. No explanations are currently available for the contrast between performance in testing and observed Northridge earthquake behavior.

Design Considerations

Research basis: Limited testing of the load-deflection behavior of tie-down devices used for diaphragm anchorage to uphill foundations has been conducted by Xiao and Xie (2002). See Cobeen, Russell and Dolan (2004) for discussion of the use and limitations of this information.

PRIMARY DIAPHRAGM
ANCHOR

(E) FOUNDATION
WALL

SECONDARY DIAPHRAGM
ANCHOR

(E)
LIGHT-FRAMED
WALL

PRIMARY
DIAPHRAGM
ANCHOR

(E) LEDGER

(E) UPHILL FOUNDATION

CONCRETE TIE-BEAM
*Slopes with grade in line
with primary anchor.*

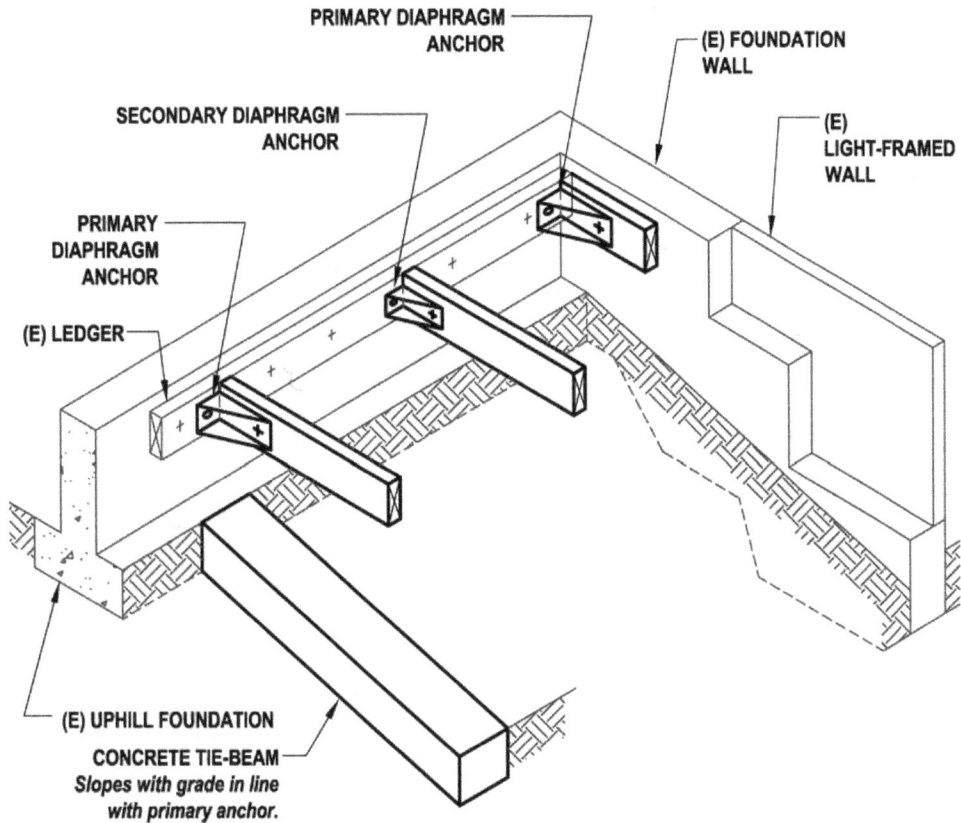

Note. Diaphragm anchor prevents existing ledger from pulling off at uphill foundation.

**Figure 5.4.5-2: Anchorage of Floor Diaphragm Framing
to Uphill Foundation in a Hillside Dwelling**

The configuration tested is similar to Figure 5.4.5-3A. Figure 5.4.5-3B is another commonly
used configuration. Care has to be taken to make the steel angle stiff enough to protect the
framing connection to the uphill foundation.

Alternate bracing approaches: Steel concentric braced frames have sometimes been used in lieu
of primary anchors at exterior stepped foundation walls (the right hand end wall in Figure 5.4.5-
2). When this approach has been used, there is often only a single diagonal brace member at each
foundation line, acting in tension for seismic loads towards the hill and in compression for
seismic loads away from the hill. This does not conform to code requirements for braced frame
design in which a balance of tension and compression resistance is required. If this approach is
taken, conservatism in estimating brace and anchorage forces is recommended, to avoid
premature failure and compensate for limited energy dissipation capacity.

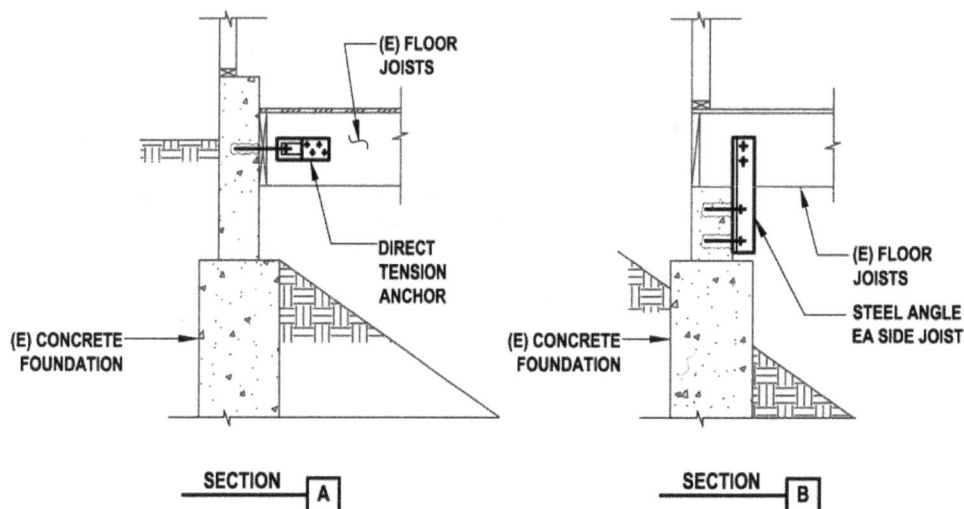

Figure 5.4.5-3: Connections for Anchorage to Uphill Foundation

Detailing Considerations

The objective of anchoring to the uphill foundation is to protect the ledger or foundation sill plate connection to foundation or foundation wall. These connections can experience brittle cross-grain tension failure at very small deflections. As a result, a very stiff primary or secondary anchor connection is needed to mitigate this failure. Stiff, direct axial connections should be favored over connections that allow movement; for example, the direct connection in Figure 5.4.5-3A would provide better protection against damage, while Figure 5.4.5-3B might flex to result in damage to the foundation sill connection but still prevent collapse. Testing has not been performed to determine what level of deformation is acceptable for the varying details that can occur at the uphill foundation.

Cost and Disruption Considerations

Because the majority of the work is intended to be in the crawl-space area under the dwelling, little disruption is generally caused by this rehabilitation work.

Construction Considerations

Construction on steep hillsides can be very difficult. In the extreme case, chemical grouting to stabilize loose soils may be required to keep the hillside from deteriorating during construction. At the end of construction, care should be taken to remove all soil that is in contact with wood framing.

Proprietary Concerns

There are no proprietary concerns with this rehabilitation technique, other than the use of proprietary connectors as part of the assemblage.

5.4.6   Rehabilitate Chimney

Deficiency Addressed by Rehabilitation Technique

This rehabilitation technique addresses inadequate component detailing associated with unreinforced and unbraced masonry chimneys. Damage to masonry chimneys has occurred in virtually every moderate to major United States earthquake. A falling hazard can be created if portions of the chimney break free.

Description of the Rehabilitation Technique

Techniques for mitigating the hazards posed by unreinforced and unbraced chimneys include:

> Removal of the chimney and fireplace
> Removal of the chimney and replacement with light-framing
> Filling of the chimney
> Anchorage of the chimney to the building

Complete removal of the masonry chimney and fireplace is the only method that will ensure elimination of the potential for damage or falling hazard. The chimney and fireplace can be removed without replacement or with replacement by well-anchored light-framing surrounding a factory-built fireplace and flue. All other rehabilitation measures mitigate rather than removing hazards.

Recommendations for removal of the masonry chimney and replacement with light-framing are published by the City of Los Angeles (2000) and California OES and FEMA (OES and FEMA, 2000). The transition to light-frame construction is shown to occur either at the top of the firebox or at a specified minimum dimension below the roof level. The farther down the chimney is removed, the more areas of potential damage are eliminated. A concrete bond beam is provided at the top of the remaining masonry. The bond beam is doweled into the existing masonry to remain and allows cast-in anchors for attachment of the light-framing above. Attention to maintaining required clearances to combustible materials is important at the transition and above. Anchorage of the flue is provided per manufacturer installation instructions. Anchorage of the light-frame enclosure to the building at floor, ceiling and roof levels is required. The OES and FEMA publication also illustrates replacement of an unreinforced masonry chimney with a code-conforming reinforced masonry chimney. The transition between existing and new construction should be carefully evaluated if this rehabilitation approach is chosen.

Figure 5.4.6-1 illustrates a possible scheme for filling in a vulnerable chimney with reinforced concrete. Reinforcing is placed in the chimney down to the smoke chamber. Most fireplace geometries will make it impractical to extend the reinforcing down further. Ties and spacers are recommended to hold the reinforcing at adequate clearances off of the flue wall so that bond is adequate to develop the reinforcing. Wheel-type spacers, sometimes put on tie or spiral reinforcing in drilled-pier foundations, could help with placement. Figure 5.4.6-1 shows the concrete extending to the damper location. Where possible, reinforcing and filling the fire-box would improve the strength and continuity of the infill. Anchorage of the chimney to floor, roof and ceiling levels needs to be provided in conjunction with chimney infilling. Filling the chimney will reduce the falling hazard of an unreinforced chimney, by providing strength and stiffness continuity at the commonly seen weak points (roof line and transitions in width). Filling the chimney may reduce, but is unlikely to eliminate damage. This rehabilitation measure is most often used for buildings of historical significance where there is a strong desire to maintain the current appearance. In some cases the height of very tall chimneys are reduced prior to filling with concrete. Use on chimneys already in poor condition due to deterioration or foundation movement is not recommended. Placement of grout between the flue liner and masonry is also recommended where this grout is completely missing or has significant gaps.

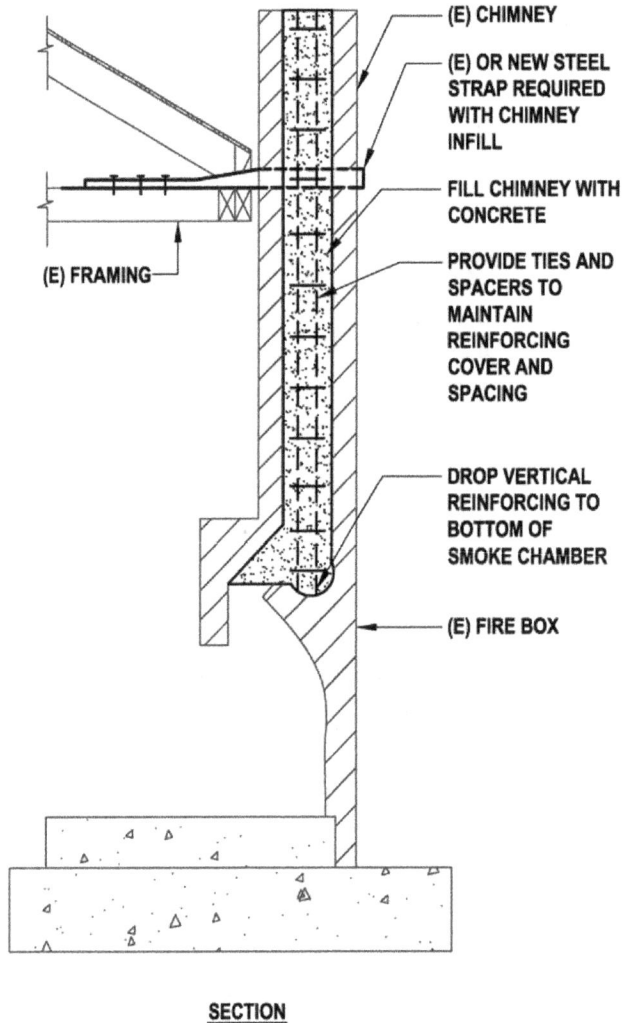

SECTION

Figure 5.4.6-1: Infill and Bracing of Masonry Chimney

Figure 5.4.6-2 illustrates anchorage of an exterior masonry chimney to floor, roof and ceiling framing. This detail is an adaptation of prescriptive information for new construction in the IBC (ICC, 2003a) and the *Masonry Fireplace and Chimney Handbook* (Amrhein, 1995). The steel strap is intended to keep the chimney from falling away from the building. In order to do this, the strap must be anchored into existing floor and roof framing with a capacity and load path adequate to resist forces from the chimney. Anchorage to wall studs or a single framing member will not accomplish this. It is often difficult and disruptive to anchor far enough into the building to develop required capacity. Figure 5.4.6-2 is intended for small to medium size chimneys common in single-family residences. Large and irregularly configured chimneys require additional consideration.

Design Considerations

Research basis: No research applicable to these rehabilitation techniques has been identified. Earthquake reconnaissance reports provide a limited record of earthquake performance of rehabilitation techniques.

Cautions: Some in the earthquake engineering community recommend against rehabilitation measures involving unreinforced masonry chimney anchorage to light-frame buildings on the basis that the anchorage is unlikely to eliminate earthquake damage. Indeed, damage and occasionally partial collapse of anchored chimneys have been seen in past earthquakes. The inherent difference in stiffness between masonry chimneys and fireplaces and light-frame construction is a likely contributor, along with widely varying adequacy of anchorage detailing and installation. The potential hazard posed by an unreinforced and/or unanchored chimney and the ability to reduce the hazard using one or more rehabilitation techniques need to be weighed for each building under consideration. Other practical measures to reduce life-safety threats due to unreinforced chimneys include limiting activities (interior as well as exterior) in the immediate vicinity of the chimney and fireplace and placing wood structural panel sheets on ceiling rafters alongside the chimney to slow down any portions falling to the interior (ABAG, 2005).

Variations in existing chimney conditions: Either careful evaluation of the existing chimney construction or worst-case assumptions regarding construction are suggested. Even when chimneys would have been required by buildings codes to be grouted and reinforced, it is common to find chimneys ungrouted, poorly grouted and unreinforced.

Foundations: In areas of poor soils, the weight of the chimney and firebox can result in higher settlement, and sometimes differential settlement, leading to leaning. Foundation problems need to be resolved before other rehabilitation measures are considered.

Detailing Considerations

Anchorage of a strap or other tie to an existing masonry chimney should be avoided where possible and otherwise approached with caution. Expansion anchors cause splitting tensile stresses that can result in cracking of the masonry. Adhesive anchors change properties under elevated temperatures that might be experienced during use of the chimney.

STEEL COLLAR AND
TENSION /
COMPRESSION STRUT
AT TALL CHIMNEY

A2 B

STEEL STRAP AT
CEILING OR ROOF
FRAMING

(E) MASONRY
CHIMNEY AND
FIRE PLACE

STEEL STRAP AT
FLOOR FRAMING

SECTION / ELEVATION   A1

(E) EXTERIOR
WALL

(E) FRAMING
PARALLEL TO
EXTERIOR WALL

STRAP EXTENDING 4
JOIST OR RAFTER
BAYS MINIMUM
NAIL TO BLOCKING

BLOCKING
BELOW STRAP

STEEL STRAP

GAP TO
COMBUSTIBLE FRAMING

7 DIAMETERS MINIMUM
(10 TO 12 PREFERRED)
TO FIRST BOLT OR
LAG SCREW

STRAP WITH BOLTS
OR LAG SCREWS TO
TIE MEMBER

BLOCKING
ALONGSIDE TIE

TIE MEMBER.
NAIL TO JOIST
OR RAFTER

PLAN DETAIL
FRAMING PARALLEL TO EXTERIOR WALL   A2

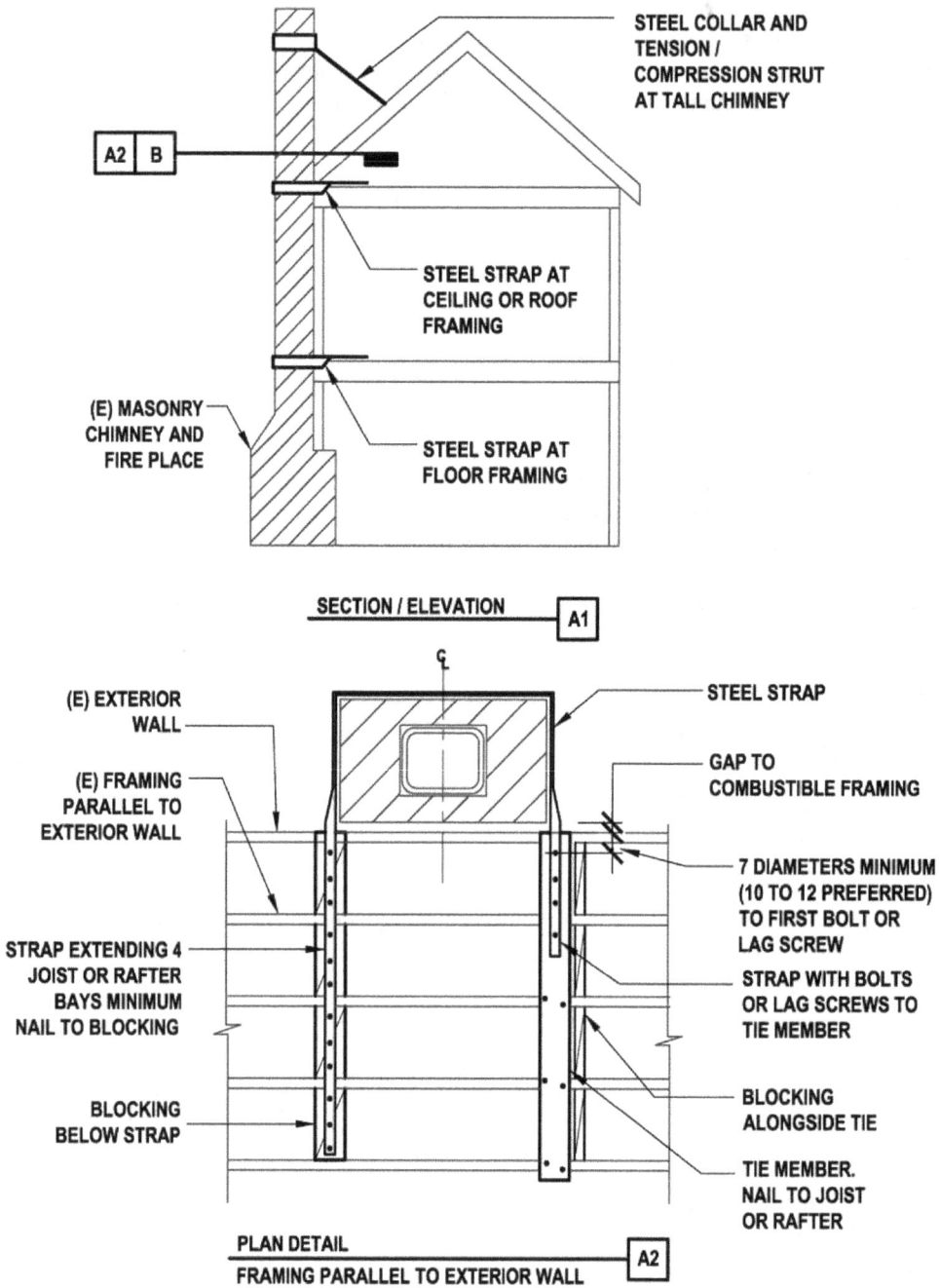

Figure 5.4.6-2A: Bracing of Masonry Chimney

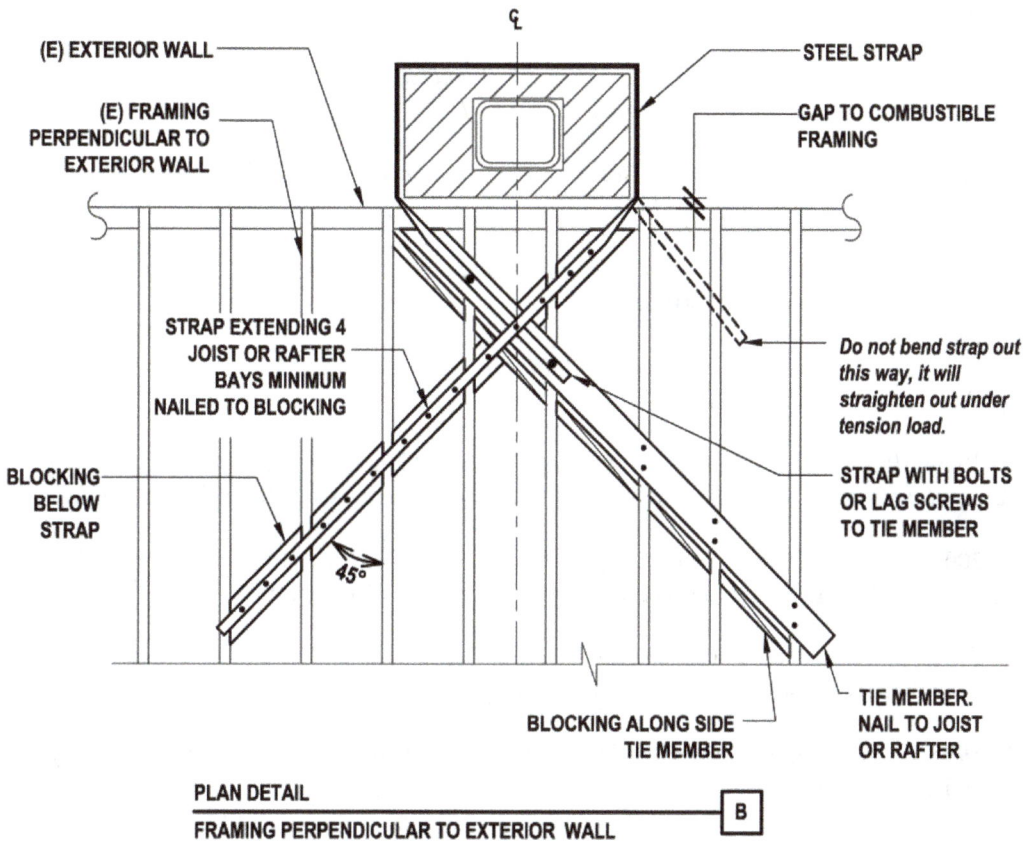

Figure 5.4.6-2B: Bracing of Masonry Chimney

Cost, Disruption and Construction and Construction Considerations

Any penetrations of the building exterior walls or roof need to be properly detailed for water resistance.

Proprietary Concerns

There are no proprietary concerns with this rehabilitation technique.

5.5   References

ABAG, 2005, Association of Bay Area Governments website: www.abag.org.

ACI, 2005, *Building Code Requirements for Reinforced Concrete and Commentary*, ACI 318, American Concrete Institute, Farmington Hills, MI.

AF&PA, 2003, *Plank and Beam Framing for Residential Buildings*, American Forest & Paper Association, Washington, D.C.

AF&PA, 2005a, *National Design Specification for Wood Construction, ASD/LRFD*, American Forest & Paper Association, Washington, D.C.

AF&PA, 2005b, *Special Design Provisions for Wind and Seismic, ASD/LRFD*, American Forest & Paper Association, Washington, D.C.

Amrhein, J., 1995, *Residential Masonry Fireplace and Chimney Handbook, Second Edition*, Masonry Institute of America, Torrance, CA.

APA, 1999a, *Wood Structural Panel Shear Walls* (Research Report 154), APA The Engineered Wood Association, Tacoma, WA.

APA, 1999b, *Research Report 158, Preliminary Testing of Wood Structural Panel Shear walls Under Cyclic (Reversed) Loading*, APA The Engineered Wood Association, Tacoma, WA.

APA, 2003, *Shear Wall Lumber Framing: Double 2x's vs. Single 3x's at Adjoining Panel Edges* (Report T2003-22), APA The Engineered Wood Association, Tacoma, WA.

ATC, 1976, *A Methodology for Seismic Design and Construction of Single-Family Dwellings*, ATC 4, Applied Technology Council, Redwood City, CA.

ATC, 1995, *Cyclic Testing of Narrow Plywood Shear Walls*, ATC R-1, Applied Technology Council, Redwood City, CA.

Chai, Y.H., T.C. Hutchinson and S. M. Vukazich, 2002, *Seismic Behavior of Level and Stepped Cripple Walls*, CUREE Publication No. W-17, Consortium of Universities for Research in Earthquake Engineering, Richmond, CA.

City of Los Angeles, 2002, "Voluntary Earthquake Hazard Reduction in Existing Hillside Buildings (Chapter 94)," *City of Los Angeles Building Code*, Los Angeles, CA.

City of Los Angeles, 2000, *Reconstruction and Replacement of Earthquake Damage Chimneys* (P/BC 2001-70), City of Los Angeles Department of Building and Safety, Los Angeles, CA.

City of Los Angeles & SEAOSC, 1996, *Findings and Recommendations of the Hillside Buildings Subcommittee of the City of Los Angeles Department of Building and Safety & Structural Engineers Association of Southern California Task Force on Evaluating Damage From the Northridge Earthquake, Final Report*, City of Los Angeles, CA.

Cobeen, K., 2004, "Recent Developments in the Seismic Design and Construction of Woodframe Buildings," *Earthquake Engineering From Engineering Seismology to Performance Based Engineering*, CRC Press, Boca Raton, FL.

Cobeen, K., J. Russell and J.D. Dolan, 2004, *Recommendations for Earthquake Resistance in the Design and Construction of Woodframe Buildings*, CUREE Publication No. W-30b, Consortium of Universities for Research in Earthquake Engineering, Richmond, CA.

Dolan, J.D., 2003, "Wood Structures," *Earthquake Engineering Handbook*," CRC Press, Boca Raton, FL.

EERI, 1996, "Supplement C to Volume II - Northridge Earthquake of January 17, 1994 Reconnaissance Report Volume 2," *Earthquake Spectra*, Earthquake Engineering Research Institute, Oakland, CA.

FEMA, 2000, *Prestandard and Commentary for the Seismic Rehabilitation of Buildings*, FEMA 356, Federal Emergency Management Agency, Washington, D.C.

FEMA, 2003, *NEHRP Provisions for New Buildings and Other Structures*, FEMA 450, Federal Emergency Management Agency, Washington, D.C.

Gatto, K. and C.M. Uang, 2002, *Cyclic Response of Woodframe Shearwalls: Loading Protocol and Rate of Loading Rate Effects*, CUREE Publication No. W-13, Consortium of Universities for Research in Earthquake Engineering, Richmond, CA.

Heine, C., 1997, *Effect of Overturning Restraint on the Performance of Fully Sheathed and Perforated Timber Framed Shear Walls*, Virginia Polytechnic Institute and State University, Blacksburg, VA.

ICBO, 1997a, *Uniform Building Code*, International Conference of Building Officials, Whittier, CA.

ICBO, 1997b, *Uniform Code for Building Conservation*, International Conference of Building Officials, Whittier, CA.

ICBO, 2001, *Guidelines for the Seismic Retrofit of Existing Buildings*, International Conference of Building Officials, Whittier, CA.

ICC, 2003a, *International Building Code*, International Code Conference, Country Club Hills, IL.

ICC, 2003b, *International Existing Building Code*, International Code Conference, Country Club Hills, IL.

ICC, 2005, *2003 International Existing Building Code (IEBC) Commentary*, International Code Conference, Country Club Hills, IL.

Isoda, H., B. Folz, and A. Filiatrault, 2002, *Seismic Modeling of Woodframe Index Buildings*, CUREE Publication No. W-12, Consortium of Universities for Research in Earthquake Engineering, Richmond, CA.

Kolba, A., 2000, *The Behavior of Wood Shear Walls Designed Using the Diekmann's Method and Subjected to Static In-plane Loading*, Marquette University, Milwaukee, WI.

Mahaney, J. A. and B. E. Kehoe, 2002, *Anchorage of Woodframe Buildings, Laboratory Testing Report*, CUREE Publication No. W-14, Consortium of Universities for Research in Earthquake Engineering, Richmond, CA.

Mosalam, K., C. Machado, K-U. Gliniorz, C. Naito, E. Kunkel and S. Mahin, 2002, *Seismic Evaluation of an Asymmetric Three-Story Woodframe Building*, CUREE Publication No. W-19, Consortium of Universities for Research in Earthquake Engineering, Richmond, CA.

OES and FEMA, 2000, *Guidelines to Strengthen and Retrofit Your Home Before the Next Earthquake*, State of California Governor's Office of Emergency Services, Sacramento, CA and Federal Emergency Management Agency, Washington, D.C.

Pardoen, G., A. Waltman, R. Kazanjy, E. Freund, and C. Hamilton, 2003, *Testing and Analysis of One-Story and Two-Story Walls Under Cyclic Loading*, CUREE Publication No. W-25, CUREE, Richmond, CA.

Salenikovich, A., 2000, *The Racking Performance of Light-Frame Shear Walls*, Virginia Polytechnic Institute and State University, Department of Wood Science and Forest Products, Blacksburg, VA.

SEAOC (Structural Engineers Association of California), 1999, *Recommended Lateral Force Requirements and Commentary,* Seventh Edition, Structural Engineers Association of California, Sacramento, CA.

von Winterfeldt, D., N. Roselund, and A. Kitsuse, 2000, *Framing Earthquake Retrofitting Decisions: The Case of Hillside Homes in Los Angeles*, PEER Report 2000-03, Pacific Engineering Research Center, Richmond, CA.

Xiao Y. and L. Xie, 2002, *Seismic Behavior of Base-Level Diaphragm Anchorage of Hillside Buildings*, CUREE Publication No. W-24, Consortium of Universities for Research in Earthquake Engineering, Richmond, CA.

Chapter 6 - Building Type W1A: Multistory, Multi-Unit Residential Woodframes

6.1 Description of the Model Building Type

Building Type **W1A** is similar to Building Type **W1** in use of woodframe wall, floor and roof construction, but includes large multi-family, multistory buildings. In **W1A** buildings, second and higher stories are almost exclusively residential use, while the first story can include any combination of parking, common areas, storage, and residential units. Post and beam framing often replaces bearing walls in non-residential areas. Multi-family residential buildings with commercial space at the first story are included in building Type **W1A** due to similar building characteristics. Lateral forces in **W1A** buildings are primarily resisted by wood diaphragms and shear walls. Figure 6.1-1 provides an illustration of this building type.

Wood joist floors with sheathing
or plywood at roof and floors

Parking sometimes
located on ground floor
with post and beam
support

Wood stud exterior and
interior bearing walls

Figure 6.1-1: Building Type W1A: Multistory Multi-Unit Residential Woodframes

This chapter addresses **W1A** buildings where the first story walls are of woodframe construction. This includes both multistory woodframe buildings supported at grade and the multistory woodframe portion of buildings with concrete or masonry walls at one or more lower stories. The stories with concrete or masonry walls represent building types other than **W1A**, and they are addressed by other chapters in this document.

Variations in the **W1A** building type can include a combination of multi-family residential use and the hillside building configuration discussed in Section 5.4.5. For this combination, rehabilitation measures from this chapter and Section 5.4.5 are applicable.

Design Practice

W1A buildings including apartment and condominium buildings, residential hotels, motels, and residential use over commercial space are very common in the current building stock, with some dating back to the early 1900s or earlier. While many **W1A** buildings constructed in the 1980s and later will have had a partially or fully engineered design, the majority of older **W1A** buildings will not. Case studies of California tuckunder buildings constructed in the 1970s (Schierle, 2001) indicate that a check of first-story walls for in-plane shear capacity was common, shear wall overturning was not considered, and bracing of upper stories commonly relied on prescriptive construction provisions. Steinbrugge, Bush and Johnson (1996) chronicled changes in California design practice of multi-family residential buildings since the 1960s. In some regions, these buildings are currently constructed using prescriptive codes.

Walls

Wall bracing materials include the same range discussed for **W1** buildings. Checks of first floor shear capacity in California tuckunder apartment buildings led to the use of wood structural panel sheathing without overturning anchorage in some first story walls in the 1960s and 1970s. Cripple walls, also discussed with the **W1** building type, are common in **W1A** buildings up until the 1950s.

Floor and Roof Diaphragms

Floor and roof diaphragms include the same materials as the **W1** building type, however plank and beam systems are rare in **W1A** buildings.

Foundations

Foundation types and issues for **W1** buildings are also applicable to **W1A** buildings. Of note, the gravity dead and live loads in **W1A** buildings can be significantly higher than in **W1** buildings.

Identification and Performance of Vulnerable Buildings

Several **W1A** building vulnerable configurations have become prominent in literature and discussion because of collapses or near collapses of lowest woodframe stories in the Loma Prieta and Northridge earthquakes. While these vulnerabilities are important for the **W1A** building type, they are not the only deficiencies that require consideration. See Section 6.3 for a systematic discussion of seismic deficiencies.

The discussion of these prominent vulnerable configurations requires a common understanding of terminology. In addition, a brief review is provided of documents that discuss performance, identification and rehabilitation provisions for vulnerable stories in **W1A** buildings.

W1A buildings, regardless of design approach, gain much of their strength and stiffness from bracing and finish materials on exterior walls and interior walls between and within residential units. This is true whether or not these walls are identified as shear walls. Where residential use occurs in multiple stories, it is common for residential unit layouts to be similar at each story, providing substantially uniform story strength and stiffness. Where the lowest story includes uses such as parking, common areas, commercial use, etc., the amount of exterior and interior wall is reduced, often resulting in significantly reduced story strength and stiffness. At the same time, the lowest story experiences the highest earthquake demands.

The terms weak story and soft story are used for this condition in which a story has less strength or stiffness than the story above. Concentration of deformation demand is understood to occur in a soft story. Inadequate strength and story failure may occur in a weak story. These would be identified as global strength and stiffness deficiencies for purposes of this chapter. Exact definitions of what constitutes a soft or weak story vary, as do opinions as to when soft and weak stories become vulnerable enough to recommend rehabilitation. Little research is available to assist in identifying when these configurations pose a hazard to life.

Where parking occurs in all or a portion of the lowest woodframe story, significant openings in the exterior walls are generally provided in order to allow access to the parking. Often there is little or no interior wall in the parking area. The term tuckunder parking (named due to the parking being tucked under the residential units) is used for this type of building configuration. Tuckunder parking buildings with woodframe walls at the parking story will often have a soft story and a weak story. Occasionally, parking only exists in a very small portion of the building plan area, and it does not significantly affect the story.

An open front building occurs when at any story level there is little or no bracing in one or more exterior walls. The term open front is a misnomer in that the open exterior wall can occur at any side of the building. Woodframe buildings are generally considered to have flexible diaphragms, and as a result bracing elements are generally provided at or near each edge of the diaphragm, most often at exterior walls. When an open front occurs, the diaphragm is required to transmit forces to other wall lines by rotation, creating torsional building behavior. This behavior is particularly critical when an exterior wall is provided at upper stories but discontinued in the first woodframe story, as this creates a significant discontinuity in the load path at the lowest story. Open front buildings often have tuckunder parking, but can also have commercial and other uses. Open front buildings will often but not always also have soft and weak stories at the open front story. Addition of vertical elements at the open front is the most direct rehabilitation approach to open front buildings. In buildings studied to date, capacities in the direction perpendicular to the open front have also been significantly lower that required by current codes and may also require rehabilitation.

What the terms soft story, weak story, tuckunder building and open front building all have in common is that they are identifying buildings that have potentially vulnerable stories due to deficient global or local strength or stiffness. In most cases, the vulnerable story is the lowest woodframed story.

Appendix Chapter 4 of the *International Existing Building Code* (IEBC) (ICC, 2003b) and Chapter 4 of the *Guidelines for Structural Rehabilitation of Existing Buildings* (GSREB) (ICBO, 2001) contain identical provisions for hazard reduction in **W1A** buildings. These provisions identify a broad range of multistory woodframe buildings as vulnerable based on:

Open front conditions (defined by IEBC as diaphragm cantilever in excess of that permitted by the applicable building code)
A weak wall line (defined by IEBC as story strength less than 80% of the strength of the story above), or
A soft wall line (defined by IEBC as not meeting story drift limits)

The IEBC provisions require evaluation and retrofit, including resisting elements from the diaphragm above the soft, weak, or open front story to the foundation-soil interface. Design is to be in accordance with the current building code except use of 75% of the code base shear is permitted. Specific rehabilitation measures are not detailed (with the exception of a prescriptive rehabilitation for limited building configurations); however, additional requirements for shear wall rehabilitation are included. The IEBC evaluation provisions create the challenge of calculating strength and stiffness for a variety of current and archaic finish materials not generally considered to be part of the lateral force-resisting system. Some guidance on strength and stiffness can be found in FEMA 356 (FEMA, 2000) and the AF&PA Wind and Seismic Supplement (AF&PA, 2005). Focus on the vulnerable first story may be lost in the calculation process. The IEBC also creates the challenge of identifying a wide range of buildings as potentially vulnerable, going well beyond open front and tuckunder configurations observed to be vulnerable to date. No guidance is given in judging relative hazard. If using IEBC Appendix Chapter 4, a commentary to the GSREB (ICC, 2005a) and ICC proposed changes (ICC, 2005b) are important additions.

The City of San Jose has developed several documents that assist in identification of vulnerable **W1A** buildings. *The Apartment Owner's Guide to Earthquake Safety* (Vukazich, 1998) uses a procedure based on ATC–21 rapid screening provisions in a broad approach to identifying vulnerable buildings and suggests shear wall enhancement and addition of steel moment frames as primary rehabilitation measures. *Practical Solutions for Improving the Seismic Performance of Buildings with Tuckunder Parking* (Lizundia and Holmes, 2000) illustrates rehabilitation techniques for three model building types, primarily using shear wall enhancement and steel moment frames. Rehabilitation measures address both life-safety and limited down time objectives. The focus of life safety measures is the first woodframed story. Work for limited down time objectives extends into upper stories.

A joint task force of the City of Los Angeles Department of Building Safety and the Structural Engineers Association of Southern California prepared the report *Wood Frame Construction Report and Recommendations* (City of Los Angeles & SEAOSC, 1994), which contains a series of observations and recommendations for multi-family residential construction based on performance in the Northridge earthquake. Issues include

Poor performance of gypsum wallboard and stucco bracing, attributed in part to high values given to these materials in past Los Angeles codes,
Poor performance of plywood shear walls, attributed to core gaps (gaps in the center ply of three-ply plywood) and slender walls,
Poor performance of tie-downs, attributed to design and installation problems, and

Excessive drift at steel columns and excessive building rotation, attributed to lack of drift checks on steel columns used as lateral-force-resisting elements.

Details and photos of observed damage are provided.

Finally, the CUREE-Caltech Woodframe Project included testing and analytical studies of open-front buildings and retrofits, summarized in Topical Discussion J (Cobeen, Russell and Dolan, 2004). One observation of note is that walls perpendicular to the open front suffered the greatest damage and degradation in testing and analysis, due to combined direct and torsional loading. Simultaneous earthquake loading in both horizontal directions should be evaluated in open-front buildings. Rehabilitation measures studied and recommended for use include:

Steel moment frames (designed as special moment frames per building code requirements or at $R = 1$) at the open front in combination with enhancement of other first story walls, and
A longitudinal wall near the building center of mass designed to carry the entire building base shear.

The CUREE research found that soft first stories are very common in woodframe construction and do not necessarily create a hazard.

Among these documents, there is currently no widely accepted definition of the point at which soft, weak and open-front stories become vulnerable to damage or constitute a life-safety hazard. The first story is the primary focus of evaluation and rehabilitation in most **W1A** buildings, and it is generally acceptable to reduce hazard through rehabilitation of the first story without improvement to upper stories. Steel moment frames and added or enhanced shear walls are the primary rehabilitation measures recognized in these documents.

6.2   Seismic Response Characteristics

Like the **W1** buildings, the dynamic response of **W1A** buildings is short period, and inelastic behavior is primarily concentrated in the vertical wall elements rather than the diaphragms. The first woodframed story will generally drift significantly more than upper stories and experience higher damage as a result. Configurations with open fronts have been seen to respond with significant torsional behavior as well as weak story behavior.

6.3   Common Seismic Deficiencies and Applicable Rehabilitation Techniques

While similar in construction to the **W1** building type, damage to **W1A** buildings has been more significant in areas of strong ground motion. Notably, the damage to finish and bracing materials and residual drift have been significant enough that re-occupancy of numerous buildings has not been permitted. Full and partial collapse of open front or tuckunder parking **W1A** buildings has occurred in recent earthquakes and resulted in loss of life in one building complex in the Northridge earthquake. The first story of these buildings was partially or completely occupied by parking; fewer and shorter bracing walls combined with archaic or heavily loaded bracing materials and rotational or torsional response contributed to vulnerability. Significant structural damage also occurred in **W1A** buildings having only residential units at the lowest story, as seen in Schierle (2001) Case Study 10, a three story residential building constructed in the early 1960s and braced with stucco and plaster over gypsum lath. See below for general discussion and Table 6.3-1 for a detailed compilation of common seismic deficiencies and rehabilitation techniques for Building Type **W1A**.

Global Strength and Stiffness

Global strength and stiffness are of particular concern in the first story of **W1A** buildings and have contributed significantly to damage in past earthquakes, sometimes accentuated by open fronts. Rehabilitation measures for global strength and stiffness include adding new vertical elements and enhancing existing elements. Common added elements are steel moment frames and added or enhanced shear walls. Steel braced frames may be added, but are not common in **W1A** buildings, since the brace would restrict access for parking or other uses.

Configuration

Although most common in **W1** buildings, some **W1A** buildings have missing or inadequately braced cripple walls. See Chapter 5 for rehabilitation techniques. Where **W1A** buildings are of large plan area, it may be necessary to add interior cripple walls and new interior foundations. It is common to enhance or add cripple walls in **W1** buildings without specifically accounting for overturning behavior in the stories above. Caution should be exercised in taking this approach with **W1A** buildings due to the larger size and weight. In addition, where uplift anchorage is being provided in stories above, the load path must be carried through the cripple wall to the foundation.

Torsional irregularities due to open fronts are prevalent and of significant concern in **W1A** buildings. Open fronts are often in the first story, and they combine with weak and soft story behavior. Where open fronts occur in tuckunder buildings, continued use of the first story parking often dictates that this deficiency be mitigated by the addition of steel moment frames. Wood shear walls and steel braced frames are alternate measures. It is important that walls perpendicular to the open front also be evaluated and enhanced, as these can be significantly deficient also.

Load Path

Adequate load path connection is a concern for **W1A** buildings, particularly so in first stories, which are likely to experience the majority of force and deformation demands. Many **W1A** buildings constructed in California in the 1960s and 1970s used wood structural panel sheathing in the first story, but did not have overturning detailing. Testing suggests that significant reductions in shear wall strength and stiffness can occur when overturning detailing is not provided. Likewise, many **W1A** and **W2** buildings are braced with diagonal lumber sheathing without overturning anchorage. Addition of overturning anchorage to these buildings could potentially greatly improve performance.

Related to the overturning load path, in **W1A** buildings where upper story shear walls are discontinued in lower stories, beams and posts providing vertical support at shear wall ends are potentially vulnerable. Instances of rehabilitation of members supporting shear walls in **W1**, **W1A** and **W2** buildings are very limited to date. This is because **W1A** building retrofits have focused on first story vulnerability, because earthquake damage to date has not shown this to be

Table 6.3-1: Seismic Deficiencies and Potential Rehabilitation Techniques for W1A Buildings					
Deficiency		Rehabilitation Technique			
Category	Deficiency	Add New Elements	Enhance Existing Elements	Improve Connections Between Elements	Reduce Demand	Remove Selected Components
Global Strength	Insufficient in-plane wall strength	Wood structural panel shear wall [6.4.2] Steel braced frame [7.4.1] Steel moment frame [6.4.1]	Enhance woodframe shear wall [6.4.2]	Shear wall uplift anchorage and compression posts [6.4.4]	Replace heavy roof finish with light finish	
Global Stiffness	Insufficient in-plane wall stiffness	Wood structural panel shear wall [6.4.2] Steel braced frame [7.4.1] Steel moment frame [6.4.1]	Enhance woodframe shear wall [6.4.2]	Shear wall uplift anchorage and compression posts [6.4.4]		
Configuration	Weak story, missing or weak cripple wall	Add woodframe cripple wall Add continuous foundation and foundation wall	Enhance woodframe cripple wall [5.4.4]			
	Open front	Wood structural panel shear wall [6.4.2] Proprietary wall Steel moment frame [6.4.1]	Enhance woodframe shear walls perpendicular to open front [6.4.2]			
Load Path	Inadequate shear anchorage to foundation			Anchorage to foundation [5.4.3]		
	Inadequate detailing for shear wall overturning		Enhance framing supporting shear wall [6.4.3]	Shear wall uplift anchors and compression posts [6.4.4]		
	Inadequate shear transfer in wood framing			Enhance load path for shear [5.4.1], [6.4.5]		

Table 6.3-1: Seismic Deficiencies and Potential Rehabilitation Techniques for W1A Buildings					
Deficiency		**Rehabilitation Technique**			
Category	**Deficiency**	**Add New Elements**	**Enhance Existing Elements**	**Improve Connections Between Elements**	**Reduce Demand**	**Remove Selected Components**
Load Path (continued)	Inadequate collectors to shear walls		Enhance existing collector	Add collector [6.4.5], [7.4.2]		
Component Detailing	Unreinforced & unbraced chimney		Infill chimney [5.4.6] Brace chimney [5.4.6]		Reduce unsupported chimney height [5.4.6]	Remove chimney [5.4.6]
Diaphragms	Inadequate in-plane strength and/or stiffness		Enhanced diaphragm [22.2.1]		Replace heavy roof finish with light finish	
	Inadequate chord capacity		Enhance chord members and connections [22.2.2]			
	Excessive stresses at openings and irregularities		Enhance diaphragm detailing			
	Re-entrant corners		Enhance diaphragm detailing			
Foundations	See Chapter 23					
[ ] Numbers noted in brackets refer to sections containing detailed descriptions of rehabilitation techniques.						

a critical weakness in woodframe construction and because rehabilitation of these supports can be difficult and expensive.

Shear transfer into and out of shear walls and other vertical elements must be adequate in order for the vertical element to fully contribute to building performance. While systematic evaluation may identify insufficient shear transfer at any story, shear transfer in first story walls is of particular concern due to reductions in the amount of shear wall and increases in unit loads. As in the **W1** building, adequate anchorage to the foundation is a high priority rehabilitation measure.

Component Detailing

Damage to unreinforced masonry chimneys has occurred in practically every earthquake to date. Approaches to rehabilitation include bracing, reducing height, infilling or removing. See Chapter 5.

Diaphragm Deficiencies

Although diaphragm deficiencies have not been seen as a significant contributor to damage to date, systematic evaluation can identify this as a deficiency. Rehabilitation measures include enhancing existing diaphragms through added fastening, blocking, and overlaying. Detailing can also be added at openings and re-entrant corners. See Chapter 22.

6.4 Detailed Description of Techniques Primarily Associated with This Building Type

6.4.1 Add Steel Moment Frame

Deficiency Addressed by Rehabilitation Technique

This rehabilitation technique addresses insufficient global or local strength or stiffness through the addition of steel moment frames. This rehabilitation technique is particularly beneficial in buildings with open fronts due to tuckunder parking, because the use of moment frames permits continued use of parking stalls. It is similarly beneficial for other buildings where continued use does not allow the addition of shear walls.

Description of the Rehabilitation Technique

This rehabilitation technique most commonly involves the addition of steel moment frames immediately adjacent to existing beams and columns, at or near a first story open front. Moment frames are less commonly added in other locations and in stories above the first story.

Figure 6.4.1-1A illustrates an elevation of a typical single-bent steel moment frame added immediately in front of existing beams and columns. Such frames might be added at every second or third framing bay across the building front. Moment frames can be brought to the job site in a complete beam plus two-column bent or in two L-shaped pieces with a field-bolted splice at beam mid-span. The use of two L-shaped pieces allows the critical beam to column connections to be welded in the fabrication shop with better access and quality control. The height required to tilt the frame into place is the factor most commonly governing whether frames are fabricated in one or two pieces. A new foundation will often be required to support the moment frame. This can either be an isolated footing at each end or a continuous footing.

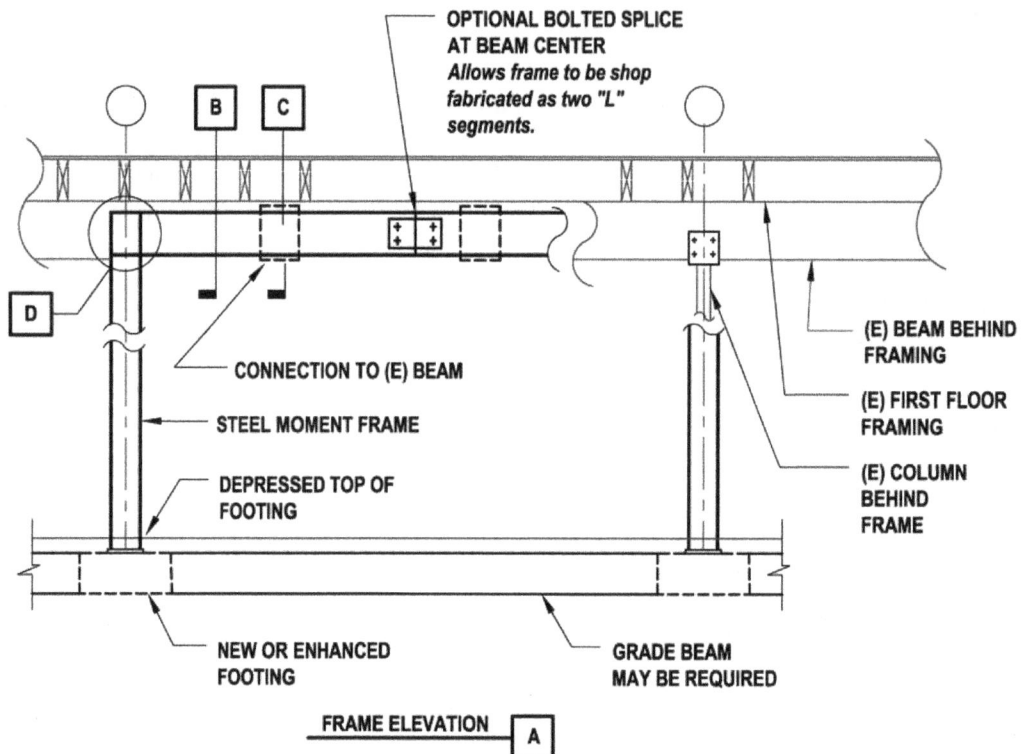

Figure 6.4.1-1A: Elevation of Steel Moment Frame in W1A Building

Footing placement will generally require the shoring of the upper stories and full or partial removal of existing footings. Transfer of earthquake load from the diaphragm above to the steel moment frame will commonly involve a collector that runs the full length of the open front and a series of connections from the collector to the steel moment frame.

Figures 6.4.1-1B, 6.4.1-1C and 6.4.1-1D illustrate possible connections. See discussion of collectors and shear transfer in the *Design Considerations* section. A number of detailing considerations discussed in Section 5.4.1 are applicable to frame connection to the existing wood building. In particular, detailing must accommodate shrinkage and possible swelling of wood, and alternate fasteners to existing sheathing may be needed.

This rehabilitation measure is not intended to address systems of steel columns cantilevered from the foundation without moment connections to a beam at the top. This cantilevered column system should be used with caution due to the difficulty of quantifying and limiting the many potential sources of rotation and deflection and to inadequate knowledge of post-elastic system behavior.

SHEATHING TO FRAMING
FASTENING. SEE FIG. 5.4.1-6

(E) SHEATHING

Nailed
plywood shim
at top of 3x
sill where
required for
bearing.

BLOCKING

(E) OR NEW
BLOCKING.
VERIFY / ADD
FASTENING
EACH END
BLOCKING

3X SILL AND
WELDED
STEEL STUDS

(E) FLOOR
FRAMING

FLAT CLIP

(E) BEAM

(E) BEAM

CONNECTION TO
(E) BEAM BEYOND
SEE C

SECTION   B1

SECTION   B2

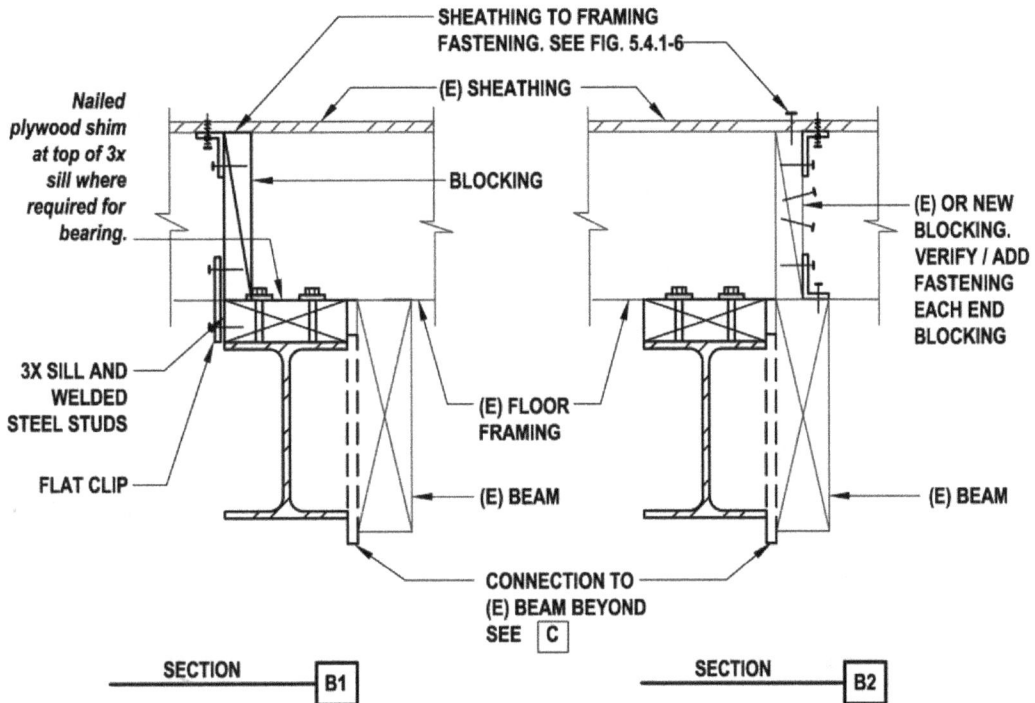

Figure 6.4.1-1B: Shear Transfer Between Moment Frame Beam and Diaphragm

Design Considerations

Research basis: Research specifically addressing steel moment frames in woodframe buildings includes: *Seismic Evaluation of an Asymmetric Three-Story Woodframe Building* (Mosalam et al., 2002) and *Improving Loss Estimation for Woodframe Buildings* (Porter et al., 2002). Results from these studies are also discussed in Cobeen, Russell, and Dolan (2004).

Moment frame design criteria: Chapter 8 of this document addresses steel moment frame rehabilitation in buildings where steel moment frames are the primary lateral force-resisting system. In contrast, when used for rehabilitation of **W1A** buildings, steel moment frames will generally only be used in one story and along one building line. The response modification factor of the woodframe building above makes use of either an ordinary or intermediate moment frame a logical choice for the first story of a multistory **W1A** building. Limitations addressing use in light-frame buildings have been in a state of flux. The most current seismic design provisions, ASCE 7-05 (ASCE, 2005) and AISC Seismic (2005), permit:

> Single story ordinary moment frames (OMF) for new buildings in Seismic Design Category (SDC) D and E, to a height of 65 feet, provided dead load tributary to the roof does not exceed 20 psf and tributary wall dead load does not exceed 20 psf

C2

(E) 1ST FLOOR
FRAMING

TO BEAM TOP
AND BOTTOM
FLANGE

(E) BEAM

VERTICAL SLOT
IN BEAM WEB

STEEL MOMENT
FRAME BEAM

STEEL
PLATE

Note: Preferred connection.
Connection had minimal slip.
See B for balance.

BOLTS IN
VERTICAL
SLOTTED HOLES
WITH PLATE
WASHERS

SECTION
LOW-SLIP CONNECTION   C1

ELEVATION
LOW-SLIP CONNECTION   C2

(E) 1ST FLOOR
FRAMING

THRU BOLT WITH
CUT WASHERS

WOOD FILLER

(E) BEAM

Note: Not preferred connection.
Connection had excessive slip.
See B for balance.

SECTION
HIGHER-SLIP CONNECTION   C3

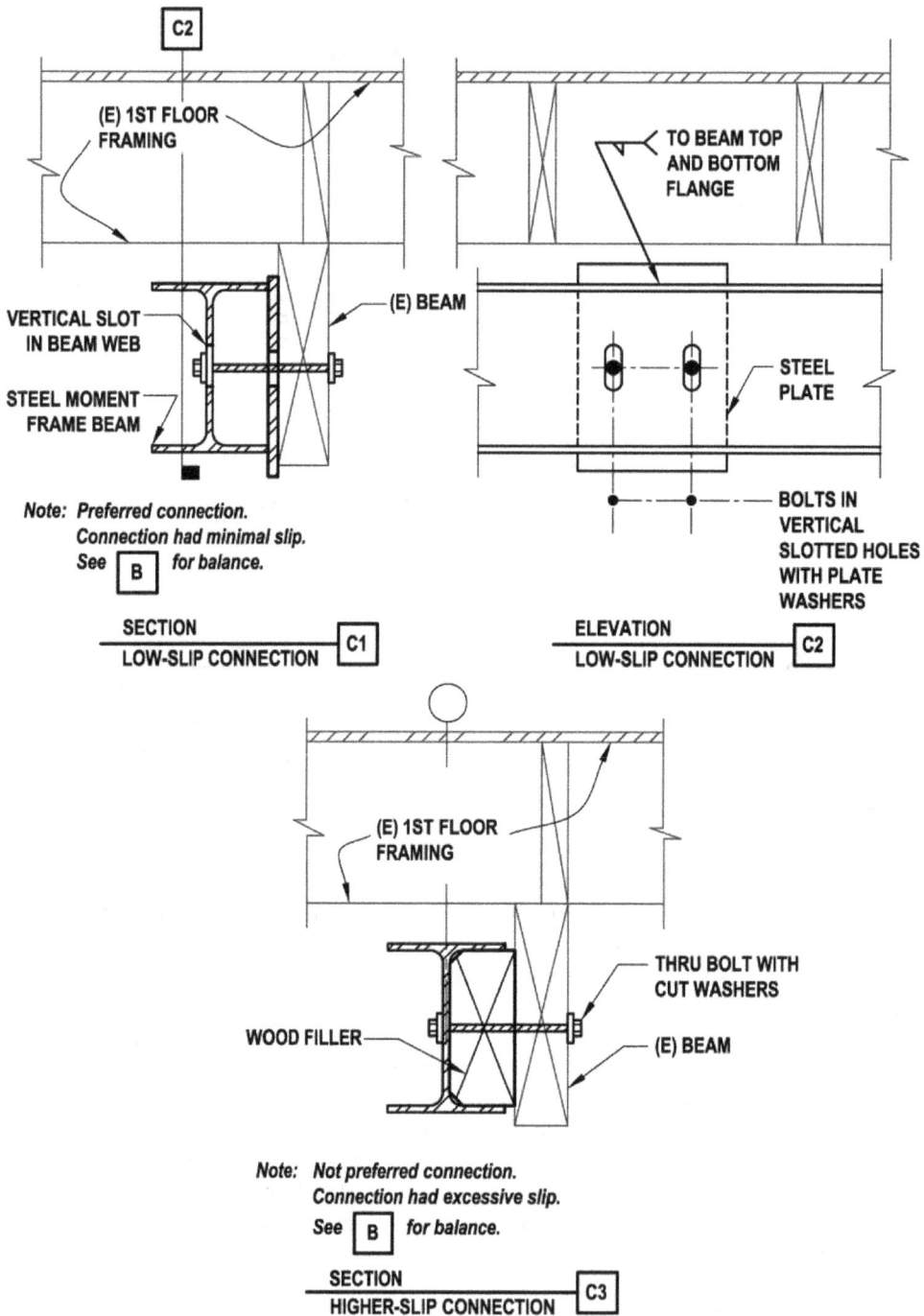

Figure 6.4.1-1C: Shear Transfer from Moment Frame Beam to Collector

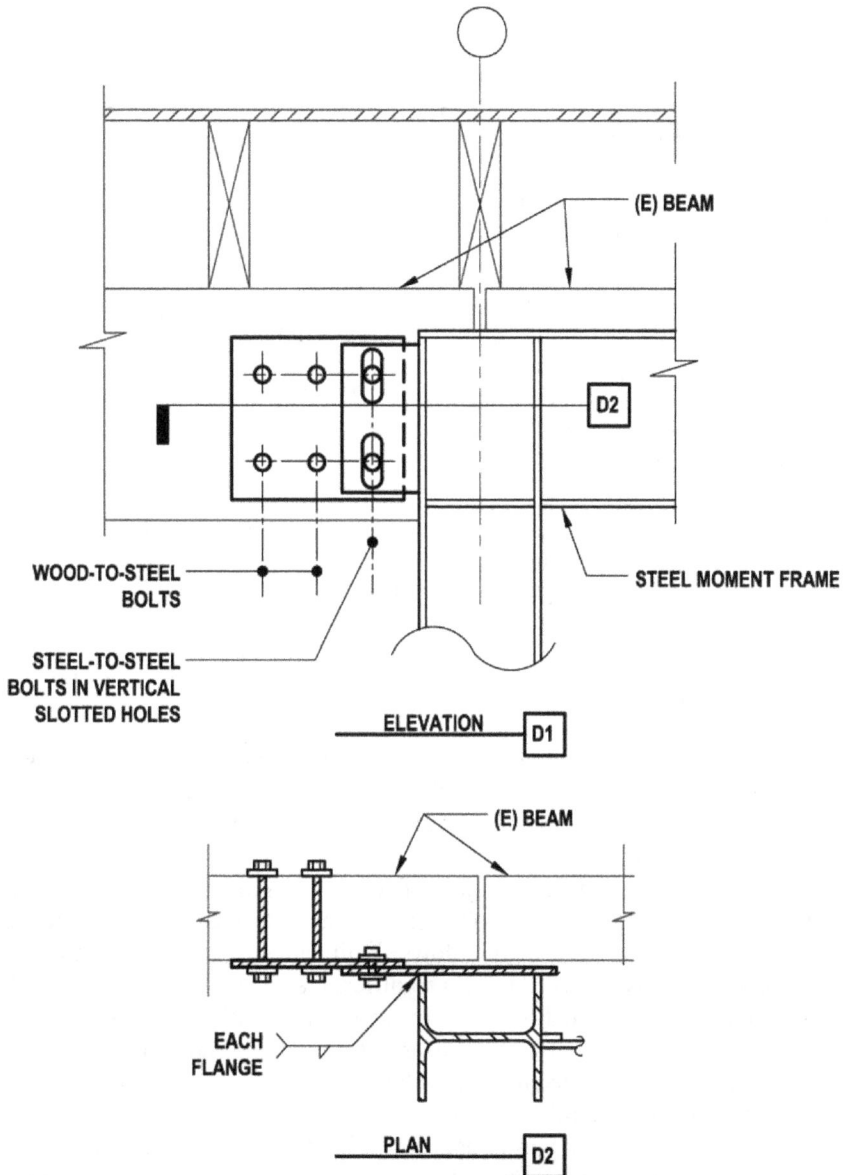

Note: Out-of-plane bracing of steel frame is required at column tops
and may be required at reduced beam sections.

Figure 6.4.1-1D: Shear Transfer from Moment Frame Beam to Collector

OMFs for new buildings in SDC D and E in light frame construction up to a height of
35 feet, with roof and floor dead load to tributary to the frame not exceeding 35 psf and
wall dead load tributary to the frame not exceeding 20 psf
Intermediate moment frames (IMFs) in SDC D up to a height of 35 feet
IMFs in SDC E up to a height of 35 feet with tributary floor and roof dead load not
exceeding 35 psf and tributary wall dead load not exceeding 20 psf

A three-story **W1A** building will generally just meet the height and weight limits to allow use of
an OMF. This allows the choice of OMF, IMF or special moment frame (SMF). While SMFs are
always acceptable, the response modification factor must not be taken as greater than for the
lateral force-resisting system above (typically wood shear wall), and use of pre-qualified welded
joints may require use of steel beam and column sizes larger than acceptable.

Because the limitations for use of moment frames in light-frame construction have been in a state
of flux, a number of organizations and jurisdictions have developed local guidance for design
and rehabilitation. Among these are:

Provisions used by the City of Santa Monica with the response modification factor set as
one (used in CUREE Woodframe Project research)
Draft guidelines by the SEAOSC Steel Ad Hoc Committee (SEAOC, 2002) addressing
up to two-story buildings and recommended reduced drift and quality assurance measures
Draft procedures by the ICC Peninsula Chapter (2004) addressing design procedures and
quality assurance measures

The need for these guidelines in addition to the latest design standards requires review. One of
the recommendations made in the guidelines is that moment frame drift be limited to less than
required by code in recognition of the lesser ductility of the connections.

Shear transfer and collector detailing: Provision of adequate strength and stiffness for shear
transfer from the building wood framing into the steel moment frame is key to improved building
performance. Where the shear transfer detail allows significant slip, undesirable building
deflection will occur. This was observed to be a significant issue in the CUREE-Caltech
Woodframe Project testing of moment frames (Mosalam et al., 2002) (Cobeen, Russell, and
Dolan, 2004). Figure 6.4.1-1C3 illustrates the shear transfer detail used in a simplified moment
frame. The shear transfer was designed using tributary seismic forces, but without consideration
of overstrength or the force that could be developed by the system. The connection used wood
filler pieces and through bolts, and it was intended to reflect common design practice. Excessive
slip developed between the wood beam and the filler. At peak capacity, the slip accounted for
40% of the total system drift, and the bolts cut long slots into the beam and fillers.

Figure 6.4.1-1C1, based on a Rutherford & Chekene detail for the CUREE testing, shows a shear
transfer detail used for the special moment frame tested in the shake table tuckunder building.
Two significant differences occur in this detail. First, the shear transfer connection was designed
to develop the capacity of the diaphragm above; and second, the wood-to-wood connection was
replaced with a lower-slip wood-to-steel connection. Although the forces seen by the frame were

moderate, the connection resulted in less slip, suggesting that better control of building drift would result.

Although the first (Figure 6.4.1-1C3) connection could be improved by design using overstrength forces, the second (Figure 6.4.1-1C1) connection approach is recommended. It is further recommended that the approach of using overstrength forces and limiting slip be applied to other shear force transfer connections, including those shown in Figures 6.4.1-1B and 6.4.1-1D.

Steel moment frame design and detailing: Design and detailing of steel moment frames used in rehabilitation should be in accordance with the most recent edition of IBC and AISC provisions.

Moment frame column bases: Columns in the CUREE testing used base plate details that are commonly considered to provide pinned conditions. This was done to minimize the moment demand put on the foundation, keeping foundation rehabilitation to a minimum, and to keep inelastic behavior in places where performance could be more easily predicted. The column base behavior during testing corresponded well to the assumed near-pinned condition, with little or no deterioration of the base plate connection seen. The use of pinned column base detailing is recommended.

Lateral bracing of columns: Bracing at the beam top and bottom flange elevations is required at the moment frame columns. For the CUREE testing, steel angle braces were provided between bottom flange continuity plates and wood floor joists.

Lateral bracing of beam flanges: Continuous bracing of the moment frame beam top flange is generally easily accomplished by the addition of a bolted nailer and connection to new or added framing, as shown in Figure 6.4.1-1B. Provisions for SMFs may require the bracing of the beam bottom flange just beyond the plastic hinge zone if bracing was included in prequalification testing. Bracing forces that are easily accommodated in steel construction can be more of a significant detailing issue in woodframe rehabilitation, depending on how far the bracing force is developed into the wood framing system. As a minimum, the brace member and its connection at either end should develop required forces.

Addition of moment frames in upper stories: Where moment frames are added in upper stories, provision for a load path to the foundation is required. The load path should be designed using the forces that can develop in the frame using overstrength, force-controlled action, or target displacement approaches.

Detailing Considerations

Accommodation of wood shrinkage: Figure 6.4.1-1 details use vertical slotted holes in the steel side plates to accommodate wood shrinkage (or expansion) and possible vertical movement due to deformation of the steel moment frame beam. This approach should be used at any location where steel side plates are placed against wood framing, provided the connection is only intended to transfer horizontal forces. See Section 5.4.1 for further discussion of wood shrinkage issues.

Cost and Disruption Considerations

It is very unlikely for the addition of a steel moment frame to be the least expensive or quickest way to rehabilitate for global or local strength or stiffness. The steel moment frame requires the involvement of multiple building trades: fabrication in a steel fabrication shop and site assembly by steel workers, in addition to foundation and framing work at the job site. The addition or enhancement of shear walls will be less expensive. In buildings were the addition of shear walls is not acceptable, however, the addition of a steel moment frame does provide a reasonable and common rehabilitation approach.

Construction Considerations

Plumbing, HVAC or electrical lines may be running in the floor framing in the vicinity of steel moment frame locations. Either accommodation in the structural design or relocation of utilities may be necessary. Job site welding of steel members requires adequate access and special ventilation measures in enclosed buildings. Welding of steel members in the vicinity of woodframe construction can be a significant fire hazard and should only be undertaken by experienced welders and only when absolutely necessary. Smoldering droppings from on-site welding and cutting have repeatedly caused structure fires. Welding should always be done by certified welders using approved welding techniques in compliance with building code welding and special inspection requirements.

Proprietary Concerns

There are no proprietary concerns with this rehabilitation technique.

6.4.2   Add New or Enhance Existing Wood Shear Wall

Deficiency Addressed by Rehabilitation Technique

This rehabilitation technique addresses insufficient global or local strength or stiffness though the addition of or enhancement of vertical elements of the lateral force-resisting system. In **W1A** buildings, stories with inadequate global first story strength and first story open fronts have been vulnerable in past earthquakes. Rehabilitation of shear walls perpendicular to the open front is often necessary.

Description of the Rehabilitation Technique

This rehabilitation technique involves the addition of a shear wall (framing and sheathing) or enhancement of an existing shear wall by the addition of sheathing, the addition of sheathing fastening, or a wood structural panel overlay.

Added shear walls: When new shear wall framing and sheathing are being added, the most difficult design issue is mobilizing dead load to resist uplift due to shear wall overturning. Design for transfer of overturning forces to the supporting soils requires an understanding of the existing foundation configuration. Added shear walls can then be located to specifically make use of or avoid existing foundations.

Figure 6.4.2-1 shows a shear wall located so that it can use the dead load carried by a building column to resist uplift at the left hand side and an existing bearing wall foundation at the right

(E) FOOTING

(E) STEEL COLUMN

SHEAR WALL

(E) FOOTING

TIE-DOWN TO
(E) FOOTING

C

B

(E) SLAB-ON-GRADE

PLAN VIEW
SUPPORT ON (E) FOOTING — A

STEEL PLATE
WASHER

PRESSURE
TREATED SILL

3/4"
MINIMUM
CLEAR

ADHESIVE OR
EXPANSION BOLT

(E) Slab must be
thick enough to
allow adequate
bolt anchorage.

SECTION — B

(E) COLUMN

EDGE NAIL

TYP.

BOLTS TO STUD
AND STEEL PLATE
WASHER

STUD

STEEL CHANNEL

PLAN DETAIL — C

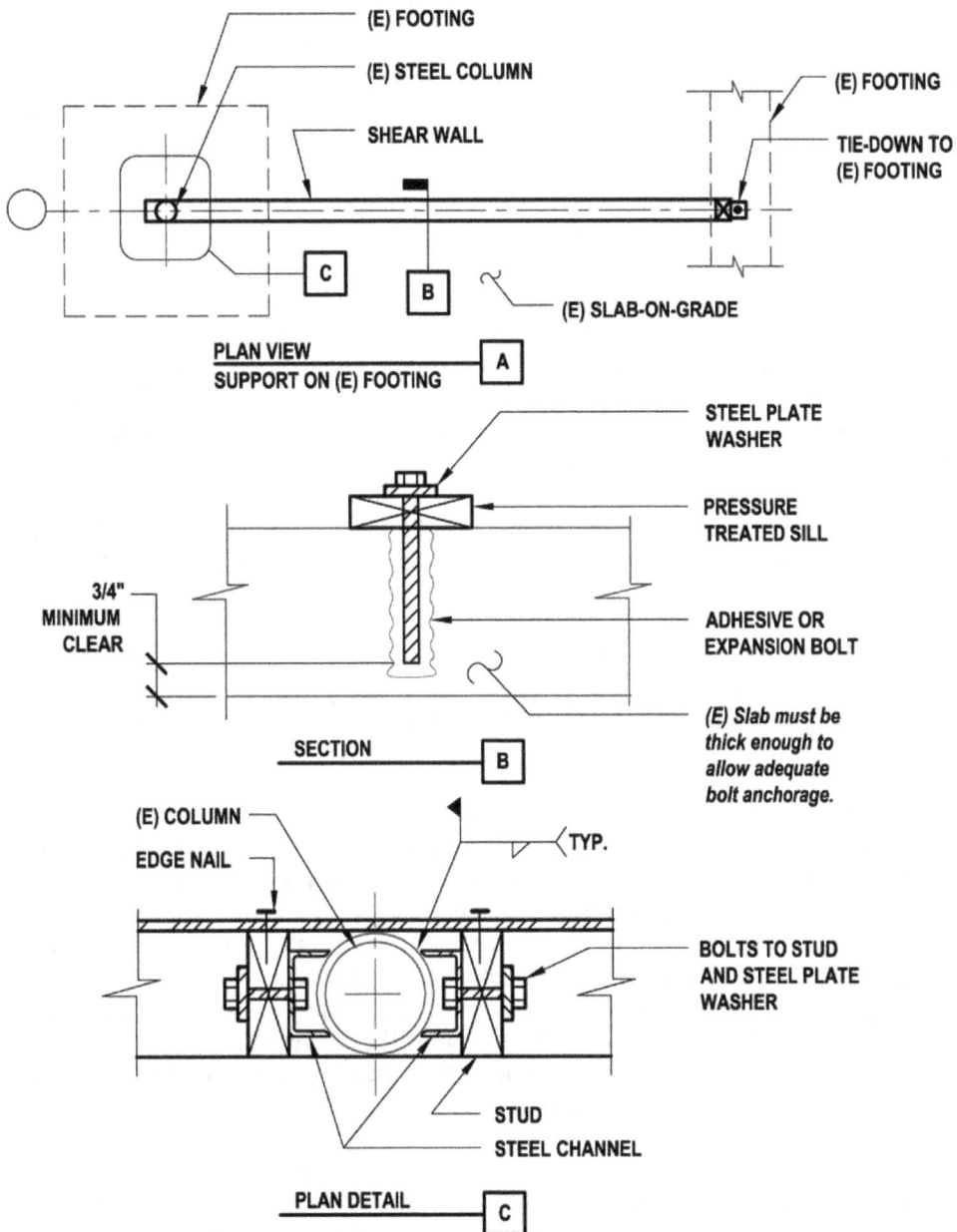

Note: This approach is only applicable where the slab-on-grade is thick
enough to permit installation of anchor bolts, and where there will not be
regular moisture exposure (such as garage floor). Do not use powder-
driven fasteners.

Figure 6.4.2-1: Added Shear Wall Supported on Existing Foundation and Slab

hand side. The existing foundations need to be checked for adequate dead load resistance and adequate capacity to resist both up and down forces within material and soil strengths. At the left hand side, the existing column connection to the foundation needs to be capable of picking up the footing and surrounding slab. Use of this detail is limited not only by the adequacy of the foundation for overturning forces, but also the adequacy of the slab for shear anchorage. The slab must be thick enough to allow the installation of expansion bolts or adhesive anchors for anchor bolts. This starts being possible at a slab thickness of about four inches and is best with a slab of five inches or greater. Use of powder-driven fasteners for shear transfer to the slab is not recommended. Testing has found that these anchors fail prematurely under cyclic loads (Mahaney and Kehoe, 2002; and Cobeen, Russell, and Dolan, 2004).

Figure 6.4.2-2 shows a shear wall supported on a new strip footing. The new footing runs between and is doweled into existing footings at each end, allowing the dead load of the existing footing to resist overturning. The addition of a new footing allows new anchor bolts to be cast-in, greatly simplifying shear anchorage. It also allows the addition of a curb to help reduce decay exposure in areas like garages that might have water exposure.

Figure 6.4.2-3 shows a shear wall added away from any existing footings. A large pad-type footing will be needed to provide enough dead load to resist overturning forces.

Enhanced shear walls: Section 5.4.1 provides a detailed discussion of enhancing shear walls by the addition of structural sheathing to walls currently braced with finish materials. This discussion is equally applicable to **W1A** building, and it is also applicable when it is decided to remove existing wood structural panel sheathing and replace it with new sheathing of higher capacity.

Other approaches to enhancing shear wall capacity include the overlaying of new wood structural panel sheathing over existing sheathing and addition of fastening (added nails or staples) to existing sheathing. Figure 6.4.2-4A illustrates the addition of nails to increase shear wall capacity. New nails do not need to be added between every existing nail pair. It is acceptable to space them out to every second or third nail pair, as long as the average over two to three feet meets the needed spacing. It is desirable to distribute added nails as evenly as possible over the height of the wall. Too many nails can reduce performance: additional detailing requirements may be triggered, the wall overstrength will be increased, and demand on anchorages will be increase. Additionally, if not symmetrically placed, added nails can reduce the capacity of the shear wall (Cobeen, Russell and Dolan, 2004).

Figure 6.4.2-4B illustrates the addition of staples. Staples are placed with their long direction parallel to the stud longitudinal direction in order to maintain edge distance in the stud and sheathing. It has been noted that workers placing staples have very little feel for whether the staple penetrates the stud, or is off the stud and only penetrates the sheathing (called a "shiner"). For this reason, careful attention to staple placement is required. This is only an acceptable approach when very modest increases in capacity are required, such that changes in detailing are not required (load path connections into and out of the shear wall, etc.). To date practice has been to waive the requirement for 3x framing at abutting panel joints when stapled shear walls are used. This is because the staples are thought to significantly reduce splitting of the wood

(E) FOOTING   FOOTING

SHEAR WALL

(E) FOOTING

B   C

DOWEL TO (E)
FOOTING

PLAN VIEW
NEW FOOTING DOWELED TO EXISTING   A

CUT (E) SLAB EDGE TO ALLOW
RESUPPORT ON CONCRETE FOOTING

RETAIN (E) SLAB
REINFORCING
WHERE POSSIBLE

CAST-IN ANCHOR BOLT WITH
STEEL PLATE WASHER

CURB AT PARKING AREA

PRESSURE TREATED SILL

DESIGN FOOTING SIDE

ACTUAL FOOTING SIDE

CONCRETE FOOTING BOTTOM
ELEVATION NOT HIGHER THAN
EXISTING FOOTING

STIRRUPS WHERE REQUIRED
FOR SHEAR

SECTION   B

CLEAN SURFACE OF
(E) FOOTING

SHEAR FRICTION DOWEL WITH
ADHESIVE ANCHOR

LOCATE DOWEL NEAR
MID-HEIGHT FOOTING

DEVELOPMENT
LENGTH

ADDED
FOOTING   (E) FOOTING

SECTION / ELEVATION   C

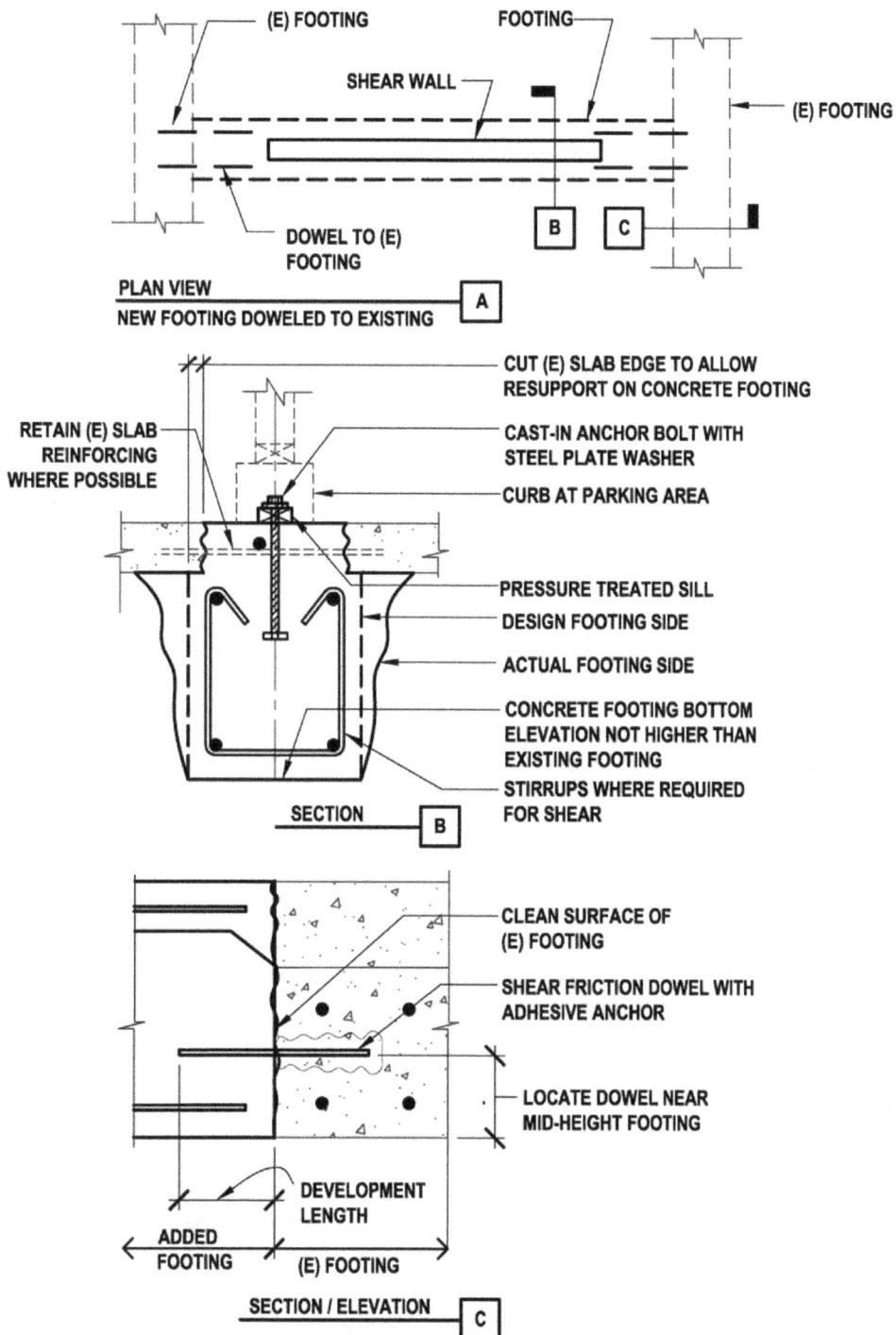

Figure 6.4.2-2: Added Shear Wall Supported by New and Existing Footings

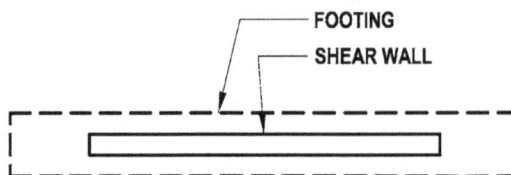

Figure 6.4.2-3: Added Shear Wall Supported on a New Footing

framing, greatly reducing the likelihood of stud failure. See further discussion in the *Design Considerations* section.

Figure 6.4.2-5 illustrates use of shear wall wood structural panel overlay over existing wood structural panel sheathing. The figure illustrates the staggering of panel edges so that edge nailing of abutting panel edges on the inside and outside sheathing layers do not occur on the same framing member. Adequacy of overlay sheathing nail penetration into the framing member needs to be verified. This may be a problem where "short" sheathing nails are used, but not likely if full length common nails are used. At shear wall boundary members, both the inside and overlay sheathing need to be fastened to the boundary member. This may require the addition of a new boundary member at this location. One set of nails should not be relied on to fasten both sheathing layers. This approach has some potential issues, discussed in the *Design Considerations* section.

Another possible use of an overlay is over existing lumber sheathing. See *Design Considerations* section for discussion.

Design Considerations

Research basis: A significant amount of research for new shear walls can be considered applicable to this use. See Section 5.4.1. Testing of stapled shear walls has been conducted by APA (1999), Zacher and Gray (1985) and Pardoen (2003). Testing of sheathing-to-framing connections with staples, wood screws, and nails using two sheathing layers has been conducted by Fonseca et al., (2002). Limited testing of plywood overlays of plywood diaphragms has been conducted by APA (1999).

Foundation design: The foundations, new or existing, have to be capable of resisting imposed forces. In Figure 6.4.2-1, the existing foundations need to be checked for both adequate dead load resistance and adequate capacity to resist up and down forces within material and soil

6 NAILS IN 2 FEET EQUALS

4" ON CENTER AVERAGE

(E) SHEATHING EDGE NAILING

(E) SHEATHING

ADDED NAILING TO GIVE SPECIFIED SPACING WHEN AVERAGED OVER 2 TO 3 FEET. DISTRIBUTE AS EVENLY AS POSSIBLE

(E) STUD

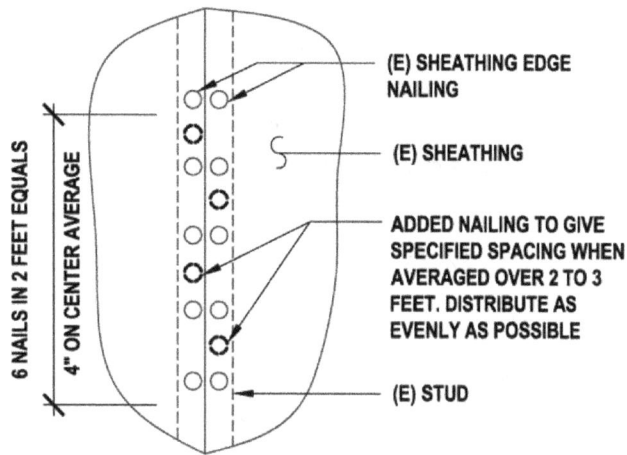

ELEVATION
RENAILING

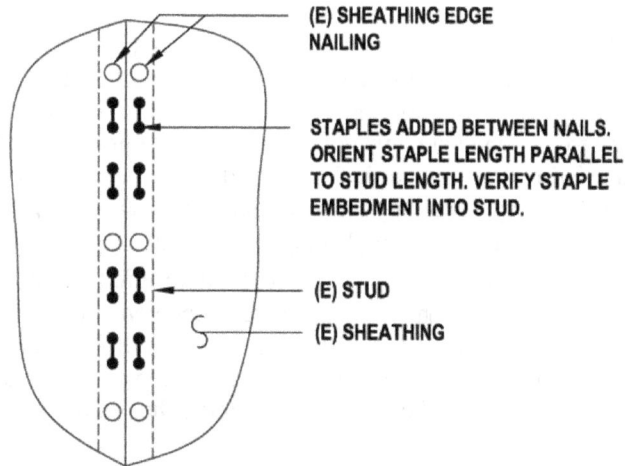

A

(E) SHEATHING EDGE NAILING

STAPLES ADDED BETWEEN NAILS. ORIENT STAPLE LENGTH PARALLEL TO STUD LENGTH. VERIFY STAPLE EMBEDMENT INTO STUD.

(E) STUD

(E) SHEATHING

ELEVATION
ADDED STAPLES

B

Note: These details are only applicable for modest increases in shear wall capacity that do not increase detailing requirements.

Figure 6.4.2-4: Enhanced Shear Wall Sheathing Fastening

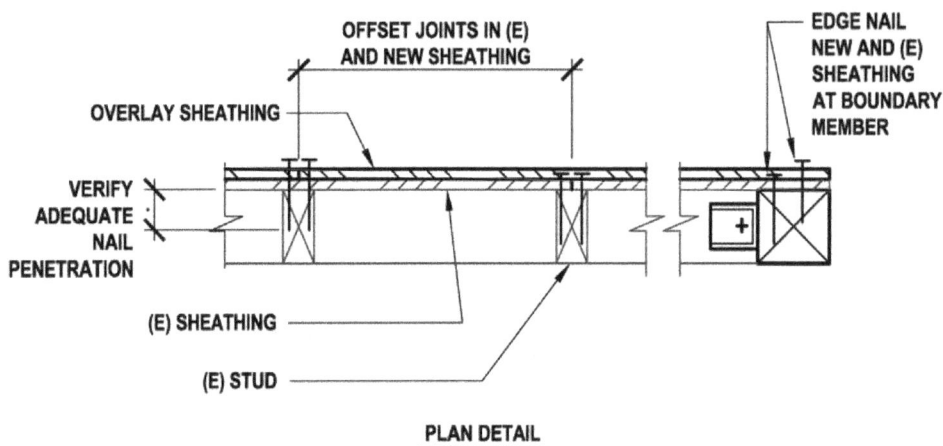

Note: See design considerations for cautions in use of overlays.

Figure 6.4.2-5: Enhanced Shear Wall With Sheathing Overlay

strengths. At the left hand side, the existing column connection to the foundation needs to be capable of picking up the footing and surrounding slab. In Figure 6.4.2-2, the new footing needs to be specifically designed for the loading; use of a typical footing section and reinforcing may not be adequate. The existing footings need to be checked for capacity to mobilize overturning resistance and to distribute downward reactions to the supporting soils. At the interface between the new and existing footings, vertical uplift and downward reactions are generally transferred through rebar doweling. Generally this is designed as a shear-friction connection, with the face of the existing footing cleaned; roughening the concrete surface to reduce the   factor below 1.0 is seldom practical, so a   of 1 is generally used in design. In order to develop shear friction, the yield strength of the reinforcing needs to be developed on either side of the interface. Embedment depths to develop the reinforcing are generally available from the adhesive anchor manufacturer. If dowels are installed too close to the top or bottom of the footing, spalling can occur. Locating dowels near the center of the footing height reduces avoids spalling issues.

Stapled shear walls: Use of stapled fastening of shear wall sheathing has been studied as a desirable approach to enhancement of existing shear walls for rehabilitation. Testing by Zacher and Gray (1985) found that use of staples avoided splitting of the framing members, making it possible to achieve higher capacities without adding in 3x studs at abutting panel edges. Stapled shear walls tested Pardoen, et al. (2003) show behavior indistinguishable from equivalent nailed shear walls. Testing of stapled connections by Fonseca et al., (2002) shows adequate load and deflection behavior, suggesting them to be equally acceptable. All of the staples tested eventually experienced fatigue failure, but this was after significantly more cycles than required by the loading protocol. When staples are being used to increase the capacity of existing shear walls, enough staples should be provided to carry the entire design shear. This is because the load-deflection behavior of the staples can be expected to be different than existing nails due to the very different fastener shank diameter. Stapled shear wall allowable design values are provided in the IBC (ICC, 2003a).

Wood screw shear wall fastening: Wood screws are occasionally used for fastening of shear wall sheathing to wood framing. The very limited research available suggests that there are concerns with using this attachment type. In testing by Mahin (1980s), the brittle fatigue failure of cut-thread wood screws was first noted. The screws failed at the transition from a full shank to a cut shank, this coincided with the framing to sheathing transition in the wall. This failure was repeated by Fonseca et al. (2002) when screw length was chosen to give minimum embedment. An increase in screw length to three inches significantly reduced but did not eliminate fatigue failure. Testing of rolled thread wood screws has not been identified.

Shear wall overlay over wood structural panel sheathing: There are two primary reasons for using an overlay rather than removing existing sheathing and putting in new. One is to avoid the expense of removing material, the other is to make use of the capacity already provided and reduce thickness of added sheathing. The downside of using an overlay is that observation and modification of framing and framing connections is not possible. Overlay of wood structural panel sheathing has been used in past rehabilitation projects; however, concerns arise that deserve consideration. The deflection of shear walls under load involves the rotation of the sheathing panel as the wall framing racks. The primary energy dissipation method is through bending of sheathing nails due to the different deflection pattern of the sheathing and framing. The addition of an overlay with staggered edges will theoretically put significant deformation demands on nails being driven in two different deformation patterns (one by each sheathing layer). Available testing on fasteners in overlay conditions (Fonseca et al., 2002) showed a significant increase in fatigue failure of nails. APA (2000) investigated plywood overlays at the end of plywood diaphragms as a means of increasing shear capacity. Slow stepped loading without load reversals was used, and the overlay was found to successfully increase capacity. Because definitive information about performance of shear wall overlays is not available, caution in using this approach is recommended.

Shear wall overlay over straight lumber sheathing: Straight lumber sheathing is generally flexible enough and of low enough capacity that when overlayed, the behavior of the wood structural panel sheathing can govern. This makes it acceptable to overlay straight sheathing; however, there is no benefit from the sheathing remaining, other than reduced work due to removal. Where removal of the straight sheathing is possible, it is preferred. Only the capacity of the wood structural panel sheathing should be relied upon. It is recommended that edge nailing of wood structural panel sheathing be through straight sheathing into framing in all cases, since reduced embedment could lead to reduced overstrength capacity due to nail withdrawal. Special attention needs to be paid to developing shear transfer to boundary members, since nailing must be through straight sheathing to the boundary member framing behind.

Shear wall overlay over diagonal lumber sheathing: The load-deflection behavior and fastener deformation patterns of diagonal lumber sheathing and wood structural and sheathing are considerably different, raising questions about the behavior resulting from the combination of the two. Due to lack of information, use is not recommended without a detailed study of behavior.

Mixing of shear wall deformation capacities: Designers are particularly cautioned against using shear wall systems or enhancements with deformation capacities less than the balance of the story or building (i.e. less than the two percent of story height drift permitted by current codes for ordinary occupancy structures). Because the building or story deformation demand or target displacement will be largely determined by the rest of the vertical elements, introduction of a stiffer element with limited deformation capacity could result in premature failure.

Detailing Considerations

Shear walls separating parking areas from residential areas may be part of fire-rated assemblies. Any fire rating needs to be maintained in the rehabilitation work. When wood structural panel sheathing is applied over gypsum wallboard, increased nail sizes are required by the building code. Because cyclic testing has not been conducted for sheathing applied over gypsum wallboard, the implications for drift are not known. Testing of gypsum wallboard has shown crushing of the gypsum, with cycled loading resulting in slotting of the wallboard and significant slip. The same behavior may lead to increased deflection where wood structural panel sheathing is applied over gypsum wallboard sheathing.

Cost and Disruption Considerations

The primary cost of enhancing existing shear walls comes from the disruption of the occupants and the removal of finishes to gain access to the structural walls. The cost of materials and connections is generally minor is comparison. As a result, it is preferable to keep the variation in sheathing, nailing, and connections to a minimum, making execution of the work as simple as possible. Planning on removal and replacement of existing sheathing can facilitate project schedule by minimizing the need to address unexpected existing sheathing conditions while construction is in progress. Other design and detailing measures that can make execution of the work more predictable are encouraged.

See Section 6.4.1 for discussion of field welding cautions.

Construction Considerations

As in new construction, it can be a challenge to assure that rehabilitation measures are constructed with the fastener (nail, staple, screw, etc.) type and size that has been assumed in design and construction documents. Use of improper type and size often results in reduced rehabilitation measure capacity. Most nails are placed with nail guns. Most gun nails are ordered by diameter and length. Indications of type and pennyweight continue to be misleading. The only way to verify that required fasteners are being used is to measure them with calipers or a similar device. Fasteners connecting sheathing to framing should not be overdriven (not break the face ply of the sheathing). Where overdriving occurs, fastener capacity may be reduced up to 40%.

Often plumbing, HVAC or electrical lines will be running in the floor framing in the vicinity of shear walls. This is particularly problematic where they cross over the shear wall at critical locations for shear or overturning transfer. Some disruption in the transfer of shear into the top of a shear wall will generally need to be accommodated, typically this means that there are a number of joist bays in which blocking and clips can not be installed. Within residential units, relocation of utilities is often not an option. In other areas, relocation of utilities may be more practical.

Proprietary Concerns

There are no proprietary concerns with this rehabilitation technique other than the use of adhesive anchors as part of the assemblage.

6.4.3   Enhance Framing Supporting Shear Wall

Deficiency Addressed by Rehabilitation Technique

This rehabilitation technique addresses inadequate beams, posts, and their interconnection supporting vertical overturning forces from ends of discontinued upper story shear walls. The primary focus is support of existing shear walls, but the discussion applies equally to support of enhanced shear walls.

Description of the Rehabilitation Technique

This rehabilitation measure involves the addition or supplementing of beams, posts, beam-to-post connections and post-to-foundation connections to support discontinued upper story shear walls.

Figure 6.4.3-1 illustrates the addition of new supports and connections where an upper story shear wall is added or enhanced. This figure shows a new beam, post and foundation system being added. Ideally, the posts would be added immediately under the shear wall ends; however, the layout of the first story will often dictate other support locations. The beam, post, beam-to-post and post-to-foundation connections must be designed for overstrength or special seismic load combinations is using ASCE 7 or IBC, or as force-controlled members per FEMA 356. Either approach will amplify the demand on these members and connections. Overturning anchorage of the shear wall is addressed in Section 5.4.1 and Figure 5.4.1-11C. Shear transfer at the wall base is addressed in Figure 5.4.1.9. Where overturning forces from the wall are significant, wood beam sizes may prove too large to be practical, in which case a steel beam may be needed. Where a steel beam is used, use of steel columns may also be practical and provide stronger and stiffer beam-to-column connections. Where an existing beam exists but is not adequate, the addition of new steel channels on either side of the beam can provide a practical solution. See Figure 6.4.3-2. Attention is needed to adequate load transfer into and out of the channels, including end supports and uplift connections.

Design Considerations

Research basis: No research applicable to the rehabilitation measure has been identified.

History: The failure of concrete columns supporting the Olive View Hospital during the 1971 San Fernando earthquake dramatically demonstrated the significant demands placed on members supporting discontinued bracing systems; however, this was not commonly considered in design of woodframe buildings until the 1997 *NEHRP Provisions* (FEMA, 1998) and 1997 UBC (ICBO, 1997) when special requirements for supporting members were expanded from columns to beams, columns and connections, and explicit application to woodframe was noted. The requirement of design for expected forces for new construction is now included in ASCE 7 (ASCE, 2005) and the IBC (ICC, 2003a), for regions of high seismic hazard, but not other regions. As a result, most buildings will not have been designed considering expected forces ($_0$ overstrength or special seismic load combinations). A systematic evaluation in accordance with

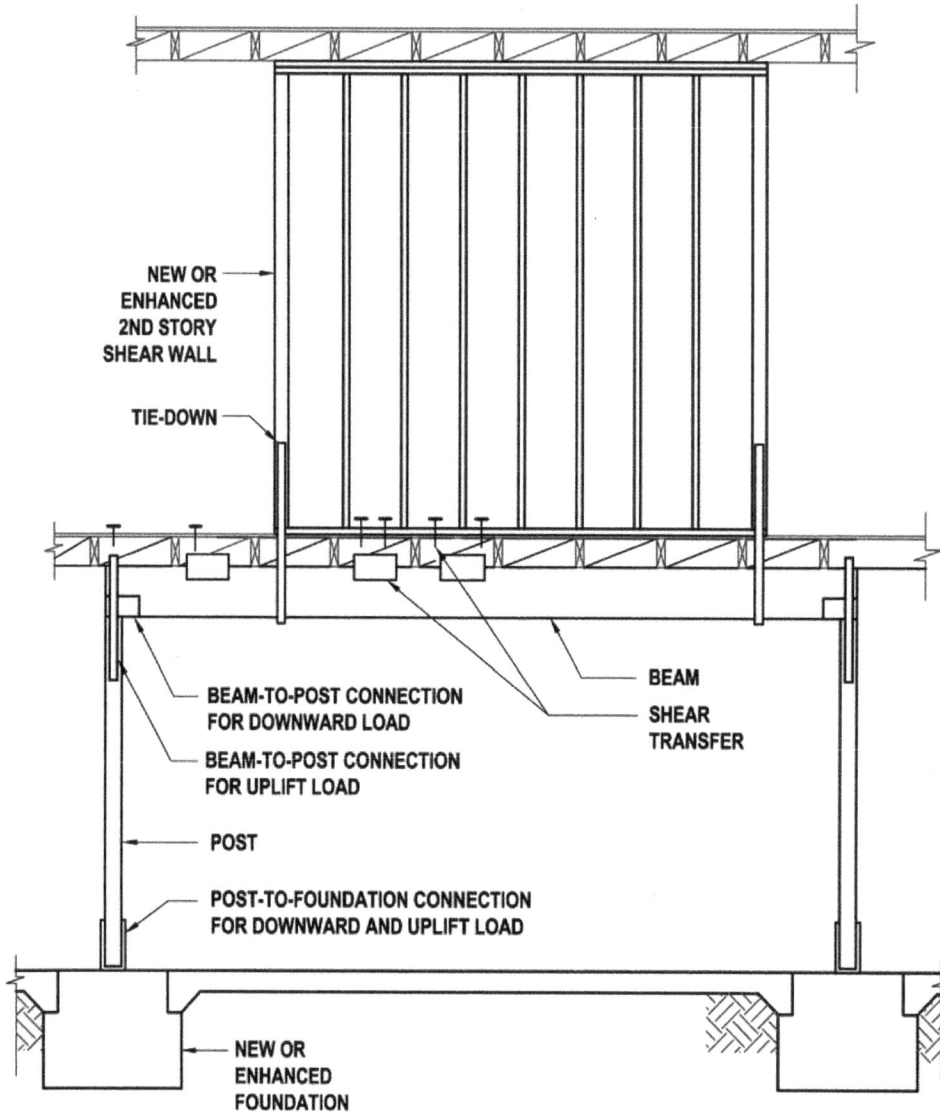

NEW OR
ENHANCED
2ND STORY
SHEAR WALL

TIE-DOWN

BEAM-TO-POST CONNECTION
FOR DOWNWARD LOAD

BEAM-TO-POST CONNECTION
FOR UPLIFT LOAD

POST

POST-TO-FOUNDATION CONNECTION
FOR DOWNWARD AND UPLIFT LOAD

BEAM

SHEAR
TRANSFER

NEW OR
ENHANCED
FOUNDATION

Note: *Evaluation of beam, post, beam-to-post and post-to-foundation connections is for force-controlled
actions; design is for over-strength load combinations.*

SHEAR WALL ELEVATION

Figure 6.4.3-1: Enhanced Overturning Support for Upper Story Shear Wall

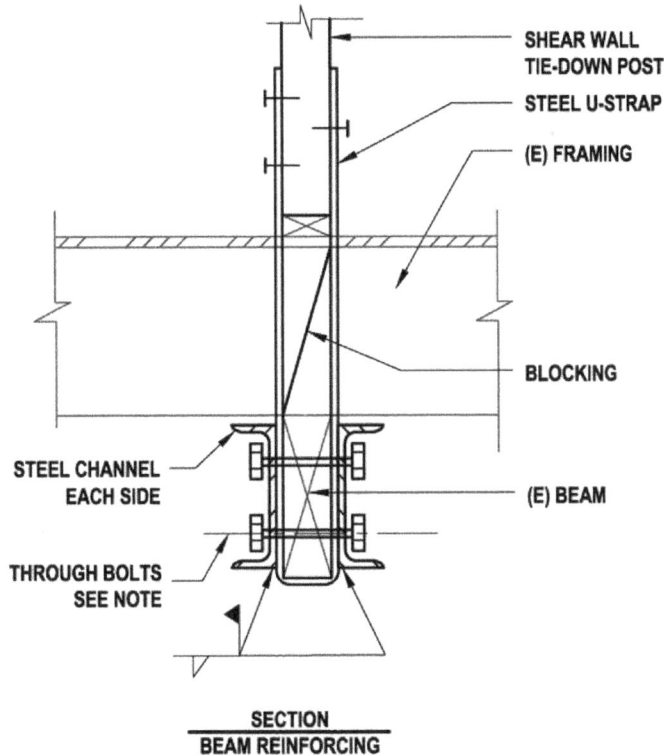

SECTION
BEAM REINFORCING

Note: A direct connection between the tie-down and channels is preferable to connecting to the existing wood beam using through bolts in perpendicular to grain loading. The same is true for connection of the beam end to a supporting post. Where transfer using bolts is necessary, the reduced effective depth of the wood beam for shear must be considered.

Figure 6.4.3-2: Enhanced Beam Supporting Discontinued Shear Wall

FEMA 356 requires that these supporting members be evaluated as force-controlled, with forces coming from 1.5 times the yield strength of the supported wall. This will have the same or a more critical effect than design per ASCE 7 and IBC requirements. As a result, support upper story shear walls will most likely be identified as a deficiency. As discussed in Section 6.3, however, rehabilitation for this deficiency has seldom occurred to date.

Support in crawl spaces: Where vertical support is needed for interior first story walls above crawlspaces with post and pier floor systems and spread footings, the easiest and least expensive rehabilitation is the addition of new foundation to support the shear wall. This is best accomplished by addition of blocking under the shear wall, fastening of a pressure treated sill with pre-placed anchor bolts, and casting of the concrete footing to the underside of the foundation sill. Access and ventilation openings in the new foundation may be required.

Detailing Considerations

See Section 5.4.1 for discussion of wood framing issues, applicable to floor blocking and added beams. Any time wood and steel members are connected to each other, the detailing needs to accommodate wood change in dimension with moisture content (either shrinkage or expansion). Figure 6.4.3-2 provides one example of where this must be considered. An existing wood beam inside of a conditioned building would be anticipated to have very little dimensional change, while a new beam or a beam with exposure to weather or humidity could have significantly more. In Figure 6.4.1-1D dimensional change was accommodated through the use of slotted holes. In Figure 6.4.3-2 it is important that the holes in the steel strap and channels not be slotted. Oversized holes in the wood beam could be used to accommodate dimensional change.

Cost/Disruption

This rehabilitation measure will require simultaneous access to the story with the shear wall and the story below. Significant areas of ceiling will need to be removed to access work. The ceiling in the garage of a **W1A** or **W2** building may be plaster rather than gypsum wall board and may be part of a fire-rated assembly separating the garage area from the residential units. Any fire rating would have to be maintained in the rehabilitation work.

Construction Considerations

Often plumbing, HVAC or electrical lines will be running in the floor framing in the vicinity of shear walls. This is particularly problematic where they cross under the shear wall at critical locations for shear or overturning transfer. In some cases it is practical to accommodate these utilities in the structural design. In other cases relocation of utilities may be more practical.

Welding of steel members requires adequate access and special ventilation measures in enclosed buildings. Welding of steel members in the vicinity of woodframe construction can be a significant fire hazard and should only be undertaken by experienced welders. Welding should always be done by certified welders using approved welding techniques in compliance with building code welding and special inspection requirements.

Proprietary Concerns

There are no proprietary concerns with this rehabilitation technique.

6.4.4   Enhance Overturning Detailing in Existing Wood Shear Wall

Deficiency Addressed by Rehabilitation Technique

This rehabilitation technique addresses inadequate or missing load path detailing for uplift and downward forces at the ends of shear walls, between shear wall and foundation, or between upper story and lower story shear walls. The uplift load path may have inadequate or missing tie-down devices and detailing. The compression load path may have inadequate compression capacity in the wall framing or through the floor framing depth.

Description of the Rehabilitation Technique

Where there is a calculated net uplift force at the ends of shear walls, proprietary tie-down connectors are fastened to the wall framing and foundation to resist the uplift forces. The tie-down connectors may be fastened to existing framing or new framing. They may be used in combination with existing shear wall sheathing (generally on the exterior face of exterior walls), new sheathing on the interior face, or new sheathing on the exterior face. Tie-down vertical bolts are generally fastened to existing foundations with adhesive anchors.

Except for very lightly loaded walls, tie-downs are generally needed to develop the in-plane strength and stiffness of wood structural panel and diagonally sheathed shear walls, as discussed in the *Design Considerations* section. Tie-downs may potentially be used on stucco shear walls, but are seldom used on gypsum wallboard shear walls due to the low capacity.

Figure 6.4.4-1A illustrates a shear wall elevation with commonly used tie-down connectors for a slab-on-grade condition. Figure 6.4.4-1B illustrates fastening to develop a load path between the tie-down connector and the shear wall sheathing edge nailing. For sheathing and framing conditions other than those shown, similar fastening must be provided to complete the load path.

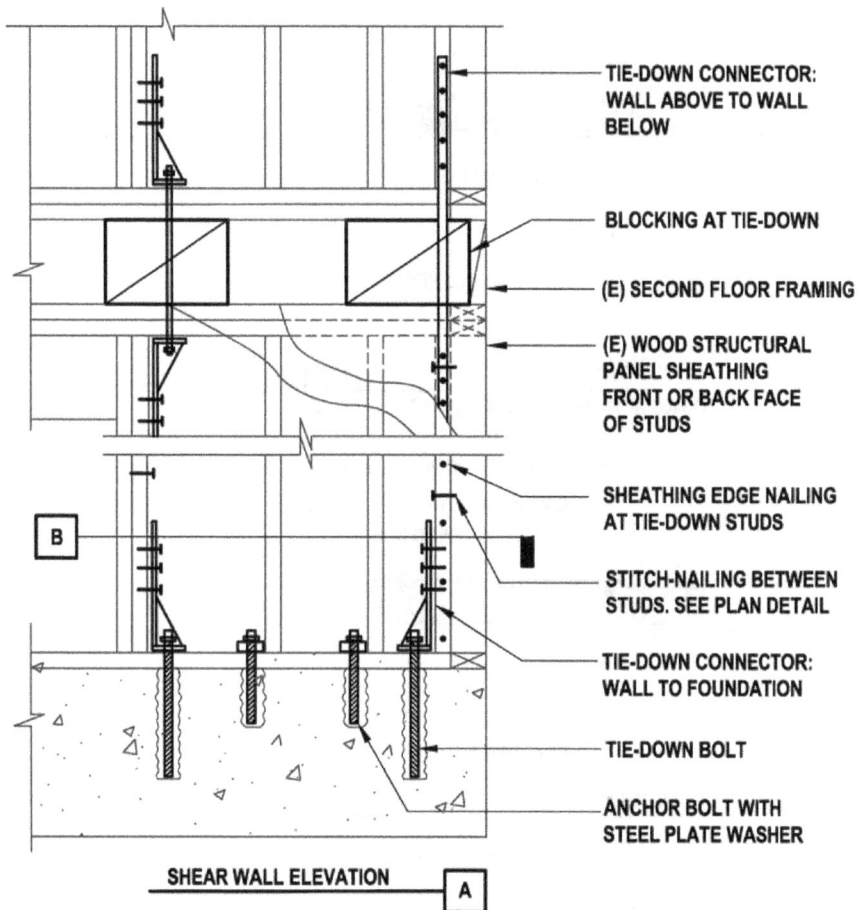

TIE-DOWN CONNECTOR: WALL ABOVE TO WALL BELOW

BLOCKING AT TIE-DOWN

(E) SECOND FLOOR FRAMING

(E) WOOD STRUCTURAL PANEL SHEATHING FRONT OR BACK FACE OF STUDS

SHEATHING EDGE NAILING AT TIE-DOWN STUDS

STITCH-NAILING BETWEEN STUDS. SEE PLAN DETAIL

TIE-DOWN CONNECTOR: WALL TO FOUNDATION

TIE-DOWN BOLT

ANCHOR BOLT WITH STEEL PLATE WASHER

SHEAR WALL ELEVATION   A

Figure 6.4.4-1A: Shear Wall Elevation with Enhanced Overturning Detailing

(E) SHEATHING EDGE NAILING

STITCH NAILING BETWEEN EXISTING STUDS

(E) SHEATHING

TIE-DOWN
CONNECTOR

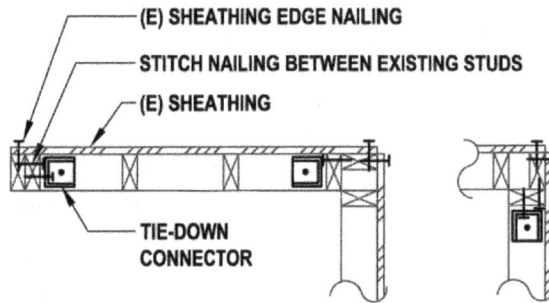

PLAN DETAIL
USING (E) EXTERIOR SHEATHING — B1

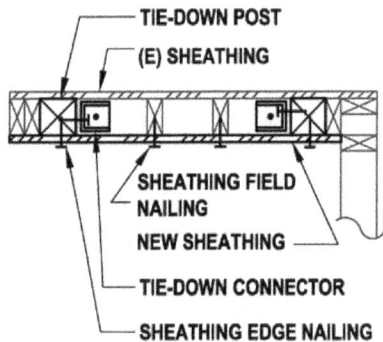

TIE-DOWN POST

(E) SHEATHING

SHEATHING FIELD
NAILING

NEW SHEATHING

TIE-DOWN CONNECTOR

SHEATHING EDGE NAILING

PLAN DETAIL
USING NEW INTERIOR SHEATHING — B2

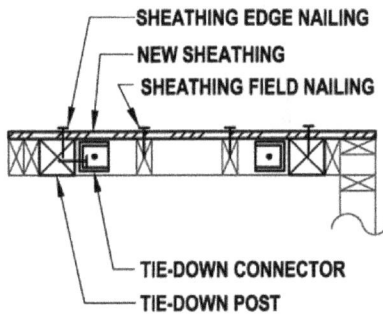

SHEATHING EDGE NAILING

NEW SHEATHING

SHEATHING FIELD NAILING

TIE-DOWN CONNECTOR

TIE-DOWN POST

PLAN DETAIL
USING NEW EXTERIOR SHEATHING — B3
(EXTERIOR FINISHES ARE REMOVED)

Figure 6.4.4-1B: Framing Fastening for Overturning Load Path

TIE-DOWN

(E) FLOOR JOIST

BLOCKING AT
TIE-DOWN IF
ACCESSIBLE

(E) FOUNDATION
SILL

(E) FOUNDATION

TIE-DOWN
ANCHOR
CONTINUOUS
THREADED ROD

ELEVATION   C1

TIE-DOWN

(E) FLOOR JOIST

BLOCKING AT
TIE-DOWN IF
ACCESSIBLE

(E) CRIPPLE WALL

(E) FOUNDATION
SILL

(E) FOUNDATION

TIE-DOWN
ANCHOR
CONTINUOUS
THREADED ROD

ELEVATION   C2

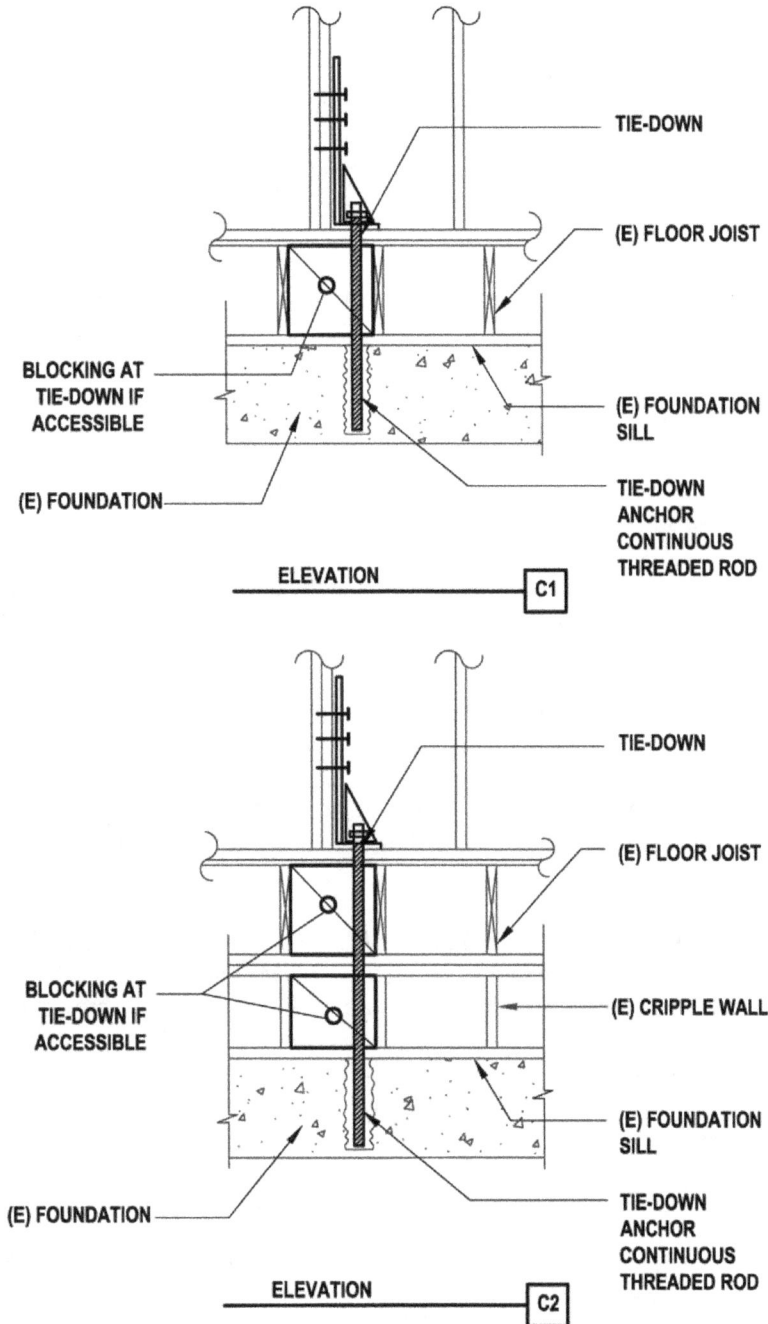

Figure 6.4.4-1C: Tie-down Details at Alternate Base Conditions

Tie-down connectors in first story walls above woodframed floors require detailing modifications; Figure 6.4.4-1C illustrates anchorage to the foundation in locations with a frame floor and a framed floor plus cripple wall. For both of these conditions, there is generally not enough height to install tie-down connectors in the floor framing or cripple wall space, so the tie-down is installed in the first story wall and the tie-down bolt is extended through the joist and cripple wall height to anchor into the foundation. Occasionally cable or rod tie-down systems running the full height of one or more stories will be used in lieu of the tie-down brackets shown in the figures.

Design Considerations

Research basis: Research results comparing in-plane strength and stiffness with and without tie-downs are summarized in Cobeen, Russell and Dolan (2004). Applicable research includes Mahaney and Kehoe (2002), Salenikovich (2000), Ni and Karacabeyli (2000), Salenikovich and Dolan (1999) and Fischer et al. (2001). The drop in shear wall capacity without tie-downs varies as a function of wall length and wall axial loading. Strength reductions up to approximately 80% (20% retained strength) were observed without tie-downs. Reductions in wall stiffness varied, but in general mirrored the drop in strength. Unless rehabilitation of uplift capacity is provided, the reduced strength and stiffness needs to be accounted for in building evaluation.

Adequacy of tie-down post or studs: The stud or post that the tie-down connector is fastened to must be designed to carry required tension and compression forces. Calculations of tension capacity must consider any reduction in the post/stud net section, such as would occur at bolted tie-downs. Where multiple stories contribute tension or compression forces to a post/stud, the full accumulated force must be considered. Single 2x studs should be carefully evaluated before they are used as tie-down studs and should be limited to appropriate loads. Where existing framing members are not adequate, new tie-down posts can be added if fastening is provided to complete the load path (see *Detailing Considerations* section). Where multiple 2x studs are to form a built-up post, it is important that stitch nailing between studs be adequate to develop the wall shear capacity.

In addition, tie-down connectors are believed to create flexure as well as tension in the post/stud being connected (Pryor, 2002). Where bare posts have been tested alone (no sheathing, wall framing), the flexure has been seen to cause both failure of the post and pull-through of bolts connecting the tie-down to the post (Nelson, 2005, and Nelson and Hamburger, 1999). The stud or post should be checked for combined tension and flexure. The type of tie-down chosen can reduce the flexure. Use of tie-downs fastened with nails or wood screws rather than bolts avoid net section reduction at the bolts and reduce possible slip. This type has been favored in California since the Northridge earthquake. Alternately, bolted tie-downs can be placed symmetrically on each side of a stud or post to minimize flexure.

Tie-down bracket devices developed by manufacturers since the Northridge earthquake have also tended to be stiffer, minimizing deformation within the bracket device. The stiffer tie-down reduces the portion of wall drift generated by uplift at the tie-down. In addition, less uplift at the wall end should reduce the likelihood of foundation sill plate splitting because sill uplift is also restrained. Stiffer tie-downs are recommended to the extent practical, as reduced wall drift should translate into less damage. Tie-downs that might have brittle failures at expected earthquake loads should be avoided.

Tie-down design criteria: FEMA 356 (FEMA, 2000) identifies fasteners used to transfer forces from wood to wood or wood to metal as being deformation-controlled actions. When coupled with several relatively high *m*-factors for static procedures, this can result in less fastening being required by FEMA 356 than the current building codes. At the same time, the shear wall sheathing fastening is identified as the desired location of inelastic behavior, which suggests that shear wall overturning restraint should be force-controlled and more fastening provided. It is recommended that current building code requirements be used for FEMA 356's Basic Safety Objective. For a higher performance objective, a capacity-based approach is suggested.

Foundation anchor type and installation: Discussion of foundation anchor type and installation can be found in Section 5.4.3. Anchorage of tie-down tension bolts to existing foundations will almost exclusively use adhesive anchors, which have more compatible capacities and allow more convenient installation. To date, it has been common to install the adhesive anchor straight down into the footing, or at a very slight angle if required for access. The capacity based on adhesive bond can be taken from manufacturer information. In the past, concrete anchorage design methods in the UBC (ICBO, 1997) have allowed calculation of the concrete pull-out capacity based on an assumed failure surface. New provisions in ACI 318 Appendix D (ACI, 2005) will not allow tie-down anchorage using current configurations. With typical anchors centered at 1-3/4 inch from the edge of concrete, required cover cannot be met, the seismic load requirement that steel rather than concrete control is difficult to meet, and side blow-out tends to restrict calculated capacities. Although this appendix chapter excludes adhesive anchors, it is difficult to consider rehabilitation anchorages acceptable that would not be acceptable for new cast-in-place connections. One possible alternative is to angle the tie-down rod in the concrete to get better cover and reduce calculated side blow-out. Although some proprietary cast-in anchors use this configuration, testing for rehabilitation use has not occurred.

Adequacy of foundation: Tie-down connectors should be attached to substantial existing footings that have the shear and flexural capacity to mobilize required resistance. Alternately, new footings or footing reinforcement can be provided. Addition of tie-down connectors at isolated footings or unreinforced masonry footings should receive very careful design consideration.

Detailing Considerations

Vertical shear load path: It is important that a load path be provided between the tie-down connector and a stud or post that has adequate fastening to the structural sheathing. Where new shear wall sheathing is provided, the tie-down connector is installed on a post/stud that receives sheathing edge nailing over the entire wall height (Figures 6.4.4-1B2 and 6.4.4-1B3). Where existing panel sheathing is being used, it is necessary to install the tie-down at an existing post/stud with sheathing edge nailing (Figure 6.4.4-1B1). Additional nailing may be required to maintain a load path between the tie-down post/stud and the post/stud with sheathing edge nailing, as seen at the right hand side of Figure 6.4.4-1B1. The nail size and spacing will need to be calculated to match the shear wall capacity. Where this is not possible, the structural sheathing should be exposed at the tie-down locations in order to provide adequate nailing into the tie-down member. Use of adhesive attachment of the tie-down post/stud to the structural sheathing should not be used as part of this load path because the stiffness of this sheathing to framing connection is not compatible with expected slippage between the sheathing and framing during shear wall racking.

Vertical compression load path: When shear wall uplift is occurring at one end of a shear wall, a downward reaction is occurring at the other end. A load path to transmit this compression through the wood framing to the foundation is generally provided at the same location as the tie-down. Often, compression blocking is added in the floor framing depth to provide full bearing of the post/stud on the top and bottom plates, as shown in Figures 6.4.4-1A and 6.4.4-1C. It is important that dry framing be used; otherwise, shrinkage is likely to make the blocking ineffective. See the Section 5.4.1 discussion of shrinkage.

Tie-down connectors: Tie-down connectors are almost exclusively proprietary. Connector types used for retrofit include brackets, straps, and occasionally full-height rod or cable systems. Where possible, it is preferred to not mix the connector types within a shear wall. All connectors should be installed in accordance with the manufacturer's recommendations and applicable ICC Evaluation Services report recommendations.

Where straps are used, the manufacturer specified capacity of the strap is dependent on the number of fasteners (nails or screws) installed at each end of the strap. It is important that the required fasteners are provided between the strap and the wall studs. Nails into the floor framing or top and bottom plates should not be counted toward the required amount. The length of the strap must be adjusted to allow installation of the proper number of fasteners into the studs. This should be clearly specified on the tie-down strap detail.

Tie-down bolts: The vertical bolt between the tie-down bracket and the foundation, or between the bracket in a story above and below, is usually all-thread rod.

Anchorage to the foundation: It is most common to use adhesive anchors for anchorage of the vertical tie-down bolt to the foundation. The calculation of the required anchorage depth must take into account the edge distance to the near face of the foundation and the foundation capacity. It is often desirable to lengthen the embedment into the foundation beyond that required by the adhesive anchor manufacturer, in order to better mobilize the foundation capacity. Adhesive anchors must be installed in accordance with the manufacturer's recommendations and the applicable ICC Evaluation Services report recommendations.

Cost/Disruption

Rehabilitation of woodframe shear walls often occurs while the building is still being occupied. This generally involves phased construction and moving furniture from room to room ahead of the work. This slows down the work, but can be less expensive and disruptive for the occupants than relocating them. When the building will be occupied a choice is sometimes made to do all of the work from the building exterior, keeping the interior as functional as possible, or completely from the building interior, avoiding opening of the building finishes. This choice greatly affects design and detailing, so it should be made very early in the design process.

Where existing shear wall sheathing has adequate shear capacity, it may be possible to selectively open interior finishes to install tie-down connectors, greatly limiting the disruption to the occupants. If locations of sheathing edge nailing are well known, it may be possible to only open up a space one stud bay wide and several feet high at each wall end. More likely, however, it will be necessary to open up the stud bay for the full wall height to provide adequate interconnection of framing members. Often shear transfer connections will also need to be provided, requiring the opening of a strip of wall finish along the base of the wall and another strip of ceiling at the wall top.

Construction Considerations

It is not uncommon for significant variation to occur in the framing detailing of existing buildings. It is important that conditions be observed during construction of rehabilitation measures, and details be modified for as-built conditions. This is most effectively done by scheduling time between opening of finishes and start of installation for the engineer to observe conditions and provide needed guidance.

Proprietary Concerns

There are no proprietary concerns with this rehabilitation technique other than the use of proprietary connectors and adhesives as part of the assemblage.

6.4.5   Enhance Shear Transfer Detailing

Deficiency Addressed by Rehabilitation Technique

This rehabilitation technique addresses detailing for transfer of shear into and out of shear walls.

Description of the Rehabilitation Technique

The addition or enhancement of shear walls is not of value unless shear forces can be transferred into and out of the wall. Section 5.4.1 addresses a wide variety of shear transfer details for the top and bottom of shear walls where the existing wall top plates will serve as collector elements. This will be applicable in most instances in **W1** and **W1A** buildings. Where new shear walls are added, however, it is likely that new collector elements will be needed. In **W1A** buildings, shear walls are generally well distributed and resist moderate loads. This section discusses collectors and shear transfer for moderate loads in new shear walls. Sections 7.4.1 and 7.4.2 address addition of collectors for new vertical elements with significant strength and stiffness, including steel braced frames and concrete and masonry walls. These elements are most often added in **W2** buildings.

In a **W1A** building, the capacity of the roof of floor diaphragms are very likely to be less than the capacity of added or enhanced shear walls. The collector needs to extend well beyond the length of the shear wall, as a minimum engaging adequate diaphragm length to resist forces. Ideally a collector would extend for the entire length of the diaphragm being supported.

Figure 6.4.5-1 illustrates collectors transferring load into the top of a new or enhanced shear wall. Figures 6.4.5-1A and 6.4.5-1B show new or existing framing parallel to the wall used as a collector. Detail A assumes that fastening of the diaphragm sheathing to framing exists or can be provided. Load transfer to the diaphragm can occur over the length of the new or existing

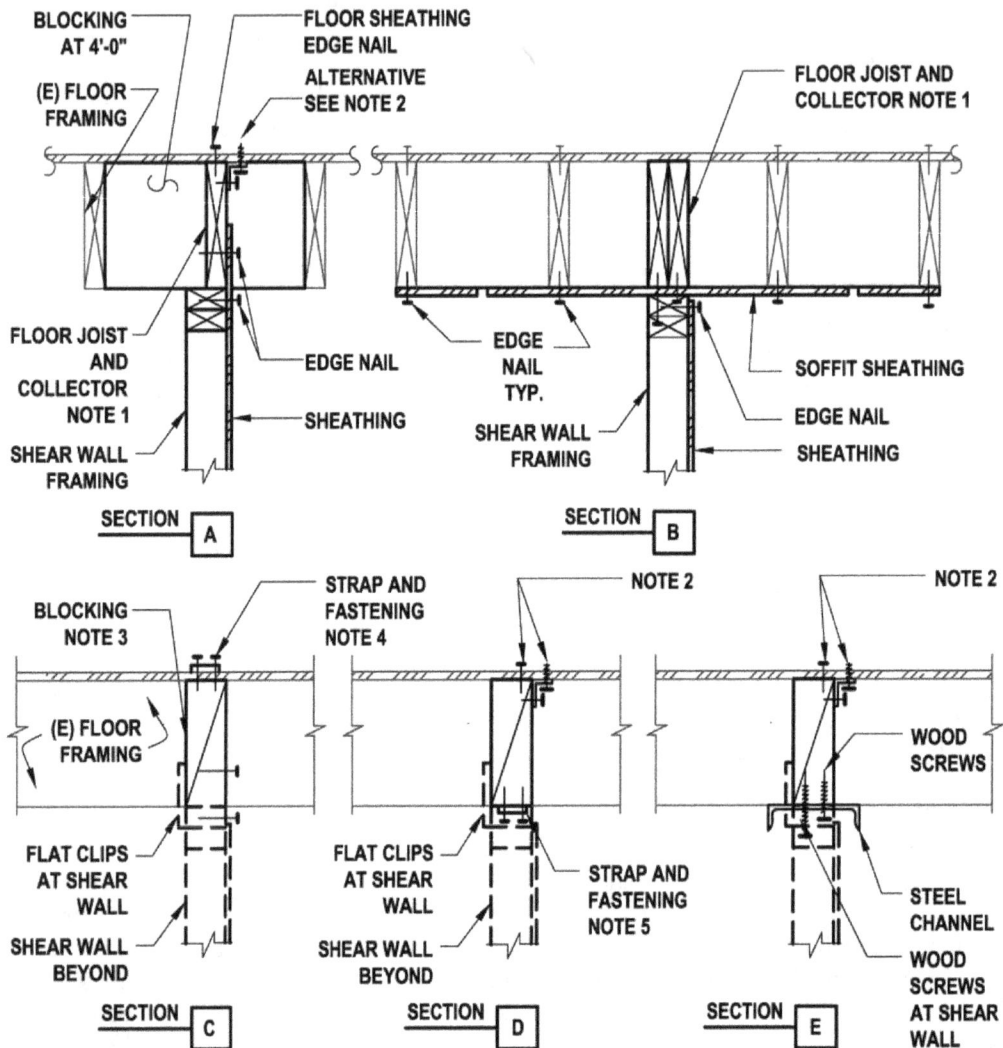

NOTES:

1. Where existing framing falls at new wall location use existing joist, otherwise add joist/collector. Where existing or added floor joists act as the collector, provide tension and compression connection at all breaks in floor joists over the required collector length.

2. Provide sheathing to framing shear transfer per Figure 5.4.1-6 or ceiling soffit per Detail B.

3. Provide blocking of adequate width to receive strap nailing. Predrilling of nail holes may be required to prevent splitting of 2x blocking. Fasten blocking each end using end nails or toe nails.

4. Run strap continuous across top of shear wall and required collector dimension in each direction. Provide splice detail for strap unless straps are provided in rolls of adequate length. Over the length of the shear wall, strap fastening must be adequate to transfer entire shear wall force into strap. Over the balance of the strap length, fastening must be adequate to transfer unit shear into the diaphragm.

5. Lap strap with shear wall top plates and fasten to develop collector force. Provide fastening adequate to transfer diaphragm unit shear over balance of strap length.

Figure 6.4.5-1: Collector Details

framing member without any splices being required. Where additional length of attachment to the diaphragm is required, splicing of the collector framing in accordance with Figure 6.4.5-2 is needed. Often framing in older buildings has a significant lap length over interior supports, sometimes making a direct nailed, screwed or bolted connection between existing members possible. Where bolts are used, detailed attention is needed to provide required bolt end and edge distances and spacing. Alternate splice approaches include steel straps and plates.

Where collector member connection to the diaphragm above is not practical, a wood structural panel soffit, as shown in Detail B can be used to transfer load from the collector to the diaphragm. A minimum soffit width of four feet will generally ensure that at least one row of diaphragm edge nailing is engaged. For large unit shears, additional soffit width and length can distribute loads further.

Figures 6.4.5-1C through 6.4.5-1E illustrate collector details where the existing framing is perpendicular to the wall. Because continuous framing is not available to act as a collector, steel straps or sections are used. Straps will generally be assumed to only carry tension loads. Blocking, already provided for shear transfer is assumed to carry compression loads. Blocking needs to have a tight fit in order to minimize deformation. Detailing is needed if splices will occur in the collector. Figure 6.4.5-3 illustrates an elevation of a collector where framing is perpendicular to the shear wall, corresponding to Figures 6.4.5-1C or 6.5.4-1D.

Design and Detailing Considerations

Research basis: Testing of shear transfer connections between wood structural panel diaphragms and shear walls below was conducted by Ficcadenti et al. (2004). No research applicable to steel straps and blocking for collectors has been identified.

Deformation in the collector: In order to be the most effective, the deformation of a collector should be as compatible as possible with the roof diaphragm it is attached to. Generally roof diaphragms in **W1A** buildings will be short span and quite stiff, suggesting that a stiff collector is preferable. This can be best achieved through use of existing framing members of as long lengths as possible, as illustrated in Figure 6.4.5-2. Splices in the collector members should also be reasonably stiff, as slip at the splice could result in tension in the diaphragm.

Where framing runs perpendicular to the framing direction, there is sometimes little choice but to use steel straps for tension and blocking or framing members for compression, as shown in Figure 6.4.5-3. Unless the straps are reasonably stiff and blocking is installed tight, significant deformation could occur in the collector, resulting in limited efficiency for transferring loads. Although this type of collector is used commonly in new construction and rehabilitation, little is know about its effectiveness and resulting building performance. Conversely, however, significant distress in diaphragms in **W1A** buildings has only been seen at significant changes in geometry such as re-entrant corners (Schierle, 2002). If collectors are to be installed, it is recommended that they be made as stiff and tight-fitting as possible. Sizing of the collector member using overstrength forces or as force-controlled actions will help keep the collector stiff, therefore increasing likely performance.

CLIP ANGLE FOR SHEAR TRANSFER

(E) SHEATHING

*Improved shear transfer to (E)
sheathing may be required.*

B

BOLTS

COLLECTOR
USING (E)
FRAMING

SHEAR WALL

(E) WALL

Note: *See wood design standards for
required bolt edge and center
spacing.*

ELEVATION
SHEAR WALL COLLECTOR   A

7 BOLT DIAMETERS
MINIMUM. INCREASE
WHERE POSSIBLE

(E) FRAMING

THROUGH BOLTS

(E) WALL BELOW

PLAN
MODERATE LOAD COLLECTOR SPLICE   B1

ADDED FRAMING

FASTENING
CALCULATED TO
TRANSFER
COLLECTOR LOAD

(E) FRAMING

(E) FRAMING

ADDED
FRAMING

(E) WALL BELOW

TIE-DOWNS
*Alternate connection approaches
include steel side plates.*

PLAN
HIGH LOAD COLLECTOR SPLICE   B2

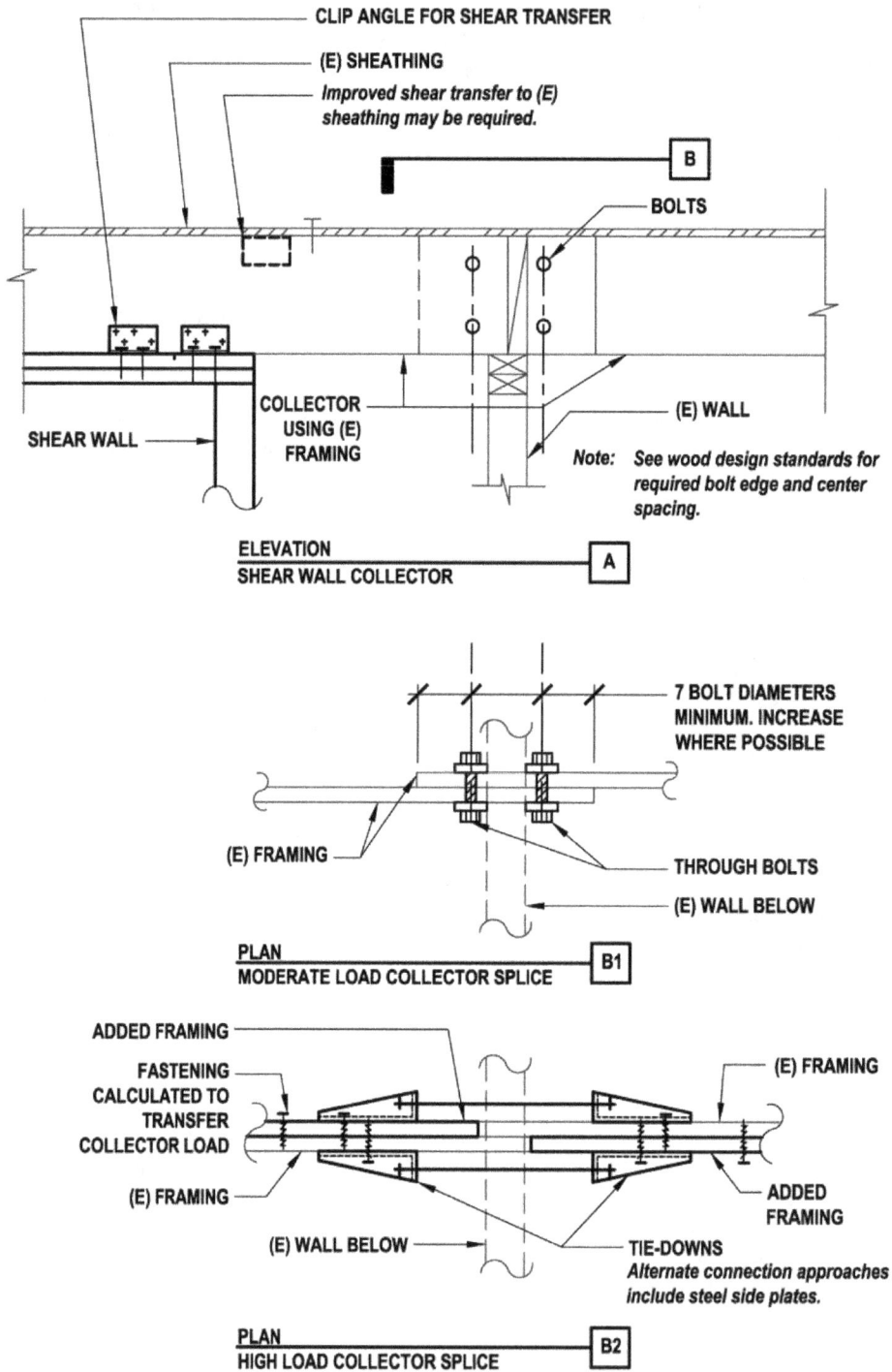

Figure 6.4.5-2: Collector Using Existing Framing Parallel to the Shear Wall

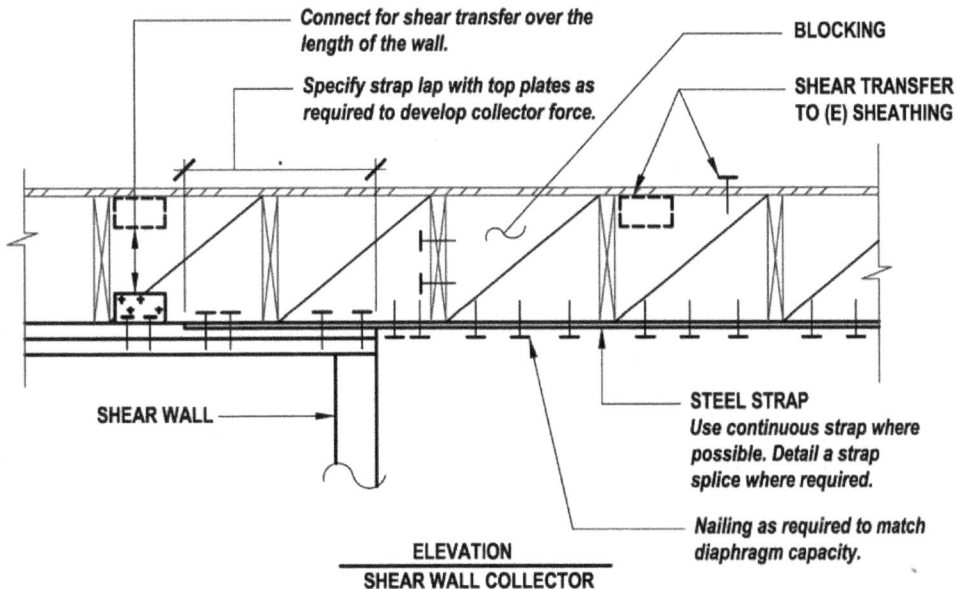

Figure 6.4.5-3: Collector Using Added Blocking at Framing Perpendicular to Shear Wall

Where the top of the existing diaphragm sheathing cannot be accessed for additional sheathing nailing, sheathing added at the ceiling soffit can help distribute forces into the existing diaphragm, as shown in Figure 6.4.5-1B.

Cost and Disruption Considerations

Removal of existing floor or roof finishes to nail diaphragm sheathing into new collector members can be both costly and disruptive. It is, however, going to provide the most predictable performance and is recommended for highly loaded walls and where a performance objective higher than life-safety is intended. Other fastening methods can be calculated and detailed, however not enough is known about their ability to perform adequately.

Construction Considerations

Tight fit of framing, blocking and straps is critical to limiting deformation and improving performance of the collectors and shear transfer connections.

Proprietary Concerns

There are no proprietary concerns with this rehabilitation technique other than the use of proprietary connectors as part of the assemblage.

6.5   References

ACI, 2005, *Building Code Requirements for Reinforced Concrete and Commentary*, ACI 318, American Concrete Institute, Farmington Hills, MI.

AF&PA, 2005, *Special Design Provisions for Wind and Seismic, ASD/LRFD*, American Forest & Paper Association, Washington, D.C.

AISC, 2005, *Seismic Provisions for Steel Buildings*, American Institute of Steel Construction, Chicago, IL.

APA, Revised 2000, *Research Report 138, Plywood Diaphragms*, APA The Engineered Wood Association, Tacoma, WA.

APA, Revised 1999, *Research Report 158, Preliminary Testing of Wood Structural Panel Shear Walls Under Cyclic (Reversed) Loading*, APA The Engineered Wood Association, Tacoma, WA.

ASCE, 2005, *Minimum Design Loads for Buildings and Other Structures* (ASCE 7), American Society of Civil Engineers, Reston VA.

City of Los Angeles and SEAOSC, 1994, *Wood Frame Construction Report and Recommendations*, City of Los Angeles and Structural Engineers Association of Southern California, Los Angeles, CA.

Cobeen, K., J. Russell, and J.D. Dolan, 2004, *Recommendations for Earthquake Resistance in the Design and Construction of Woodframe Buildings*, CUREE Publication No. W-30, Consortium of Universities for Research in Earthquake Engineering, Richmond, CA.

FEMA, 2000, *Prestandard and Commentary for the Seismic Rehabilitation of Buildings* (FEMA 356), Federal Emergency Management Agency, Washington, D.C.

FEMA, 1998, *NEHRP Recommended Provisions for Seismic Regulations for New Buildings and Other Structures* (FEMA 302), Federal Emergency Management Agency, Washington, D.C.

Ficcadenti, S., E. Freund, G. Pardoen, and R. Kazanjy, 2004, *Cyclic Response of Shear Transfer Connections Between Shear Walls and Diaphragms in Woodframe Construction*, CUREE Publication No. W-28, Consortium of Universities for Research in Earthquake Engineering, Richmond, CA.

Fischer, D., A. Filiatrault, B. Folz, C.M. Uang, and F. Seible, 2001, *Shake Table Tests of a Two-Story Woodframe House*, CUREE Publication No. W-06, Consortium of Universities for Research in Earthquake Engineering, Richmond, CA.

Fonseca, F., S. Rose and S. Campbell, 2002, *Nail, Screw and Staple Fastener Connections*, CUREE Publication No. W-16, Consortium of Universities for Research in Earthquake Engineering, Richmond, CA.

ICBO, 1997, *Uniform Building Code*, International Conference of Building Officials, Whittier, CA.

ICBO, 2001, *Guidelines for Seismic Retrofit of Existing Buildings* (GSREB), International Conference of Building Officials, Whittier, CA.

ICC, 2003a, *International Building Code*, International Code Council, Country Club Hills, IL.

ICC, 2003b, *International Existing Building Code* (IEBC), International Code Council, Country Club Hills, IL.

ICC, 2005a, *2003 International Existing Building Code (IEBC) Commentary*, International Code Conference, Country Club Hills, IL.

ICC, 2005b, *Proposed Changes to the 2003 IEBC*, International Code Council, Country Club Hills, IL.

ICC Peninsula Chapter, 2004, *Simplified Design Procedure for Structural Steel Ordinary Moment-Resisting Frame* (OMRF) 9/3/04 draft), ICC Peninsula Chapter, CA.

Lizundia, B. and W. Holmes, 2000, *Practical Solutions for Improving the Seismic Performance of Buildings with Tuckunder Parking*, prepared for the City of San Jose Department of Housing and Office of Emergency Services, San Jose, CA.

Mahaney, J. A. and B. E. Kehoe, 2002, *Anchorage of Woodframe Buildings, Laboratory Testing Report*, CUREE Publication No. W-14, Consortium of Universities for Research in Earthquake Engineering, Richmond, CA.

Mahin, S., 1980s, personal communication.

Mosalam, K., et al., 2002, *Seismic Evaluation of an Asymmetric Three-Story Woodframe Building*, CUREE Publication No. W-19, Consortium of Universities for Research in Earthquake Engineering, Richmond, CA.

Nelson, R., 2005, personal communication.

Nelson, R. and R. Hamburger, 1999, "Hold-down Eccentricity and the Capacity of the Vertical Wood Member," *Building Standards*, November-December 1999, International Conference of Building Officials, Whittier, CA.

Ni, C. and E. Karacabeyli, 2000, "Effect of Overturning Restraint on Performance of Shear Wall," *Proceedings of the World Conference on Timber Engineering*, University of British Columbia, Vancouver, CA.

Pardoen, G., A. Waltman, R. Kazanjy, E. Freund, and C. Hamilton, 2003, *Testing and Analysis of One-Story and Two-Story Walls Under Cyclic Loading*, CUREE Publication No. W-25, Consortium of Universities for Research in Earthquake Engineering, Richmond, CA.

Porter, K., J. Beck, H. Seligson, C. Scawthorn, L.T. Tobin, R. Young and T. Boyd, 2002, *Improved Loss Estimation for Woodframe Buildings*, CUREE Publication No. W-18, Consortium of Universities for Research in Earthquake Engineering, Richmond, CA.

Pryor, S. 2002, *The Effect of Eccentric Overturning Restraint in Complete Shear Wall Assemblies*, Simpson Strong-Tie Company, Inc., Dublin, CA.

Rutherford & Chekene, 2000, *Seismic Rehabilitation of Three Model Buildings with Tuckunder Parking: Engineering Assumptions and Cost Information*, prepared for the City of San Jose Department of Housing and Office of Emergency Services, May.

Salenikovich, A., 2000, *The Racking Performance of Light-Frame Shear Walls, Virginia Polytechnic Institute and State University*, Blacksburg, CA.

Salenikovich, A. and J. Dolan, 1999, "Effects of Aspect Ratio and Overturning Restraint on Performance of Light-Frame Shear Walls Under Monotonic and Cyclic Loading," *Proceedings of the Pacific Timber Engineering Conference, 1999*, Volume 3, New Zealand Forest Research Institute, Rotorua, New Zealand.

Schierle, G.G., 2002, *Northridge Earthquake Field Investigations: Statistical Analysis of Woodframe Damage*, CUREE Publication No. W-09, Consortium of Universities for Research in Earthquake Engineering, Richmond, CA.

Schierle, G. G. editor, 2001, *Woodframe Project Case Studies*, CUREE Publication No. W-04, Consortium of Universities for Research in Earthquake Engineering, Richmond, CA.

SEAOC (Structural Engineers Association of California), 2002, *Recommended Guidelines for Ordinary Moment Frames (OMF) for Buildings*, drafted by the SEAOSC Steel Ad-Hoc Committee for the SEAOC Seismology Committee (August 5, 2002 version), Structural Engineers Association of California, Sacramento, CA.

Seismic Safety Commission, State of California, 1994, *1994 Northridge Earthquake Buildings Case Studies Project* (Proposition 122; Product 3.2 SSC 94-06), Sacramento, CA.

Steinbrugge, J., V. Bush, and J. Johnson, 1996, "Standard of Care in Structural Engineering Woodframe Multiple Housing," *Proceedings, 1996 Convention, Structural Engineers Association of California*, Structural Engineers Association of California, Sacramento, CA.

Vukazich, S., 1998, The *Apartment Owner's Guide to Earthquake Safety*, City of San Jose, San Jose, CA.

Zacher, E. and R.G. Gray, 1985, "Dynamic Tests of Woodframed Shear Panels," *Proceedings, 1985 Convention, Structural Engineers Association of California*, Structural Engineers Association of California, Sacramento, CA.

Chapter 7 - Building Type W2: Woodframes, Commercial and Industrial

7.1   Description of the Model Building Type

Building Type **W2** consists of commercial, institutional, and smaller industrial buildings constructed primarily of wood framing. Most **W2** buildings have first floor slab-on-grade construction; however, woodframe floors supported on foundation walls or cripple walls occur. The upper floor and roof framing consist of wood joists and can include wood or steel trusses, beams, and columns. Post and beam framing is common at interior and at storefronts or garage openings. Lateral forces are resisted by woodframe diaphragms and shear walls. In older buildings, steel rod bracing systems may also be used in place of diaphragms. In newer buildings, wood shear walls are sometimes used in combination with isolated concrete or masonry shear walls or steel braced frames or moment frames. Figure 7.1-1 provides one illustration of this building type.

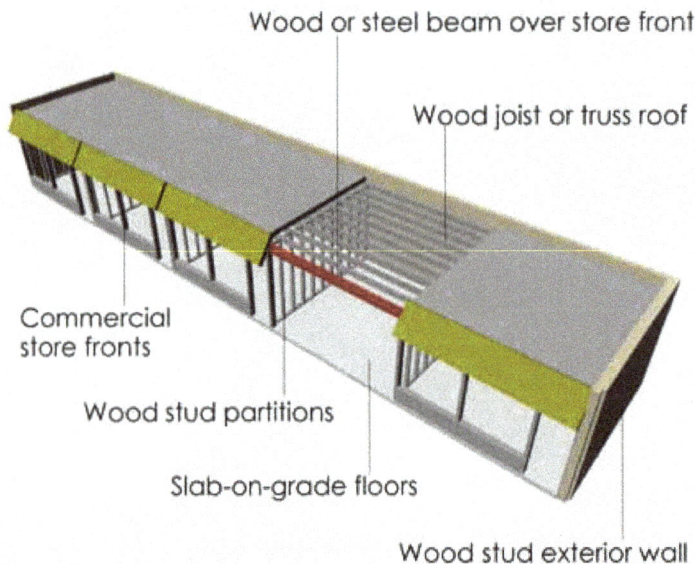

Wood or steel beam over store front

Wood joist or truss roof

Commercial store fronts

Wood stud partitions

Slab-on-grade floors

Wood stud exterior wall

Figure 7.1-1: Building Type W2: Woodframes, Commercial and Industrial

Design Practice

Design practice for **W2** buildings can include no design, design per conventional constructions provisions, engineered gravity design and conventional construction bracing, and engineered gravity and lateral design. More **W2** buildings are likely to have an engineered gravity design than **W1** and **W1A** buildings because the framing systems often fall beyond conventional construction provisions. Lateral bracing of multistory **W2** buildings in accordance with conventional construction provisions was permitted by the UBC (ICBO, 1994) through the 1994 edition. Construction of single-story **W2** buildings using conventional construction provisions is still permitted under the IBC (ICC, 2003).

In California, woodframe school buildings constructed in the 1950s included engineered lateral-force-resisting systems with tie-down anchors and diagonal lumber or plywood sheathed shear walls and diaphragms (Jephcott and Hudson, 1974). Commercial construction in California would generally be anticipated to match school construction, with a time lag. It is anticipated that most **W2** buildings constructed today will have engineered gravity and lateral designs.

Walls and Other Vertical Elements

While wall bracing materials can include the same range discussed for **W1** buildings, the use of diagonal lumber sheathing or wood structural panel sheathing is much more likely in **W2** buildings. The use of overturning anchorage would be varied between the 1950s and 1970s, but common in engineered buildings from the 1980s on.

W2 buildings often have significantly fewer interior bracing walls than **W1** and residential portions of **W1A** buildings. School buildings have moderate room sizes. Commercial buildings with simple geometries are often braced only at the building perimeter, creating large open rooms. Interior bracing may be added for more complex geometries. In newer commercial buildings, concrete or masonry shear walls, steel moment frames, or steel braced frames are sometimes used at the street front or as interior bracing walls in order to maximize the occupant's or user's ability to see across the retail or office space. These vertical elements are used specifically because needed bracing capacity can be provided by much shorter element lengths than with woodframe shear walls. Inclusion of these vertical elements requires additional attention to force distribution, potential torsional irregularities, and collectors adequate to transmit lateral loads to the elements.

Cripple walls, also discussed with the **W1** building type, sometimes occur in **W2** buildings. See Chapter 5.

Floor and Roof Diaphragms

Floor and roof diaphragms include the same materials as the **W1** building type; however, plank and beam systems are rare in **W2** buildings. Significant in **W2** buildings is the occurrence of longer diaphragm spans and more complicated roof diaphragm configurations. The longer spans should result in larger force and deformation demands in the diaphragms and more out-of-plane movement of walls following the diaphragm deflection. More complicated roof configurations require attention to boundary members at diaphragm edges and vertical offsets in chords and collectors.

7.2   Seismic Response Characteristics

Many **W2** buildings, like Building Types **W1** and **W1A** are short period with inelastic behavior concentrated in the vertical elements. Some **W2** buildings (primarily single story), however, have long-span diaphragms, creating the possibility of high stresses, inelastic behavior, and high deformation in the diaphragm.

7.3 Common Seismic Deficiencies and Applicable Rehabilitation Techniques

Very little information has been published on the earthquake performance of **W2** buildings. Earthquake reconnaissance report discussions of wood buildings have tended to focus on residential rather than commercial and light industrial uses. One exception to this is an exhaustive and detailed review school building performance in the 1971 San Fernando earthquake (Jephcott and Hudson, 1974). Only occasional and generally moderate damage is reported to have occurred in one-story and two-story woodframe school buildings. This is consistent with observations of schools made following the Northridge earthquake (EERI, 1996). For schools, however, nonstructural damage was reported to be significant.

While reports of damage are scarce, the construction materials and demands are essentially the same as in **W1** and **W1A** buildings, and many of the vulnerabilities and damage types should be expected to be similar. In fact, the generally larger building size and fewer interior walls should make **W2** buildings more vulnerable than **W1** or **W1A**. This was true in a case study of the Satellite Student Union Center at California State University, Northridge (Schierle, 2001), where significant damage to finish materials occurred. In addition, the **W2** category includes buildings such as churches that often have very irregular building configurations, which should make them susceptible to damage.

See below for general discussion and Table 7.3-1 for a detailed compilation of common seismic deficiencies and rehabilitation techniques for the Building Type **W2**.

Global Strength and Stiffness

Global strength and stiffness can be of concern in **W2** buildings. This is particularly true where use of the first story results in few structural and nonstructural walls and open fronts. Rehabilitation is commonly addressed by the addition or enhancement of wood shear walls, or the addition of steel moment frames, steel braced frames or concrete or masonry shear walls. Where **W2** buildings are large in plan area, it may become practical to introduce a steel braced frame to resist lateral loads. The braced frame can resist higher loads than wood shear walls and lighter moment frames, allowing concentration of lateral loads into fewer and shorter bracing elements. This increases the level of force in the collector and at the base resisting shear and overturning. Occasionally concrete or masonry shear walls are used for rehabilitation; this must be done with caution however, because the weight of the wall will increase seismic forces perpendicular to the wall and attention to wall anchorage is required.

Configuration

The open-front torsional irregularities and weak cripple wall configuration deficiencies introduced in **W1** and **W1A** buildings are equally applicable to **W2** buildings. In addition, mixing of lateral force systems in **W2** buildings can lead to torsional irregularities. Where shear walls are mixed with other vertical bracing elements, care should be taken in evaluating the distribution of lateral forces and deformations. Torsional irregularity may have contributed to damage to the CSU Northridge Satellite Student Center (Schierle, 2001). Common measures for rehabilitation of torsional irregularities include the addition of steel moment frames, wood shear walls, steel braced frames and concrete or masonry shear walls.

Table 7.3-1: Seismic Deficiencies and Potential Rehabilitation Techniques for W2 Buildings						
Deficiency		Rehabilitation Technique				
Category	Deficiency	Add New Elements	Enhance Existing Elements	Improve Connections Between Elements	Reduce Demand	Remove Selected Components
Global Strength	Insufficient in-plane wall strength	Wood structural panel shear wall [6.4.2] Steel braced frame [7.4.1] Steel moment frame [6.4.1]	Enhance woodframe shear wall [6.4.2]	Uplift anchorage and compression posts [6.4.4]	Replace heavy roof finish with light finish	
Global Stiffness	Insufficient in-plane wall stiffness	Wood structural panel shear wall [6.4.2] Steel braced frame [7.4.1] Steel moment frame [6.4.1]	Enhance woodframe shear wall [6.4.2]	Uplift anchorage and compression posts [6.4.4]		
Configuration	Weak story, missing or weak cripple wall	Wood structural panel shear wall [5.4.3], [6.4.2] Add woodframe cripple wall Add continuous foundation and foundation wall	Enhance woodframe shear wall [6.4.2] Enhance woodframe cripple wall [5.4.4]			
	Torsional irregularity including open front	Wood structural panel shear wall [6.4.2] Proprietary wall Steel moment frame [6.4.1] Concrete or masonry wall	Enhance woodframe shear wall [6.4.2]			
Load Path	Inadequate shear anchorage to foundation			Anchorage to foundation [5.4.3]		
	Inadequate overturning anchorage			Uplift anchors and compression posts [6.4.4]		

Table 7.3-1: Seismic Deficiencies and Potential Rehabilitation Techniques for W2 Buildings						
Deficiency		**Rehabilitation Technique**				
Category	**Deficiency**	**Add New Elements**	**Enhance Existing Elements**	**Improve Connections Between Elements**	**Reduce Demand**	**Remove Selected Components**
Load Path (continued)	Inadequate shear transfer in wood framing			Enhance load path for shear [5.4.1], [6.4.5]		
	Inadequate collectors to vertical elements		Enhance existing collector [7.4.2]	Add collectors [6.4.5], [7.4.2]		
Diaphragms	Inadequate in-plane strength and/or stiffness		Enhanced existing diaphragm [22.2.1]		Replace heavy roof finish with light finish	
	Inadequate chord capacity		Enhance chord members and connections [22.2.2]			
	Excessive stresses at openings and irregularities		Enhance diaphragm detailing			
	Re-entrant corners		Enhance diaphragm detailing			
Foundations	See Chapter 23					
[ ] Numbers note in brackets refer to sections containing detailed descriptions of rehabilitation techniques.						

W2 buildings with inadequate cripple wall bracing and foundation anchorage are just as vulnerable as similar **W1** and **W1A** buildings. Rehabilitation of these deficiencies is recommended to be highest priority. For **W2** buildings, use of an engineered rather than prescriptive design for cripple wall bracing and bolting is recommended.

Load Path

The load path deficiencies in **W2** buildings are much the same as **W1** and **W1A** buildings. Rehabilitation measures typically involve fasteners and connectors to resist shear and overturning, and addition of collectors. As in **W1** buildings, anchorage to the foundation is a high priority for rehabilitation in **W2** buildings.

Diaphragm Deficiencies

W2 buildings can have highly irregular diaphragms, with vertical offsets, folded plates, and saw-tooth configurations. Rehabilitation of chords and collectors is key to adequate performance of irregular diaphragms. Rehabilitation enhancing the capacity of the diaphragm is discussed in Chapter 22.

7.4   Detailed Description of Techniques Primarily Associated with This Building Type

7.4.1   Add Steel Braced Frame (Connected to Wood Diaphragm)

Deficiency Addressed by Rehabilitation Technique

This rehabilitation technique addresses inadequate global or local strength or stiffness through the addition of a new steel braced frame element.

Description of the Rehabilitation Technique

Figure 7.4.1-1A illustrates an elevation of a steel braced frame added in a **W2** building. This type of element is generally introduced because it can provide needed bracing capacity in a short element length. The resulting highly loaded element will almost always require the addition of a significant collector to transfer load into the top and the addition of a significant new foundation to transmit forces to the supporting soils. Because existing foundations and collectors would not likely be adequate, it is common to add these elements in an area clear of existing foundations and beams.

Figures 7.4.1-1B1 and 7.4.1-1B2 illustrate unit shear transfer over the length of the braced frame for framing perpendicular and parallel to the frame.

Figures 7.4.1-1C1 and 7.4.1-1C2 illustrate collector members and their connection to framing perpendicular and parallel to the frame. These details are discussed further in Section 7.4.2.

Figure 7.4.1-2 illustrates a two-story steel braced frame added in a **W2** building. Significant in this detail is that the second floor is opened up allowing the braced frame to run continuous over the two-story height. Installing a separate frame at each story would lead to unmanageable connection details; the two-story configuration provides the strongest and stiffest solution. Shear

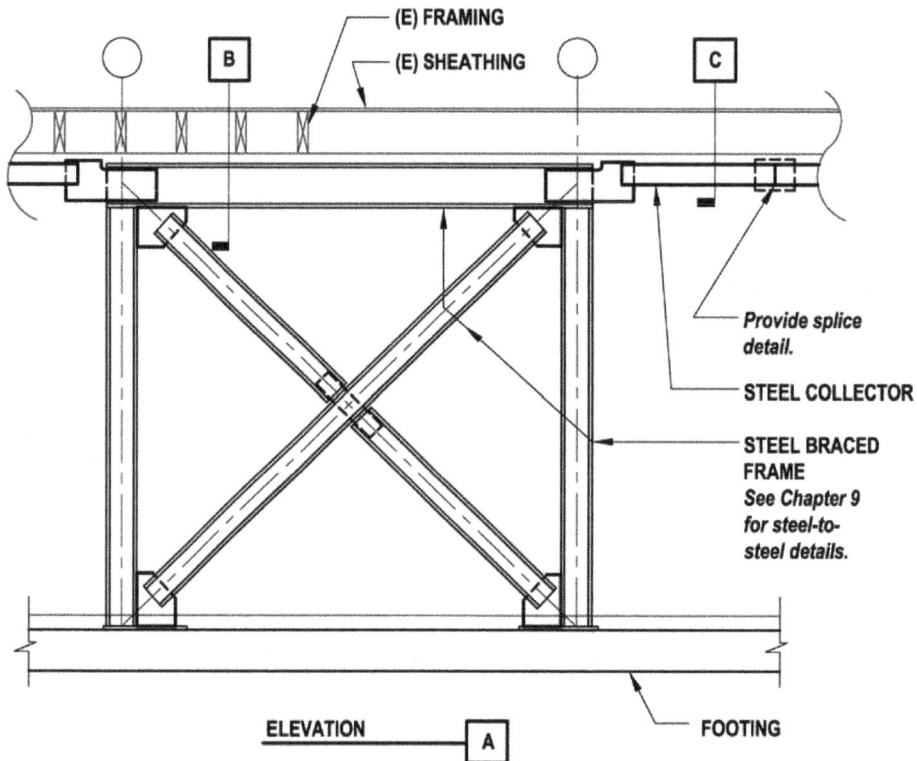

Figure 7.4.1-1A: Steel Braced Frame Added in W2 Building

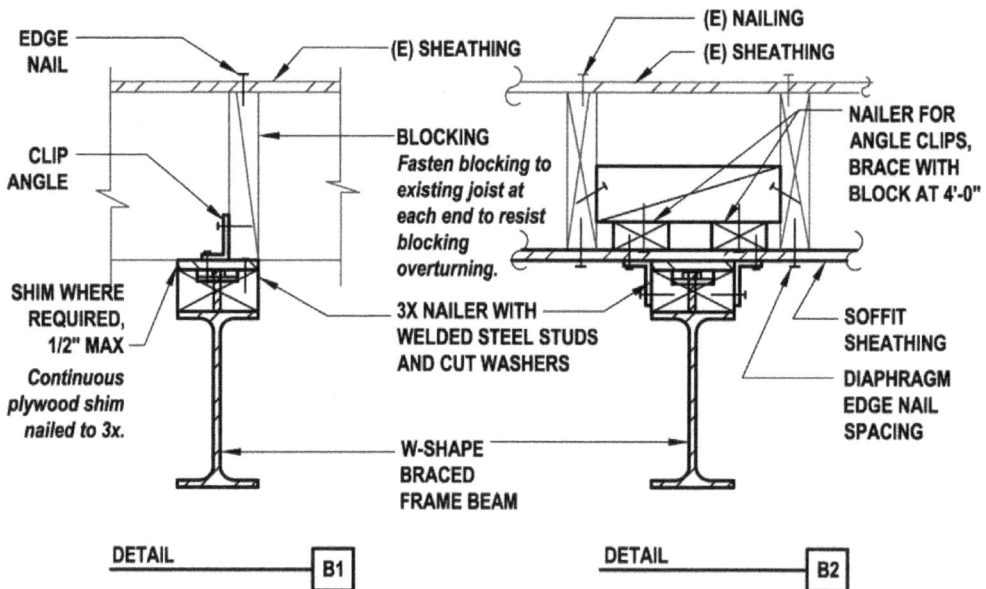

Figure 7.4.1-1B: Shear Transfer and Collector for Steel Braced Frame

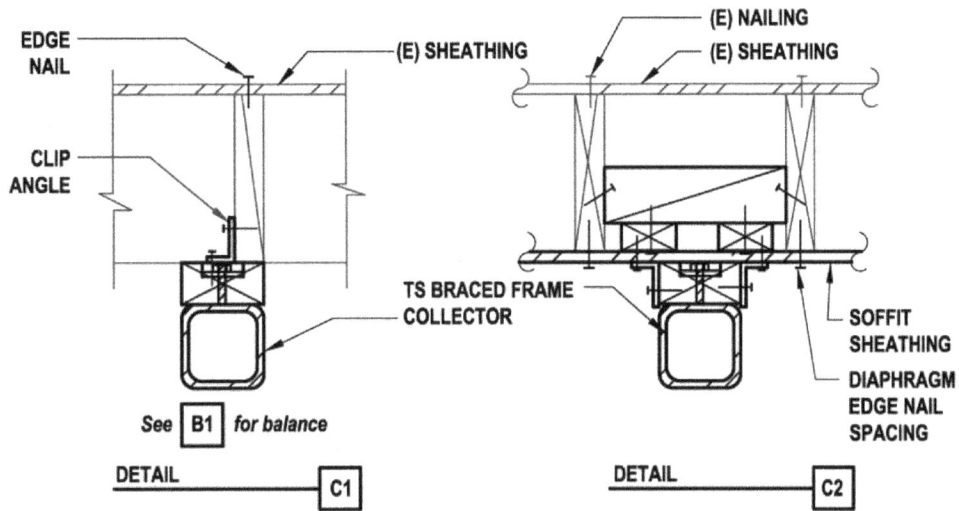

Figure 7.4.1-1C: Shear Transfer and Collector for Steel Braced Frame

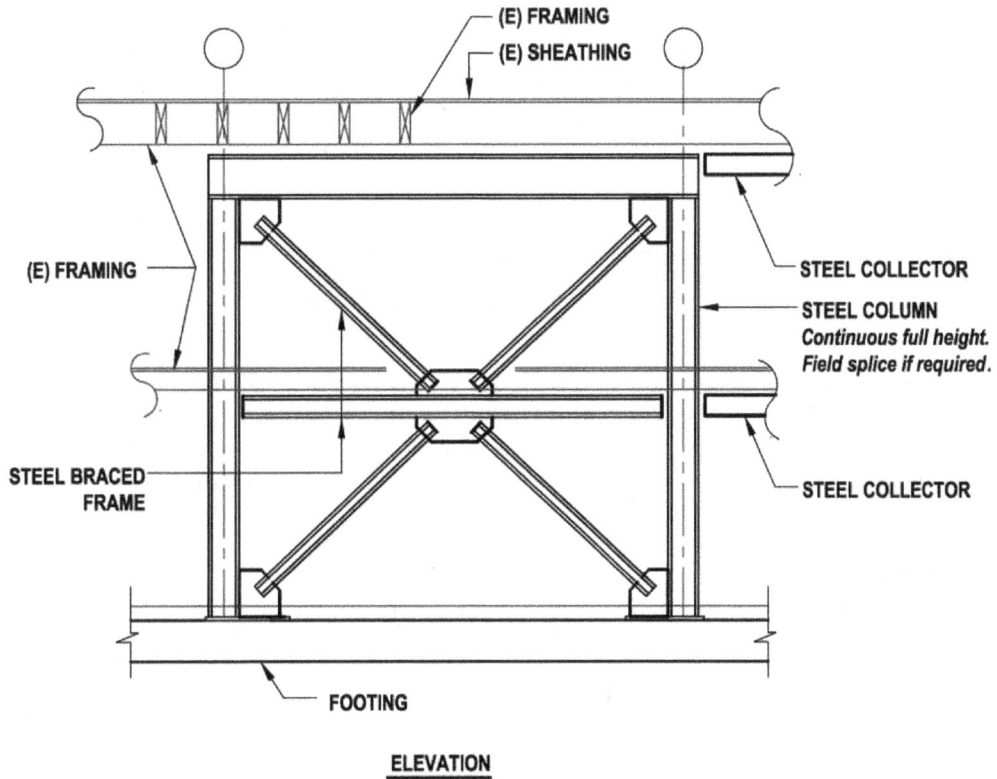

Figure 7.4.1-2: Two-Story Steel Braced Frame in a W2 Building

transfer, collectors, and frame member bracing need to be provided at the second floor as well as the roof.

Design Considerations

Research basis: See Chapter 9 of this document for detailed discussion of steel braced frames. No research applicable to steel braced frames in wood buildings has been identified.

Foundation: The cost of the new braced frame foundation will be a significant part of the cost of this rehabilitation measure. Generally, the dead load available to resist foundation uplift will be minimal, so a large and heavy foundation is often needed. In some cases, it may be necessary to resort to drilled piers or helical anchors to provide uplift resistance.

Detailing Considerations

See Chapter 9 for discussion of detailing of the steel braced frame. See Section 5.4.1 for discussion of basic issues related to rehabilitation of woodframe structure, including wood shrinkage and splitting. These issues are pertinent to load path connections for attachment of steel frames and collectors into the existing woodframe structure.

Steel Connections: Details for connections within the steel braced frame and from the frame to the collector will need to give careful consideration to access for field assembly and field welding. Out-of-plane bracing will be required at the top of the columns and as required by the AISC *Seismic Provisions for Structural Steel Buildings* (AISC, 2005) along the length of the beam. See Section 6.4.1 for discussion of similar connection at a steel moment frame.

Cost and Disruption Considerations

The addition of the steel frame will be disruptive to the area immediately surrounding the frame. Field welding is very difficult to avoid when adding steel braced frames to existing buildings, but should be minimized as much as possible. See Section 7.4.2 for discussion of interruption due to the steel collector.

Construction Considerations

Placement of the steel columns is one of the significant construction challenges, particularly in a multistory frame as shown in Figure 7.4.1-2. It may be desirable to place the steel columns and then cast the foundation and/or slab concrete around them. This allows the depth of the footing to be used for maneuvering the steel column. See Section 6.41 for discussion of field welding issues.

Proprietary Concerns

There are no proprietary concerns with this rehabilitation technique.

7.4.2   Provide Collector in a Wood Diaphragm

Deficiency Addressed by Rehabilitation Technique

This rehabilitation technique addresses provision of collectors to new high capacity vertical elements in wood diaphragm buildings. This technique is primarily intended for use with steel braced frames, as discussed in Section 7.4.1, but would also be applicable to a collector for a new concrete or masonry shear wall in a **W2** building. Sections 6.4.5 and 5.4.1 discuss shear transfer and collectors for woodframe vertical elements with low to moderate loads.

Description of the Rehabilitation Technique

Figures 7.4.1-1C1 and 7.4.1-1C2 illustrate a steel collector member added perpendicular or parallel to existing diaphragm framing. Fastening is provided for shear transfer between the collector and the diaphragm. As discussed in Section 7.4.1, it is assumed that this collector is added in a location away from existing continuous framing members. See Section 6.4.5 for collector alternatives where this is not the case.

In Figure 7.4.1-1C1 new blocking is added, and nailing is provided between the diaphragm sheathing and blocking and between the blocking and a wood nailer. The nailer is attached to the collector with welded steel studs. A 3x framing member is shown as the collector because it gives better bolt values and because it provides enough depth to allow counter sinking of the washer if required.

In Figure 7.4.1-1C2, a wood structural panel ceiling soffit is used to distribute collector forces into the existing diaphragm.

Design Considerations

Research basis: No research applicable to installation of steel braced frames in a **W2** building have been identified; however Section 6.4.1 discussion of shear transfer and collector detailing for steel moment frames (Mosalam et al., 2002) (Cobeen, Russell, and Dolan, 2004) is similar to this technique.

Design demand: The primary deformation in a wood diaphragm should occur as nail slip in the fasteners attaching the sheathing to the framing. The most effective collector will allow this slip to occur, while not adding other sources of significant deformation. This is why the collector is illustrated as a substantial steel section rather than a light steel strap. The collector and connections would require design for overstrength forces in accordance with current building codes. Similarly, use of force-controlled actions would be appropriate when using rehabilitation guidelines. Design to avoid yielding of the steel section is recommended.

Nailing of the existing sheathing to the framing or blocking should not be increased beyond what is required for design level forces or deformation-controlled actions. Note, however, that design level forces may require two rows of diaphragm edge nailing at the collector member, as the sum of the unit shear from two sides may be up to twice the diaphragm unit shear capacity. It is highly recommended that roof sheathing or floor finish materials be removed to allow the sheathing edge nailing to be installed from the top of the existing sheathing (Figure 5.4.1-6A). The other fastening and connections between the steel collector and existing roof sheathing should be designed for amplified forces to the extent possible in order to limit deformation. Connection alternatives shown in Figures 5.4.1-6B, 5.4.1-6C and 5.4.1-6D are not recommended for this level of demand.

Detailing Considerations

Extent of collector: It is recommended that the collector be extended for the full dimension of the diaphragm wherever practical. If the collector is stopped short of the end, the change in diaphragm shear and therefore, deformation may occur at the collector end.

Detailing of splices: Collector member splice locations should be planned, and splice details should be developed. Collector interruptions at beams may make logical splice locations. In Detail C2, it may be possible to move the collector up into the area between joists, thus avoiding collector breaks at beams. Where this is the case, ideal splice locations may be a few feet away from beams. Collector compression forces should be considered in splice design.

Tolerances in existing floor framing: It can generally be expected that there will be some unevenness in the underside of the existing floor framing in the areas where the steel beam and collectors are to be added. It is best to anticipate and include in detailing shimming or other approaches to dealing with this tolerance. Detailing should show locations where shimming is acceptable, set upper limits on acceptable shimming, and adjust fastener capacity or length to account for reduced fastener penetration when shimming is provided.

Cost/Disruption

The addition of collectors at the underside of roof or floor framing can be quite disruptive in buildings that are in use because of the extent of the work; however with adequate planning the work can generally be installed quickly. Where quick work is desirable, the ceiling should be removed for observation of existing conditions over the full extent of the collector prior to steel fabrication. In a one-story building, disruption of occupants can be reduced by installing the collector member on the roof top. This will require removal and replacement of roofing, and adjustment of roof drainage if drainage is altered by the added collector member.

Construction Considerations

Welding of steel studs: Threaded steel studs should be welded to the steel collector in a fabrication shop with periodic special inspection. Field welding of the studs is discouraged due to a lower level of control and fire hazard. Smaller fabrication shops may not have fusion welding equipment for attachment of the studs. A fillet weld around the stud perimeter is acceptable when used with wood nailers. Slight routing of the wood nailer may be required to accommodate the weld.

Proprietary Concerns

There are no proprietary concerns with this rehabilitation technique.

7.5   References

AISC, 2005, *Seismic Provisions for Steel Buildings*, American Institute of Steel Construction, Chicago, IL.

Cobeen, K., J. Russell, and J.D. Dolan, 2004, *Recommendations for Earthquake Resistance in the Design and Construction of Woodframe Buildings*, CUREE Publication No. W-30, Consortium of Universities for Research in Earthquake Engineering, Richmond, CA.

ICBO, 1994, *Uniform Building Code*, International Conference of Building Official, Whittier, CA.

ICC, 2003, *International Building Code*, International Code Council, Country Club Hills, IL.

Jephcott, D. K. and D. E. Hudson, September 1974, *The Performance of Public School Plants During the San Fernando Earthquake*, Earthquake Engineering Research Laboratory, California Institute of Technology, Pasadena, CA.

Mosalam, K., et al., 2002, *Seismic Evaluation of an Asymmetric Three-Story Woodframe Building*, CUREE Publication No. W-19, Consortium of Universities for Research in Earthquake Engineering, Richmond, CA.

Schierle, G. G. editor, 2001, *Woodframe Project Case Studies*, CUREE Publication No. W-04, Consortium of Universities for Research in Earthquake Engineering, Richmond, CA.

Chapter 8 - Building Types S1/S1A: Steel Moment Frames

8.1   Description of the Model Building Type

Building Type **S1** consists of an essentially complete frame assembly of steel beams and columns. Lateral forces are resisted by moment frames that develop stiffness through rigid connections of the beam and column created by angles, plates, and bolts, and/or by welding. Moment frames may be developed on all framing lines or only in selected bays. It is significant that no structural walls or steel braces are provided. Floors are cast-in-place concrete slabs or concrete fill over metal deck. These buildings are used for a wide variety of occupancies such as offices, hospitals, laboratories, and academic and government buildings. Figure 8.1-1 shows an example of this building type.

Building Type **S1A** is similar but has floors and roofs that act as flexible diaphragms such as wood or untopped metal deck. One family of these buildings is older warehouse or industrial buildings, while another more recent use is for small office or commercial buildings in which the fire rating of concrete floors is not needed.

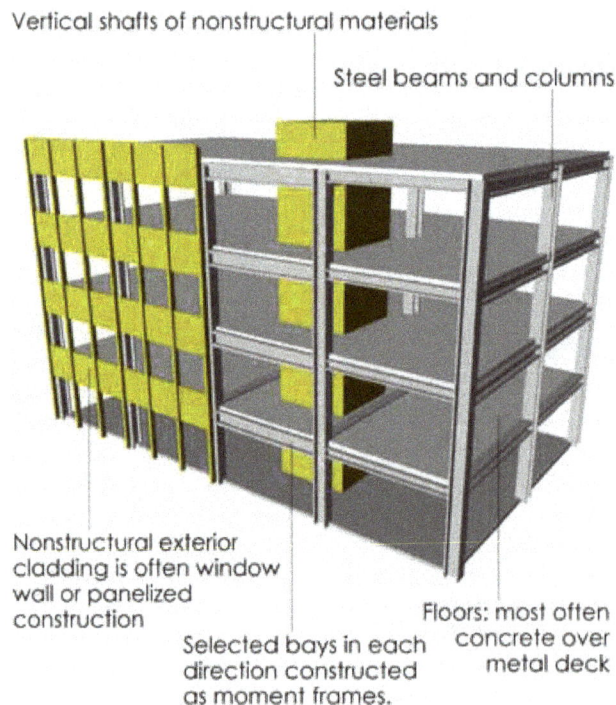

Vertical shafts of nonstructural materials

Steel beams and columns

Nonstructural exterior cladding is often window wall or panelized construction

Selected bays in each direction constructed as moment frames.

Floors: most often concrete over metal deck

Figure 8.1-1: Building Type S1: Steel Moment Frames

Variations Within the Building Type

The use of structural steel in building construction started in the last decades of the 19th century. Lateral load resistance was initially provided by masonry infill. Additional resistance may have been provided by the encasement of the steel members in concrete, though these encased members were not designed with composite considerations. In the 1920s, the use of riveted connections introduced steel moment frames. Beam flanges and webs were joined to the columns by structural shapes, most commonly T-sections. Rivets had low strength and ductility, which limited the overall capacity of the frames. In the 1960s, high strength bolts replaced the rivets. Bolts were faster to install and permitted larger clamping forces, increasing the rigidity of the frames. Connections became smaller when cover plates bolted to the beam flanges and welded to the columns replaced T-sections in the 1950s. By the 1960s, beam flanges and webs were welded directly to the columns to create fully restrained connections, initiating the welded steel moment frame (WSMF) construction era. Shear tabs bolted to the beam webs and welded to the columns later replaced welded beam webs. These welded-flange and bolted-web connections were used extensively in the 1970s and 1980s and are now known as pre-Northridge connections. After the 1994 Northridge earthquake, it was concluded that these connections did not provide enough plastic rotational capacity for most seismic applications (FEMA, 2000c). Several types of connections have been developed and tested to address the flaws in the pre-Northridge connections. The newer connections are designed to develop the full moment capacities of the beams and provide large inelastic rotation capacity.

Floor and Roof Diaphragms

Diaphragms associated with this building type could be either rigid or flexible. The typical rigid diaphragm found in modern buildings consists of structural concrete fill on metal deck. Diaphragm forces transfer to the frames through shear studs welded to the beams. Older steel buildings that were constructed before metal decks were commonly used may have concrete slabs or masonry arches that span between the beams. Flexible diaphragms include bare metal deck or metal deck with nonstructural fill. These are frequently used on roofs that support light gravity loads. Decks could be connected to the steel members with shear studs, puddle welds, screws, or shot pins. The steel members also act as chords and collectors for the diaphragm.

Foundations

There is no typical foundation for this building type. Foundations can be of any type including spread footings, mat footings, and piles, depending on the characteristics of the building, the lateral forces, and the site soil. Spread footings are used when lateral forces are not very high and a firm soil exists. For larger forces and/or poor soil conditions, a mat footing below the entire structure is commonly used. Pile foundations are used when lateral forces are extremely large or poor soil is encountered. The piles can be either driven or cast-in-place. Vertical forces are distributed to the underlying soil through a combination of skin friction between the pile and soil and/or direct bearing at the end of the pile; lateral forces are resisted primarily through passive pressure on the vertical surfaces of the pile cap and piles.

8.2   Seismic Response Characteristics

Steel moment frame buildings are generally flexible, but subject to large interstory drifts. The ductility of these buildings is achieved through yielding and plastic hinging of beams and/or shear yielding of column panel zones at beam-column connections. This inelastic behavior allows moment frames to sustain many cycles of loading and load reversals. Historically, it was believed that large plastic rotations could be developed without significant strength degradation. Up until the Northridge earthquake, the performance of steel moment frame buildings was also believed to far exceed that of masonry and concrete buildings based on observations from previous earthquakes.

The Northridge earthquake exposed severe deficiencies in WSMF connections. A significant number of the frames inspected after the earthquake exhibited visible cracking in the beam flange-to-column flange welds. In a few rare cases, the flanges completely fractured and the damage extended into either the shear tab or the column panel zone. Newer buildings, which relied on deeper beams with thick flanges and less redundancy, were discovered to be even more susceptible to this type of damage. In retrospect, a review of data from past earthquakes indicates that WSMF buildings rarely received close inspection following the event, and often these buildings were overlooked due to the more obvious damage in other types of structures (FEMA, 2000d). This oversight also applies to older steel buildings with riveted or bolted connections. These older buildings, though designed as moment frames, may have suffered limited damage due to the interaction of the steel frames with infill walls and concrete cores. Lastly, if steel damage was discovered after an earthquake, engineers often attributed this to poor construction quality (FEMA, 2000d).

8.3   Common Seismic Deficiencies and Applicable Rehabilitation Techniques

See Table 8.3-1 for deficiencies and potential rehabilitation techniques particular to this system. Selected deficiencies are further discussed below by category.

Global Strength

The lack of global strength is caused by insufficient frame strength, resulting in excessive demands on the existing frames. Yielding or fracturing of the beams, columns, and/or connections could lead to excessive drifts. As a result, the building could be deemed irreparable after an eathquake.

Global Stiffness

Moment frames are much more flexible than other types of lateral force-resisting systems. Their flexibility could lead to excessive building drifts and interstory drifts. This is likely to cause structural damage to the connections and nonstructural damage to the partitions and cladding. Additional concerns include P- effects and pounding with adjacent buildings. Common rehabilitation measures include strengthening the existing frames or providing new vertical lateral force-resisting elements.

Configuration

Soft story conditions occur when stiffness from one floor to the next changes abruptly. This is common at ground floors of commercial and office buildings with tall first stories. It could also occur at mid-heights of five-story to fifteen-story story tall buildings that have not been designed for higher mode effects and near field motions.

Table 8.3-1: Seismic Deficiencies and Potential Rehabilitation Techniques for S1/S1A Buildings						
Deficiency		**Rehabilitation Technique**				
Category	**Deficiency**	**Add New Elements**	**Enhance Existing Elements**	**Improve Connections Between Elements**	**Reduce Demand**	**Remove Selected Components**
Global Strength	Insufficient frame strength	Moment frame Braced frame [8.4.1] Concrete/masonry shear wall [8.4.2] Steel plate shear wall [8.4.8]	Strengthen beams [8.4.3], columns [8.4.3], and/or connections [8.4.6], [8.4.9]		Seismic isolation [24.3] Supplemental damping [24.4]	
Global Stiffness	Excessive drift	Moment frame Braced frame [8.4.1] Concrete/masonry shear wall [8.4.2] Steel plate shear wall [8.4.8]	Strengthen beams [8.4.3], columns [8.4.3], and/or connections [8.4.6], [8.4.9]		Supplemental damping [24.4]	
Configuration	Soft story	Moment frame Braced frame [8.4.1] Concrete/masonry shear wall [8.4.2] Steel plate shear wall [8.4.8]				
	Re-entrant corner	Moment frame Braced frame [8.4.1] Concrete/masonry shear wall [8.4.2] Collector [8.4.4]	Enhance detailing [8.4.3], [8.4.4]			
Load Path	Missing collector	Add collector [8.4.4]				
	Inadequate shear, flexural, and uplift anchorage to foundation		Embed column into a pedestal bonded to other existing foundation elements [8.4.5]	Provide steel shear lugs or anchor bolts from base plate to foundation [8.4.5]		

Table 8.3-1: Seismic Deficiencies and Potential Rehabilitation Techniques for S1/S1A Buildings

Deficiency		Rehabilitation Technique				
Category	Deficiency	Add New Elements	Enhance Existing Elements	Improve Connections Between Elements	Reduce Demand	Remove Selected Components
Load Path (continued)	Inadequate out-of-plane anchorage at walls connected to diaphragm			Tension anchors [16.4.1]		
Component Detailing	Inadequate capacity of beams, columns, and/or connections		Enhance beam-column connection [8.4.6] Add cover plates or box members [8.4.3] Provide gusset plates or knee braces [9.4.1] Encase columns in concrete			
	Inadequate capacity of panel zone		Provide welded continuity plates [8.4.6] Provide welded stiffener or doubler plates [8.4.6]			
	Inadequate capacity of horizontal steel bracing	Provide additional secondary bracing [9.4.2]	Strengthen bracing elements Reduce unbraced lengths	Strengthen connections		
Diaphragms	Inadequate in-plane strength and/or stiffness	Collectors to distribute forces [8.4.3], [8.4.4] Moment frame Braced frame [8.4.1] Concrete/masonry shear wall [8.4.2] Steel plate shear wall [8.4.8]	Concrete topping slab overlay Wood structural panel overlay at flexible diaphragms [22.2.1] Strengthen chords [8.4.3], [8.4.4], and [22.2.2]	Add nails at flexible diaphragms [22.2.1]		
	Inadequate shear transfer to frames			Provide additional shear studs, anchors, or welds [22.2.7]		

Table 8.3-1: Seismic Deficiencies and Potential Rehabilitation Techniques for S1/S1A Buildings

Deficiency		Rehabilitation Technique				
Category	Deficiency	Add New Elements	Enhance Existing Elements	Improve Connections Between Elements	Reduce Demand	Remove Selected Components
Diaphragms (continued)	Inadequate chord capacity	Add steel members or reinforcement [8.4.3], [8.4.4]				
	Excessive stresses at openings and irregularities	Add reinforcement [8.4.3] Provide drags into surrounding diaphragm [8.4.4]				Infill opening [22.2.4], [22.2.6]
Foundations	See Chapter 23					
[ ] Numbers noted in brackets refer to sections containing detailed descriptions of rehabilitation techniques.						

Load Path

Load path deficiencies in steel moment frame buildings include inadequate collectors and frame anchorage to foundations. In Type **S1** buildings, seismic forces transfer from the diaphragm to the frame through shear studs welded to collectors or directly to the frame beams. The collectors or the connections to the frame may be too weak and insufficient to transfer these forces. Connections from columns to a base plate or pile cap have to resist shear, flexural and potentially uplift forces. Connections that cannot develop these frame forces do not allow the frame to develop its full capacity.

Component Detailing

The most common detailing deficiencies in steel moment frames are related to beam-column connections. Pre-Northridge welded moment connections consist of complete penetration flange welds and bolted or welded shear tabs. These connections were previously thought to be ductile but were found to fracture at small plastic rotations or even under elastic loads. Rehabilitation techniques primarily focus on forcing the yielding and plastic hinging to occur away from the joint to reduce stresses on the welds. This can be achieved with the use of reduced beam sections (RBS) or strengthening the section of a beam adjacent to the beam-column joint. These techniques are presented in Section 8.4.6 and discussed in detail in *AISC Design Guide 12* (Gross et al., 1999) and FEMA 351 (FEMA, 2000b). Before welded moment connections became common, connections were either riveted or bolted. These types of connections are prone to net section fracture.

Panel zones were previously thought to provide excellent ductility and strain hardening and were given increased predicted shear strength in the 1988 Uniform Building Code (FEMA, 2000c). The increased strength also meant larger inelastic demands in the panel zones. As a result, panel zones are subjected to yielding or buckling before adjacent members fully develop their capacities. Studies performed after the Northridge earthquake found that large panel zone deformations could increase the potential for connection failures (FEMA, 2000c).

Welded column splices are vulnerable to fracture when subjected to large tensile loads (FEMA, 2000b), since they generally use partial penetration groove welds and thus, are not designed for the full capacity of the smaller column.

Diaphragm Deficiencies

Common diaphragm deficiencies include insufficient in-plane shear strengths, inadequate chords, and excessive stresses at openings. Causes for these deficiencies could be due to lack of slab or fill thickness, lack of reinforcing steel in the slabs, insufficient connections to chord elements, and poor detailing at openings. See Chapter 22 for common rehabilitation techniques.

Foundation Deficiencies

Foundations that are inadequate do not develop the full capacity of the lateral force-resisting system. Their deficiencies result from insufficient strengths and sizes of footings, grade beams, pile caps, and piles. See Chapter 23 for common rehabilitation techniques.

8.4 Detailed Description of Techniques Primarily Associated with This Building Type

8.4.1 Add Steel Braced Frame (Connected to an Existing Steel Frame)

Deficiency Addressed by Rehabilitation Technique

Moment frame buildings that are insufficient to resist lateral forces or too flexible to control building drifts can be converted into braced frame buildings.

Description of the Rehabilitation Technique

The seismic performance of a building may be improved by adding braces to existing welded or riveted steel moment frames. Braces can be added without substantially increasing the mass of the building. Various concentrically braced frame (CBF) configurations should be considered, though some tend to perform much better than others in earthquakes. In addition, systems that meet the provisions for special concentrically braced frames (SCBF) are expected to exhibit stable and ductile behavior in large earthquakes. See Chapter 9 for a discussion of different CBF configurations and their behaviors. Moment frames are not commonly converted to eccentrically braced frames (EBF) due to complicated design and detailing issues that would be encountered. Different brace types can be used, including W-shapes, hollow structural sections (HSS), steel pipes, double angles, double channels, double HSS, and buckling-restrained braces.

Design Considerations

Adding braced frames to a moment frame building increases its stiffness considerably. The upgraded structure should be evaluated for higher lateral and overturning forces accordingly. The primary system performance issues are those associated with CBF systems, in which the most vulnerable elements are the braces and their end connections. Thorough knowledge of the existing material behaviors and strengths are necessary to ensure that new and existing elements to interact in the desired manner. Other design issues include the following:

Research basis: No references directly addressing the addition of steel braced frames to moment frame buildings have been identified.

Brace locations: Preference of brace locations should be given to existing moment frame bays to utilize the strengths of existing members, connections, and foundations. If this is not possible and braces are added at other locations, the existing moment frames should be considered when forces are distributed to the lateral force-resisting system. If the building drifts are large enough, deficiencies in the moment frames may still require mitigation whether the frames are included in the new lateral force-resisting system or not.

Brace selection: Use compact and non-slender sections whenever possible to avoid premature fracturing or buckling of the braces during post-yield behavior. Issues related to conventional structural shapes are discussed in Chapter 9. Two particular brace types not common in older braced frame buildings are double HSS section and buckling-restrained braces. Double HSS sections can be used in configurations similar to double angles or channels (Lee and Goel, 1990). They provide reduced fit-up issues and smaller width-to-thickness ratios compared to a single HSS, resulting in increased energy dissipation capacity. The other type of brace, used in a buckling-restrained braced frame (BRBF), is typically used in new buildings but has also been used successfully in new BRBF systems in existing buildings. One example of a brace used in a BRBF consists of a steel core inside a casing, which consists of a hollow structural section (HSS) infilled with concrete grout. Proprietary materials separate the steel core and concrete to prohibit bonding between the two materials. There are other buckling-restrained braces that do not use grout or additional separating agents between the steel and grout. The main advantage of these braces is the ability of the casing to restrain the buckling of the steel core without providing any additional axial force resistance beyond the capacity of the steel core. Provisions for new building BRBF design are included in the *NEHRP Recommended Provisions* (FEMA, 2003) and the *AISC Seismic Provisions for Structural Steel Buildings* (AISC, 2005b).

Nonstructural issues: The addition of braces to an existing structure changes the architectural character of the building. Braces in exterior frames will be visible in buildings with clear glazing. At interior bays, braces have to be configured to avoid obstruction of existing corridors, doorways, and other building systems. Braces are also commonly exposed and incorporated into the interior architecture. For this particular option, note that gusset plates designed in accordance with the *AISC Seismic Provisions* for SCBF can be fairly large and should be discussed with the architect and tenants. If the braced frames are hidden in partition walls, the architect should be aware that these walls will be thicker than typical walls. Beams that are increased in size and new collectors may affect nonstructural components by reducing clear floor heights. These components typically include suspended ceilings, pipes, conduits, and ducts. Coordination with the architect and other trades should not be overlooked or underestimated.

Detailing Considerations

In addition to obtaining the latest drawings for the building including as-built drawings, if available, and conducting comprehensive field surveys, the following issues should be noted:

Connections: Gusset plates provide greater tolerances for the installation of braces than directly attaching the braces to the frame members. It may be impossible to completely eliminate welding at a braced frame connection in a seismic rehabilitation. Continuity and doubler plates on columns, stiffener plates on beams, brace-to-gusset plate connections, and most gusset plate-to-frame member connections all require welding. A typical fully welded connection appropriate for use in SCBF is shown in Figure 8.4.1-1. If the brace capacity is governed by out-of-plane buckling, stable post-buckling behavior can be achieved by allowing the gusset plate to develop restraint-free plastic rotations. AISC recommends an offset of two times the plate thickness along the brace centerline, measured from the end of the brace to a line perpendicular to the nearest point on the gusset plate constrained from out-of-plane rotation. If gaps are provided between the gusset plate and concrete fill, then only the beam or column can act as constraints, as is the case in Figure 8.4.1-1. On the other hand, if concrete is placed directly against the gusset plate, then the slab can also act as a constraint.

For low to moderate seismic applications, Figure 8.4.1-2 shows a more compact connection. An example of a W-shape used as a brace and welded directly to the beam and column is shown in Figure 8.4.1-3. The two latter connections do not allow for restraint-free plastic rotations out-of-plane and should be used primarily in situations where in-plane buckling of the braces govern or

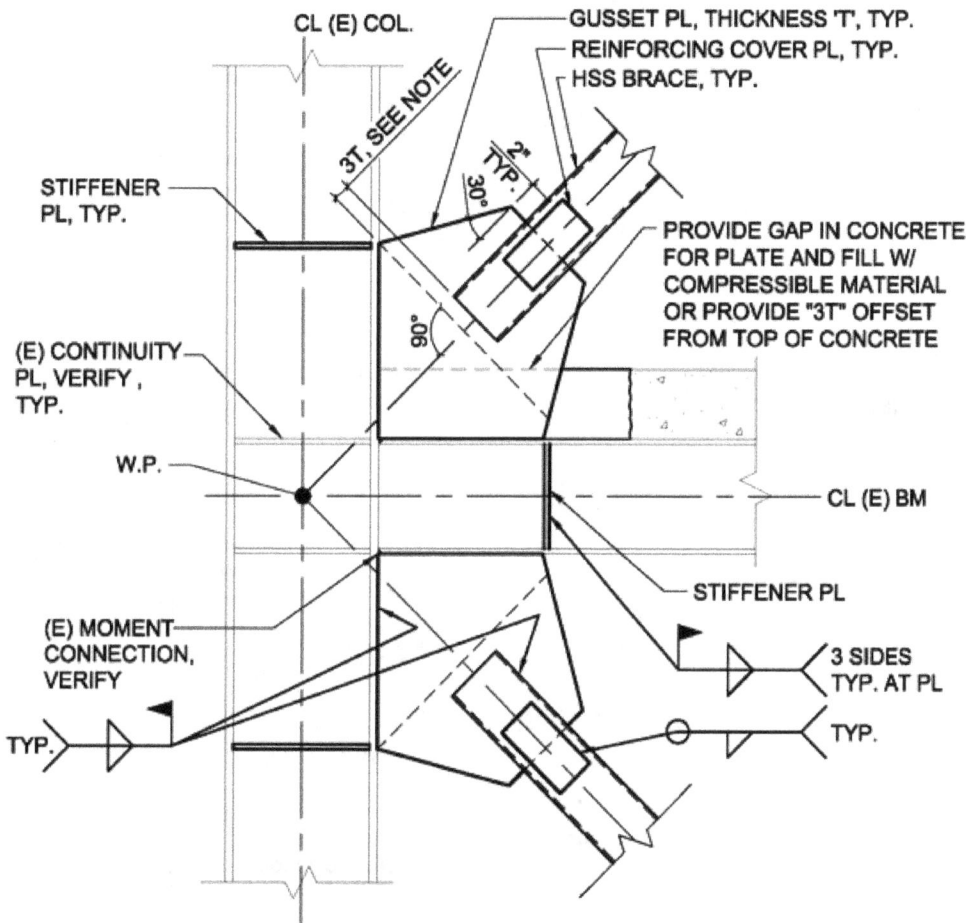

Figure 8.4.1-1: HSS Brace at Existing Beam-Column Connection in SCBF

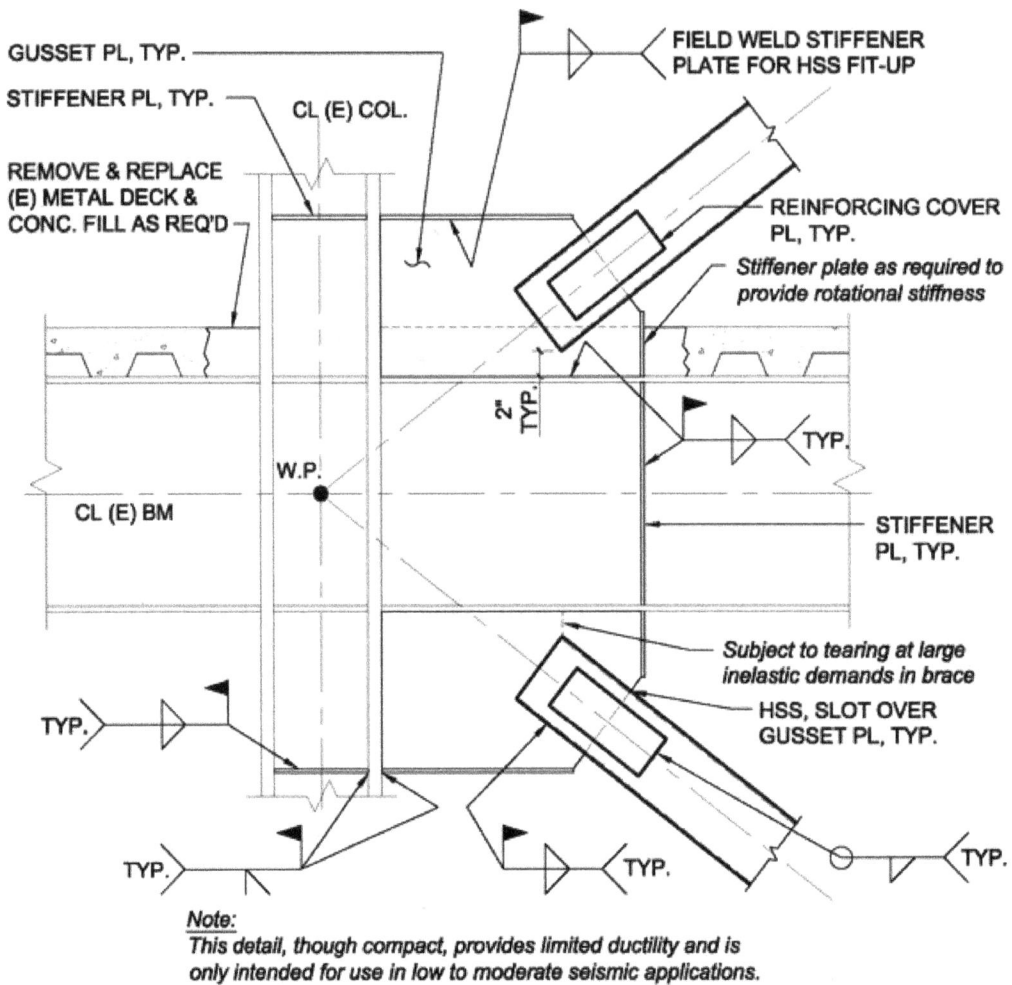

Figure 8.4.1-2: HSS Brace at Existing Beam-Column Connection in Ordinary CBF

ductility demands are low. In addition, the connection shown in Figure 8.4.1-3 requires extensive field welding and could present fit-up issues if field dimensions are not verified on a case-by-case basis.

Buckling-restrained braces are typically bolted to gusset plates. Their bolt configurations at these connections allow for very small tolerances. Double angle and double channel braces can be easily bolted to a gusset plate but space restrictions may limit the number of bolts and subsequently, the strength of the connection.

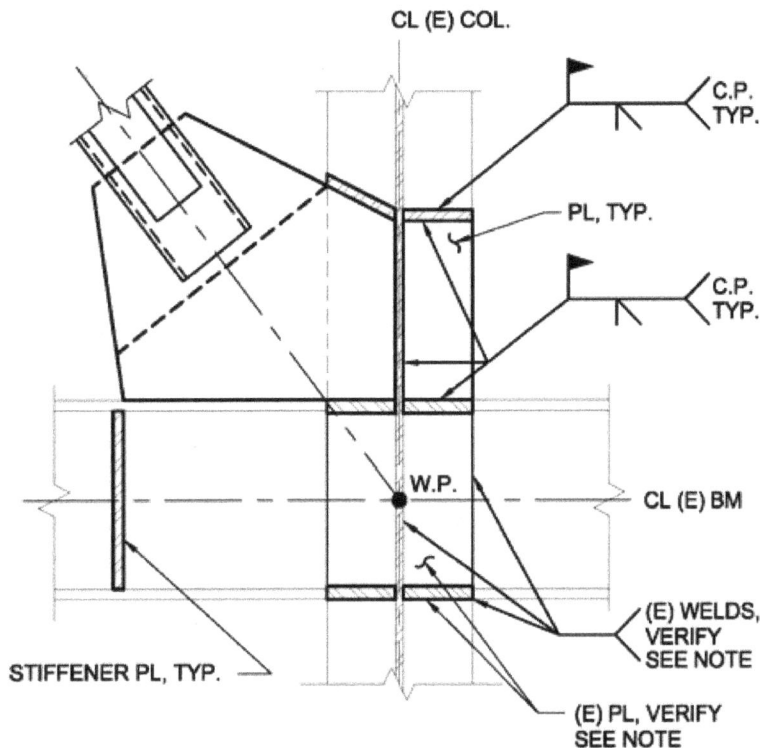

Figure 8.4.1-3: HSS Braced at Existing Beam-Column Connection

It may not always be practical or possible to locate a work point at the intersection of the centerlines of the beam and column, e.g. deep beams. This may deemed to be acceptable if the eccentricity is included in the connection design.

Built-up brace members: While double angles, double channels, and double HSS offer advantages for installation, more stringent criteria apply to these members when used in SCBF. These include stitch spacing, member compactness, and strength of the stitches (AISC, 2005b).

Reinforcing cover plates: HSS and pipe braces are subject to net section fracture at the gusset plate slots (Uriz and Mahin, 2004, and Yang and Mahin, 2005). This brittle failure mode can be eliminated by adding reinforcing cover plates to the sides of the HSS without the slots, as shown in Figure 8.4.1-1 and Figure 8.4.1-2. For pipes, the reinforcing plates can be oriented at right angles to the pipe and appear like stiffeners.

Cost/Disruption

This system could be cost-effective when compared to other alternatives for upgrading steel moment frame buildings. Costs will be less when existing moment frames are converted into braced frames to take advantage of the existing strength and stiffness of the frame members, connections, and foundations. Designs that are simple and details that are not overly complicated will also minimize costs.

Costs can also be reduced if disruption is minimal during construction. Installing braces at the perimeter frames reduces logistical issues associated with working in confined spaces and temporary removal of the nonstructural elements. Noise associated with this type of work is loud and disturbing to the tenants if the building is occupied while the work is being performed.

Construction Considerations

The engineer's involvement during the construction phase is critical during a seismic rehabilitation. The design of the retrofit scheme must not neglect the construction phase and should consider these issues at a minimum:

Welding issues: A work environment in which the welder can perform quality welds is critical. This includes an environment with adequate space to properly operate welding equipment, adequate lighting, and a stable work platform for overhead welds, all while ensuring worker health and safety. Additional considerations include venting of welding fumes and fire protection.

Removal of existing nonstructural elements: Exterior cladding and interior partitions must often be removed to deliver and install the braces to their final locations. Connection modifications at the roof level may warrant the removal of roofing and waterproofing. Installation of the connection to the underside of beams, column continuity and doubler plates, and beam stiffeners, will affect ceilings, lights, and other mechanical/electrical/pluming components. Almost all steel buildings will have some form of fireproofing around the members and connections. With older buildings, asbestos may be present in the fireproofing, which could require costly abatements to expose the members. Older buildings with items of historical significance may require additional coordination and effort to ensure that these items are removed and restored properly. Buildings constructed in the early 20th century commonly had steel members encased in concrete.

Removal of existing structural elements: Slabs and metal decks must be chipped and cut away to install connections. Slabs oriented perpendicular to the beams require temporary shoring. Temporary openings in the slab not only require shoring but also consideration for how the openings will be closed. Beams are rarely removed but if unavoidable, the slabs supported by the beams need shoring and the columns supporting the beams may require bracing.

Construction loads: Loads from construction activities vary and may be either temporary or permanent. Temporary loads could include the weight of construction equipment and patterned loading on perimeter framing when heavy cladding is temporarily removed. Permanent loads could be induced if connection stiffness is modified, such as converting a simple shear tab support into a fully rigid moment connection, and when members are temporarily removed, causing forces to redistribute.

Proprietary Concerns

Braces used in BRBF are proprietary. There are a limited number of manufacturers of the braces used in BRBF.

8.4.2   Add Concrete or Masonry Shear Wall (Connected to An Existing Steel Frame)

Deficiency Addressed by Rehabilitation Technique

Moment frames buildings that are insufficient to resist lateral forces or too flexible to control building drifts can be strengthened and stiffened by adding shear walls. The shear walls may be used alone as the new lateral force-resisting system or in conjunction with the moment frames.

Description of the Rehabilitation Technique

Shear walls can add considerable strength and stiffness to a structure. These walls could be considered as an alternative to adding braced frames if the existing beams and columns are incapable of resisting forces in the braced frame system. Concrete shear walls can be placed using conventional formwork or shotcrete can be used instead if skilled operators for placing shotcrete walls are available. Masonry shear walls, though typically weaker than their concrete counterparts, have the advantage of not requiring formwork when filling the cell cavities with grout.

Design Considerations

The addition of a shear wall to an existing steel frame forms a composite shear wall system. The horizontal shear forces are resisted by the wall elements and the vertical overturning forces are primarily resisted by the steel columns that become boundary elements. This system is almost always controlled by shear due to the substantial flexural strength provided by the steel column boundary elements. Though shear cracking and yielding of a wall is not as ductile as flexural hinging at the base of the wall, limiting the wall to small drifts prevents early loss of strength and stiffness degradation. Other design issues are as follows:

Research basis: No references directly addressing the addition of shear walls to moment frame buildings have been identified.

Design forces: The forces on the structure could increase significantly when this mitigation technique is employed due to the increased stiffness and mass of the walls. The entire structure should be reanalyzed, including all components that were previously determined to have sufficient capacity to resist the forces on the more flexible structure. In-plane forces due to the self-weight of the walls remain in the walls and do not increase demands on other elements. The out-of-plane forces, on the other hand, must be transferred through the diaphragms and collectors to other walls or frames. The design of the wall foundations depends on the magnitude of the forces and the soil properties. If overturning forces are greater than the available counteracting building mass or the soil is poor, some type of pile foundation will be necessary; otherwise, strip footings may suffice.

Wall locations: Preference of shear wall locations should be given to existing bays of moment frames to utilize the layout of the existing collectors and the strengths of the existing members and connections. If this is not possible and shear walls are added at other locations, the existing moment frames should be considered when forces are distributed to the lateral force-resisting system. Deficiencies in the moment frames may still require mitigation whether the frames are included in the lateral system or not.

Continuous walls vs. frames with infill walls: A continuous wall that encases the existing frame beams and columns, shown in Figures 8.4.2-1 and 8.4.2-2, respectively, provides the most robust and simple design. It may be desired to use a system that utilizes independent wall panels that appear similar to older style steel frame buildings with infill walls. The detailing and behavior would be much different from the older buildings though. The wall panels in each bay do not encase the beams and columns, as shown in Figures 8.4.2-3 and 8.4.2-4, respectively. One advantage of this system is that it may allow the elimination of shoring by not having to remove the metal deck though some slab concrete removal may still be required. Another advantage is that beams that have existing penetrations and the elements penetrating the beams would not be affected. The main drawback is that this system may be limited to use in buildings with lower seismic forces. The wall strengths would be limited by the number of studs that can be installed and the capacity of the beam webs. Another option is to encase the beams but not the columns.

Coupled walls: Beams in existing bays of moment frames that are located between new walls will behave like coupling beams, particularly if the bay is relatively short. These beams, subject to high shears and moments, require ductile detailing. The level of coupling between the walls is a function of the stiffness of the beams, which can be determined by a computer model of the system.

Nonstructural issues: The addition of walls to an existing structure changes the architectural character of the building. Walls at the exterior will be visible in buildings with clear glazing. At interior bays, walls have to be configured to avoid obstruction of existing corridors, doorways, and other building systems. Walls can be exposed and incorporated into the interior architecture or hidden in partition walls. The architect should be aware that these walls will be thicker than typical partition walls. Encased beams and new collectors affect nonstructural components by reducing clear floor heights. These components typically include suspended ceilings, pipes, conduits, and ducts. Encased columns reduce usable floor space. Coordination with the architect and other trades should not be overlooked or underestimated.

Detailing Considerations

In addition to obtaining the latest drawings for the building including as-built drawings, if available, and conducting comprehensive field surveys, the following issues should be noted:

Connection to existing frame: Positive connection should be provided to the existing frame to transfer seismic loads to the walls and to provide overturning resistance. Welded shear studs or reinforcing bars are typically used. In much older buildings where cast iron members still exist, it is difficult to weld to cast iron due to its high carbon content. Instead, holes for the reinforcing steel bars may have to be drilled in the members to rely on direct bearing for the load paths.

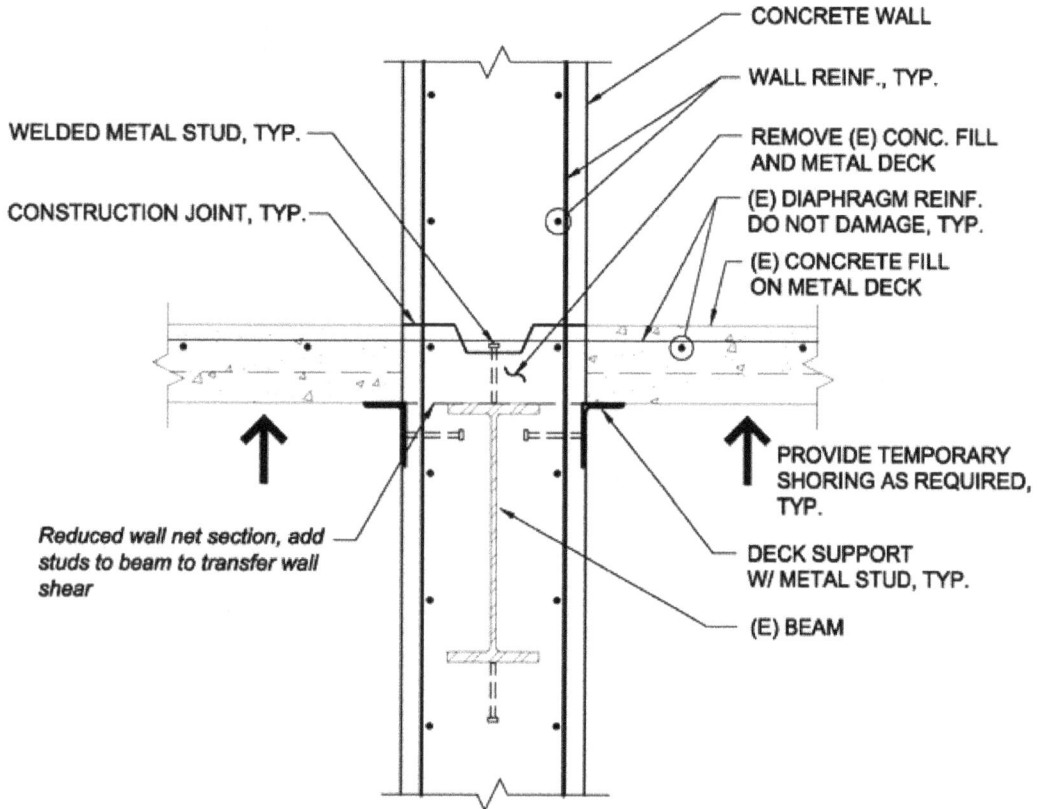

Note:
Offset wall to one side of beam if additional room required for vibrating equipment or for shotcrete construction.

Figure 8.4.2-1: Cast-in-Place Concrete Wall at Existing Beam

Wall reinforcing: Reinforcing steel in the walls should meet all ACI 318 (ACI, 2005) requirements, including the seismic provisions. Columns with positive connections to the walls could be counted as boundary reinforcing steel limited by the strength of the shear connections. Congestion issues may be encountered at the ends of the walls due to the presence of the columns, confinement steel, and anchorage for horizontal wall reinforcing steel. It may be necessary to drill holes through the columns for the horizontal steel.

CONCRETE WALL

ALTERNATE ENDS OF
CROSSTIES

*Note space required for bar
bends and termination*

HORIZ. REINF., TYP.

*Avoid specifying studs at inside
face of flanges to allow use of
standard stud installing
equipment*

(E) COLUMN

VERT. REINF., TYP.

DRILL HOLE IN COL. FLANGE
(Specify maximum size)

*As an alternative,
terminate wall horizontal
reinforcement on this side
of column to avoid drilling
holes in existing column,
not recommended for
special reinforced
concrete structural walls*

WELDED METAL STUD, TYP.

WALL REINF., TYP.

PLAN

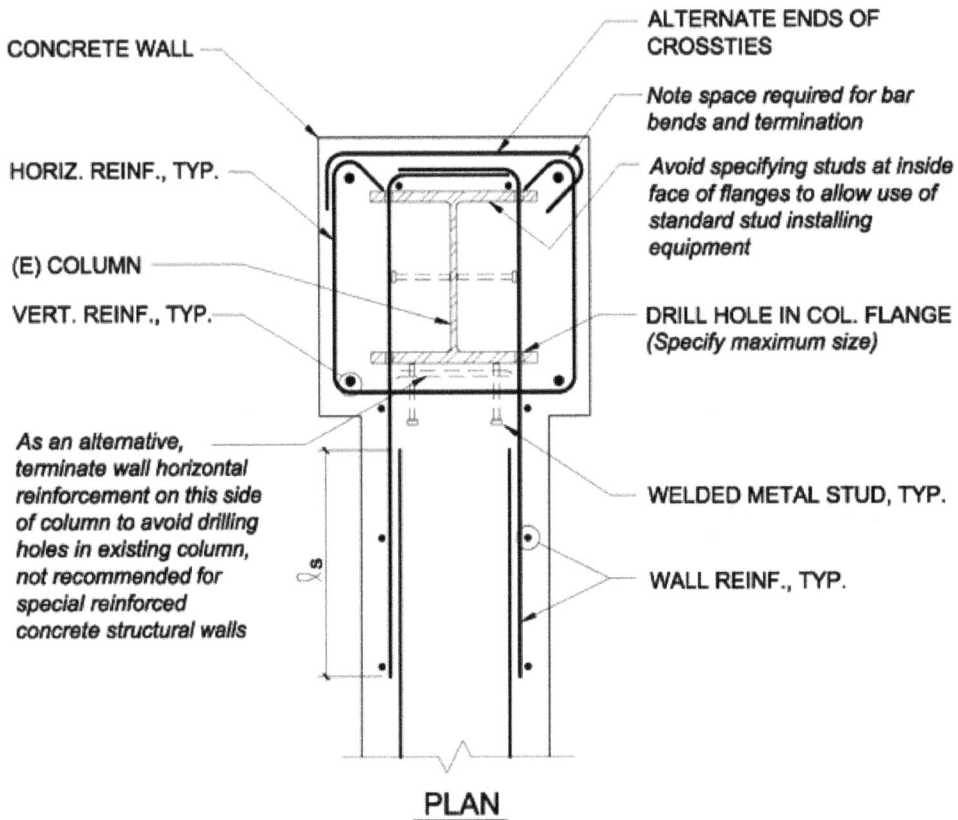

Figure 8.4.2-2: Cast-in-Place Concrete Wall Encasing Existing Column

Mechanical couplers: Consider using reinforcing steel couplers at areas of congestion. Note that diameters of the couplers could be as much as twice that of the bars and must meet ACI cover requirements for reinforcing steel.

Effective wall thickness: At beam flanges, the wall thickness could be reduced considerably, which compromises the shear strength of the wall. There are several ways to add the shear strength of the wall at these locations. Shear studs can be added to the beams to transfer some forces through the beam web. Thickness of the wall can be increased at the beam. Extra shear reinforcing steel can be placed in the wall provided that ACI limits for reinforcing steel are not exceeded.

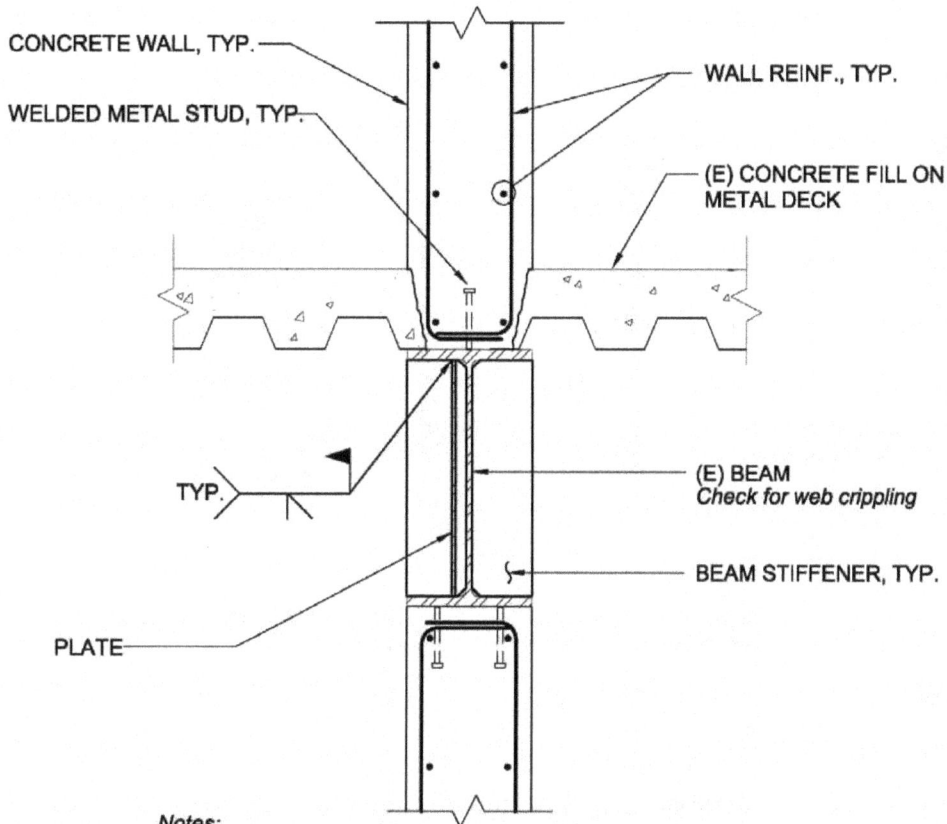

CONCRETE WALL, TYP.

WALL REINF., TYP.

WELDED METAL STUD, TYP.

(E) CONCRETE FILL ON
METAL DECK

TYP.

(E) BEAM
Check for web crippling

BEAM STIFFENER, TYP.

PLATE

Notes:

1. *Detail is appropriate for shotcrete construction and more problematic for cast-in-place concrete construction due to concrete placement and vibrating challenges.*

2. *Shear in wall is transferred entirely through studs.*

3. *For deck perpendicular to beam condition, number of metal studs limited by deck flute spacing.*

Figure 8.4.2-3: Discontinuous Wall at Existing Beam

Wall connections at slabs: To avoid damaging the slab reinforcing steel critical for transferring diaphragm forces to the shear walls, the concrete slab could be chipped away without damaging the slab steel. This still allows for a monolithic wall construction with a construction joint most likely located at the top of slab. If the shear forces in the walls are sufficiently low, the entire slab may be preserved and merely roughened at the construction joint. Holes can be drilled in the slab to allow vertical reinforcing steel to pass through.

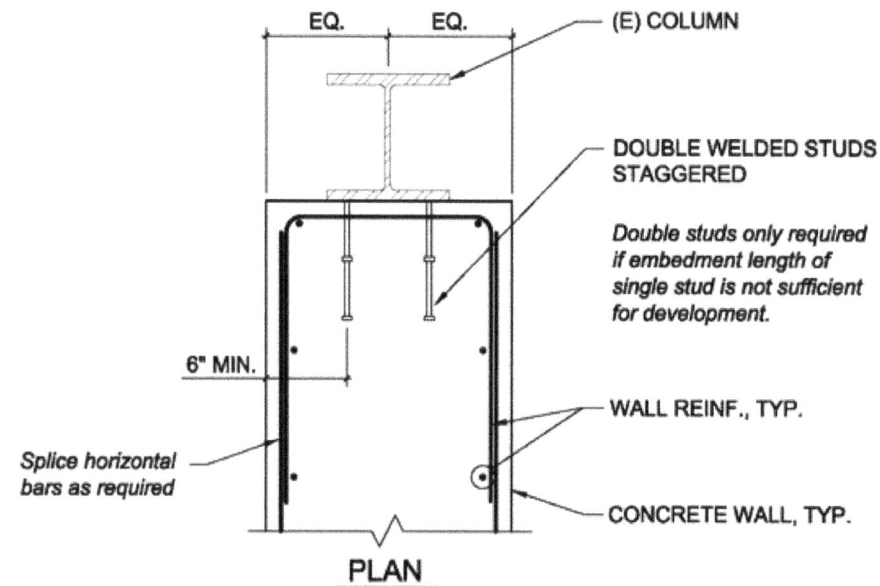

Figure 8.4.2-4: Wall at Existing Column

Frame with infill walls: Since the purpose of this technique is to create a shear wall system, the detailing should meet this goal by providing continuous load paths for both the laterally induced shear and overturning forces. Welded shear studs or reinforcing bars should be provided along all wall panel edges to achieve this. The horizontal shear force transfers from one wall to another through the studs on the top and bottom flanges and the webs of the beams at each floor. Additional plates could be welded between flanges to reduce stresses in beam webs. The columns become the boundary elements of the shear walls. Their effectiveness in providing overturning resistance may be limited by the number of shear studs or reinforcing bars that can be installed. Consider welding one stud on top of another to increase the embedment depth to avoid pullout failure.

Offset walls: Walls do not necessarily have to be centered on the existing beams and columns. By offsetting a wall and using the beam and column webs as the edge of the wall, shoring and metal deck removal only has to be performed on one side of the frame. An offset wall also lends itself to shotcrete construction, while walls with centered steel members have to be cast-in-place since it is not possible to place shotcrete on both sides of a steel member.

Cost/Disruption

Adding concrete walls to an existing building is costly, though some savings can be found in the wall construction. Shotcrete walls are typically cheaper and faster to construct than conventional concrete walls due to the savings in materials and labor associated with formwork. Cost savings can be even greater if shotcrete is applied against an existing wall at a stair or elevator and mechanical shafts. CMU walls are generally less expensive than either shotcrete or cast-in-place concrete. New foundations are almost always required for new walls and could be extremely costly if deep foundations, such as drilled piers, are added.

Installing new walls is disruptive to the occupants because of the noise and vibrations associated with construction. Even if tenants are relocated to parts of the building where the work is not being performed, vibrations associated with cutting, chipping, and drilling of concrete can transmit through the structure. The disruption can be reduced somewhat if the walls are installed at the perimeter.

Construction Considerations

The engineer's involvement during the construction phase is critical during a seismic rehabilitation. The design of the retrofit scheme must not neglect the construction phase and should consider the issues below at a minimum.

Shotcrete walls: The quality of a shotcrete wall is highly dependent on the skill of the nozzle operator. Building codes typically require preconstruction test panels constructed and reinforced similarly to the actual walls. This allows for inspection of the finished product, sawcutting to verify the quality of the shotcrete, and coring to determine strength. Additional test panels for overhead joints should also be requested to allow for cores to be taken to inspect the surface preparation and the joint bond. A series of ACI 506 publications provide general information, specifications, certification of nozzle operators, and evaluation of shotcrete (ACI, 1991, 1994, 1995a, 1995b, 1998).

Congestion: Congestion issues may be encountered at the ends of the walls due to the presence of the columns, confinement steel, and anchorage for horizontal wall reinforcing steel. Consider using mechanical couplers in heavily reinforced boundary elements and walls where lap splices are impractical.

Concrete/shotcrete placement: The type of construction should be determined during the design phase because details for cast-in-place concrete and shotcrete construction are different. It may not always be possible to confidently use either type of construction; and, thus, the details creating such situations should be avoided. During construction, some locations require particular attention when shotcrete and concrete are being placed to ensure that all gaps are filled – tops of wall panels, k-region of steel sections, and construction joints. Figure 8.4.2-5 shows a pilaster at a steel column where shotcrete shadowing restrictions require the use of cast-in-place construction for the pilaster but still allow shotcrete construction away from the pilaster. See Section 21.4.5 for additional discussion concerning concrete and shotcrete construction.

Removal of existing nonstructural elements: Exterior cladding and interior partitions may have to be removed to deliver concrete formwork and other equipment. Walls that encase beams at the roof require temporary removal of the roofing and waterproofing. Installation of the walls will affect ceilings, lights, and other mechanical/electrical/pluming components. Nonstructural elements located within the frame bays being strengthened have to be moved for construction of the walls. Permanent relocation may be desired for some of these elements to minimize openings in the new walls. Older buildings with items of historical significance require additional coordination and effort so that these items are not damaged when temporarily removed and are restored properly.

Removal of existing structural elements: See general discussion in Section 8.4.1.

Construction loads: See general discussion in Section 8.4.1.

Proprietary Concerns

Mechanical couplers for reinforcing steel are proprietary.

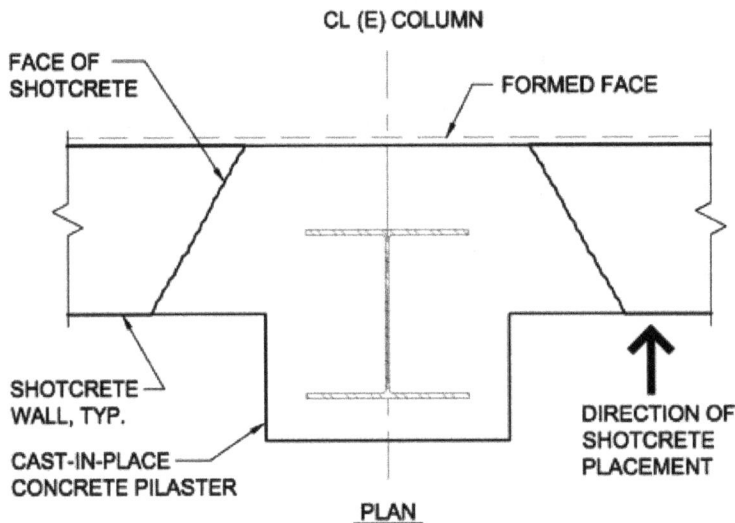

Notes:
1. Wall reinforcing steel not shown for clarity.
2. Shotcrete shadowing restrictions at steel column prevents use of shotcrete for pilaster.

Figure 8.4.2-5: Combined Shotcrete and Cast-In-Place Construction

8.4.3   Add Steel Cover Plates or Box Existing Steel Member

Deficiency Addressed by Rehabilitation Technique

Frame members that are inadequate to resist the seismic demands are strengthened with cover plates or by adding side plates to W-shapes to create box sections. This reduces axial and flexural stresses in beams and columns and could also be used to increase the shear strengths of these members.

Description of the Rehabilitation Technique

Cover plates are welded to the outside of existing flanges, which in effect, increase the flange areas. When strengthening a beam, a cover plate is more commonly attached only to the underside of the beam since the presence of an existing diaphragm makes it difficult and costly to weld to the top of the beam or attach side plates to box the beam. However, for strengthening non-composite beams, top cover plates are virtually indispensable. Conversely for columns, plates are typically welded to the sides of W-shapes because it is much more effective to increase the axial and flexural capacities by converting a column to a box section and the shear capacity by essentially providing two additional webs.

Design Considerations

Thorough knowledge of the existing material behaviors and strengths are necessary for the new and existing elements to interact in the desired manner. If welding to the existing components, the carbon equivalent of these components need to be verified through as-built records or new testing to determine their weldability; see FEMA 356 (FEMA, 2000e) for further discussion of this issue. Other design issues include the following:

Research basis: No references directly addressing the addition of steel cover plates to existing frame members have been identified.

Beams: The flexural strength of a beam can be improved by welding cover plates to the bottom if there is a composite slab present. If there is only a bare metal deck, a cover plate on only one side of the beam may not be very effective. However, it could be useful for strengthening beams with large axial forces, primarily collector members.

Columns: The overall capacity of a column is determined by axial-flexural interaction. Boxing a column decreases its slenderness and, therefore, increases its axial and flexural capacities. Whether the areas of the new plates can be directly included in computing these capacities depends on their continuity and detailing at a beam-column joint. Except for one side of the exterior columns, beams framing into the columns at each floor will disrupt the continuity of the new plates.

Foundations: Where cover plates are added to the columns at their base, a reevaluation of the foundation system is warranted. It is not uncommon for frame columns to develop plastic hinges at their bases and thus, the increase in demand on the foundation may be greater than intended. The strength and stiffness of the base plate and the anchor rods should be evaluated and upgraded accordingly. A more in-depth discussion of column connections to the foundation is provided in Section 8.4.5. Foundation upgrades are not common with this technique, but if required, refer to the chapter on foundations.

Nonstructural issues: The beam and column upgrades hardly take up any additional space but minor modifications and temporary relocations may be required for some architectural and M/E/P elements.

Detailing Considerations

In addition to obtaining the latest drawings for the building including as-built drawings, if available, and conducting comprehensive field surveys, the following issues should be noted:

Beams: When a section of beam is strengthened, the cover plate typically attaches to the bottom flange with fillet welds, as shown in Figure 8.4.3-1. The length and size of the welds determine the shear flow between the plate and the existing beam, similar to interactions between other composite elements.

Columns: Side plates used to create a box section can be attached to the column flanges with CJP, PP, or fillet welds, as shown in Figure 8.4.3-2. The fillet weld is the simplest to implement since it does not require additional preparations like the other welds, particularly the use of backing plates and beveling of the welded edges. The type of connection used may limit the shear flow between the plates and the existing column.

Existing floors: Whether the columns are upgraded with cover plates or box sections, the new plates have to terminate at the existing beams and slabs. Though it is possible to achieve continuity by using additional plates or sections if necessary, this would create detailing and construction complexities and should be avoided. Furthermore, excessive welding around a beam-column joint could create undesirable residual stresses in the joint.

Cost/Disruption

Schemes that involve slab removal, work around a connection, and foundation work are costly. As typical with seismic upgrades, cost and disruption is minimized when schemes are kept simple.

Construction Considerations

See Section 8.4.1 for general discussion of welding issues, removal of existing nonstructural and structural elements, and construction loads.

Proprietary Concerns

There are no known proprietary concerns with this technique.

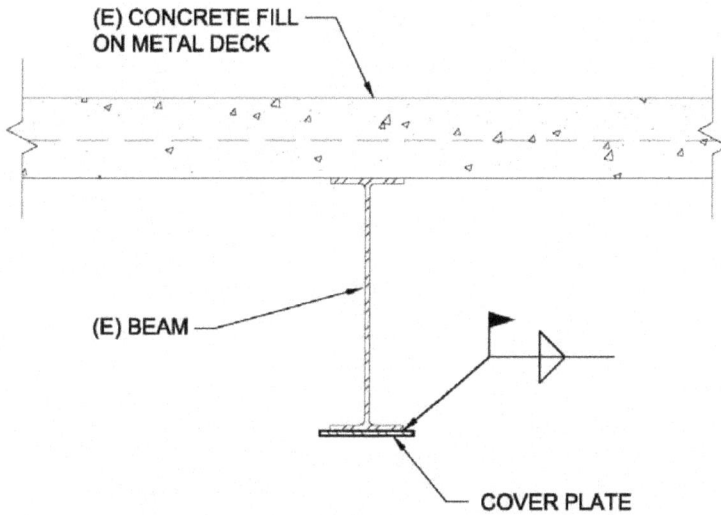

Figure 8.4.3-1: Cover Plate at Existing Beam

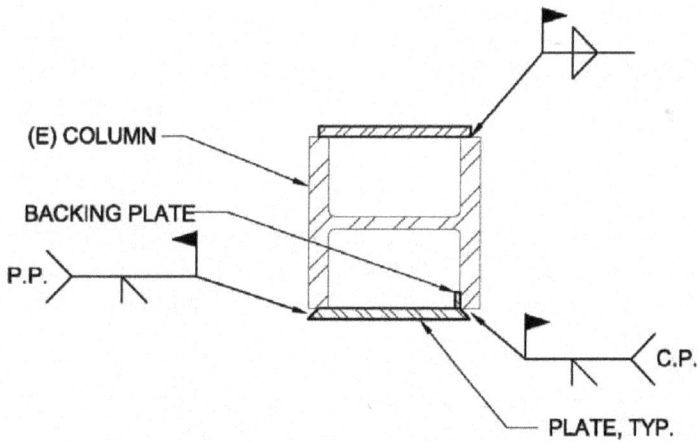

Note:
Welds shown indicate alternate possibilities of plate attachment.

Figure 8.4.3-2: Box Section at Existing Column

8.4.4   Provide Collector in a Concrete Fill on Metal Deck Diaphragm

Deficiency Addressed by Rehabilitation Technique

Beams that are inadequate to transfer collector loads are strengthened.

Description of the Rehabilitation Technique

Plates can be welded to various portions of the existing beam. Cover plates, discussed in Section 8.4.3, are commonly attached to the underside of beams. To attach plates to top flanges, it may be simplest to orient the plates vertically and weld to the underside of the flanges, as shown in Figure 8.4.4-1, or similarly on top of the bottom flanges. Given all the possible options, the plate locations may be dictated by the eccentricities, continuity options at a beam-column joint, or the presence of nonstructural elements.

Design Considerations

Though the purpose of this technique is to strengthen a beam axially, it also changes the flexural properties of the beam by increasing its stiffness. It should be verified that this does not have unintended consequences, such as converting the beam into a frame member. Eccentricities in collectors are typically neglected. However, deep collectors and/or collectors with large forces could be subject to significant moments from these eccentricities. The eccentricities could be reduced if plates are attached to the top flanges but requires other detailing considerations at a beam-column joint.

Detailing Considerations

The new plates should attach to the columns with complete joint penetration (CJP) welds. Consider replacing the welds in the existing collector with high notch toughness welds in a high seismic region. Continuity plates directly aligned with the collector elements at a beam-column joint should be used as much as possible but may be offset if all eccentricities and their effects on the joint are considered.

Cost/Disruption

This technique is relatively inexpensive compared to other types of steel frame upgrades. The most expensive part of the structural upgrade is associated with the work at the beam-column connections. The nonstructural disruptions to the architectural and M/E/P elements are typical.

Construction Considerations

See Section 8.4.1 for general discussions of welding issues, removal of existing nonstructural and structural elements, and construction loads.

Proprietary Concerns

There are no known proprietary concerns with this technique.

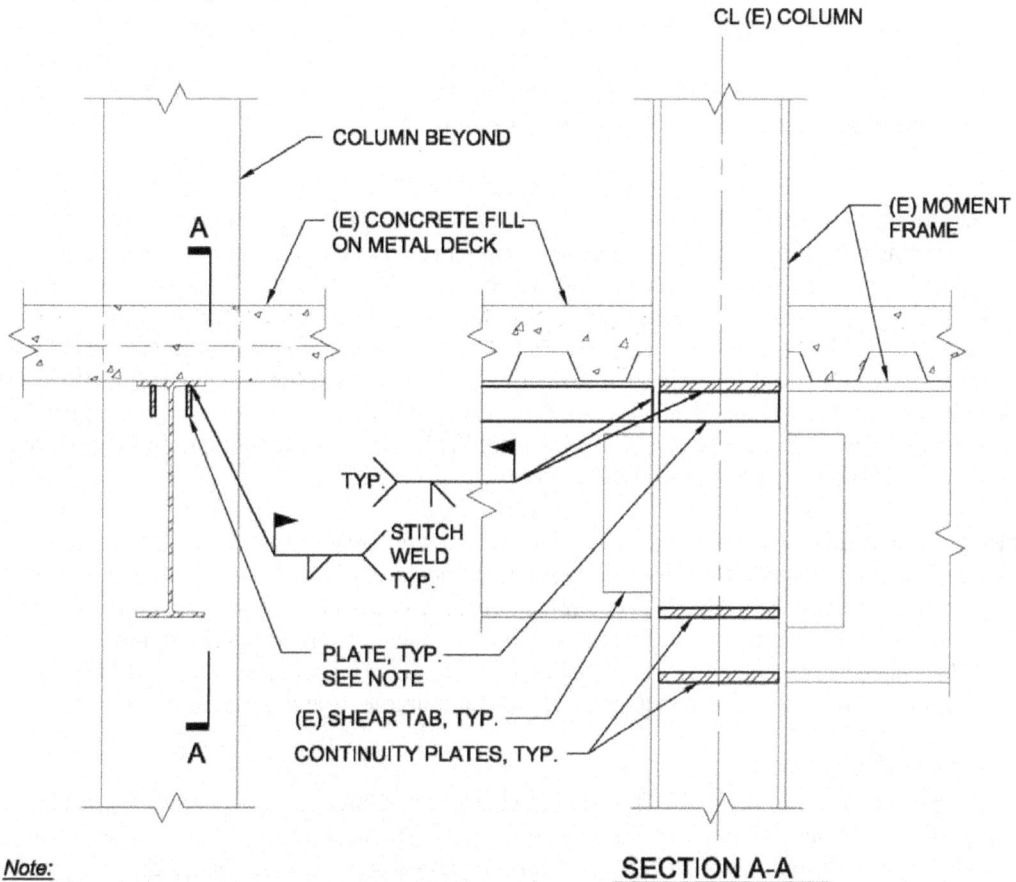

Figure 8.4.4-1: Plate Collectors at Existing Beam

8.4.5   Enhance Connection of Steel Column to Foundation

Deficiency Addressed by Rehabilitation Technique

Frame columns are subject to axial (including possible tension), flexural, and shear forces. To
this end, columns with inadequate anchorage to the foundation limit the capacity of a frame. The
columns could be part of an existing lateral force-resisting system that do not meet current
standards or part of an upgraded system with larger forces resulting from increased stiffness.

Description of the Rehabilitation Technique

Two methods are common for enhancing the column connection to the foundation. First, modifications at the base plate could include the addition of anchor rods, welding shear lugs to the base plate, and/or enlargement of the base plate. The other method is to encase the column in a concrete pedestal. It is also possible to use both of these methods together. The foundation system itself is not addressed here but is discussed in Chapter 23.

Anchor rods can be used to resist tensile forces due to uplift and flexure. Holes drilled in the existing footing for these rods can be filled with a nonshrink high strength grout or chemical adhesive. Shear lugs may be used to transfer shear forces into the foundation. If the existing base plate is not large enough to accommodate the new rods or lugs, new plates can be welded to the existing plate to enlarge its area. This is also necessary if the allowable bearing stress is exceeded due to increased column compression. The increased base plate size leads to greater shear and flexural forces in the plate. It is likely that the plate will not be thick enough to resist these forces. Thus, stiffeners that act as supports can be welded to the base plate to reduce the plate forces. The stiffeners also provide additional load paths from the column to the base plate. Figure 8.4.5-1 shows some of these modifications.

Base plate modifications may not always be practical or possible. Instead, the column can be encased in a concrete pedestal above the footing, as shown in Figure 8.4.5-2. Shear forces would transfer through direct bearing of the column against the pedestal. The pedestal could also be used to transfer uplift and flexural forces by relying on the existing base plate and other mechanisms, such as welded shear studs along the column. The pedestal itself should be detailed as a reinforced concrete column that meets all ACI requirements and seismic provisions.

Design Considerations

Thorough knowledge of the existing material behaviors and strengths are necessary for the new and existing elements to interact in the desired manner. A complete presentation of column base plate design can be found in AISC *Design Guide 1* (Fisher and Kloiber, 2005). Key design issues include the following:

Research basis: No references directly addressing the upgrade of connections of steel columns to foundations have been identified.

Shear lugs: Columns that are in compression offer shear resistance through friction below the base plate. AISC recommends different friction coefficients depending on the location of the base plate with respect to the top of concrete. Additional shear resistance is required when the compressive force is not great enough or there is tension in the column. Though not common in retrofit applications, shear lugs, or simply plates welded to the bottom of a base plate, have the advantage of shallow embedment compared to anchor rods. The lugs also receive confinement from the base plate above and simplify the design by allowing anchor rods to only resist tension. The design of a shear lug is a function of the allowable bearing stress in the surrounding grout and flexural forces in the lug.

Anchor rods: Tension in an anchor rod is developed through bond along its length and/or direct bearing through a hook, bolt head, or nut at the end of the rod. Though it is possible to drill a hole large enough in an existing footing to accommodate a bolt head or nut, manufacturers of grouts and adhesives sometimes limit the size of the a hole for a given rod diameter in which its product can be used as an infill. In this case, the tensile force has to be developed entirely through bond. For bond development, the rod has to be threaded or some other means of mechanical anchorage has to be provided along the rod. Anchor rods are not typically used to transfer shear to the foundation because the mechanism for shear transfer is difficult to define and still subject to debate. If unavoidable, several factors should be considered, including bending of the rod through the oversized hole in the base plate and shear and tension interaction on the rod. If the column shear force is being distributed between both new and existing rods, note that the holes for the existing rods are likely to be oversized and welded plate washers should be verified or provided.

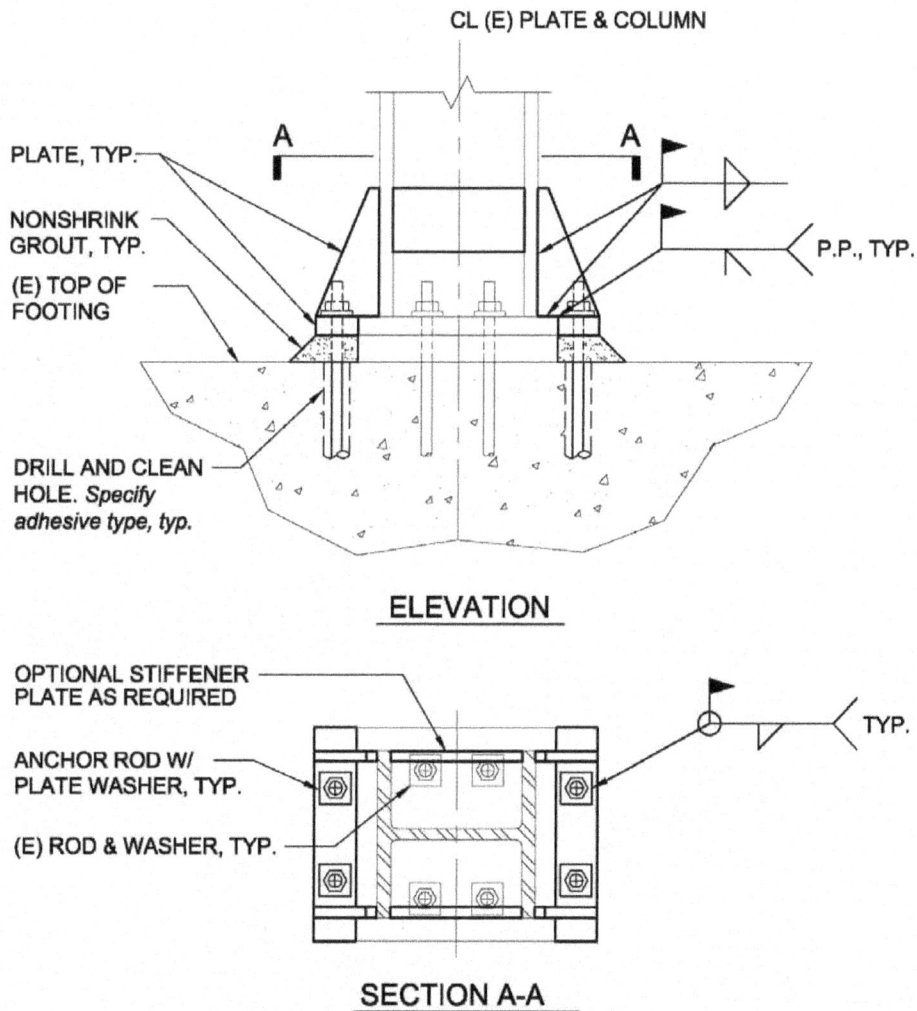

Figure 8.4.5-1: Modified Base Plate to Increase Uplift Capacity

CL (E) COLUMN

A ———   ——— A

(E) BASE PLATE
ROUGHEN SURFACE
(E) TOP OF FOOTING

Specify minimum embedment

ANCHOR ROD

ELEVATION

CONCRETE PEDESTAL

WELDED SHEAR STUD, TYP.

HORIZONTAL REINF., TYP.

ALTERNATE ORIENTATION OF
CROSSTIES

VERTICAL REINF., TYP.

SECTION A-A

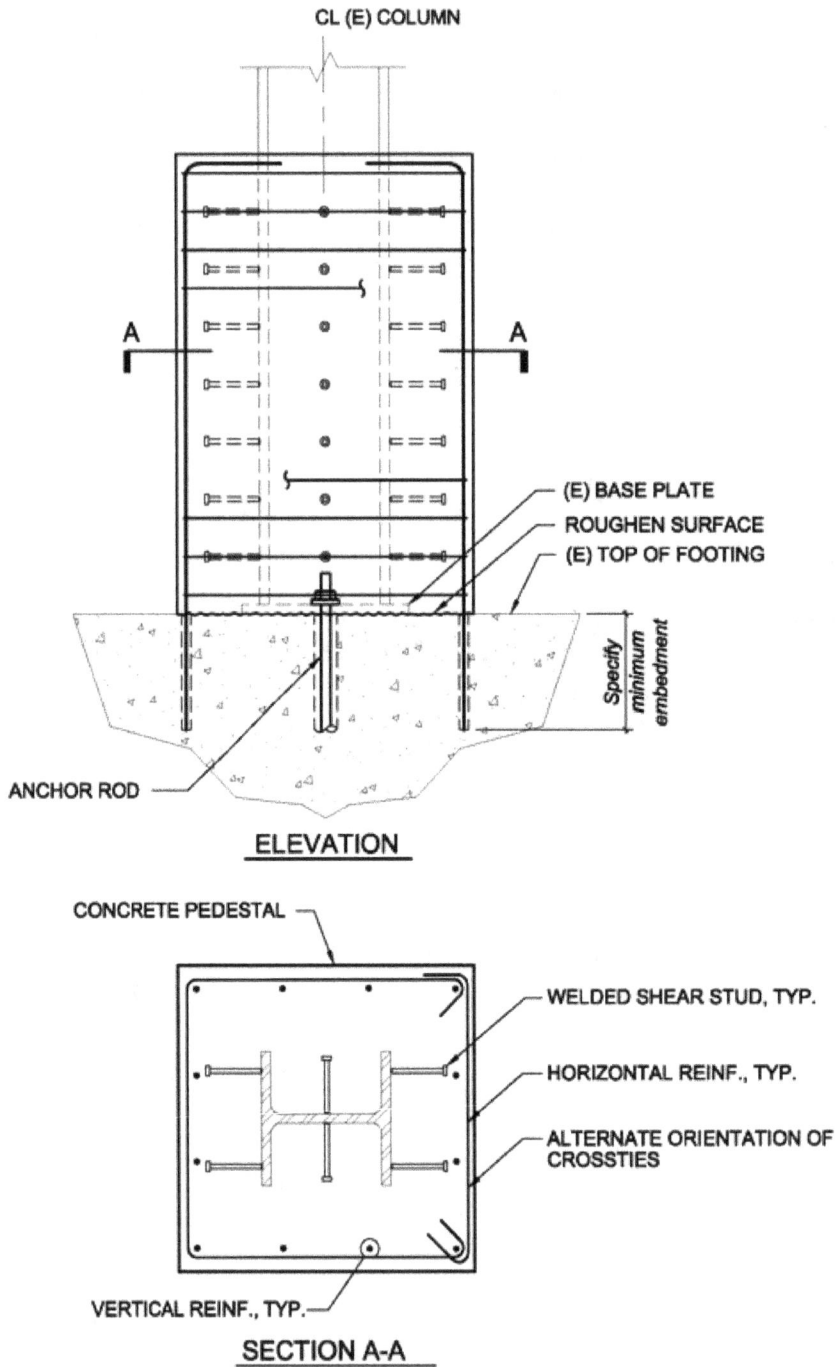

Figure 8.4.5-2: Concrete Pedestal at Existing Column

Stiffeners: The simplest way to reduce flexural stresses in a base plate is to add stiffeners that essentially provide multiple supports along the base plate. These stiffeners also assist in transferring uplift and flexural forces to the base plate and reduce the weld stresses at the base of a column.

Nonstructural issues: Base plates are typically hidden in the slab and thus, the upgrade could remain hidden if it only involves adding anchor rods. Pedestals, on the other hand, require additional space and could present aesthetic issues. However, basements are often used for parking, storage, or as equipment locations, where aesthetic considerations may be negligible.

Detailing Considerations

In addition to obtaining the latest drawings for the building including as-built drawings, if available, and conducting comprehensive field surveys, the following issues should be noted:

Base plates: If the existing base plate can accommodate new anchor rods or lugs, burning holes in the plate is usually acceptable. Otherwise, enlargement of an existing base plate is accomplished by welding new plates to the existing plate. Only partial penetration (PP) welds can be used since the lack of backing precludes using complete joint penetration (CJP) welds unless concrete is removed to allow placement of backing plates. The depth of the PP weld limits the effective thickness of a plate. Also of note is that AISC relaxes its typical edge distance provisions if there is no lateral load on a base plate. The only recommendation is that enough edge distance remains such that the drill or punch does not drift when a hole is made. *AISC Design Guide 1* suggests that one quarter of an inch is enough to meet this condition.

Anchor rods: Detailing differs depending on if an anchor rod is designed to resist small shear forces and tension or tension only. For the latter, an oversized hole in the base plate and a thick plate washer over the hole is considered adequate. Since oversized holes are intended to allow for inaccuracies in anchor rod placement when concrete footings are placed, it should be possible to use smaller holes if the holes are being drilled. In addition to the design considerations mentioned above, anchor rods with shear forces require welding heavy plate washers to the base plate. The washers should have close-fit holes to minimize the movement required to engage all of the rods together. Nuts are sometimes welded to the washers but the mechanical properties of the nuts could lead to welding complications.

Pedestals: Reinforcement of the pedestal is similar to a reinforced concrete column. Vertical bars provide tensile and flexural strength while hoops or ties provide confinement. At the top of footing, the shear is transferred through shear friction in the anchored reinforcing steel dowels. The concrete surface should be roughened, which also increases the shear friction values. If the pedestal is also used to transfer uplift and flexural forces, additional dowels are required for tension. A single dowel should not be considered to provide both shear and tensile resistance. One mechanism to transfer uplift and flexural forces from the column to the pedestal is the use of welded shear studs along the column. The size and spacing of the shear studs may determine the minimum pedestal diameter and height.

Concrete anchorage: Installation of the anchor rods and dowels requires drilling into the existing footing. Consider using a scanning device to avoid damaging the reinforcing steel in the top layer of the footing. Cores may have to be taken to confirm the concrete strength. Many commercial grouts and adhesives are available for bonding the anchor to the concrete. Verify that the selected product is appropriate for the specific anchor in a seismic application. On site inspection and testing of the anchors are mandatory since their performance rely heavily on the installation process. The grout and adhesive product vendors' ICC Evaluation Service reports provide standardized installation procedures and anchorage capacities.

Cost/Disruption

The cost of a column connection to foundation upgrade is not very expensive. However, this technique is not usually performed only by itself. Costs would be small relative to an overall lateral force-resisting system upgrade and a foundation upgrade. Disruption could be minimal since typically, there are no tenants in the basement. Even if the basement were used for tenant access, such as parking, only a few columns would be affected at a time.

Construction Considerations

The engineer's involvement during the construction phase is critical during a seismic rehabilitation. The design of the retrofit scheme must not neglect the construction phase and should consider these issues at a minimum:

Welding issues: See general discussion in Section 8.4.1.

Removal of existing nonstructural elements: This technique tends to be less disruptive of nonstructural functions compared to other techniques. However, there may still be some architectural and M/E/P elements, such as partitions and pipes adjacent to the columns, that require temporary removal. See Section 8.4.1 for discussions of fireproofing, asbestos, and concrete encasement.

Removal of existing structural elements: To access the base plate, an existing slab and slab reinforcement will probably have to be removed and replaced. Care should be taken to not damage any of the existing structural elements. If the existing base plate is being replaced entirely, a column shoring scheme has to be devised. The existing anchor rods should be cut at the top of footing and the new rods have to be relocated.

Construction loads: See general discussion in Section 8.4.1. Construction loads at the basement level are not typically a concern if a slab-on-grade is present.

Proprietary Concerns

Many grout and adhesive products are available.

8.4.6   Enhance Beam-Column Moment Connection

Deficiency Addressed by Rehabilitation Technique

Riveted, bolted, and WSMF connections are upgraded to improve their ability to withstand inelastic rotational demands and develop the plastic moment capacity of the beams.

Description of the Rehabilitation Technique

The techniques discussed in this section were developed specifically to address pre-Northridge WSMF connections. These techniques can be adapted with prudence to existing riveted and bolted connections that are found to be inadequate or to upgrade partially restrained connections. These techniques are covered thoroughly in *AISC Design Guide 12* and FEMA 351 and only briefly presented here. The reduced beam section (RBS) is the only technique that weakens the beam in flexure, which in turn, moves the plastic hinge away from the column and reduces the demand on the complete joint penetration (CJP) welds. Two other methods - welded haunch and bolted bracket – also move the hinge away from the column but strengthen the existing connection and seek to maintain the original flexural capacity of the beam. A similar method employs cover plates over the beam flanges, requiring little additional space. This method is not discussed in detail here but more information can be found in FEMA 351. Additional modifications that should be performed for each technique include adding or verifying the capacity of beam flange continuity plates across the column web and strengthening the panel zone; see the references mentioned above and *AISC Design Guide 13* (Carter, 1999) for reference.

The selection of a particular connection modification depends on specific project factors. There are advantages and disadvantages to each of the methods. While each of the three connections discussed below consistently developed a minimum plastic rotation of 0.02 radian in cyclic loading experiments, the welded haunch and bolted bracket demonstrated higher levels of performance and reliability (Gross et al., 1999). On the other hand, the RBS modification requires no additional space and reduces the beam capacity to enforce a strong-column weak-beam condition. The welded haunch is the only modification that exhibited desirable behavior in tests where the beam top flanges were left in a pre-Northridge condition (Gross et al., 1999). The bolted bracket scheme requires top flange reinforcement but eliminates field welding.

Design Considerations

Connection modifications merely ensure that the connections behave as originally intended in a moment frame. These modifications by themselves do not reduce the frame forces in an earthquake. Other structural upgrades are necessary if the frame does not have sufficient global strength or stiffness to resist the demands.

Research basis: Numerous experimental programs have been performed under the auspices of the SAC Joint Venture, the National Science Foundation, the National Institutes of Standards and Technology, and AISC. Most of the results of these tests have been summarized in *AISC Design Guide 12* and the FEMA documents listed in the references. The user is cautioned to extrapolate beyond the conditions that have been tested only with proper considerations of the differences between the tested connections and the building conditions.

Design forces: AISC enforces the strong-column weak-beam concept by specifying a minimum column-beam moment ratio in the *AISC Seismic Provisions*. The general intent is to prevent plastic hinges from developing in the columns, which could lead to a soft story. Some yielding may still occur in the columns without causing loss in frame strengths. The required flexural strength of a column is determined from several variables associated with the beam. The beam plastic moment at the critical plastic section – centerline of RBS or tip of haunch and bracket –

includes a strain hardening factor and the expected yield stress of the flanges. An additional moment is a function of the shear in the beam and the distance from the critical plastic section to the column centerline. This shear is a combination of the shear associated with the plastic moment and the gravity loads.

RBS: The design of moment frames is often governed by global and interstory drift limits instead of strength requirements. Thus, the reduction in beam strength and stiffness from an RBS could be acceptable within practical limits. If the presence of an existing slab poses construction and design issues, then the top flange may remain intact and only the bottom flange has to be cut; though minimal removal of the existing slab is still required to replace the existing top flange weld. Note that a bottom flange modification only achieves a minimal stress reduction and that both the top and bottom flange CJP welds still require replacement with high toughness weld metal. Various RBS cut shapes have been tested successfully, though the radius cut RBS minimizes stress concentrations. The critical dimension is the depth of the cut, which can also be expressed in terms of flange width reduction as a percentage. AISC recommends a maximum flange reduction of 50% as a practical limit for field modifications. To determine if a flange reduction is adequate, the maximum moment at the face of the column can be computed for the modified beam. If the ratio of this moment to the plastic moment of the beam exceeds 1.05, a bottom flange RBS modification is not recommended. Either a top flange RBS should also be provided or another method should be used.

Welded haunch: A tapered haunch welded to a beam bottom flange changes the force transfer mechanism at the connection. Analytical and experimental studies have found that the beam shear transfers through the haunch flange (FEMA, 2000b). This, in turn, reduces the shear stresses in the flange welds and increases the depth of the column panel zone. The welded haunch is the only connection out of the three presented for this technique to reach plastic rotations on the order of 0.03 radian while allowing the top flange weld to remain in a pre-Northridge condition in tests. Additional rotational capacity can be obtained by upgrading the top flange groove weld to a higher toughness, adding a cover plate, or adding another haunch, in order of increasing performance. The two latter modifications can be performed without replacing the top flange weld.

Bolted bracket: Considered an alternative to the welded haunch, this option does not require any field welding. A portion of the flange force is transferred to the bracket and reduces the weld stresses. Several bolted bracket configurations are possible and can be applied to either beam flange or both flanges for large beams. Common types of brackets include the haunch bracket, pipe bracket, angle bracket, and double angle bracket. Pipe and angle brackets are relatively compact and can be covered up by the slab for a top flange application. For bottom flange haunch only schemes, a stiff angle is still recommended at the top flange (Gross et al., 1999). The pre-Northridge weld in the top flange can remain in this case even if it is found to have defects.

Material strength: Yield strengths of existing materials are best determined from tests of samples taken from the actual members. Samples taken from flanges are preferred over the web. When samples are not available, AISC *Design Guide 12* prescribes some overstrength values for different grades of steel.

Minor axis connections: Moment connections to the minor axis of a column are not as common, but sometimes present in older moment frame construction. Their performance in the Northridge earthquake did not result in significant damage. Thus, little work has been done to study the rehabilitation of these connections. Similar concepts presented in this section may be used to rehabilitate these connections with the recognition that the columns may be weaker than the beams, which would also limit the inelastic demands on the connections. Note the governing jurisdiction may require qualification through the testing program described in FEMA 351.

Nonstructural issues: Welded haunches and bolted brackets are considerable in size. The recommended welded haunch and bolted bracket depths are approximately one third and one half times the beam depths, respectively. For a W36 beam used in a moment frame, this adds up to 18 inches to the connection depth. This makes the option of adding components to only the bottom flange more attractive, though it could still interfere with the ceiling. At the top flange, the modification would certainly have to be hidden or incorporated into an architectural feature.

Detailing Considerations

In addition to obtaining the latest drawings for the building including as-built drawings, if available, and conducting comprehensive field surveys, the following issues should be noted:

Weld filler metal matching and overmatching: Weld filler metals with greater tensile strength than the connected steel should be used. Flux cored arc welding and shielded metal arc welding electrodes that conform to E70 specifications exhibit overmatching properties compared to common steel specifications, including ASTM A36, A572 (Grade 42 and 50), A913 (Grade 50), and A992 (FEMA, 2000b).

Weld metal toughness: Tests on existing connections modified with RBS but no modifications to the existing beam flange welds showed poor performance (Gross et al., 1999). Improved connection performance was achieved for connections where the top and bottom flange welds were replaced with higher toughness weld metal. *AISC Seismic Provisions* now require weld metals with minimum Charpy V-Notch (CVN) toughness of 20 ft-lbs at -20°F and 40 ft-lbs at 70°F for demand critical welds.

Weld backing and access holes: Backing plates provided for flange welds create a notch effect and also hinder detection of weld flaws at the weld root. Backing should be removed at flanges receiving new CJP groove welds and reinforced with fillet welds. The removal of existing backing without weld replacement is an ineffective upgrade since the existing weld is still likely to have low toughness. The size and shape of the weld access hole should be configured to provide welder access and minimize stress concentrations. FEMA 351 provides a specific weld access hole detail that meets these criteria.

RBS (Figure 8.4.6-1): The distance from the face of column to the start of the RBS cut and the length of the cut, more concisely known as the offset and the chord length, respectively, should be kept small as to not allow the moment to increase significantly from the plastic hinge to the column. Yet, the distance to the start of the cut should also be large enough to allow for the flange force from the RBS to be uniformly distributed across the flange at the CJP weld. The length of the cut should also be long enough to control the inelastic strains along the RBS. The radius of the cut is a function of its depth and the chord length. See AISC *Design Guide 12* for specific recommendations and guidelines for selecting these dimensions.

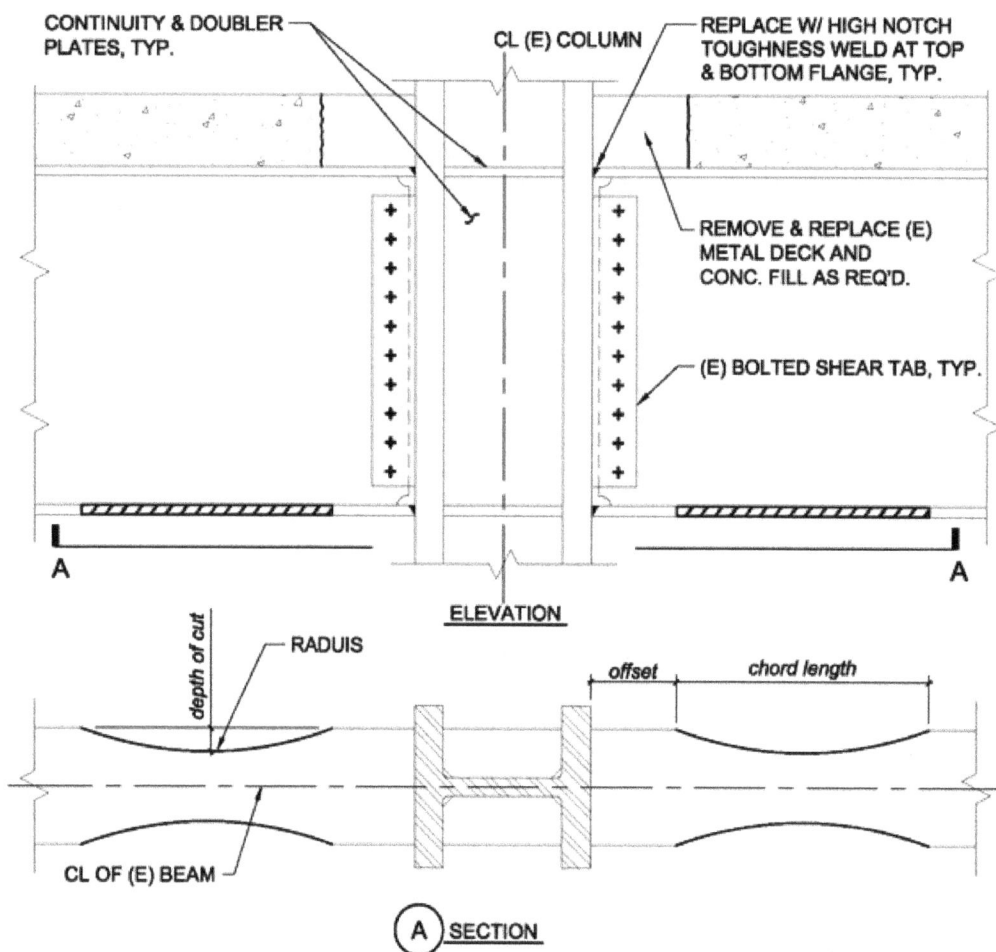

Figure 8.4.6-1: Reduced Beam Section at Bottom Flange of Existing Beam
(adapted from AISC Design Guide 12)

Welded haunch (Figure 8.4.6-2): Haunches can be cut from WT- or W-shapes. Typically, the haunch flange is groove welded to the beam and column flanges while the haunch web is fillet welded to these elements. Experimental programs have consistently used the same haunch geometry—length and depth approximately one half and one third that of the beam depth,

CL (E) COLUMN

(E) WELD MAY REMAIN IF ULTRASONIC
TESTING DOES NOT SHOW DEFFECTS,
SEE NOTE

(E) CONC. FILL ON
METAL DECK

(E) BEAM

CONTINUITY
& DOUBLER
PLATES, TYP.

TYP.

STIFFENER
PL., EA. SIDE

(E) BOLTED
SHEAR TAB

TYP.

REPLACE (E) WELD
& REMOVE BACKING
TAB, TYP.

WT SHAPE

Note:

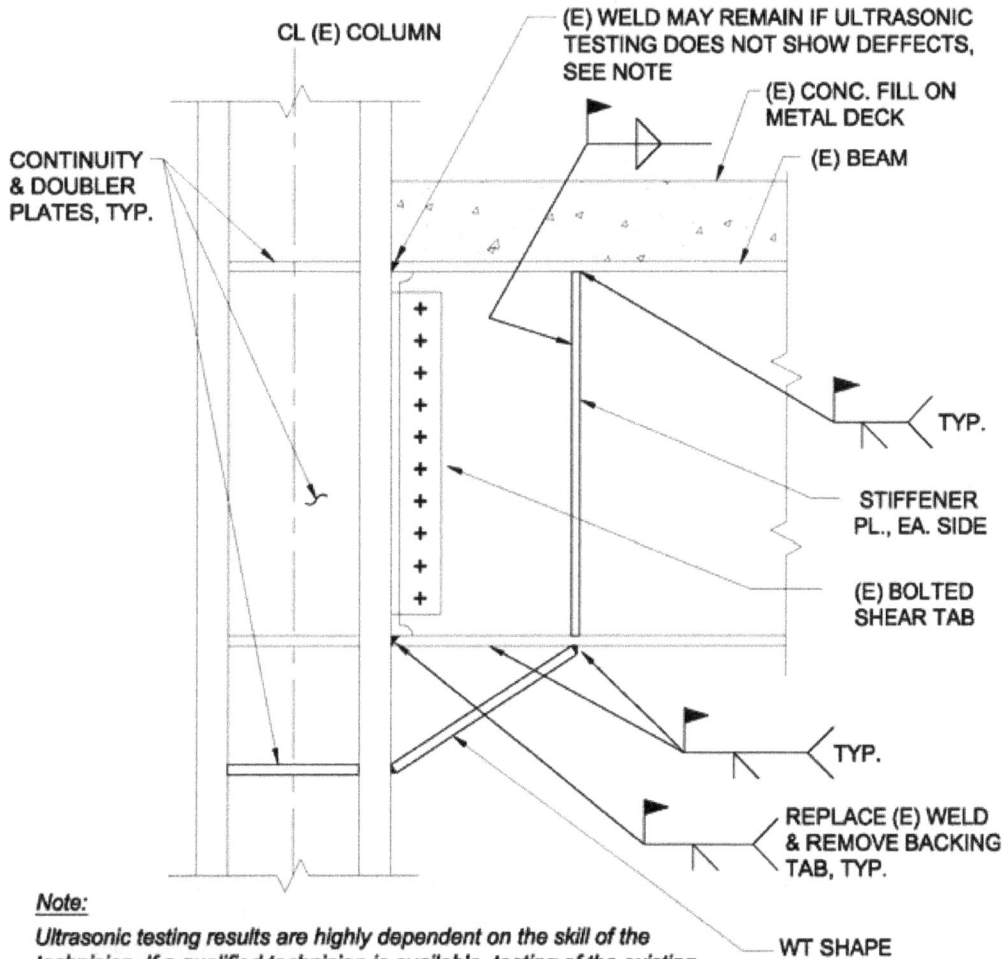

Ultrasonic testing results are highly dependent on the skill of the technician. If a qualified technician is available, testing of the existing weld at the top flange can be performed from below. Otherwise, chipping of the existing slab may be required to perform the test.

Figure 8.4.6-2: Welded Haunch at Bottom Flange of Existing Beam
(adapted from AISC Design Guide 12)

respectively. The haunch flange is sized to resist most of the shear force from the beam. The haunch web size can be established to achieve equilibrium. At the intersection of the beam and haunch flanges, web stiffeners should be provided for the beam. Similarly at the column, add continuity plates to align with the haunch flange.

Bolted bracket: The preferred configuration for this modification is a haunch bracket at the bottom flange and an angle bracket at the top flange that can be hidden within the slab. If additional reinforcement is necessary at the top flange, angles can be added below the top flange on each side of the web, as shown in Figure 8.4.6-3. Close-fit holes minimize slip in this connection. Tests have demonstrated that these connections are essentially fully rigid and can reach large plastic rotations while allowing the pre-Northridge welds at the flanges to remain (Gross et al., 1999). AISC recommends neglecting the existing welds and designing each bracket for the entire flange tension force. The bottom flange haunch bracket will be slightly larger than the welded haunch for the same beam size. The top flange angle bracket may require several rows of bolts along the beam flange and has to be designed for the prying force on the column flange. Due to limitations in the available L-shapes, the angle bracket may have to be cut from a W-shape.

Cost/Disruption

Connection modifications are locally very disruptive. Modification of a connection typically requires access from two floors to perform the work on each flange. Noise associated with this type of work will spread and disrupt tenants on other floors unless the work is done during off-hours. The RBS is probably cheaper than the other two modifications since it requires the least amount of material and labor. With older buildings, there may be asbestos present in the fireproofing around the steel members, which could require costly abatements to expose the connections.

Construction Considerations

The engineer's involvement during the construction phase is critical during a seismic rehabilitation. The design of the retrofit scheme must not neglect the construction phase and should consider these issues at a minimum:

Welding/bolting issues: See general discussion in Section 8.4.1. The bolted bracket modification can be performed without any field welding. Primary issues associated with the bolted bracket consist of typical field bolting issues such as set up, fit-up, and alignment.

Removal of existing nonstructural elements: The connection work will affect ceilings, lights, and other mechanical/electrical/pluming components. Connection modifications at the roof level may warrant removal of the roofing and waterproofing. See Section 8.4.1 for discussions of fireproofing, asbestos, and concrete encasement.

Removal of existing structural elements: See general discussion in Section 8.4.1.

Construction loads: See general discussion in Section 8.4.1.

Proprietary Concerns

There are several proprietary connections that have been developed to upgrade pre-Northridge connections, generally sharing similar design intents. Some of these connections briefly presented in FEMA 351 include the Side Plate connection system, the Slotted Web connection, and one particular type of the bolted bracket connection. The reader should contact the licensors of these technologies for more information.

8.4.7   Enhance Column Splice

Deficiency Addressed by Rehabilitation Technique

Strengthen welded or bolted column splices that do not meet the detailing and minimum design strength requirements in *AISC Seismic Provisions*.

Description of the Rehabilitation Technique

Numerous options exist for upgrading a column splice. The approach and the level of strengthening depend on the type of lateral system since special moment frames have different requirements from other systems in the *AISC Seismic Provisions*. It also depends on if the existing splice is welded or bolted and if field welding is permitted. Field welding may be a necessity in some cases since the *AISC Seismic Provisions* do not allow bolts and welds to share loads on the same faying surface.

Most existing welded splices are likely to be complete joint penetration (CJP) welded or partial joint penetration (PP) welded. CJP welds require beveled transitions to avoid stress concentrations. A beveled transition can be constructed in the field by building up the weld over the thicker column flange or grinding away a portion of the flange. PP welds inherently possess stress concentrations at the unwelded portion of the joint and thus, have higher strength requirements, but do not require beveled transitions. PP welds can be strengthened by welding plates or stiffeners across the splice. When field welding is impractical or undesirable, it may be possible to use bolted plates at an existing welded splice if the bolts are designed to resist the entire load.

The rehabilitation method for a bolted splice depends on the controlling design mode for the splice – whether the splice is governed by net section fracture of the column, yielding of the bolts, or gross yield or net section fracture of the splice plate. Bolts that govern the existing splice capacity could be replaced with stronger bolts, such as replacing A325 with A490. The upgrade may also be as simple as replacing bolts that have threads included in the shear plane with longer bolts. Alternatively, more bolts or larger bolts could be provided, but both would require more extensive field work. Splice plates with insufficient strength can be replaced or new plates could be provided on the opposite side of the existing plate, which would require longer bolts. The existing plates may also be extended by welding new plates to the ends.

Design Considerations

Considerations for a column splice include the justification of a load path, preservation of symmetry to avoid eccentric loads, and deformation compatibility if welds and bolts are both used. In addition to the requirements in the *AISC Seismic Provisions*, the *AISC Steel Construction Manual* (AISC, 2005c) has a general discussion of column splices and contains typical details of W-shape splices. Information can be found regarding filler plate sizes for different column depths and erection tolerances. Important design issues include the following:

Research basis: No references directly addressing the upgrade of existing column splices have been identified.

Design strength: The *AISC Seismic Provisions* specify a minimum strength requirement for all splices as a function of the expected yield strength and the flange area of the column. In addition, PP welded splices have higher strength requirements due to the stress concentration manifested in the crack-like notch at the unwelded side of a flange. Lastly, column splices in special moment frames have to be designed for the full flexural strength of the smaller column and their web splices require a shear strength of twice the plastic flexural strength of the column divided by the story height.

PP welded splices: For flange thickness differences between the columns up to 1/8", steel shims can be fillet welded to the inside of the flanges. Greater than 1/8", it may be more practical to attach a plate to the flanges using CJP welds.

Plate locations: Plates can be extremely versatile whether they are welded or bolted to the columns. Bolted plates are commonly placed at the outside of flanges, where extra filler plates can be used to make up differences in flange thicknesses. Filler plates are not necessary at the inside of flanges, where a flush surface is already provided if the same nominal depth columns are present. Plates can also be welded from one flange tip to another, which provides increased shear strength and stability across the joint. See Figure 8.4.7-1.

Nonstructural issues: The column splices hardly take up any additional space but minor modifications and temporary relocations may be required for some architectural and M/E/P elements.

Detailing Considerations

In addition to obtaining the latest drawings for the building including as-built drawings, if available, and conducting comprehensive field surveys, the following issues should be noted:

Welded splices: Large shrinkage strains could develop when welding heavy sections. Bolted or fillet welded plate slices should be considered as an alternative if there is a possibility of a brittle weld fracture.

Bolted splices: If new holes are drilled in the existing column or splice plate, tolerances of the existing holes should be verified to ensure that bolts will be loaded evenly. The capacity of the net section should also be checked. See Figure 8.4.7-2 for a sample detail.

Web splices: Column webs in moment frames that use bolted web splices require plates or channels on both sides to reduce eccentricities. This should be considered for column webs in other lateral force-resisting systems though it is not a specific requirement in the *AISC Seismic Provisions*.

Cost/Disruption

The cost of implementing this technique is highly dependent on the level of modification performed to the existing splice. The level of disruption is typical for that of steel frame upgrades, involving vacancy of space, removal of all nonstructural elements around the column, and significant noise.

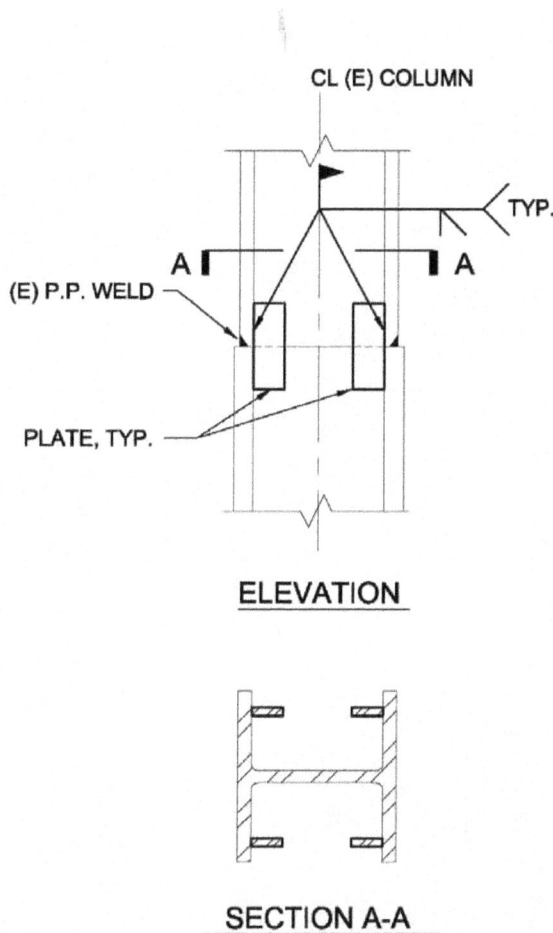

Figure 8.4.7-1: Welded Splice Upgrade at Existing Column

Construction Considerations

The engineer's involvement during the construction phase is critical during a seismic rehabilitation. The design of the retrofit scheme must not neglect the construction phase and should consider these issues at a minimum:

Welding issues: See general discussion in Section 8.4.1. Though weld fractures were not found at column splices after the Northridge earthquake, an environment in which the welder can perform quality welds is critical.

Removal of existing nonstructural elements: See general discussion in Section 8.4.1. Depending on the location of the splice, the upgrade could affect ceilings, lights, and other mechanical/electrical/pluming components.

CL (E) COLUMN

PLATE

FILLER PLATE

BOLTS AS REQUIRED

A   A

(E) PLATE

(E) WEB SPLICE PLATE

ELEVATION

SECTION A-A

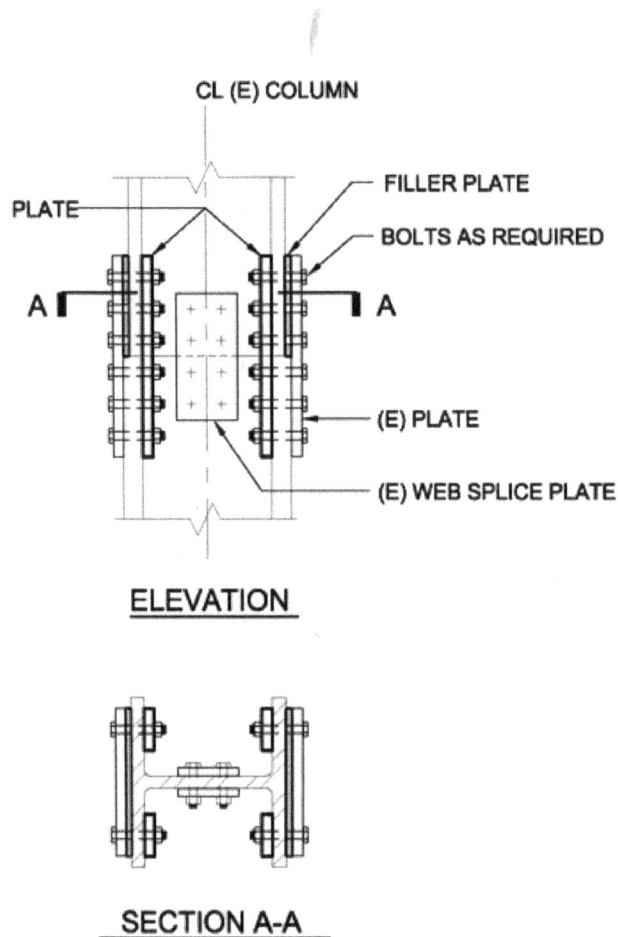

Figure 8.4.7-2: Bolted Splice Upgrade at Existing Column

Removal of existing structural elements: Due to the critical nature of columns, the removal of existing welds or bolts at a column should be minimized. Column alignment and stability should be maintained at all times.

Construction loads: See general discussion in Section 8.4.1. Typically, welding on a loaded column should not create a safety issue, although stability during construction should always be considered. At a minimum, see section in *AISC Steel Construction Manual* (2005c) on column splices and the *AISC Code of Standard Practice for Steel Buildings and Bridges* (AISC, 2005a) for other construction considerations.

Proprietary Concerns

There are no known proprietary concerns with this technique.

8.4.8   Add Steel Plate Shear Wall (Connected to An Existing Steel Frame)

Deficiency Addressed by Rehabilitation Technique

Moment frames buildings that are insufficient to resist lateral forces or too flexible to control building drifts can be strengthened and stiffened by adding steel plate shear walls (SPSW), as shown in Figure 8.4.8-1A. The shear walls may be used alone as the new lateral force-resisting system or in conjunction with the moment frames.

Description of the Rehabilitation Technique

Shear walls can add considerable strength and stiffness to a structure. It should be considered as an alternative to adding braced frames if additional stiffness is required. Compared to concrete shear walls, steel plate shear walls are lighter and add less seismic mass to a structure. They also take up less space and may be more economical to construct, particularly in taller buildings where the costs of delivering formwork and pumping concrete are significant. The behavior of this system is analogous to braced frames that rely on tension-only braces, as well as plate girders whereby tension fields develop along the diagonals. However, neither of these examples completely characterizes the behavior of an SPSW. The beams and columns in an SPSW behave as boundary elements that are subject to a complex array of forces and require a considerable amount of stiffness to develop the capacity of the steel panels. Large strut forces are imposed on the beams while columns are subject to axial, flexural, and shear forces. Thus, preference of shear wall locations should be given to existing bays of moment frames, in which the members and connections are less likely to require modifications for use in the SPSW system. The energy dissipating mechanisms in an earthquake include tension yielding and eventual tearing of the diagonal tension fields, shear yielding of the plates, compressive buckling of the plates along the diagonal compression fields, and slipping of the bolted connections to the fin plates when used.

In an existing building, fin plates at the wall boundaries would be field welded to the beams and columns first. Large steel panels would then be welded or bolted to these fin plates, as illustrated in Figure 8.4.8-1B. Panel splices could also be welded or bolted in the field. If welded splices are used, it is recommended to provide full penetration welds with the backing bars removed after welding. Openings are acceptable if stiffeners are provided around the edges. The panels themselves may be unstiffened or stiffened. Provisions for the design of this system have been included in the 2005 *AISC Seismic Provisions*. AISC is also in the process of publishing a design guide for unstiffened SPSW.

Design Considerations

Adding shear walls to a moment frame building increases its stiffness considerably. The upgraded structure should be evaluated for higher lateral and overturning forces accordingly. This system is almost always controlled by shear due to the substantial flexural strength provided by the steel column boundary elements.

Research basis: No references directly addressing the addition of SPSW to moment frame buildings have been identified. However, if the inelastic deformations can be limited to the steel plates, the seismic performance of the strengthened structure is not expected to differ from new

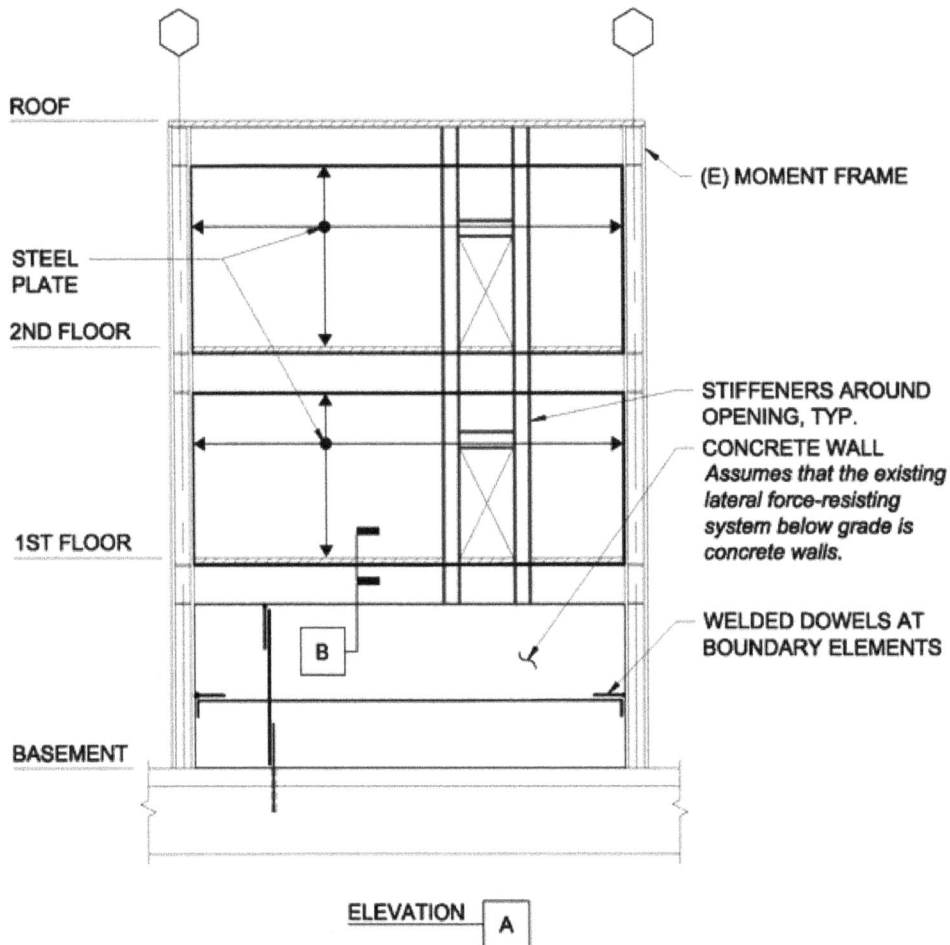

ROOF

(E) MOMENT FRAME

**STEEL
PLATE**

2ND FLOOR

**STIFFENERS AROUND
OPENING, TYP.**

CONCRETE WALL
*Assumes that the existing
lateral force-resisting
system below grade is
concrete walls.*

1ST FLOOR

B

**WELDED DOWELS AT
BOUNDARY ELEMENTS**

BASEMENT

ELEVATION   A

Figure 8.4.8-1A: Unstiffened Steel Plate Shear Wall

buildings utilizing SPSW. A summary of the major research on this system as well as its seismic
performance in past earthquakes can be found in Astaneh-Asl (2001).

Unstiffened vs. stiffened walls: Stiffened walls tend to exhibit higher shear strength though both
types of walls can be expected to exhibit ductile behavior. Buckling of the steel plates in
unstiffened walls allow tension fields to develop and resist the lateral forces. Stiffeners, such as
plates or channels, can be welded to the steel panels to prevent buckling of steel plates. These
walls are more likely to yield in shear instead of developing tension fields. The use of stiffeners
also permits the panels themselves to be thinner than panels in unstiffened walls. The panels in
stiffened wall participate in resisting the overturning forces because buckling does not occur in
the panels. Consequently, overturning forces in unstiffened walls are primarily resisted by the
columns.

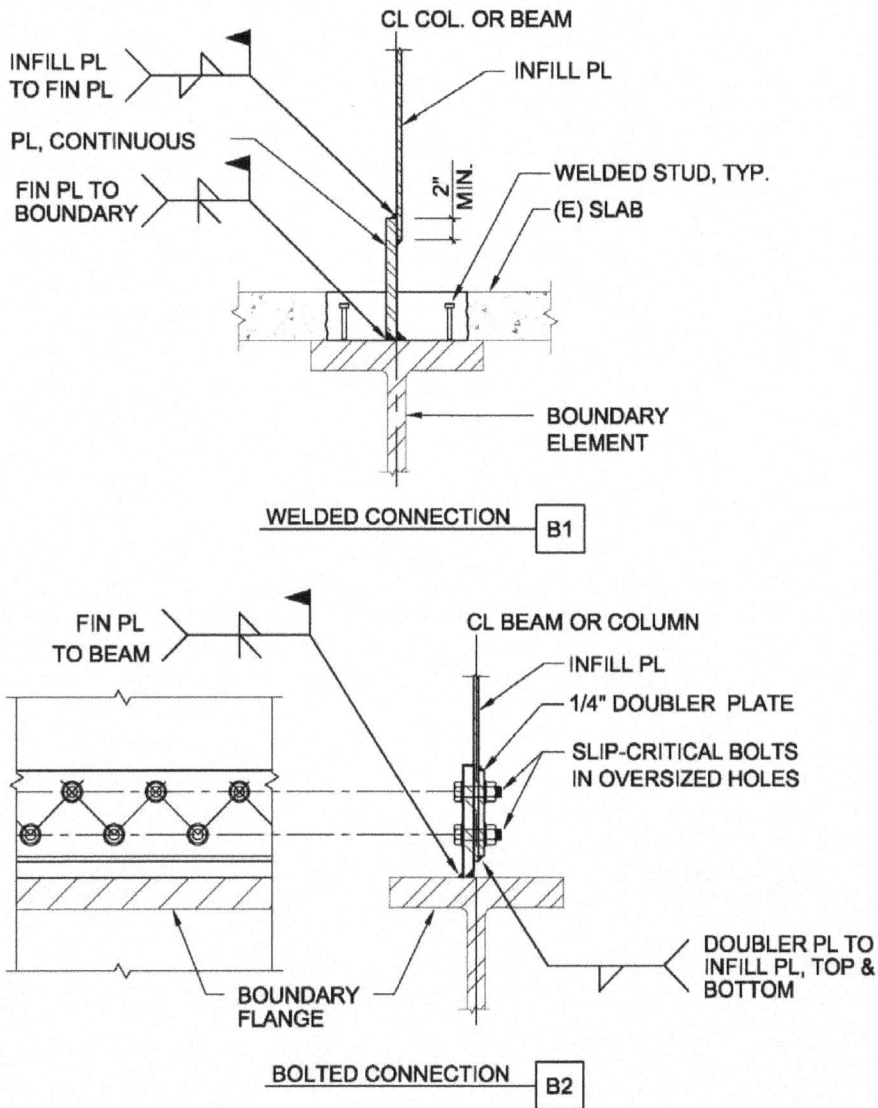

Figure 8.4.8-1B: Fin Plate Connection Options

Nonstructural issues: The addition of walls to an existing structure changes the architectural character of the building. Walls at the exterior will be visible in buildings with clear glazing. At interior bays, walls have to be configured to avoid obstruction of existing corridors, doorways, and other building systems. Walls can be exposed and incorporated into the interior architecture or hidden in partition walls.

Detailing Considerations

In addition to obtaining the latest drawings for the building including as-built drawings, if available, and conducting comprehensive field surveys, the following issues should be noted:

Steel connections: Welds from the fin plates to beams and columns should develop the capacity of the steel panels through full penetration welds or fillet welds on both sides of the plates. To allow for construction and field tolerances, the steel panels can be lapped with the fin plates and connected using fillet welds along both edges of the lap.

Connections at slabs: To avoid damaging the slab reinforcing steel critical for transferring diaphragm forces to the shear walls, the concrete slab could be chipped away without damaging the slab steel and individual fin plates could be placed between the reinforcing steel. The diaphragm forces are transferred to the wall through shear studs on the beams. Alternatively, if continuous fin plates are required and/or the shear studs are inadequate, the reinforcing steel in could be cut and welded directly to the fin plates. In this case, forces transfer from the slab into the wall through shear friction.

Cost/Disruption

As always, the cost of a strengthening scheme depends on the project and its unique requirements. There are no issues with the SPSW system that is known to cost significantly more than adding braced frames or concrete shear walls to a moment frame building. Unstiffened walls are cheaper and less labor intensive than stiffened walls. New foundations are almost always required for new walls and could be extremely costly if deep foundations, such as drilled piers, are added.

Installing new walls is disruptive to the occupants because of the noise and vibrations associated with construction. Even if tenants are relocated to parts of the building where the work is not being performed, vibrations associated with cutting, chipping, and drilling of concrete as well as the installation of steel panels can transmit through the structure. The disruption can be reduced somewhat if the walls are installed at the perimeter.

Construction Considerations

See Section 8.4.1 for general discussions of welding issues, removal of existing nonstructural and structural elements, and construction loads.

Proprietary Concerns

There are no known proprietary concerns with this technique.

8.4.9   Convert an Existing Steel Gravity Frame to a Moment Frame

Deficiency Addressed by Rehabilitation Technique

This technique addresses moment frames buildings that only require slight gains in strength and/or stiffness. The extent of connection modification depends on the seismic hazard and the type of moment frame.

Description of the Rehabilitation Technique

Converting existing gravity frame connections to moment frame connections does not increase the strength or stiffness of a structure significantly unless a large majority of the gravity frame column connections are made moment resisting. The strength and stiffness gain are also limited because the existing beams and columns used in the gravity frame are typically much lighter than the moment frame members. However, these members could also be strengthened as part of the rehabilitation scheme. This technique has less impact on the architectural character of the building than adding braced frames or shear walls.

The simplest method for implementing this technique is through the addition of welded flange cover plates from the beam to the column without any modification of the bolted shear tab. The beam flanges remain unattached from the column since the gap typically exceeds the maximum permitted root opening size for full penetration welds. This method should only be used with Ordinary Moment Frame applications.

In high seismic regions where moment frames have to meet Special Moment Frame requirements, a more sophisticated connection upgrade that forces the beam yielding to occur away from the connection should be provided. This could range from adding welds to the bolted shear tab for the method described above to welding top and bottom haunches from the beam to the column. Several methods are presented in Section 8.4.6 as well as FEMA 350 (FEMA, 2000a) and FEMA 351 (FEMA, 2000b). The level of upgrade ultimately depends on the expected ductility demand and the performance objective.

Design and Detailing Considerations

The strong-column weak-beam concept should still be a primary design consideration. All detailing issues related to the design of moment frame connections need to be considered, including weld filler metal matching, weld metal toughness, removal of weld backing, and column flange and web reinforcing. For Ordinary Moment Frames, joint reinforcing should be limited in order to permit some yielding, as to not place excessive demands on the flange to column connections. For Special Moment Resisting Frames, other requirements such as width-thickness limitations, lateral bracing requirements, etc., should be checked to be in accordance with the *AISC Seismic Provisions*. In most cases, it is expected that the existing gravity beam and column configurations will be such that it will be difficult to meet the requirements for Special Moment Frames without other major modifications. As a result, it is expected that in most cases, the Ordinary Moment Frame requirements would apply.

Cost/Disruption

These issues are discussed in Section 8.4.6 for moment connections. Connection upgrades are typically less disruptive than adding braced frames or shear walls. The costs could vary depending on if existing moment frame connections are also being upgraded and the total number of connections being modified.

Construction Considerations

See Section 8.4.6 for general discussions of issues related to moment connections.

8.5   References

ACI, 1991, *Guide to Certification of Shotcrete Nozzlemen*, ACI-506.3R, American Concrete Institute, Farmington Hills, MI.

ACI, 1994, *Guide for the Evaluation of Shotcrete*, ACI-506.4R, American Concrete Institute, Farmington Hills, MI.

ACI, 1995a, *Guide to Shotcrete*, ACI-506R, American Concrete Institute, Farmington Hills, Michigan.

ACI, 1995b, *Specification for Shotcrete*, ACI-506.2, American Concrete Institute, Farmington Hills, MI.

ACI, 1998, *State of the Art Report on Fiber Reinforced Shotcrete*, ACI-506.1R, American Concrete Institute, Farmington Hills, MI.

ACI, 2005, *Building Code Requirements for Reinforced Concrete and Commentary*, ACI 318, American Concrete Institute, Farmington Hills, MI.

AISC, 2005a, *Code of Standard Practice for Steel Buildings and Bridges*, American Institute of Steel Construction, Chicago, IL.

AISC, 2005b, *Seismic Provisions for Structural Steel Buildings*, American Institute of Steel Construction, Chicago, IL.

AISC, 2005c, *Steel Construction Manual, 13th Edition*, American Institute of Steel Construction, Chicago, IL.

Astaneh-Asl, A., 2001, *Seismic Behavior and Design of Steel Shear Walls*, Structural Steel Educational Council, Moraga, CA.

Carter, C.J., 1999, AISC Design Guide 13, *Stiffening of Wide-Flange Columns at Moment Connections: Wind and Seismic Applications*, American Institute of Steel Construction, Chicago, IL.

FEMA, 2000a, *Recommended Seismic Design Criteria for New Steel Moment-Frame Buildings*, FEMA 350, Federal Emergency Management Agency, Washington, D.C.

FEMA, 2000b, *Recommended Seismic Evaluation and Upgrade Criteria for Existing Steel Moment-Frame Buildings*, FEMA 351, Federal Emergency Management Agency, Washington, D.C.

FEMA, 2000c, *State of the Art Report on Connection Performance*, FEMA 355D, Federal Emergency Management Agency, Washington, D.C.

FEMA, 2000d, *State of the Art Report on Past Performance of Steel Moment-Frame*

Buildings in Earthquakes, FEMA 355E, Federal Emergency Management Agency, Washington, D.C.

FEMA, 2000e, *Prestandard and Commentary for the Seismic Rehabilitation of Buildings*, FEMA 356, Federal Emergency Management Agency, Washington, D.C.

FEMA, 2003, *NEHRP Recommended Provisions for Seismic Regulations for New Buildings and Other Structures*, FEMA 450, Federal Emergency Management Agency, Washington, D.C.

Fisher, J.M. and L.A. Kloiber, 2006, AISC Design Guide 1, *Base Plate and Anchor Rod Design*, American Institute of Steel Construction, Chicago, IL.

Griffis, L.G., 1992, AISC Design Guide 6, *Load and Resistance Factor Design of W-Shapes Encased in Concrete*, American Institute of Steel Construction, Chicago, IL.

Gross, J.L., Engelhardt, M.D., Uang, C.M., Kasai, K., and N.R. Iwankiw, 1999, AISC Design Guide 12, *Modification of Existing Welded Steel Moment Frame Connections for Seismic Resistance*, American Institute of Steel Construction, Chicago, IL.

Lee, H. and S.C. Goel, 1990, "Seismic Behavior of Steel Built-up Box-Shaped Bracing Members, and Their Use in Strengthening Reinforced Concrete Frames," Report No. UMCE 90-7, University of Michigan, Ann Arbor, MI.

Uriz, P. and S. Mahin, 2004, *Summary of Test Results for UC Berkeley Special Concentric Braced Frame No. 1 (SCBF-1) Draft*, Version 1.1, Department of Civil and Environmental Engineering, University of California, Berkeley, CA.

Yang, F. and S. Mahin, 2005, *Limiting Net Section Fractures in Slotted Tube Braces*, Structural Steel Educational Council, Moraga, CA.

Chapter 9 - Building Types S2/S2A: Steel Braced Frames

9.1   Description of the Model Building Type

Building Type **S2** consists of a frame assembly of steel beams and columns. Lateral forces are resisted by diagonal steel members placed in selected bays. Floors are cast-in-place concrete slabs or concrete fill over metal deck. These buildings are typically used for buildings similar to steel moment frames, although more often for low-rise applications. Figure 9.1-1 shows an example of this building type.

Building Type **S2A** is similar but has floors and roof that act as flexible diaphragms such as wood or untopped metal deck. This is a relatively uncommon building type and is used primarily for small office or commercial buildings in which the fire rating of concrete floor is not needed.

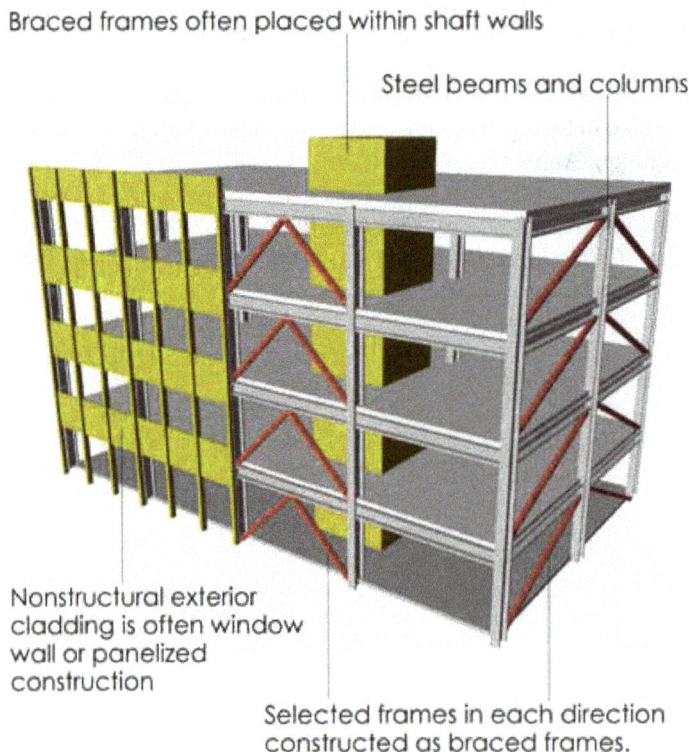

Figure 9.1-1: Building Type S2: Steel Braced Frames

Variations Within the Building Type

The two principal types of braced frame configurations are concentrically braced frames (CBFs) and eccentrically braced frames (EBFs). In a CBF, the centerlines of members that meet at a joint all intersect at a single point. These frames behave as vertical truss systems by transferring lateral loads primarily through axial loading of beams, columns, and braces. During earthquakes, inelastic behavior is typically limited to the braces and connections. Common CBF configurations include diagonal bracing, X-bracing (or 2-story X-bracing), V-bracing (or inverted-V-bracing), and K-bracing. The diagonal braces in an EBF are offset at joints such that link beams separate the ends of braces from columns or other braces. Inelastic behavior is concentrated in the links while all members outside of the links remain elastic or near elastic. Link beams can be located adjacent to a column or at the center of a beam. Common types of braces include W-shapes, hollow structural sections (HSS), steel pipes, double angles, and double channels.

Braces are either welded directly to the beams and columns or welded to gusset plates. It is standard practice on the West Coast to weld gusset plates to the beams and columns. Away from the West Coast, the plate is more commonly welded to the beam and bolted to the column with a pair of angles or a WT-shape. Beam-column connections vary depending on whether there is a brace at the joint or not. Connections range from simple shear tabs to fully welded moment connections.

Floor and Roof Diaphragms

Diaphragms associated with this building type may be either rigid or flexible. The typical rigid diaphragm found in modern buildings consists of structural concrete on metal deck. Diaphragm forces transfer to the frames through shear studs welded to the beams. Older steel buildings that were constructed before metal decks were commonly used may have concrete slabs or masonry arches that span between the beams. Flexible diaphragms include bare metal deck or metal deck with nonstructural fill. These are frequently used on roofs that support light gravity loads. Decks could be connected to the steel members with shear studs, puddle welds, screws, or shot pins. The steel members also act as chords and collectors for the diaphragm.

Foundations

There is no typical foundation for this building type. Foundations can be of any type, including spread footings, mat footings, and piles, depending on the characteristics of the building, the lateral forces, and the site soil. Spread footings are used when lateral forces are not very high and a firm soil exists. For larger forces and/or poor soil conditions, a mat footing below the entire structure is commonly used. Pile foundations are used when lateral forces are extremely large or poor soil is encountered. The piles can be either driven or cast-in-place. Vertical forces are distributed to the underlying soil through a combination of skin friction between the pile and soil and/or direct bearing at the end of the pile; lateral forces are resisted primarily through passive pressure on the vertical surfaces of the pile cap and piles.

9.2   Seismic Performance Characteristics

Braced frames are generally considered to be stiff systems in the elastic range. Their nonlinear response depends on their ability to redistribute forces between bays and drifts between stories.

Braces and connections in CBF undergo large inelastic deformations in tension and compression into the post-buckling range. Ductility of CBF systems in past earthquakes have been limited by local failures of braces and connections. It is thought that CBF systems that are properly designed and detailed can possess ductility in excess of that previously assigned to these systems (AISC, 2005). Yet, recent experimental testing at UC Berkeley found that special concentrically braced frames (SCBF), designed with an inverted-V configuration using the 1997 *AISC Seismic Provisions*, fractured the HSS braces after only a few cycles of loading in the inelastic range (Uriz and Mahin, 2004). Also, the damage concentrated at the level that first experienced brace buckling, resulting in weak story response. As of this writing, these results are still under investigation.

EBF systems approach the higher performance levels of structural systems such as buckling-restrained braced frames and fluid or friction damped frames (Horne et al., 2001). By limiting nonlinear action to the link beams, the post-yield behavior of the system and the maximum demand on the frame can be better predicted. The ductility of an EBF system is dependent on the length and detailing of the link beams and whether shear or flexural yielding governs their inelastic response.

9.3   Common Seismic Deficiencies and Applicable Rehabilitation Techniques

Undesirable behaviors of CBF systems that have been observed in past earthquakes include fracture of connection elements, fracture of braces, and local buckling of braces. These failures, in turn, cause excessive demands on other elements in the system and lead to overall frame failure. See Table 9.3-1 for deficiencies and potential rehabilitation techniques particular to this system. Selected deficiencies are further discussed below by category.

Global Strength

The lack of global strength to resist the seismic demands is a direct result of weak frames. In some cases, the existing beams and columns in a frame may possess enough capacity to accommodate upgrades to the braces and connections. If the beams and columns cannot make this accommodation, braces can be added to other bays to create new frames and thereby, reduce the demand on the existing frames. This would be more easily achieved for CBF systems than for EBF systems, which have special link beam requirements (AISC, 2005).

Global Stiffness

Braced frames are extremely stiff in the elastic range. Concerns with global stiffness occur in the inelastic range when braces are prone to buckling and cause a loss of stiffness. Limited post-elastic stiffness is potentially provided by the frame and non-frame columns in the system (Tang and Goel, 1987; Hassan and Goel, 1991).

Configuration

Concentrically braced frames: K-bracing and inverted-V-bracing configurations exhibit undesirable behavior when a brace buckles. The remaining tension brace at a joint imparts an unbalanced force onto the column or beam. This is particularly hazardous for frames with K-bracing since it may cause a column to fail entirely. In frames that use inverted-V-bracing, the unbalanced force is additive to gravity loads on beams.

Table 9.3-1: Seismic Deficiencies and Potential Rehabilitation Techniques for S2/S2A Buildings						
Deficiency		**Rehabilitation Technique**				
Category	Deficiency	Add New Elements	Enhance Existing Elements	Improve Connections Between Elements	Reduce Demand	Remove Selected Components
Global Strength	Insufficient frame strength	Braced frame [8.4.1] Concrete/masonry shear wall [8.4.2] Steel plate shear wall [8.4.8]	Strengthen braces [9.4.2], beams [8.4.3], columns [8.4.3], and/or connections [9.4.1]		Seismic isolation [24.3] Supplemental damping [24.4]	
Global Stiffness	Excessive drift	Braced frame [8.4.1] Concrete/masonry shear wall [8.4.2] Steel plate shear wall [8.4.8]	Strengthen braces [9.4.2], beams [8.4.3], columns [8.4.3], and/or connections [9.4.1]		Supplemental damping [24.4]	
Configuration	Soft story	Braced frame [8.4.1] Concrete/masonry shear wall [8.4.2] Steel plate shear wall [8.4.8]				
	Re-entrant corner	Braced frame [8.4.1] Concrete/masonry shear wall [8.4.2] Collector [8.4.4]	Enhance detailing [8.4.3], [8.4.4]			
Load Path	Missing collector	Collector [8.4.4]				
	Inadequate shear, flexural, and uplift anchorage to foundation		Embed column into a pedestal bonded to other existing foundation elements [8.4.5]	Provide steel shear lugs or anchor bolts from base plate to foundation [8.4.5]		
	Inadequate out-of-plane anchorage at walls connected to diaphragm			Tension anchors [16.4.1]		

Table 9.3-1: Seismic Deficiencies and Potential Rehabilitation Techniques for S2/S2A Buildings						
Deficiency		**Rehabilitation Technique**				
Category	Deficiency	Add New Elements	Enhance Existing Elements	Improve Connections Between Elements	Reduce Demand	Remove Selected Components
Component Detailing	Inadequate capacity of braces and/or connection	Replace braces [9.4.1]	Increase area of braces [9.4.2] Make braces composite elements [9.4.2] Improve b/t ratios [9.4.2]	Add bolts and welds [9.4.1] Increase size of gusset plates [9.4.1]		
	Inadequate capacity of beams, columns, and/or connections		Add cover plates or box members [8.4.3] Provide gusset plates or knee braces [9.4.1]	Provide gusset plates [9.4.1]		
	EBFs not conforming to current standards		Check current EBF design standards [9.4.2]			
	Inadequate capacity of horizontal steel bracing	Provide additional secondary bracing [9.4.2]	Strengthen bracing elements [9.4.2] Reduce unbraced lengths [9.4.2]	Strengthen connections [9.4.1]		
Diaphragms	Inadequate in-plane strength and/or stiffness	Collectors to distribute forces [8.4.4] Moment frame Braced frame [8.4.1] Concrete/masonry shear wall [8.4.2]	Concrete topping slab overlay Wood structural panel overlay at flexible diaphragms [22.2.1] Strengthen chords [8.4.3], [8.4.4], and [22.2.2]	Add nails at flexible diaphragms [22.2.1]		
	Inadequate shear transfer to frames			Provide additional shear studs, anchors, or welds [22.2.7]		
	Inadequate chord capacity	Add steel members or reinforcement [8.4.3], [8.4.4]				

Table 9.3-1: Seismic Deficiencies and Potential Rehabilitation Techniques for S2/S2A Buildings						
Deficiency		Rehabilitation Technique				
Category	Deficiency	Add New Elements	Enhance Existing Elements	Improve Connections Between Elements	Reduce Demand	Remove Selected Components
Diaphragms (continued)	Excessive stresses at openings and irregularities	Add reinforcement [8.4.3] Provide drags into surrounding diaphragm [8.4.4]				Infill opening [22.2.4], [22.2.6]
Foundations	See Chapter 23					
[ ] Numbers noted in brackets refer to sections containing detailed descriptions of rehabilitation techniques.						

Eccentrically braced frames: The inelastic response of a link beam is influenced by its length relative to the ratio M_p/V_p of the link section (AISC, 2005). Link beams that exhibit shear yielding have greater inelastic deformation capacity than ones that exhibit flexural yielding. *AISC Seismic Provisions* (AISC, 2005) permit shear yielding links to have four times the plastic rotation angles of flexural yielding links. Configurations with link beams adjacent to columns are susceptible to weld fractures found in pre-Northridge connections unless special detailing measures are taken to reduce the demands on these welds.

Load Path

Load path deficiencies in steel braced frame buildings include inadequate connections, collectors, and frame anchorage to foundations. Brace connections may have less strength than the braces. In Type **S2** buildings, seismic forces transfer from the diaphragm to the frame through shear studs welded to collectors or directly to the frame beams. The collectors or the connections to the frame may be too weak to transfer these forces. Connections from columns to base plates or pile caps must resist shear, flexural, and potential uplift forces. Connections that cannot develop these frame forces prevent the frame from developing its full capacity.

Component Detailing

CBF: HSS and pipe braces with high b/t ratios and other steel shapes that lack compactness are subject to local buckling or fracture after a limited number of inelastic cycles. Ductility of these braces can be improved by infilling with concrete or adding longitudinal stiffeners. Alternatively, the braces can be replaced with double HSS sections, which can be used in configurations similar to double angles or channels (Lee and Goel, 1990). HSS and pipe braces are also subject to net section fracture at the gusset plate slots (Uriz and Mahin, 2004). This deficiency can be mitigated by adding reinforcing plates to the sides of the HSS without the slots, as shown in Figure 8.4.1-1 and Figure 8.4.1-2. For pipes, the reinforcing plates can be oriented at right angles to the pipe and appear like stiffeners. Brace connections that are only designed for the axial capacity of the braces may not be adequate to generate the full strength of the braces. To achieve good post-inelastic response, all eccentricities in the connection must be considered. A brace that buckles in the plane of the gusset plates should have its end connections designed for the full axial load and flexural strength of the brace (AISC, 2005). A brace that buckles out-of-plane should ensure that the gusset plates can develop restraint-free plastic rotations without buckling.

EBF: Frames that rely on link-to-column connections have traditionally utilized similar detailing as pre-Northridge connections at beam-column joints. These connections should be reevaluated in the wake of the findings following the Northridge earthquake. Also, experimental research has found that link beams at the first floor undergo the largest inelastic deformation and have the potential to create a soft story (AISC, 2005).

Diaphragm Deficiencies

Common diaphragm deficiencies include insufficient in-plane shear strengths, inadequate chords, and excessive stresses at openings. Causes for these deficiencies could be due to lack of slab or fill thickness, lack of reinforcing steel in the slabs, insufficient connections to chord elements, and poor detailing at openings. See Chapter 22 for common rehabilitation techniques.

Foundation Deficiencies

Foundations that are inadequate do not develop the full capacity of the lateral force-resisting system. Their deficiencies result from insufficient strengths and sizes of footings, grade beams, pile caps, and piles. See Chapter 23 for common rehabilitation techniques.

9.4   Detailed Description of Techniques Primarily Associated with This Building Type

9.4.1   Enhance Braced Frame Connection

Deficiency Addressed by Rehabilitation Technique

Adequate capacities of connections are essential to the proper performance of a braced frame. Connections with insufficient strength and/or ductility to develop stable inelastic frame behavior are strengthened or replaced.

Description of the Rehabilitation Technique

Brace end connections commonly rely on additional connection elements (e.g., gusset plates), but the braces may also be directly welded to the beam and column through a moment connection. Moment connections are also used for link-to-column connections in an EBF and sometimes used for beam-to-column connections in both EBF and CBF systems. The mitigation approach is different depending on whether the existing connection relies on a connection member or a moment connection.

If the existing connection members have sufficient capacity, the most economical alternative may be to increase the connection capacity by providing additional welds or bolts. This typically only allows for a limited increase in capacity since existing brace connection configurations can rarely accommodate significant modifications. If the existing connection members have inadequate capacity, the existing configuration and accessibility need to be assessed to determine whether adding supplemental connection members or replacing the existing connection members with members of greater capacity is more economical. Supplementing the existing connection eliminates the challenges associated with removal of existing connection welds and temporary support of the braces.

The primary concern with moment connections used in braced frames is the use of low notch toughness weld metals. For braces that are expected to develop plastic hinges at their ends, consider replacing the existing welds with a high notch toughness weld metal. Beam-to-column connections in CBF do not typically experience large flexural forces and likely do not need to be upgraded. An exception occurs when the braces or their end connections fail and frame action becomes the primary mechanism for lateral force resistance; however, this is not a recommended design approach, as it does not ensure stable post-elastic behavior. In EBF configurations where the link beam is adjacent to the column, an upgrade should be considered given the large demands on the link beam and its critical nature.

Design Considerations

AISC Seismic Provisions does not permit the sharing of loads by both welds and bolts on the same faying surface. Thus, a bolted brace to gusset plate connection should only be enhanced with bolts or replaced entirely with welds. Note it is not uncommon in some regions for a gusset plate to be bolted to the column through a shear plate and welded to the beam since these are separate faying surfaces. In addition to having thorough knowledge of the existing material behaviors to ensure that the new and existing elements to interact in the desired manner, other design issues include the following:

Research basis: No references directly addressing upgrades of braced frame connections have been identified.

Design forces: Brace connections that are only designed for the axial capacity of the braces may not be adequate to generate the full strength of the braces. A brace that buckles in the plane of the gusset plates should have its end connections designed for the full axial load and flexural strength of the brace (Astaneh-Asl et al., 1986). This recommendation may be more appropriate for high seismic applications.

Bolted connections: Bolts that govern the existing connection capacity could be replaced with stronger bolts, such as replacing A325 with A490. The upgrade may also be as simple as replacing bolts that have threads included in the shear plane with longer bolts. More bolts could be added if the existing configuration allows for it. Larger bolts could be provided but this would reduce the net section capacities due to enlargement of the holes.

Welded connections: Existing fillet welds can be thickened provided the welds are not of low notch toughness weld metals or found to be inadequate through material testing. Otherwise, the existing welds should be removed and replaced with high notch toughness weld metals. A typical fully welded connection appropriate for use in SCBF is shown in Figure 8.4.1-1. For low to moderate seismic applications, Figure 8.4.1-2 shows a more compact connection. Welded brace connections to the weak axis of columns are complex and expensive; an example of this type of connection is shown in Figure 8.4.1-3.

Moment connections: As mentioned above, moment connections that are subject to large flexural forces should be upgraded with high notch toughness weld metals. This would primarily apply to braces that directly connect to beams and columns in SCBF and some OCBF in high seismic applications. EBF link beams that are adjacent to columns also fall into this category. Link beams develop large flexural forces whether shear or flexural yielding governs.

Nonstructural issues: Gusset plates designed in accordance with the *AISC Seismic Provisions* for SCBF can be fairly large and should be discussed with the architect and tenants.

Detailing Considerations

In addition to obtaining the latest drawings for the building including as-built drawings, if available, and conducting comprehensive field surveys, the following issues should be noted:

Gusset plates: A brace in a SCBF that buckles out-of-plane could form plastic hinges at midspan and in the gusset plates at each end. The gusset plates should provide restraint-free plastic rotations without buckling. *AISC Seismic Provisions* suggests a minimum distance of two times the plate thickness between the end of the brace and the assumed line of restraint to achieve this. Consider increasing this distance to three times the plate thickness to accommodate over cutting of the slots in HSS and other erection tolerances. Connections that do not allow for restraint-free plastic rotations out-of-plane should be used primarily in situations where in-plane buckling of the braces govern or ductility demands are low.

Bolted connections: If new holes are drilled in the existing brace and connection member, tolerances of the existing holes should be verified to ensure that bolts will be loaded evenly.

Weld filler metal matching and overmatching: Weld filler metals with slightly greater tensile strength than the connected steel should be used. Flux cored arc welding and shielded metal arc welding electrodes that conform to E70 specifications exhibit overmatching properties compared to common steel specifications, including ASTM A36, A572 (Grade 42 and 50), A913 (Grade 50), and A992 (FEMA, 2000b).

Weld metal toughness: AISC Seismic Provisions now require weld metals with minimum Charpy V-Notch (CVN) toughness of 20 ft-lbf at -20°F for all welds in the lateral force-resisting system. There is also an additional requirement of 40 ft-lbs at 70°F for demand critical welds.

Cost/Disruption

Connection modifications are locally very disruptive. Noise associated with this type of work will spread and disrupt tenants on other floors unless the work is done during off-hours. These modifications could be particularly costly if existing gusset plates have to be replaced.

Construction Considerations

See Section 8.4.1 for general discussions of welding issues, removal of existing nonstructural and structural elements, and construction loads.

Proprietary Concerns

There are no known proprietary concerns with this technique.

9.4.2   Enhance Strength and Ductility of Braced Frame Member

Deficiency Addressed by Rehabilitation Technique

Inadequate beams, columns, and braces are strengthened or replaced to achieve ductile frame behavior.

Description of the Rehabilitation Technique

The ductility of an existing brace can be enhanced by reducing its slenderness, which can be accomplished by decreasing its unbraced length, infilling hollow sections with concrete, or adding longitudinal stiffeners. The unbraced length of a brace can be reduced by adding secondary bracing members that are not part of the primary lateral force-resisting system. Infilling existing hollow sections with concrete can reduce the severity of local buckling (Liu and

Goel, 1988; Lee and Goel 1987). An effective width-thickness ratio for the infilled member is determined by multiplying the width-thickness ratio of the section by the factor (0.0082 x KL/r + 0.264), applicable to braces with KL/r values between 35 and 90 (Goel and Lee, 1992). Adding longitudinal stiffeners presents the least field complications; the stiffeners could consist of plates or small angle sections.

If both strength and ductility are required, new braces have to be added. Some configurations may lend themselves to schemes that allow the existing braces to remain. These include single angle, double angle, and channel braces that can be doubled; rolled sections can also be cover plated. In other cases, it is more practical to replace the existing brace with a new brace, of which numerous options exist. The increase in brace strength may require upgrades to other components of the braced frame, such as the brace connections, beams, and columns. Connection upgrades are discussed in the Section 9.4.1. In many cases, the most cost-effective alternative for increasing the capacity of the existing beams and columns in is to add cover plates or side plates to create box sections. This technique is discussed in Section 8.4.3.

Design Considerations

It would be preferable to limit the strengthening of the existing braces to the capacity of the other members of the lateral force-resisting system, including the foundations, to avoid triggering too many upgrades. Thorough knowledge of the existing material behaviors and strengths are necessary for the new and existing elements to interact in the desired manner. Other design issues include the following:

Research basis: No references directly addressing upgrades of braced frame members have been identified.

Existing brace strengthening: Significant modifications to an existing brace could trigger strengthening or redesign of its end connections. Strengthening of existing K- or inverted-V-bracing should be undertaken only after careful evaluation of the additional bending forces following the buckling of the compression bracing. Where the existing bracing in these systems is found to have inadequate capacity, the preferred solution is to replace it with a diagonal or X-bracing configuration.

Secondary bracing: A brace member is designed to resist both tension and compression forces, but its capacity for compression stresses is limited by potential buckling and is therefore less than the capacity for tensile stresses. Since the design of the system generally is based on the compression capacity of the brace, some additional capacity may be obtained by simply reducing the unsupported length of the brace by means of secondary bracing provided the connections have adequate reserve capacity or can be strengthened for the additional loads.

New brace selection: If existing braces are replaced, use compact and non-slender sections whenever possible to avoid premature fracturing or buckling of the braces during post-yield behavior. Two particular brace types not common in older braced frame buildings are double HSS sections and buckling-restrained braces. Double HSS sections can be used in configurations similar to double angles or channels (Lee and Goel, 1990). They provide reduced fit-up issues and smaller width-to-thickness ratios compared to a single HSS, resulting in increased energy dissipation capacity. The other type of brace, used in a buckling-restrained braced frame (BRBF), is typically used in new buildings but has also been used successfully in new BRBF systems in existing buildings. One example of a brace used in a BRBF consists of a steel core inside a casing, which consists of a hollow structural section (HSS) infilled with concrete grout. Proprietary materials separate the steel core and concrete to prohibit bonding between the two materials. There are other buckling-restrained braces that do not use grout or additional separating agents between the steel and grout. The main advantage of these braces is the ability of the casing to restrain the buckling of the steel core without providing any additional axial force resistance beyond the capacity of the steel core. Provisions for new building BRBF design are included in the *NEHRP Recommended Provisions* (FEMA, 2003) and the *AISC Seismic Provisions for Structural Steel Buildings*. Note that significant connection modifications may be required when braces are replaced.

Nonstructural issues: Brace modifications, when exposed, will affect the interior architecture or if hidden in partition walls, these walls may be thicker than typical walls. Beams that are increased in size affect nonstructural components by reducing clear floor heights. These components typically include suspended ceilings, pipes, conduits, and ducts. Coordination with the architect and other trades should not be overlooked or underestimated.

Detailing Considerations

In addition to obtaining the latest drawings for the building including as-built drawings, if available, and conducting comprehensive field surveys, the following issues should be noted:

Built-up brace members: While double angles, double channels, and double HSS offer advantages for installation, special criteria apply to these members when used in a SCBF. Buckling of these types of braces imposes large shear forces on the stitches. Therefore, closer stitch spacing and higher stitch strengths are required. More stringent member compactness is also necessary for ductility and energy dissipation.

Reinforcing cover plates: HSS and pipe braces are subject to net section fracture at the gusset plate slots (Uriz and Mahin, 2004). This brittle failure mode can be eliminated by adding reinforcing cover plates to the sides of the HSS without the slots, such as the ones shown in Figures 8.4.1-1 and 8.4.1-2. For pipes, the reinforcing plates can be oriented at right angles to the pipe and appear like stiffeners. Additional information regarding the design of these plates can be found in *Limiting Net Section Fracture in Slotted Tube Braces* (Yang and Mahin, 2005).

Cost/Disruption

Designs that are simple and details that are not overly complicated will minimize costs associated with this technique. This could include maximizing the use of existing members, minimizing connection upgrades, and reducing the amount of field welding.

Costs can also be reduced if disruption is minimal during construction. Installing braces at the perimeter frames reduces logistical issues associated with working in confined spaces and temporary removal of the nonstructural elements. Noise associated with this type of work is loud and disturbing to the tenants if the building is occupied while the work is being performed.

Construction Considerations

See Section 8.4.1 for general discussions of welding issues, removal of existing nonstructural and structural elements, and construction loads. Connecting members, such as gusset plates, that are not being replaced should be protected when braces are removed.

Proprietary Concerns

Braces used in buckling-restrained braced frames (BRBF) are proprietary. There are a limited number of manufacturers of the braces used in BRBF.

9.5   References

AISC, 2005, *Seismic Provisions for Structural Steel Buildings*, American Institute of Steel Construction, Chicago, IL.

Astaneh-Asl, A., Goel, S.C., and R.D. Hanson, 1986, "Earthquake-Resistant Design of Double Angle Bracing," *Engineering Journal*, Vol. 23, No. 4, American Institute of Steel Construction, Chicago, IL.

FEMA, 2000a, *Recommended Seismic Evaluation and Upgrade Criteria for Existing Steel Moment-Frame Buildings*, FEMA 351, Federal Emergency Management Agency, Washington, D.C.

FEMA, 2003, *NEHRP Recommended Provisions for Seismic Regulations for New Buildings and Other Structures*, FEMA 450, Federal Emergency Management Agency, Washington, D.C.

Goel, S.C. and S. Lee, 1992, "A Fracture Criterion for Concrete-Filled Tubular Braces," *Proceedings of the 1992 ASCE Structures Congress*, pp. 922-925, ASCE, Reston, VA.

Hassan, O. and S.C. Goel, 1991, "Seismic Behavior and Design of Concentrically Braced Steel Structures," Report No. UMCE 91-1, University of Michigan, Ann Arbor, MI.

Horne, J., Rubbo, A., and J. Malley, 2001, "AISC-LRFD Design and Optimization of Steel Eccentrically Braced Frames," *Proceedings of 2001 Convention of the Structural Engineers Association of California*, Sacramento, CA.

Lee, H. and S.C. Goel, 1990, *Seismic Behavior of Steel Built-up Box-Shaped Bracing Members, and Their Use in Strengthening Reinforced Concrete Frames*, Report No. UMCE 90-7, University of Michigan, Ann Arbor, MI.

Lee, S. and S.C. Goel, 1987, *Seismic Behavior of Hollow and Concrete-Filled Square Tubular Bracing Members*, Report No. UMCE 87-11, University of Michigan, Ann Arbor, MI.

Liu, Z.,and S.C. Goel, 1988, "Cyclic Load Behavior of Concrete-Filled Tubular Braces," *Journal of Structural Division*, Vol. 114, No. 7, ASCE, Reston, VA.

Tang, X. and S.C. Goel, 1987, *Seismic Analysis and Design Considerations of Braced Steel Structures*, Report UMCE 87-4, University of Michigan, Ann Arbor, MI.

Uriz, P. and S. Mahin, 2004, *Summary of Test Results for UC Berkeley Special Concentric Braced Frame No. 1 (SCBF-1) Draft*, Version 1.1, Department of Civil and Environmental Engineering, University of California, Berkeley, CA.

Yang, F. and S. Mahin, 2005, *Limiting Net Section Fractures in Slotted Tube Braces*, Structural Steel Educational Council, Moraga, CA.

Chapter 10 - Building Type S4: Steel Frames with Concrete Shear Walls

10.1   Description of the Model Building Type

Building Type **S4** consists of an essentially complete frame assembly of steel beams and columns. The floors are concrete slabs or concrete fill over metal deck. These buildings feature a significant number of concrete walls effectively acting as shear walls, either as vertical transportation cores, isolated in selected bays, and/or as a perimeter wall system. The steel column and beam system may act only to carry gravity loads or may have rigid connections to act as a moment frame to form a dual system. This building type is generally used as an alternate for steel moment or braced frames in similar circumstances. These buildings will usually be mid- or low-rise. Figure 10.1-1 shows an example of this building type.

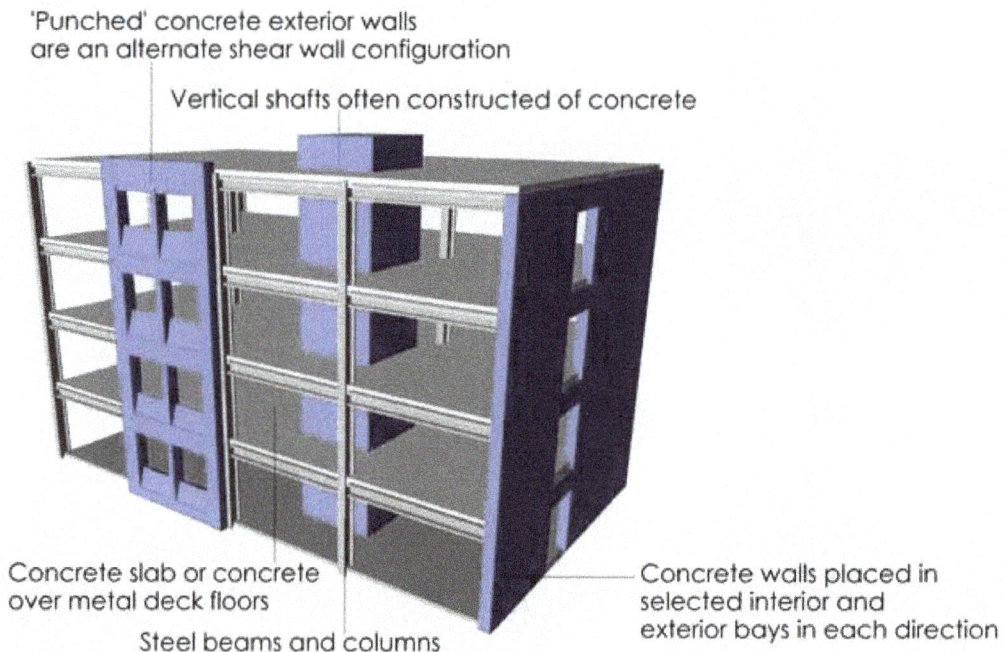

'Punched' concrete exterior walls are an alternate shear wall configuration

Vertical shafts often constructed of concrete

Concrete slab or concrete over metal deck floors

Steel beams and columns

Concrete walls placed in selected interior and exterior bays in each direction

Figure 10.1-1: Building Type S4: Steel Frames with Concrete Shear Walls

10.2   Seismic Performance Characteristics

In older buildings, the steel frame carries only gravity loads while all lateral loads are resisted by the concrete shear walls. In modern buildings, both lateral systems work together in proportion to relative rigidity. Generally, except in tall buildings, these systems tend to behave more like shear wall structures due to the much greater stiffness of the walls. The contribution of the steel moment frame to the lateral capacity of the building is a function of the number of frames and the detailing of the beam-column joints. See performance characteristics described in Section 14.2 for concrete shear wall buildings and Section 8.2 for steel moment frame buildings.

10.3   Common Seismic Deficiencies and Applicable Rehabilitation Techniques

See deficiencies and techniques described in Section 13.3 and 14.3 for concrete shear wall buildings and Section 8.3 for steel moment frame buildings.

10.4   Detailed Description of Techniques Primarily Associated with This Building Type

See recommended techniques in Section 13.4 for concrete shear wall buildings and Section 8.4 for steel moment frame buildings.

10.5   References

See references in Section 13.5 and 14.5 for concrete shear wall buildings and Section 8.5 for steel moment frame buildings.

Table 10.3-1: Seismic Deficiencies and Potential Rehabilitation Techniques for S4 Buildings						
Deficiency		Rehabilitation Technique				
Category	Deficiency	Add New Elements	Enhance Existing Elements	Improve Connections Between Elements	Reduce Demand	Remove Selected Components
Global Strength	Insufficient in-plane wall shear strength	Concrete/masonry shear wall [8.4.2] Braced frame [8.4.1]	Concrete wall overlay [21.4.8] Fiber composite wall overlay [13.4.1] Steel overlay		Seismic isolation [24.3] Reduce flexural capacity [13.4.4]	
	Insufficient flexural capacity	Concrete/masonry shear wall [8.4.2] Braced frame [8.4.1]	Add or enhance chords			
	Insufficient frame strength	Moment frame Braced frame [8.4.1] Concrete/masonry shear wall [8.4.2]	Strengthen beams [8.4.3], columns [8.4.3], and/or connections [8.4.6]		Seismic isolation [24.3]	
Global Stiffness	Excessive drift	Concrete/masonry shear wall [8.4.2] Braced frame [8.4.1] Moment frame	Strengthen beams [8.4.3], columns [8.4.3], and/or connections [8.4.6] Concrete wall overlay [21.4.5]			
	Inadequate capacity of coupling beams	Concrete/masonry shear wall [8.4.2] Braced frame [8.4.1]	Strengthen beams [13.4.2] Improve ductility of beams [13.4.2]			Remove beams
Configuration	Discontinuous walls	Concrete/masonry shear wall [8.4.2]	Enhance existing column for overturning loads	Improve connection to diaphragm [13.4.3]		Remove wall
	Soft story	Concrete/masonry shear wall [8.4.2] Braced frame [8.4.1]				
	Re-entrant corner	Moment frame Braced frame [8.4.1] Concrete/masonry shear wall [8.4.2] Collector [8.4.4]	Enhance detailing [8.4.3], [8.4.4]			

10-3

Table 10.3-1: Seismic Deficiencies and Potential Rehabilitation Techniques for S4 Buildings						
Deficiency		Rehabilitation Technique				
Category	Deficiency	Add New Elements	Enhance Existing Elements	Improve Connections Between Elements	Reduce Demand	Remove Selected Components
Configuration (continued)	Torsional layout	Add balancing walls [8.4.2], braced frames [8.4.1], or moment frames				
Load Path	Missing collector	Add collector [8.4.4]	Strengthen existing beam [8.4.3] or slab Enhance splices or connections of existing beams [8.4.4]			
	Discontinuous Walls	Provide new wall support components to resist the maximum expected overturning moment	Strengthen the existing support columns for the maximum expected overturning moment [8.4.3] Provide elements to distribute the shear into the diaphragm at the level of discontinuity [13.4.3]			
	Inadequate shear, flexural, and uplift anchorage to foundation		Embed column into a pedestal bonded to other existing foundation elements [8.4.5]	Provide steel shear lugs or anchor bolts from base plate to foundation [8.4.5]		
	Inadequate out-of-plane anchorage at walls connected to diaphragm			Tension anchors [16.4.4]		

10-4

Table 10.3-1: Seismic Deficiencies and Potential Rehabilitation Techniques for S4 Buildings					
Deficiency		Rehabilitation Technique			
Category	Deficiency	Add New Elements	Enhance Existing Elements	Improve Connections Between Elements	Reduce Demand	Remove Selected Components
Component Detailing	Wall inadequate for out-of-plane bending	Add strongbacks [21.4.3]	Concrete wall overlay [21.4.5]			
	Wall shear critical		Concrete wall overlay [21.4.5] Fiber composite wall overlay [13.4.1]		Reduce flexural capacity of wall [13.4.4]	
	Inadequate capacity of beams, columns, and/or connections		Enhance beam-column connections [8.4.6] Add cover plates or box members [8.4.3] Provide gusset plates or knee braces [9.4.1] Encase columns in concrete [8.4.2]			
	Inadequate capacity of panel zone		Provide welded continuity plates [8.4.6] Provide welded stiffener or doubler plates [8.4.6]			
	Inadequate capacity of horizontal steel bracing	Provide additional secondary bracing [9.4.2]	Strengthen bracing elements [9.4.2] Reduce unbraced lengths [9.4.2]	Strengthen connections [9.4.1]		
Diaphragms	Inadequate in-plane strength and/or stiffness	Collectors to distribute forces [8.4.4] Moment frame Braced frame [8.4.1] Concrete/masonry shear wall [8.4.2]	Concrete topping slab overlay Strengthen chords [8.4.3], [8.4.4]			

Table 10.3-1: Seismic Deficiencies and Potential Rehabilitation Techniques for S4 Buildings						
Deficiency		Rehabilitation Technique				
Category	Deficiency	Add New Elements	Enhance Existing Elements	Improve Connections Between Elements	Reduce Demand	Remove Selected Components
Diaphragms (continued)	Inadequate shear transfer to frames			Provide additional shear studs, anchors, or welds [22.2.7]		
	Inadequate chord capacity	Add steel members or reinforcement [8.4.3], [8.4.4]				
	Excessive stresses at openings and irregularities	Add reinforcement [8.4.3] Provide drags into surrounding diaphragm [8.4.4]				Infill opening [22.2.4], [22.2.6]
Foundations	See Chapter 23					
[ ] Numbers noted in brackets refer to sections containing detailed descriptions of rehabilitation techniques.						

Chapter 11 - Building Types S5/S5A: Steel Frames with Infill Masonry Shear Walls

11.1   Description of the Model Building Type

Building Type **S5** is normally an older building that consists of an essentially complete gravity frame assembly of steel floor beams or trusses and steel columns. The floor consists of masonry flat arches, concrete slabs or metal deck and concrete fill. Exterior walls, and possibly some interior walls, are constructed of unreinforced masonry, tightly infilling the space between columns and between beams and the floor such that the infill interacts with the frame to resist lateral movement. Windows and doors may be present in the infill walls, but to effectively act as a shear resisting element, the infill masonry must be constructed tightly against the columns and beams. The steel gravity framing in these buildings may include truss spandrels or knee braces on the exterior walls, or partially restrained beam-column connections in a more extensive pattern. The steel frame also is often cast in concrete for fireproofing purposes. The buildings intended to fall into this category normally feature exposed clay brick masonry on the exterior and are common in commercial areas of cities with occupancies of retail stores, small offices, and hotels. Figure 11.1-1 shows an example of this building type.

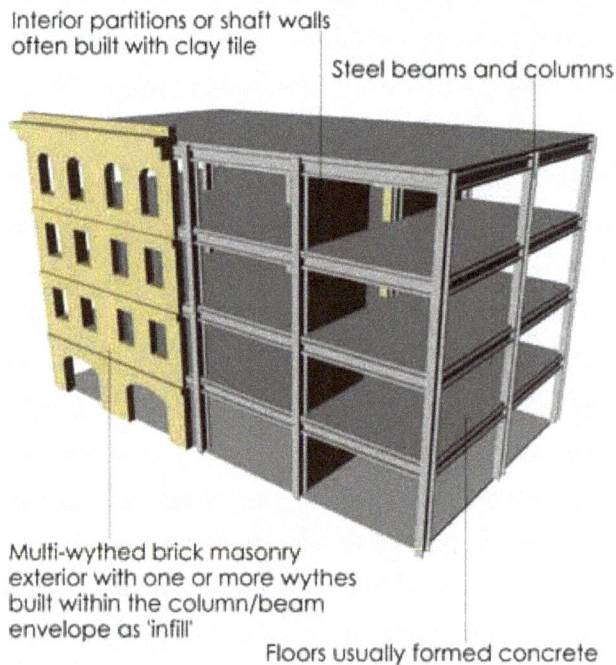

Figure 11.1-1: Building Type S5: Steel Frames with Infill Masonry Shear Walls

The **S5A** building type is similar but has floors and roof that act as flexible diaphragms such as wood, or untopped metal deck. This type of building will almost always date to the 1930 or earlier.

Variations Within the Building Type

The building type was identified primarily to capture the issues of interaction between unreinforced masonry and steel gravity framing. The archetypal building has solid clay brick at the exterior with one wythe of brick running continuously past the plane of the column and beam and two or more wythes infilled within the plane of the column and beam. The exterior wythe of clay brick forms the finish of the building although patterns of terra cotta, stone, or precast concrete may be attached to the brick or laid up within the brick. However, there can be many variations to this pattern depending on the number and arrangement of finished planes on the exterior of the building. For example, the full width of the infill wall may be located with the plane of the column and beam with a pilaster built out and around the column and a horizontal band of brick or other material covering the beam. The beam is often placed off center of the column, usually on the out-board side. In extreme conditions, the primary plane of the masonry wall may not directly engage the column at all. In these cases, strut compression must be transferred eccentrically through the masonry surrounding the column, reducing effectiveness.

In some buildings the steel frame is encased in concrete, primarily for fireproofing. This encasement is normally reinforced with mesh and may contribute to overall frame stiffness and to connection strength and stiffness of partially restrained steel connections. Importantly, at the perimeter frames, the concrete encasement forms a smooth surface at the masonry interface and probably encouraged a neater fit during construction. Concrete encasement of columns also will assist in transferring eccentric strut loading into the column-frame system

Hollow clay tile masonry may also be used as an exterior infill material. Although this material often has a very high compression strength, the net section of material available to form the compression strut within the frame will normally contribute a lateral strength of only a small percentage of the building weight. The material being brittle and the wall being highly voided, these walls may also lose complete compressive strength quite suddenly. Therefore, walls of hollow clay tile infill will probably not contribute a significant portion of required lateral resistance except in areas of low seismicity and/or when walls are arranged as infill on both the exterior and interior of the building.

More recent buildings may have unreinforced concrete block masonry configured as an exterior infill wall, with a variety of finish materials attached to the outside face of the concrete block. Similar to hollow clay tile walls, these walls may exhibit moderate to low compressive strength and brittle behavior that marginalizes their usefulness as lateral elements. In addition, hollow concrete block exterior walls often will not be installed tight to the surrounding framing, eliminating infill compression strut behavior.

Floor and Roof Diaphragms

The earliest version of this building type may include floors constructed of very shallow masonry arches spanning between steel beams. A relatively flat top surface is created with masonry rubble or light-weight cementitious fill and the floor is finished with wood sheathing. In some cases, the thrust from the arches is resisted by tie rods running perpendicular and through the steel beams. The only diaphragm action provided by such floors is the finish wood sheathing and the lateral flexibility of this system is incompatible with the stiff but brittle masonry arches.

Building Type **S5A** will have heavy timber floors with one or more layers of sheathing forming a diaphragm. The flexibility of such diaphragms will often form a seismic deficiency because, assuming no interior shear elements, the large drift at the diaphragm mid-span will damage perpendicular walls and gravity framing. Specific strengthening techniques for this building type are not covered here. For generalized strengthening of diaphragms, see Chapter 22.

Most typically, the floor and roof are cast-in-place concrete slabs spanning between beams. The concrete slab is often integral with lightly reinforced concrete surrounding each beam. This building type can also be found with metal deck and concrete floor slabs.

Foundations

There is no typical foundation for this building type. Foundations can be found of every type depending on the height of the building, the span of the gravity system and the site soil. The exterior walls are exceptionally heavy and typically will be supported by a continuous concrete footing or often a continuous concrete wall forming a basement space below.

11.2   Seismic Response Characteristics

Most steel frame infill buildings will incorporate some beam column connections with moment resistance, either from top and bottom chord truss connections, knee bracing, or partially restrained tee or angle connections. The restraint is often enhanced by cast-in-place concrete cover. The lateral strength and stiffness of these systems is difficult to assess, although some testing has been done (Roeder et al., 1996). See also Abrams (1994). Unless the perimeter infill is penetrated with large openings, the frame will be far more flexible than the infill. Therefore, both in terms of stiffness and strength, the exterior infill walls typically will form the effective lateral system for this building type. The effectiveness of the system depends on the size and extent of openings and articulation of the plane of the wall. With solid or nearly solid infill panels, strut action will be stiff and strong. As openings in panels increase in size, struts or combinations of struts cannot effectively form around the opening and the steel columns and beams will begin to work as a moment frame, with "fixity" at the beam-column joint provided by the masonry. For low and moderate intensity shaking, the exterior walls may provide adequate strength to satisfy the specified performance objective. As the shaking demand increases, the masonry will tend to crack and spall, losing stiffness and potentially creating a falling hazard. The complete steel gravity system, characteristic of this building type, is generally expected to provide sufficient stability to prevent collapse, particularly if designed for lateral resistance. However, in configurations with large height-to-width ratios, end or corner columns could fail in compression or at tension splices, potentially leading to partial collapse.

This building type is often characterized by a commercial store-front first floor with little or no infill at that level on one or more faces of the building. This condition can cause a soft story condition or a severe torsional response if open on one or two sides only. Such conditions can lead to concentration of seismic deformation at the open level, potentially leading to local P-delta failure. This open commercial story was a common feature in many buildings of this type that were shaken in the 1906 San Francisco earthquake, but there were no story-mechanisms reported. It is speculated that the soft story provided isolation for the upper stories and that the displacement demand, for reasons unknown, did not exceed the story capacity. In fact, there have been no reports of collapses or damage that suggested imminent collapse in typical U.S. multistory office-like steel infill buildings in strong ground motion. Earthquakes providing such tests include the 1906 San Francisco, the 1933 Long Beach, and to a lesser extent, the 1994 Northridge events. In general, current seismic evaluation technology does not reach the same conclusion.

11.3 Common Seismic Deficiencies and Applicable Rehabilitation Techniques

See Table 11.3-1 for deficiencies and potential rehabilitation techniques particular to this system. Deficiencies related to steel moment frames and masonry shear walls are shown in Table 5.3-1 and Table 18.3-1, respectively. Selected deficiencies are further discussed below by category.

Global Strength

The overall strength provided by the exterior walls may be insufficient to prevent serious degradation and resulting amplified displacements in the building that can lead to irreparable damage or even instability. The strength may be limited by inadequate number of panels of infill, excessive openings, or masonry weak in compressive strength. The standard approach to such deficiencies will be to add new, relatively stiff lateral force-resisting elements such as concrete shear walls or steel braced frames often located on the interior between existing columns. Concrete walls can also be added at the perimeter on the inside face of the masonry. This procedure is usually conceptualized and analyzed as a concrete shear wall rather than an infill to the frame.

Fiber composite layers also can be added to the face of masonry to enhance infill strut action. Although this technique has been tested for increasing shear strength of URM walls, little research is available directly on the effects of adding these layers to infill panels.

Unless the masonry is completely doweled or connected to supporting backing, the damage state of the masonry wall must be estimated for the expected drifts of the combined system to determine if the desired performance has been achieved.

Global Stiffness

For this building type, the methods for adding stiffness are similar to those adding strength.

Configuration

Two global configurational deficiencies are common in this building type. The first is a soft and weak story at the street level created by commercial occupancies with exterior bays with little or

Table 11.3-1: Seismic Deficiencies and Potential Rehabilitation Techniques for S5/S5A Buildings					
Deficiency		Rehabilitation Technique			
Category	Deficiency	Add New Elements	Enhance Existing Elements	Improve Connections Between Elements	Reduce Demand	Remove Selected Components
Global Strength	Inadequate length of exterior wall	Interior concrete walls [8.4.2] Interior steel braced frames [8.4.1]	Concrete wall overlay [21.4.5] Fiber composite wall overlay [21.4.6]			
	Excessive sized openings in infill panels	Interior concrete walls [8.4.2] Interior steel braced frames [8.4.1]	Infill selected openings [21.4.7] Concrete wall overlay [21.4.5] Fiber composite wall overlay [21.4.6]			
	Inadequate columns for overturning forces		Add cover plates or box members [8.4.3] Encase columns in concrete			
	Weak or deteriorated masonry	Interior concrete walls [8.4.2] Interior steel braced frames [8.4.1]	Point outside and/or inside wythes of masonry Inject wall with cementitious grout Add concrete or fiber composite overlay on exterior walls pier and/or spandrel [21.4.5], [21.4.6]			
Global Stiffness	See inadequate strength					
Configuration	Soft or weak story	Interior concrete walls [8.4.2] Interior steel braced frames [8.4.1]				

Table 11.3-1: Seismic Deficiencies and Potential Rehabilitation Techniques for S5/S5A Buildings						
Deficiency		**Rehabilitation Technique**				
Category	**Deficiency**	**Add New Elements**	**Enhance Existing Elements**	**Improve Connections Between Elements**	**Reduce Demand**	**Remove Selected Components**
Configuration (continued)	Torsion from one or more solid walls	Balance with Interior concrete walls Balance Interior steel braced frames				Remove selected infill panels on solid walls
	Irregular Plan Shape	Balance with interior concrete walls Balance with interior steel braced frames				
Load Path	Out-of-plane failure of infill due to loss of anchorage or slenderness of infill	Provide vertical strongback wall supports [21.4.3]	Concrete wall overlay [21.4.5] Fiber composite wall overlay [21.4.6]			Remove infill
	Inadequate connection of finish wythe to backing		Add interwythe tie [21.4.12]			
	Inadequate collectors	Add steel collector on surface of concrete [12.4.3] Embed or add collector in concrete floor slab [12.4.3]	Strengthen beam to column or beam to beam splices			
Component Detailing	Inadequate columns splice for tension due to uplift force induced by infill			Add splice plates Provide splice through added reinforced concrete encasement		
	Inadequate beam column connection to resist compression thrust			Strengthen connection in shear		

Table 11.3-1: Seismic Deficiencies and Potential Rehabilitation Techniques for S5/S5A Buildings						
Deficiency		**Rehabilitation Technique**				
Category	**Deficiency**	**Add New Elements**	**Enhance Existing Elements**	**Improve Connections Between Elements**	**Reduce Demand**	**Remove Selected Components**
Component Detailing (continued)	Weak or incompletely filled joint between masonry and surrounding steel components			Repair or fill voids to provide essentially continuous bearing.		
Diaphragms	Flat masonry arch diaphragm	Add diagonal steel braced diaphragm under floor [22.2.8] Remove top layers of floor construction and add concrete slab diaphragm	Add tension ties to prevent loss of arch action [22.2.8]			
Foundation	See Chapter 23					
[ ] Numbers noted in brackets refer to sections containing detailed descriptions of rehabilitation techniques.						

no infill. This deficiency can be corrected by adding selected bays of infill or by adding shear walls or braced frames at this level. The second common issue is a plan torsional irregularity created by solid masonry walls on property lines coupled with walls with many openings on street fronts. If shown by analysis to be necessary, torsional response can be minimized by stiffening the more flexible side of the building with more infill or by the addition of lateral elements. In rare cases, the solid walls can be balanced with the open side by selected removal of panels or disengagement of the infill strut action.

Load Path

The primary load path issue with this building type is to assure that the mass of the exterior walls will not become disengaged from the frame which will both prevent infill strut action as well as to create a significant falling hazard on the street below.

In-plane, the articulation of the exterior walls may result in offsets of the wall plane between floors. The presence of a complete load path and maintenance of confinement for strut formation must be reviewed in such instances.

If new lateral load-resisting elements are added, existing slab and steel beam construction may need to be strengthened to provide adequate collectors.

Component Detailing

In order to qualify as an infill lateral force-resisting element, the infill must be installed tight to the surrounding steel frame. Loose or incomplete infill can be mitigated with local patching of the masonry or by injection of cemetitious or epoxy grout. However, unless the building is gutted for remodeling purposes, this procedure will be extremely disruptive.

The detailing of the steel frame forming the confinement for the masonry is important to achieve infill strut behavior. The connection of beam to column must be capable of resisting the strut compression forces from the masonry. Many different configurations are possible, each with a different potential weakness, but the shear capacity of the beam-to-column connection is often critical. In addition, column splices may be inadequate to transfer the overturning forces created by strut action. Critical connections normally can be strengthened with steel plates.

Diaphragm Deficiencies

A wide variety of concrete diaphragms can be found in this building type. Solid slab-type floors will often provide an adequate diaphragm while joisted floors may include only a thin, poorly reinforced continuous slab with low shear capacity. The connection of slabs to exterior wall should be reviewed because dowels or other positive connections may not have been provided.

See Chapter 21 on URM construction for discussion of wood diaphragms in this type of building.

Flat masonry arch floors are problematic. The diaphragm capacity of such built up construction has not been established. Damage causing loss of arch action can create falling hazards or vertical load failures. Removal and replacement may not be feasible, either from a pure economic standpoint or due to historical preservation issues. The added weight of a new concrete slab is often difficult to accommodate, even if top layers of the existing floor are removed.

If space is available, a new steel diagonal frame diaphragm can be added underneath such floors. FRP can be layered on the masonry arches to better secure them in place. New lateral force-resisting elements can be added to minimize the need for diaphragm action.

Foundation Deficiencies

No systematic deficiency in foundations should be expected solely due to the characteristics of this building type.

Other Deficiencies

Although deterioration of material, in general, is not covered in this document, it is known that most buildings of this type have no reliable waterproofing system for the exterior steel framing, particularly the columns. Significant damage to columns from water infiltration has been noted in several cases, and this condition should be investigated before assuming that the perimeter frame is a significant lateral force-resisting element.

11.4 Detailed Description of Techniques Primarily Associated with This Building Type

Most significant recommendations listed in Table 11.3-1 are similar to techniques more commonly associated with other building types such as steel framed buildings (**S1**, **S2**, or **S4**), unreinforced masonry bearing wall buildings (**URM**), or general techniques applied to concrete diaphragms. Details concerning these techniques can be found in other chapters.

11.5 References

Abrams, D.P. (Editor), 1994, *Proceedings from the NCEER Workshop on Seismic Response of Masonry Infills*, Technical Report NCEER-94-0004, National Center for Earthquake Engineering Research, Buffalo, NY.

Roeder, C.W., Leon, R.T., and Preece, F.R., 1996, "Expected Seismic Behavior of Older Steel Structures," *Earthquake Spectra*, EERI, Vol. 12, No. 4, Oakland, CA, pgs 805-824.

Chapter 12 - Building Type C1: Concrete Moment Frames

12.1   Description of the Model Building Type

These buildings consist of concrete framing, either a complete system of beams and columns or columns supporting slabs without gravity beams. Lateral forces are resisted by cast-in-place moment frames that develop stiffness through rigid connections of the column and beams. The lateral force-resisting frames could consist of the entire column and beam system in both directions, or the frames could be placed in selected bays in one or both directions. An important characteristic is that no significant concrete or masonry walls are present, or that they are adequately separated from the main structure to prevent interaction. Some buildings of this type have frames specifically designed for lateral loads, but also have interacting walls apparently unaccounted for in the design. These buildings could be classified as moment frames and the wall interaction would immediately be considered a seismic deficiency. Alternately, these buildings could be classified as Building Type **C2f** (Shear Wall with Gravity Frames). Older concrete buildings may include frame configurations that were not designed for lateral load, but if no walls or braces are present, the frames become the effective lateral force system and should be included in this building category. Buildings of this type that include integral concrete or masonry walls on the perimeter should be considered as Building Type **C2f** or **C3**. Floors may be a variety of cast-in-place or precast concrete. Buildings with concrete moment frames are generally used for most occupancies listed for steel moment frames, but are also used for multistory residential buildings.

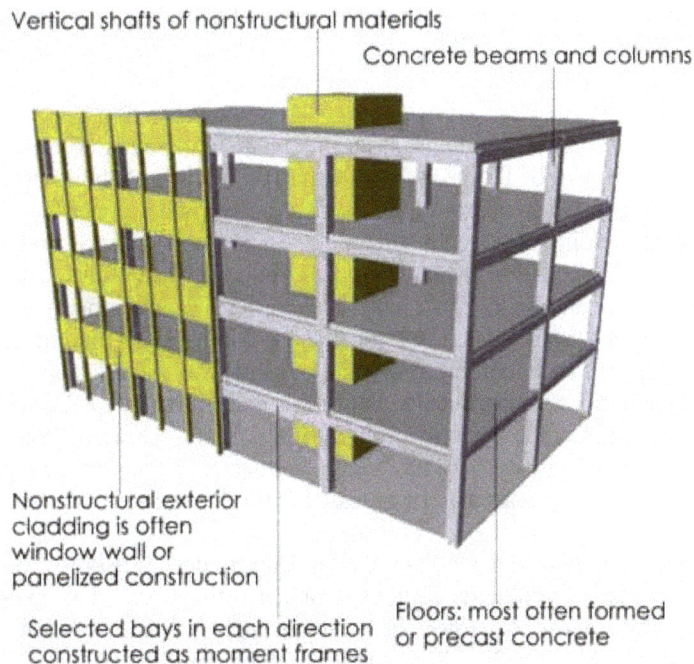

Vertical shafts of nonstructural materials

Concrete beams and columns

Nonstructural exterior
cladding is often
window wall or
panelized construction

Selected bays in each direction
constructed as moment frames

Floors: most often formed
or precast concrete

Figure 12.1-1: Building Type C1: Concrete Moment Frames

Variations Within the Building Type

The primary variation within this type is the type of frame and the number of frames included. Frames can range from column-girder systems of one bay on each face of the building to systems that employ every column coupled with two-way slabs. Frames classified by code as ductile or semi-ductile by code beginning in the late 1960s and early 1970s are far more constrained in configuration due to prescriptive rules governing girder configuration, strong column-weak beam, and limitations on joint shear.

Floor and Roof Diaphragms

The floor and roof diaphragms in this building type are essentially the same as the bearing wall system, and are almost always cast-in-place concrete. The diaphragms are stiff and strong in shear because the horizontal slab portion of the gravity system is either thick or frequently braced with joists. However, one way joist systems could be inadequate in shear in the direction parallel to the joists. Collectors are seldom in place and transfer of load from diaphragm to shear wall must be carefully considered.

Foundations

There is no typical foundation for this building category. Foundations could be found of every type depending on the height of the building, the span of the gravity system and the site soil.

12.2   Seismic Response Characteristics

This building type must be separated into older frame systems, often not even designed for lateral loads and including few, if any, features that would assure ductile behavior, and frames specifically designed to exhibit ductility under seismic loading. Rules for design of ductile concrete frames were developed during the 1960s.

Older, non-ductile frame buildings, assuming an insignificant amount of concrete or masonry walls are present, will be far more flexible than other concrete buildings, and will probably be relatively weak. Most importantly, columns are often not stronger than beam or slab system, forcing initial yielding in these key elements. In addition, unless spiral ties were used, the column will typically fail in shear before a flexural hinge can form. Buildings with these characteristics are among the most hazardous in the U.S. inventory and are in danger of collapse in ground motion strong enough to initiate shear failures in the columns. Buildings of this type that are configured such that initial hinging occurs in the floor system will exhibit stiffness and strength degradation and large drifts, but unless exceptionally weak, are far less likely to collapse. The ratio of the inherent strength of the frame—designed for lateral loads or not—compared to the seismic demand has a large influence on the performance, and frames in low and moderate seismic zones may be at less risk for this reason.

Semi-ductile frames, with some but not all of current design features for concrete frames, likely will perform better, particularly if the columns are protected by basic strength and are designed to be flexurally controlled. However, many of these early concrete frames may be excessively weak and suffer from high ductility demands which could have serious consequences if a soft or weak story is present due to architectural configuration or column layout.

Buildings with "fully ductile" frames are expected to perform well, unless vertical or horizontal configuration irregularities concentrate inelastic deformation on certain structural components.

12.3   Common Seismic Deficiencies and Applicable Rehabilitation Techniques

See Table 12.3-1 for deficiencies and potential rehabilitation techniques particular to this system. Selected deficiencies are further discussed below by category.

Global Strength

Although lack of ductility is the overwhelming deficiency for this building category, low strength may contribute to poor performance. It is difficult to add significant strength within the confines of the existing frames and most often new elements of braced frames or shear walls are added in these buildings.

Global Stiffness

See *Global Strength*.

Configuration

The most common configuration issue in this building type is a soft or weak story created by a non-typical story height. If the building is not to receive new walls or frames as part of a global retrofit, such configuration deficiencies can be minimized or eliminated with local strengthening of columns.

Load Path

There are no load path issues particular to this building type.

Component Detailing

The major deficiencies of this building type are due to inadequate component detailing, namely the structural components of the frame. Current requirements for "ductile frames" include capacity design techniques to assure flexural yielding in both girders and columns, as well as, for the most part, to limit yielding to the floor system. Retrofit procedures to obtain this ductile behavior of the frames are difficult, disruptive, and expensive, and are therefore seldom done. In high seismic zones, retrofit of these buildings is normally accomplished by adding new, stiffer lateral force-resisting elements that prevent significant ductility demand on the frames.

Some research has been completed to investigate methods of retrofit for concrete moment frames (see Section 12.4.6), and in lower seismic zones where demands over and above gravity designs are not great, local strengthening and confinement of frame elements may be practical.

Diaphragm Deficiencies

The most common diaphragm deficiency in this building type is a lack of adequate collectors. The addition of effective collectors in an existing diaphragm is difficult and disruptive. Existing strength to deliver loads to the shear walls should be studied carefully before adding new collectors.

Table 12.3-1: Seismic Deficiencies and Potential Rehabilitation Techniques for C1 Buildings

Deficiency		Rehabilitation Technique				
Category	Deficiency	Add New Elements	Enhance Existing Elements	Improve Connections Between Elements	Reduce Demand	Remove Selected Components
Global Strength	Insufficient number of frames or weak frames	Concrete/masonry shear wall [12.4.2] Steel braced frame [12.4.1] Concrete or steel moment frame Steel moment frame	Increase size of columns and/or beams [12.4.5]		Remove upper story or stories [24.2] Seismically isolate [24.3] Supplemental damping [24.4]	
Global Stiffness	Insufficient number of frames or frames with inadequate stiffness	Concrete/masonry shear wall [12.4.2] Steel braced frame [12.4.1] Concrete or steel moment frame	Increase size of columns and/or beams [12.4.5] Fiber composite wrap of gravity columns [12.4.4] Concrete/steel jacket of gravity columns [12.4.5] Provide detailing of all other elements to accept drifts		Supplemental damping [24.4]	Remove components creating short columns
Configuration	Soft story or weak story	Add strength or stiffness in story to match balance of floors				
	Re-entrant corner Torsional layout	Add floor area to minimize effect of corner Add balancing walls, braced frames, or moment frames		Provide chords in diaphragm		

Table 12.3-1: Seismic Deficiencies and Potential Rehabilitation Techniques for C1 Buildings

Deficiency		Rehabilitation Technique				
Category	Deficiency	Add New Elements	Enhance Existing Elements	Improve Connections Between Elements	Reduce Demand	Remove Selected Components
Configuration (continued)	Incidental walls failing or causing torsion	Add balancing walls, braced frames, or moment frames	Uncouple incidental walls Convert incidental walls to lateral elements walls			Remove incidental walls
Load Path	Inadequate collector	Add or strengthen collector [12.4.3]				
Component Detailing	Lack of Ductile detailing-- general		Perform selected improvements to joints [12.4.6]		Seismic isolation [24.3]	
	Lack of ductile detailing: Strong column-weak beam		Jacket columns [12.4.4]			
	Lack of ductile detailing: Inadequate shear strength in column or beam		Fiber composite wrap [12.4.4] Concrete/steel jacket [12.4.5]			
	Lack of ductile detailing: Confinement for ductility or splices		Fiber composite wrap [12.4.4] Concrete/steel jacket [12.4.5]			
Diaphragms	Inadequate in-plane shear capacity	Concrete or masonry shear wall [12.4.2] Braced frame [12.4.1] Moment frame	R/C topping slab overlay FRP overlays [22.2.5]			

Deficiency		Rehabilitation Technique				
Category	**Deficiency**	**Add New Elements**	**Enhance Existing Elements**	**Improve Connections Between Elements**	**Reduce Demand**	**Remove Selected Components**
Diaphragms (continued)	Inadequate chord capacity	New concrete or steel chord member [12.4.3]				
	Excessive stresses at openings and irregularities	Add chords [12.4.3]				Infill openings [22.2.4]
Foundations	See Chapter 23					
[ ] Numbers noted in brackets refer to sections containing detailed descriptions of rehabilitation techniques.						

Table 12.3-1: Seismic Deficiencies and Potential Rehabilitation Techniques for C1 Buildings

12.4   Detailed Description of Techniques Primarily Associated with This Building Type

12.4.1   Add Steel Braced Frame (Connected to a Concrete Diaphragm)

Deficiencies Addressed by the Rehabilitation Technique

Inadequate global shear capacity

Inadequate lateral displacement (global stiffness) capacity

Description of the Rehabilitation Technique

Addition of steel diagonal braced frames to an existing concrete moment frame building is a method of adding strength and/or stiffness to the structural system. The steel braces can be added without a significant increase in the building weight. The new braces will commonly be some configuration of concentric braced frame (CBF); it is very uncommon to use an eccentrically braced frame (EBF) due to costs and difficult detailing issues associated with the link mechanism. Any of a variety of diagonal brace configurations may be used, as well as a variety of brace member section types.   Figure 12.4.1-1 shows several common configurations. Common connections of the new brace to the existing concrete structure are shown in Figures 12.4.1-2A, 12.4.1-2B, and 12.4.1-2C.

Design Considerations

Research basis: Design of the lateral force-resisting system for the building should account for the stiffness of both the braced frame system and the existing concrete moment frames. While basic research regarding adding braced frames at the interior of a concrete moment frame building has not been identified, research in the 1980s at the University of Texas at Austin on frames at the exterior façade demonstrated the ability of the new steel braced frames to increase the deformation capacity of the non-ductile concrete frames (Jones and Jirsa, 1986). A schematic detail of the connection used in this testing is shown in Figure 12.4.1-3.

Braced frame – concrete frame interaction: Most designs of braced frame retrofits will be governed by maintaining drifts within the range of acceptability for the existing concrete elements. This can be accomplished by setting up a model that includes both the stiffness of the braced frame and of the concrete frame and meeting acceptability requirements for the displacements (or psuedo forces) in the concrete elements. Some engineers prefer to consider only the braced frames as a new lateral system, determine real drift demand for that system, and then check that drift for acceptability superimposed on the existing frame

In taller buildings, the possible incompatibility between vertical cantilever behavior of discrete braced frames and the existing moment frames must be assessed. Existing beams or slabs, if unusually thick, that frame directly into the ends of new braces may restrain the global flexural deformation of the brace and require special consideration. Finally, due to the wide variety of nonlinear behavior of braced frames that is dependent on configuration and detailing, it may be difficult to obtain an adequate understanding of overall deformation compatibility using linear methods.

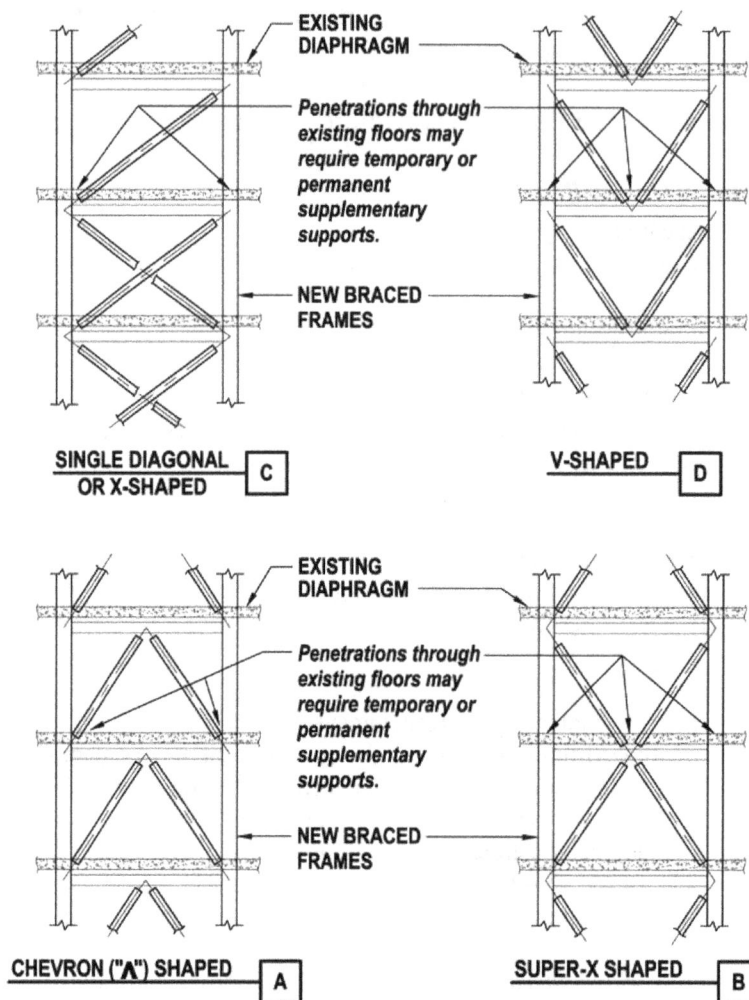

Figure 12.4.1-1: Typical Braced Frame Configurations

Braced frame location: The new braces may be located on the exterior or interior of the building. An exterior location generally allows for easier construction access and perhaps less cost, but is visible, exposed to the environment and probably will impact exterior building finishes. Braces placed parallel to the façade can be connected to the exterior faces of perimeter spandrel beams, perimeter moment frames or edges of floor and roof diaphragms relatively easily, but will most likely cross in front of some windows. Alternatively, exterior bracing may be placed as buttresses, perpendicular to the existing façade. This configuration will probably require more extensive new collectors to deliver lateral forces from the diaphragms but may allow creation of new stair or elevator shafts, or perhaps additional floor area. For projects that include expansion of or additions to the existing building, the new braces could be located in the adjacent new construction, tied to the existing building.

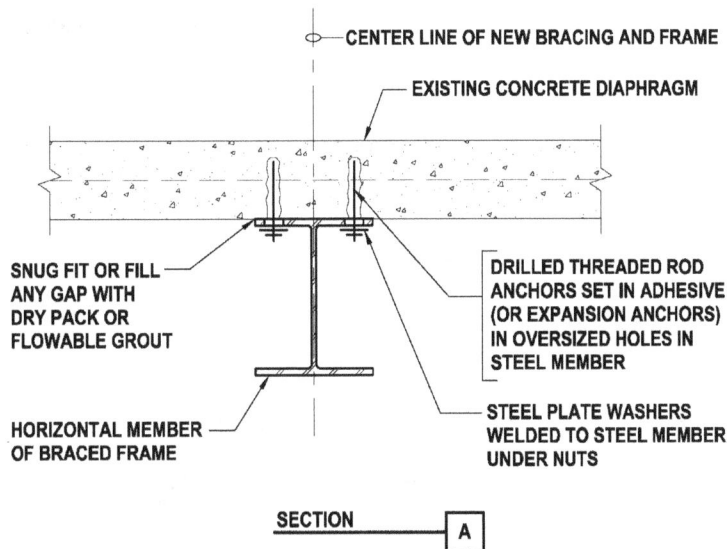

Figure 12.4.1-2A: Typical Connection to Concrete Diaphragm

Interior braces will most commonly be located along existing frame lines, particularly at moment frame bays. This will allow for best use of any existing diaphragm chords and collectors and for best moment frame – braced frame interaction. In some cases however, interior braces will be located offset from existing column-frame lines to minimize direct impact on existing structural or architectural components or to simplify the frame-diaphragm connections.

The addition of new braced frames to a building will always impact the architectural character and functional uses of the building to some degree. Selection of preferred brace locations must be made considering these issues, such as space layout, corridor locations, doorways, windows, main M/E/P distribution runs, as well as the structural or construction considerations.

Braced frame configuration and member section type: In most cases where diagonal steel braces are used to strengthen or stiffen a concrete frame building, a complete braced frame including horizontal beam and column members, as well as the diagonal braces themselves, is employed. Installation of diagonal bracing members between existing concrete columns is difficult because transfer of a large concentrated axial force from the concrete members through a localized connection with a limited number of anchors is rarely feasible. The steel columns are often continuous, passing through the floors, from foundations up to the roof or highest level required to avoid transfer of load from the steel system in and out of the concrete at each floor. In some cases, columns can be connected to adjacent concrete columns, but if the concrete column becomes part of the primary chord, reinforcing splice locations must be carefully considered.

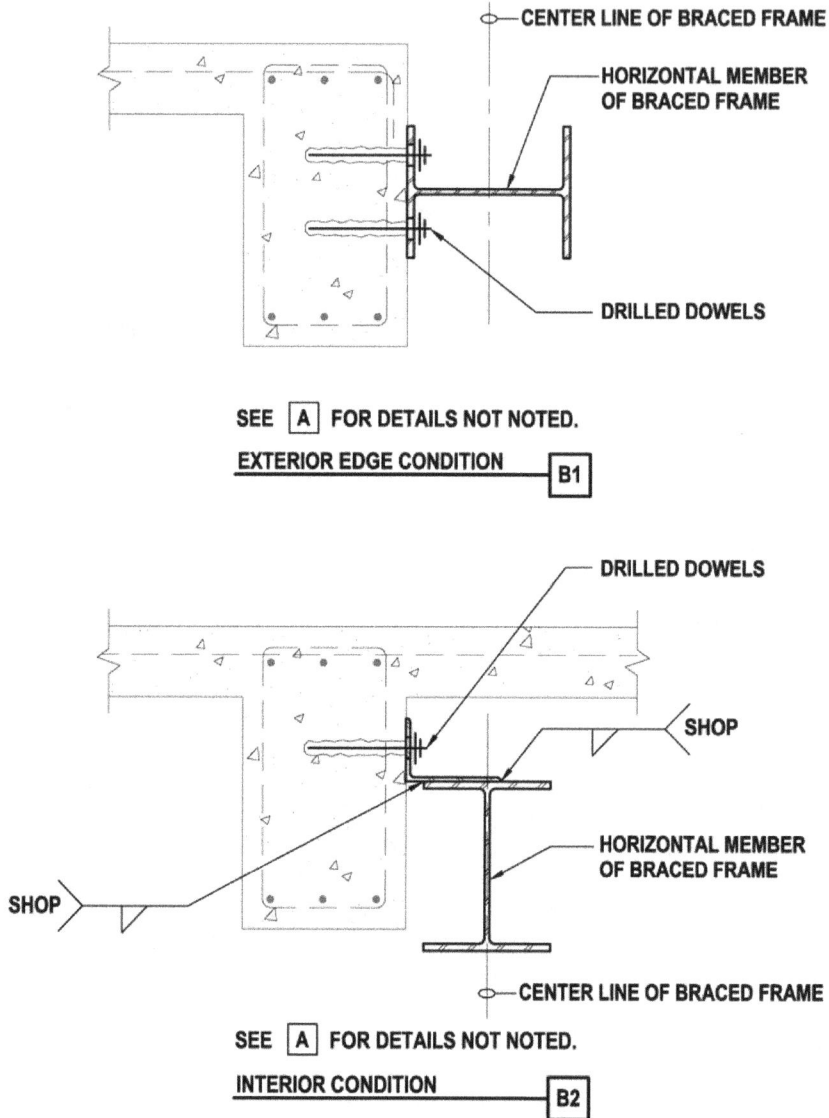

Figure 12.4.1-2B: Typical Connection to Existing Concrete Beam

New steel horizontal elements are similarly needed to facilitate the connections of the diagonal and to transfer forces from each floor into the frame. These steel elements are generally placed below the floor and roof diaphragms or adjacent to beams or spandrels. The diagonal steel braces may be placed in any of the commonly used configurations indicated in Figure 12.4.1-1;

**EXISTING CONCRETE
BEAMS OF 2-WAY MOMENT
FRAME SYSTEM ABOVE & BELOW**

DRILLED DOWELS

**SHOP OR
FIELD**

**SHOP OR
FIELD**

**CENTER LINE OF
BRACED FRAME**

SEE FIG. A FOR DETAILS NOT NOTED.

PLAN - TWO-WAY MOMENT FRAME CONNECTION C1

**EXISTING BEAMS OF ONE-WAY
MOMENT FRAME ABOVE
& BELOW**

DRILLED DOWELS

**BRACED FRAME
COLUMN**

**CENTER LINE OF
BRACED FRAME**

SEE FIG. A FOR DETAILS NOT NOTED.

PLAN - ONE-WAY MOMENT FRAME CONNECTION C2

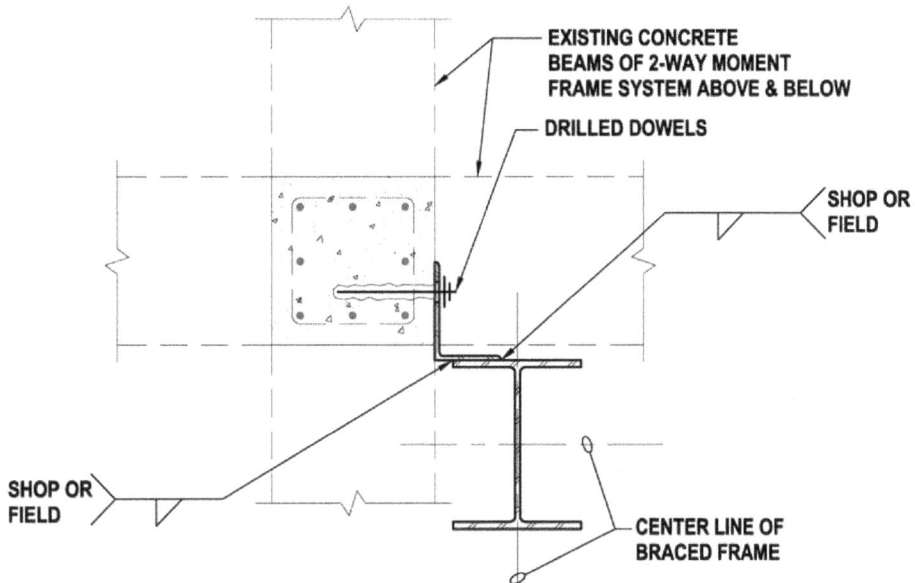

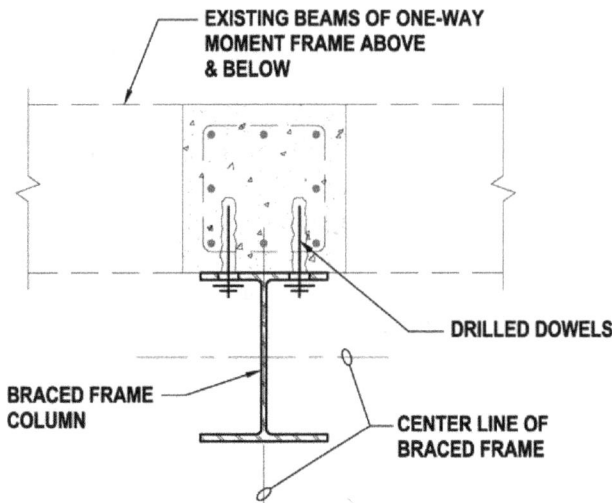

Figure 12.4.1-2C: Typical Connection to Existing Concrete Column

12-11

CL (E) COLUMN

FACE OF (E) CONC. COLUMN

MC SHAPE

TYP. TO MC AND
GUSSET PL

GUSSET PL AT EACH FLANGE

WEB TO
PL, TYP.

W - SHAPE   WT - SHAPE

TYP.

W - SHAPE

ERECTION BOLTS & PL

TYP.

ELEVATION

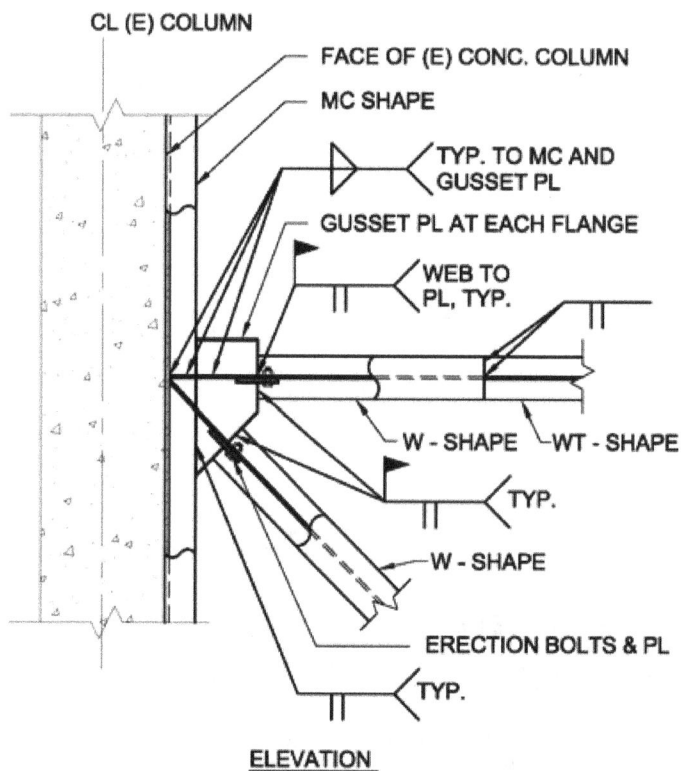

Figure 12.4.1-3: Test Specimen Connection Detail for Braced Frame

single diagonal or X-shaped, V-shaped, chevron (inverted-V) shaped, or "super-X" shaped (a combination of chevron and V braces in alternate stories forming a two-story X shape). Two-story X-bracing has the advantage over V- or inverted-V-bracing should a compression brace buckle. In the latter configurations, the remaining tension brace has an unbalanced vertical component that has to be resisted by the beam. For an X-bracing configuration, even if a compression brace buckles, the force in the remaining tension brace is transmitted directly to the tension brace on the opposite side of the beam.

Configuration will be selected based on consideration of structural issues, relative strength, stiffness and performance, as well as of several other issues including aesthetics, conflicts with doorways, corridors or windows, M/E/P systems, or the number of connections and penetrations. Column and beam members are often W-shapes, but may be other shapes such as channels or hollow structural section (HSS) tubes to improve aesthetics or to ease detailing. Diagonal members may be of any typically used sections including W-shapes, hollow (HSS) pipes or tubes, or double channels, angles or HSS tubes.

Buckling-Restrained Braced Frames (BRBFs), in which steel plates or cruciform shaped braces are surrounded by unbonded concrete in such a way as to prevent bucking of the brace, act essentially the same in tension and compression. The yield strength of a bay braced with one or more of these braces can be relatively accurately predicted. In situations where many, lightly loaded braces will be employed, sufficient global strength can be obtained by designing the braces to yield prior to yielding or other failure of the existing columns, preventing the need to retrofit the columns.

Detailing Considerations

Connection to existing concrete floor and roof diaphragms: A significant concern associated with installing a new steel braced frame in a concrete building is the connection of the beam at the top of the frame in each story to the underside of the existing concrete diaphragm overhead. The primary concern is that a relatively large shear force must be transferred from the overhead diaphragm into the new steel bracing below through a relatively localized connection using discreet anchors/bolts. The connection is generally made by one or more rows of concrete anchors as shown in Figure 12.4.1-2A. Typically, the anchors are threaded rods set in epoxy, but drilled expansion anchors may be used if they provide sufficient force transfer capacity and adequate testing to show they can resist cyclic loading. An alternate connection method, installed from the top down, consists of providing large holes in the concrete slab to expose the steel beam sufficiently to installed welded dowels to the top flange. The hole is then backfilled with cementitious or epoxy grout. In many cases however, the shear capacity of the existing concrete diaphragm is inadequate to deliver the relatively large shear force within the length of the braced frame. In those cases, a collector will be required (refer to Section 12.4.3).

Connection to existing moment frames: New braced frames are often located on or alongside of the existing moment frame lines. This generally allows for better use of the existing collectors (beams) to deliver diaphragm forces to the bracing and, perhaps, use of the existing frame columns and footings to help resist overturning and uplift forces. It is generally preferable to locate the new braces alongside of the existing moment frames instead of as an "infill" within the width of the existing concrete frame beams and columns.

For diagonal braces installed in an "infill" configuration, it is often extremely difficult to transfer large seismic forces from the surrounding concrete members through very localized connections with a limited number of discreet anchors. Also, if steel columns or vertical members are used in the "infill" frame, it is virtually impossible to provide vertical continuity from floor to floor through the existing concrete beams. Furthermore, if connections of sufficient strength can be made, the anchors must be threaded into or through the relatively densely reinforced beams and columns and, where collector strengthening is required, the added collector components will not connect directly to the new braces. In addition, physical installation and fit-up of the new braces and their connecting gusset plates often becomes significantly more difficult.

These detailing difficulties can be reduced or avoided by placing the new braced frame alongside of the concrete moment frame. In most cases, placing the new bracing alongside an exterior frame will allow the greatest ease in detailing. Bracing in this location will almost always require installation of a complete new steel braced frame instead of only the new diagonal braces themselves. In this configuration, the connection of the concrete diaphragm to steel braced frame can be made as discussed above or by installing anchors into the side of the adjacent concrete frame beam as shown in Figure 12.4.1-2B. Braced frame overturning forces are carried directly by the new steel column members. However, if concrete beam framing occurs in two directions, the new steel columns will generally need to be offset from the existing concrete column on a 45-degree diagonal to provide continuity of the new column through the floors without interference with the existing concrete beams. Overturning resistance can be obtained by connecting the new steel column to the existing concrete column (see Figure 12.4.1-2C1) and footing. For cases where the new braced frame can be placed on the exterior of the building, the new steel columns can be continuous and the connections to the adjacent concrete columns or pilasters can be made with relative ease. If diaphragm collector strengthening is required, the additional collector can be installed alongside of the existing beam line and can be connected directly to the new braced frame.

Exterior bracing at offset columns: There are some buildings where the exterior columns protrude farther out than the exterior beams. The Jones and Jirsa (1986) research can be applied in these situations, where the new steel framing is placed adjacent to the beams and in the plane of the outer portion of the protruding concrete column as shown in Figures 12.4.1-3 through 12.4.1-5. The primary challenges lie in connecting the two types of frames and delivering loads into the braced frames. As an alternative to drilling numerous holes for bolts or dowels into the concrete columns, steel lugs can be provided at each floor. In this approach, steel pipes are inserted through cores drilled through the concrete columns and filled with grout. Next, the pipes are welded to steel plates on the sides of the concrete columns, which then provide surfaces for welding to the columns of the braced frames. An example of this connection is shown in Figure 12.4.1-4. If required, horizontal forces can also be transferred directly to the braced frames through the braced frame beams. The beams are welded to steel plates, which are connected to the concrete slab or beams at the building perimeter with dowels, bolts, or lugs, as shown in Figure 12.4.1-5.

Exposed exterior braced frames require simple and clean connections that fit the architectural character of the building. Use of W-shapes for the braced frame members can eliminate gusset plates and allow direct connection of the members through complete joint penetration welds. Shop welding of the connections and on-site prefabrication of the braced frames will minimize field welding on the structure. W-shapes also simplify other architectural issues by not allowing rainwater or debris to accumulate.

Installation of additional collectors: Installation of new braced frames in a concrete frame structure, especially in one with a distributed frame system, will often result in increased diaphragm demands at the individual braces. An advantage of locating the new brace at an existing frame line is that the existing beams can then be used as a collector. However, insufficient continuity and/or laps of reinforcing steel combined with highly concentrated diaphragm demand may still require strengthening of the existing collector (refer to Section 12.4.3).

Footings: Addition of steel braced frames to an existing building will almost always require construction of new footings, or augmentation of existing ones, to resist the concentrated overturning demand. In many cases, the overturning uplift demand will require installation of tie

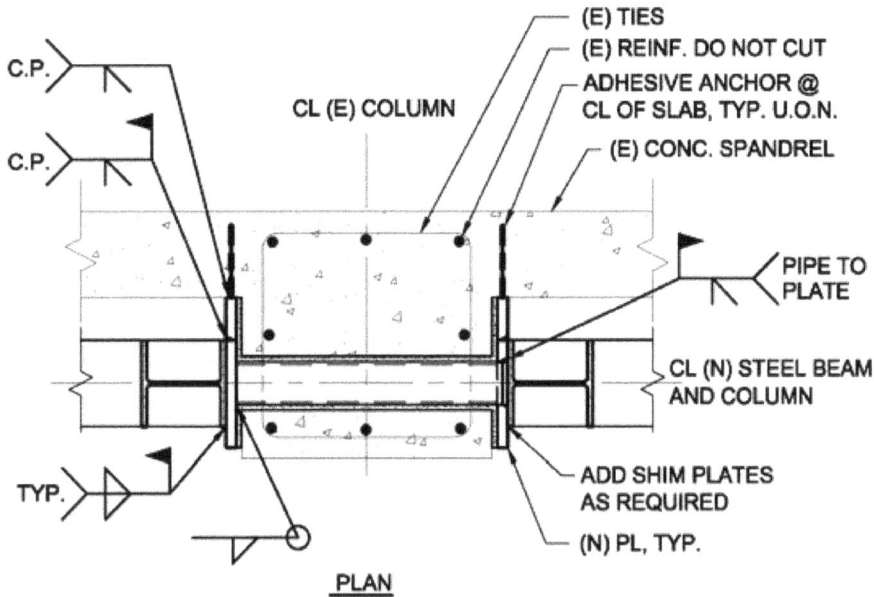

Installation Procedure:
1. Install adhesive anchor and plate on each side of column.
2. Core hole for pipe through column.
3. Install pipe with shop welded plate on one side.
4. Install plate with hole and weld to pipe.
5. Weld cap plate to pipe.
6. Weld plate attached to pipe to plate attached to adhesive anchor.
7. Fill pipe and annular space solid with nonshrink grout.

Figure 12.4.1-4: Braced Frame to Concrete Column Connection

downs. Alternatively, the new frame can be located between two existing column frame lines, instead of directly on or along one frame line, and new foundations or grade beams can be used to engage more than one existing column to resist the uplift demand.

Cost/Disruption Considerations

The cost and level of disruption associated with installation of steel bracing is generally less than that of shear walls. The number of penetrations that need to be cut through the existing concrete structure and of drilled dowels and anchors may be less than for the shear wall alternative, and the work is generally not as wet or messy. Also, it will not be necessary to prepare any existing concrete surfaces that will be in contact with new steel members. The new members are discrete and welded or bolted connections are localized. However, there will be noise and vibrations resulting from the required cutting and drilling that will make continued occupancy difficult.

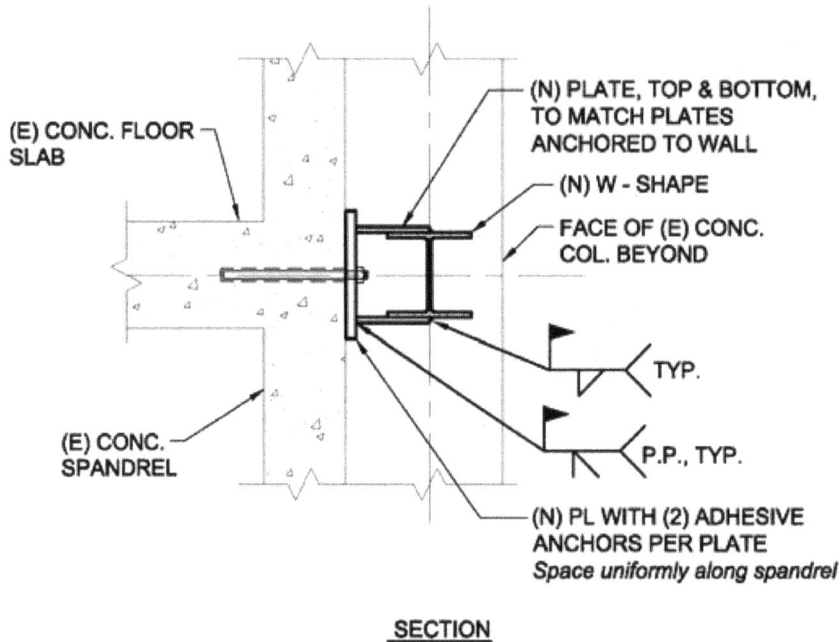

Figure 12.4.1-5: Braced Frame to Slab Connection

Invariably, some architectural and M/E/P system components will require relocation or replacement.

Construction Considerations

The primary construction consideration will be fit up and installation of the steel braces and their connections. The desire to limit the number of splice connections must be balanced against the difficulties of installing longer members such as multistory columns. Installation of the diagonal braces will require careful planning and will often require member splices in the field. Installation of drilled threaded rod or expansion anchors will require some precision and extensive use of templates and oversized holes, to assure proper fit with the steel members. In some cases, the steel members themselves could be used as the template for the anchors.

Proprietary Concerns

In general, there are no proprietary concerns related to installation of steel diagonal braced frames in a building. The one exception, however, occurs if buckling-restrained braces are used.

12.4.2   Add Concrete or Masonry Shear Wall (Connected to a Concrete Diaphragm)

Deficiencies Addressed by the Rehabilitation Technique

Inadequate global shear capacity

Inadequate lateral displacement (global stiffness) capacity

Description of the Rehabilitation Technique

Addition of shear walls to an existing concrete frame building is a common method of adding significant strength and/or stiffness to the structure. The new walls may be of cast-in-place concrete, shotcrete or fully grouted concrete masonry unit (CMU) construction.

Design Considerations

Research basis: No research focused on the overall effects of adding shear walls to existing concrete frames has been identified. The effects of surface preparation, concrete strength, and interface reinforcement on interface shear capacity between new and existing concrete were examined by Bass, Carrasquillo, and Jirsa (1985). These tests indicate that surfaces prepared with heavy sandblasting exhibit shear capacities greater than or equal to those exhibited by chipped surfaces or surfaces prepared with shear keys. Increased concrete strength resulted in increased interface shear capacity for chipped surfaces and those prepared with shear keys, but it had little effect on the shear capacity of interface surfaces prepared by sandblasting. Specimens in which drypack mortar was used exhibited a significantly smaller shear capacity than those where new concrete was cast directly against the interface. Increasing the amount or embedment depth of reinforcement across the interface resulted in greater interface shear capacity.

Frame-wall interaction: Most designs of shear wall retrofits will be governed by maintaining drifts within the range of acceptability for the existing concrete frame elements. This can be accomplished by setting up a model that includes both the stiffness of the shear walls and of the concrete frame and meeting acceptability requirements for the displacements (or psuedo forces) in the concrete frame elements. Some engineers prefer to consider only the shear walls as a new lateral system, determine the expected drift demand for that system and then check that drift superimposed on the existing frame for acceptability.

In taller buildings, the possible incompatibility between vertical cantilever behavior of discrete shear walls and the existing concrete frames must be assessed. Existing beams or slabs, if unusually thick, that frame directly into the ends of new walls may restrain the global flexural deformation of the wall and require special consideration. Two shear walls are often purposely placed in line and connected by a short beam to form a coupled shear wall system. In this system, the coupling beams are specially designed to accept significant inelastic deformations. Seldom can two such walls utilize existing beams as coupling beams due to inadequate detailing. Thus coupling beams, when employed, are installed new, as part of the system.

Frame-wall configuration: A primary design consideration is determination of whether or not the existing concrete frames may be used as an effective part of a combined system. Are the existing frame columns strong enough and/or well detailed enough to serve as the chord/boundary member of a shear wall without improvement? Are the frame beams detailed well enough to serve as coupling beams or to be incorporated into the wall itself? These considerations may limit the choices of wall-frame physical relationship: that is, should the walls be placed 1) within the plane of the existing concrete frames, 2) as vertically continuous walls alongside of, and joined to, the existing frames, or 3) as separate vertical elements independent of the frames? The first alternative is often best avoided as noted in the *Detailing Considerations* discussion below. Considering alternates 2 and 3, it must be determined if the frames are capable of becoming part of the shear wall (primarily as chord elements) or if it is beneficial to prevent direct interaction by placing the shear walls free of the existing concrete frame elements. In some cases, it is not feasible to stiffen the building into the range of acceptable deformation of the existing frames, and improvement in deformation capacity may be required in addition to the addition of new walls.

Wall location: The new walls may be placed on the exterior or interior of the building. An exterior location generally allows for easier construction access and perhaps less cost, but is visible, exposed to the environment and may impact exterior building finishes. Walls placed parallel to the façade can be connected to the exterior edges of floor and roof diaphragms or perimeter concrete frames relatively easily, but will most likely require closure or reduction in size of some windows. Alternatively, exterior walls may be placed as buttresses perpendicular to the existing façade. This configuration will probably require more extensive new collectors to deliver lateral forces from the diaphragms but may allow creation of new stair or elevator shafts, or even of additional floor area. For projects that include expansion of or additions to the existing building, the new walls could be located in the adjacent new construction.

Interior walls located along frame lines, particularly at moment frame bays will often allow for best use of any existing diaphragm chords and collectors. Beams that frame directly into the ends of new walls may behave like coupling beams as described above. In some cases, interior walls are better located offset from existing column-frame lines to minimize direct impact on existing structural or architectural components or to simplify the wall-diaphragm connections.

The addition of shear walls to a building will always impact the architectural character and functional uses of the building. Selection of preferred wall locations must be made considering these issues, such as space layout, corridor locations, doorways, windows, main M/E/P distribution runs, as well as the structural or construction considerations.

Detailing Considerations

Connection to existing concrete floor and roof diaphragms: Arguably, the most significant detail associated with installing a new shear wall in a concrete building is the connection at the top of the new wall to the underside of the existing concrete diaphragm overhead. The construction joint must be made tight, without any gapping, to facilitate transfer of shear forces from the overhead diaphragm into the new wall below and to minimize the possibility of joint slip. See the discussion under the *Research basis* section.

Typical details of this connection for a new cast-in-place concrete wall below an existing concrete flat slab are shown in Figures 12.4.2-1A and 12.4.2-1B. The vertical dowels must be sufficient to transfer forces from the existing diaphragm and from the new wall above (if it exists), to the lower wall. Shears can also be transferred across this joint with large diameter

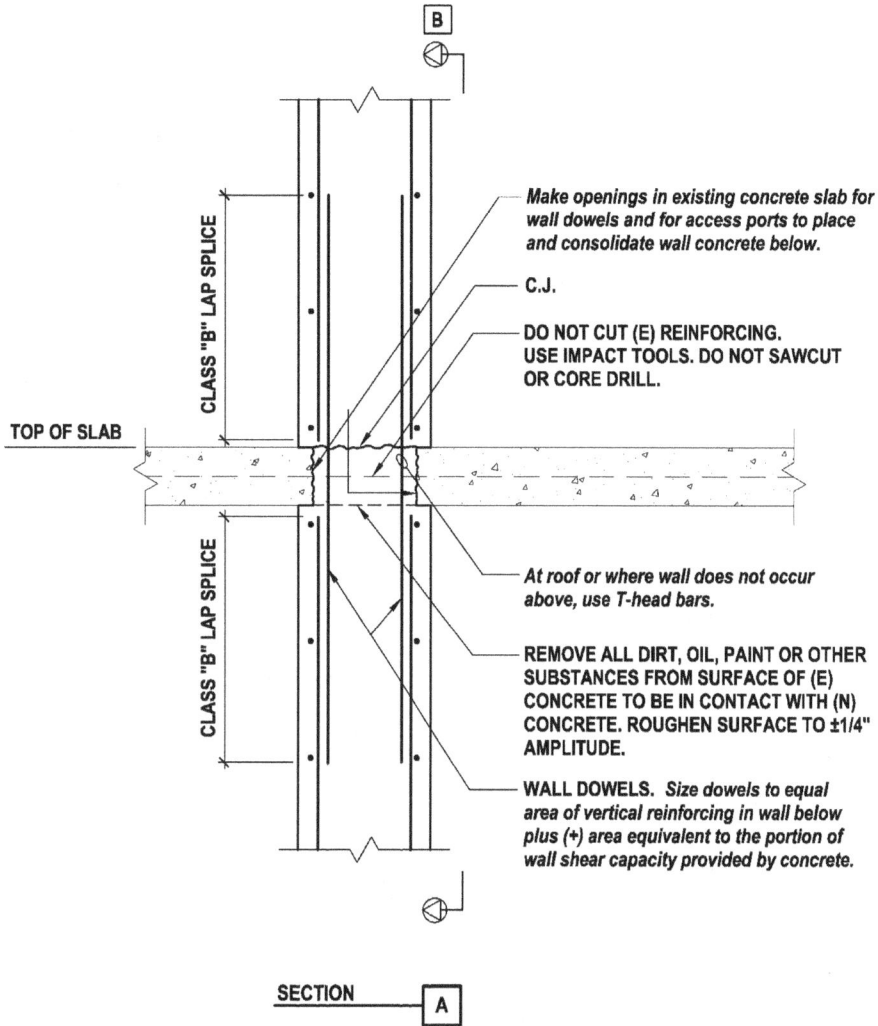

Figure 12.4.2-1A: Concrete Wall Connection to Concrete Slab

pipes or structural shapes. The holes made through the existing slab must serve not only to install the dowels, but also to allow for placement and consolidation of the wall concrete. The concrete head created by placement up to the top of slab coupled with cleaning and roughening the existing concrete contact surface by either sandblasting or chipping will provide the best joint available. The larger holes through the slab will also be more like intermittent shear keys. The holes should be drilled or made with impact tools instead of saws or core drills to avoid cutting or damaging existing slab reinforcement. Prior to cutting the holes, temporary shores may be required below the slab along each side of the row of holes. The concrete should be placed through the slab openings into the forms below, up to top of slab, to provide some head on the joint at the underside of the diaphragm.

**Set dowels with adhesive grout
or drill oversized holes
(diameter at least 1" larger
than dowel) to allow for
placement of flowable
cementifious grout.**

A

TOP OF SLAB

**6" MAX.
OFFSET**

**Chip out enlarged holes for
concrete placement and
consolidation ±4'-0" o.c. max.
Place concrete up to top of slab.
(Not needed if walls are shotcrete.)**

SECTION   B

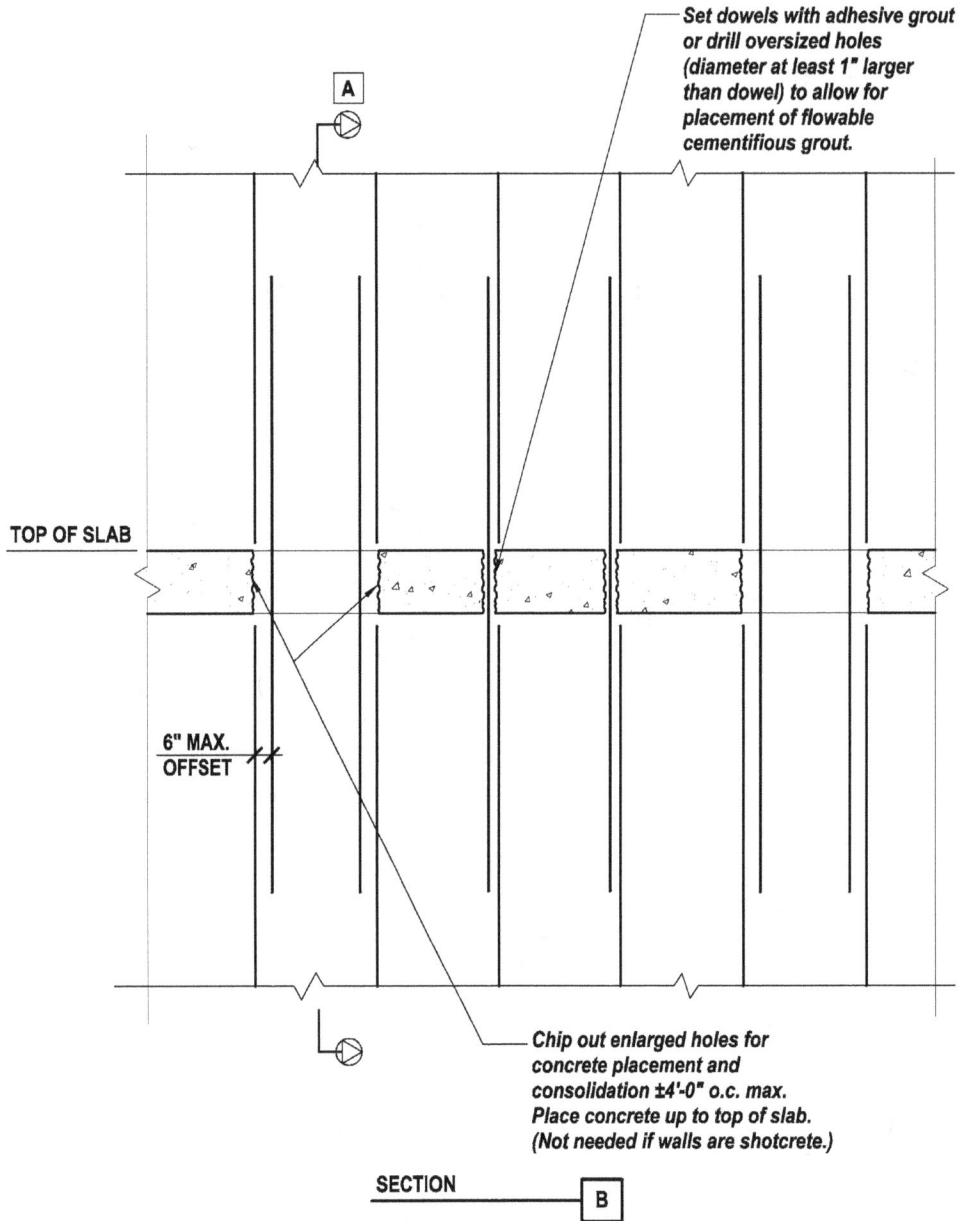

Figure 12.4.2-1B: Concrete Wall Connection to Concrete Slab – Partial Elevation View

If the new wall is shotcrete, special care is required by the nozzle operator when placing the shotcrete directly at the underside of the slab to provide a tight, well-bonded joint free of rebound or gaps. To minimize the possibility of creating a gap due to sagging, the last lift of shotcrete should be a short one. In the end, however, such a well-bonded joint is often not achieved at the slab soffit and remedial work, similar to crack-injection repairs, is likely to be needed. As an alternative, the holes through the slab needed for placement of the vertical rebar dowels could be made oversized, sufficient to allow placement of pourable, cementitious grout at the top of the shotcrete wall below, similar to the cast-in-place concrete alternative.

For CMU wall construction, the masonry units will typically be constructed up to within one or two courses of the overhead slab soffit, leaving enough of a gap to allow placement of the upper lift of grout. Preferably, the gap should be formed and grouted from above through holes in the slab similar to cast-in-place concrete alternative described above. Consolidation of the upper lift of grout should be performed through holes in the slab above. Although the gap can be dry packed from below, this is a considerably less effective alternative as confirmed by the research results noted above.

Regardless of whether the new wall is cast-in-place concrete, shotcrete or CMU, some shrinkage or sagging will probably occur creating a crack at the joint. To account for the resulting reduction in effective aggregate interlock along this joint, it may be prudent to use a lower coefficient of friction, and increase the size the vertical dowels.

Figure 12.4.2-1C shows the conditions where the existing concrete diaphragm is in a pan joist or waffle slab system instead of a flat slab. For these types of floor or roof systems, the joists or waffle ribs must be preserved to avoid shoring. However, there is likely to be more flexibility in the extent of the openings that can be made through the slab between the joists/ribs, and temporary shoring will generally not be required. Where the new wall is parallel to the joists, it is preferable to locate the wall offset from the joist as shown in the detail. The holes in the slab may be made as intermittent keys, similar to the flat slab condition discussed above, or they can be made as relatively long slots or as a continuous opening the length of the wall. Additional diaphragm to wall shear transfer capacity can be obtained by doweling into the side of the adjacent rib.

Where the new wall is perpendicular to the joists, or at a waffle slab condition, the slab can be removed between the ribs as indicated in Figure 12.4.2-1D. Since installation of continuous horizontal wall bars through the perpendicular ribs is generally not possible, installation of one or two horizontal hoop ties may be required at the upper portions of the wall between the ribs. For CMU wall construction, the masonry will stop below the joists or ribs, and the large vertical gap up to the slab, between the ribs, will be completed with poured concrete.

Connection to existing frames: New walls are often located on or alongside existing frame lines. This generally allows for better use of the existing diaphragm collectors (beams) and of the existing frame columns as wall chords or boundary elements. It is almost always preferable to locate the wall alongside the frame beams instead of as an "infill," within the width of the frame

WALL DOWELS. *Size dowels
to equal area of vertical reinforcing
in wall below plus (+) area
equivalent to portion of wall
shear provided by concrete.*

CHIP CONTINUOUS SLOT IN
EXISTING SLAB BETWEEN
JOISTS. DO NOT CUT EXISTING
REINFORCEMENT.

CLASS "B" LAP SPLICE

TOP OF SLAB

*Added drilled dowels
as required for shear
transfer from diaphragm.*

AT CONDITION PARALLEL TO
JOISTS, PROVIDE CONTINUOUS
HORIZ. WALL REINFORCEMENT
DRILLED THROUGH BEAMS.
AT CONDITION PERPENDICULAR
TO JOISTS, SEE D .

CLASS "B" LAP SPLICE

(E) JOISTS OR WAFFLE
SLAB RIBS

SECTION   C

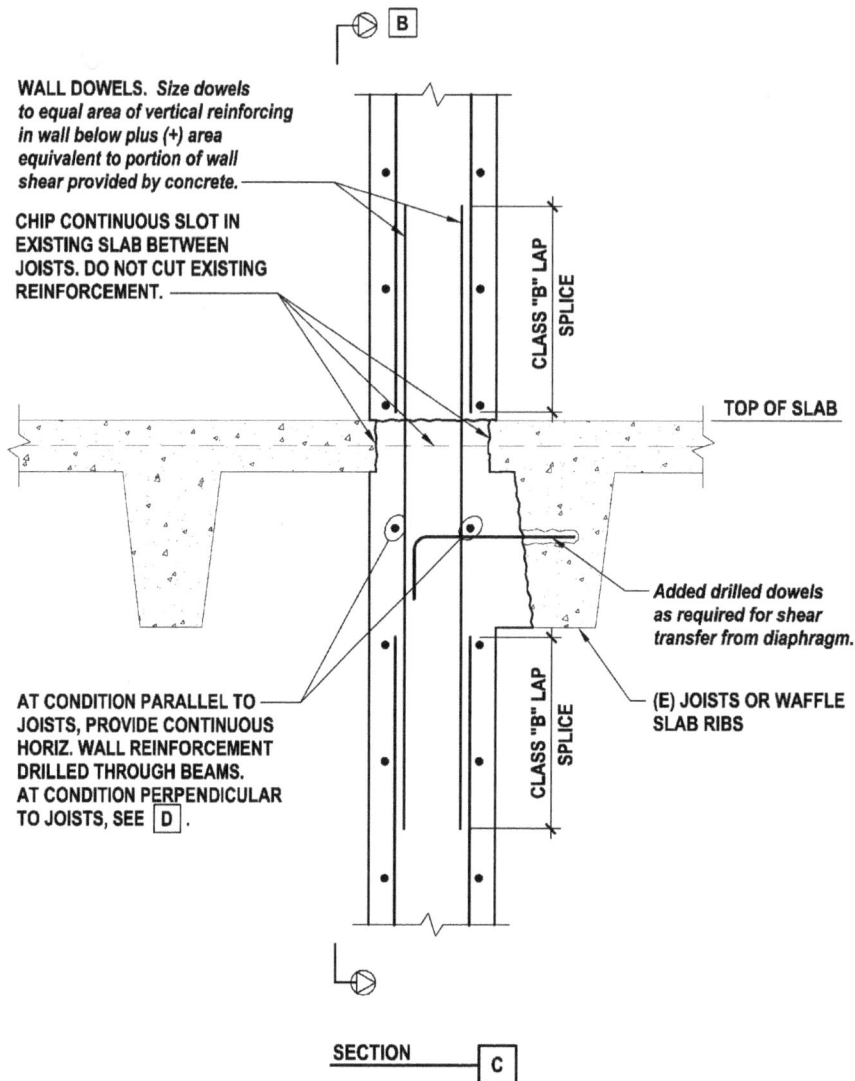

Figure 12.4.2-1C: Connection of Concrete Wall to Concrete Joists or Waffle Slab

beams and columns. When placed alongside the frame, the wall-diaphragm connections are as
discussed above, and additional shear transfer and wall chord capacity can be obtained by
doweling into the side of the beam and the column, respectively. Also, if diaphragm collector
strengthening is required, the additional collector can be installed alongside the existing beam
line and will be lead directly to the new wall. In the "infill" configuration, the vertical wall
dowels must be threaded through the relatively densely reinforced beams, concrete placement
and consolidation becomes significantly more difficult, and additional collectors do not connect
directly to the new wall.

REMOVE PORTION OF
(E) SLAB BETWEEN
CONC. JOISTS, TYP.

Typical horizontal bars not shown.

6"
MAX.

TYP. VERT. BARS
ABOVE FLOOR LEVEL

TOP OF SLAB

BOTTOM OF
RIBS BEYOND

CLASS "B"
LAP SPLICE

TYP. VERT. BARS
BELOW FLOOR LEVEL

TIES ☐ TO MATCH SIZE
& SPACING OF HORIZ.
WALL REINFORCEMENT.

WALL DOWELS. DISTRIBUTE
DOWELS EVENLY ALONG
WALL LENGTH; PLACE 3 DOWELS
MINIMUM, AT EQUAL SPACES,
BETWEEN ADJACENT (E) JOISTS.

(E) CONC. JOISTS OR
WAFFLE RIBS, TYP.

DETAIL   D

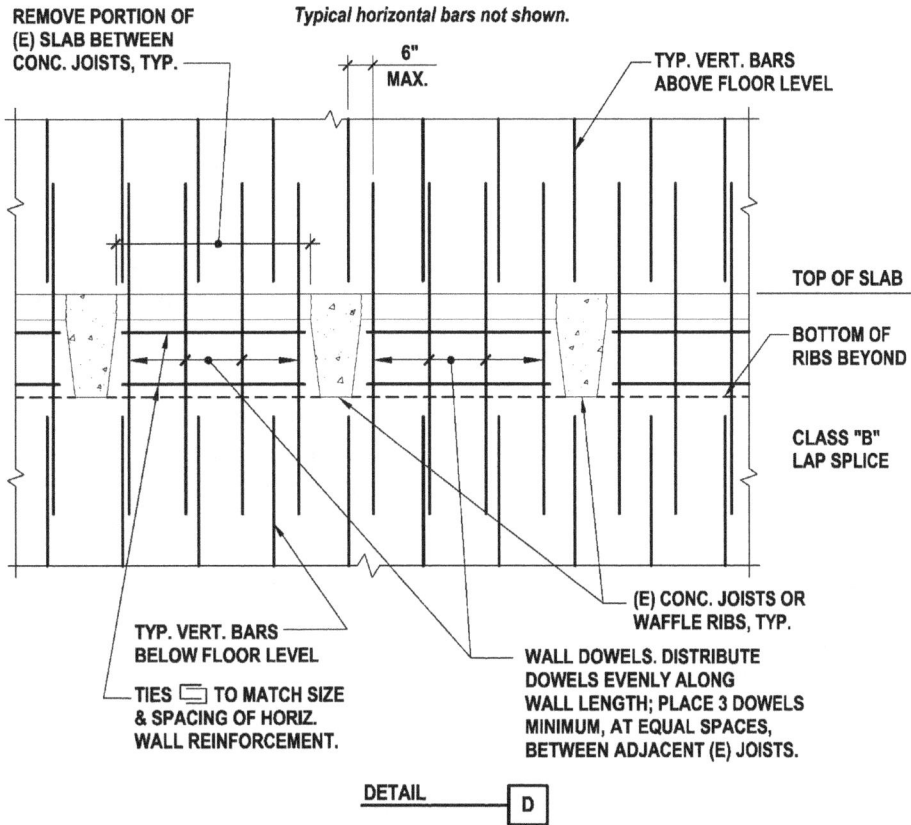

Figure 12.4.2-1D: Concrete Wall Connection to Waffle Slab – Partial Elevation View

If the existing columns have sufficient strength and appropriate reinforcement detailing, they
may be used as the wall chord or boundary element, by doweling into the column. The
effectiveness of this is limited by the amount of doweling that can be installed. In many cases,
however, the existing column will require strengthening or jacketing, or new wall chords will be
needed.

Installation of additional collectors: Installation of shear walls in a frame structure, especially in
one with a complete frame system, will result in increased diaphragm demands at the individual
walls. An advantage of locating the new wall at an existing frame line is that the existing beams
can then be used as a collector. However, insufficient rebar continuity and/or laps combined with
highly concentrated diaphragm demand may still require strengthening. Refer to Section 12.4.3.

Footings: Addition of concrete or masonry shear walls will almost always require construction
of new footings, or augmentation of existing ones, to support the added weight as well as to

resist the increased and/or concentrated overturning demand. In many cases, the overturning uplift demand will require installation of tie downs. Where the new wall is located between column frame lines, instead of directly on or along one frame line, new foundations can be used to engage more than one column to resist the uplift demand.

Cost/Disruption Considerations

In general, shotcrete walls are less expensive than cast-in-place concrete because at least one side of the wall forming is eliminated. If shotcrete can be applied against an existing wall at stair, elevator or mechanical shafts, the cost savings of shotcrete is even greater. CMU walls are generally less costly, per square foot, than either shotcrete or cast-in-place concrete walls. However, CMU walls may not provide comparable strength or stiffness, requiring the addition of more linear feet of CMU walls than either cast-in-place concrete or shotcrete walls.

Construction of new shear walls in an existing building can be very disruptive to any building occupants. Noise, vibration, and dust associated with many operations, especially cutting holes through and drilling dowels into concrete, can be transmitted throughout a concrete structure. Placing cast-in-place concrete, shotcrete or even grouted masonry is a wet process and very messy. Shotcreting in an enclosed area creates differential pressures that can spread debris beyond nominal construction barriers. Also, excavation and drilling operations and the use of mechanized and/or truck mounted equipment associated with installation of new foundations can be very disruptive.

Construction Considerations

The existing concrete surfaces to be in contact with the new concrete walls should be cleaned of all finishes, paint, dirt, or other substances and then be roughened to at least attempt to provide 1/4" minimum amplitude aggregate interlock at joints and bonded surfaces. At overhead joints where such preparations may be less effective, as discussed in the *Detailing Considerations* section above, additional dowels can be used with less roughening.

For shotcrete applications, separate trial test "panels" at the overhead joints should be included with the normal preconstruction test panels. These test joints should be cored to inspect the adequacy of the surface preparation and the joint bond. Nozzle operators should have several years experience with similar structural seismic improvement applications.

In addition to the usual concrete/shotcrete core sampling and testing, the overhead joints should be cored to allow inspection of the joint quality and determine whether or not repairs are needed.

For CMU shear walls, practical limitations on placement of wall reinforcing steel must be considered. In particular, use of "seismic comb" type of joint reinforcement (a prefabricated mesh of welded wire reinforcement used as transverse reinforcement at boundary elements of CMU walls) has often proven to be very difficult to install and the resulting rebar congestion interferes with grouting operations.

12.4.3   Provide Collector in a Concrete Diaphragm

Deficiencies Addressed by the Rehabilitation Technique

Inadequate or missing collector
Inadequate diaphragm chord capacity

Description of the Rehabilitation Technique

Addition of a new collector or strengthening of an existing collector is often needed when new steel braced frames or concrete shear walls are added to an existing building. The new collector must extend as far as necessary, often one or more bays from one or both ends of the new brace or wall, to draw the required shear demand from the existing diaphragm. The new collector will be constructed of reinforced concrete or steel, generally depending on whether the general building upgrade involves installation of new concrete shear walls or steel braced frames. The new collector will most often be installed at the underside of floor. At roofs, the collector may be placed either from below or above the roof.

In cases where the existing diaphragm chord is absent or inadequate, the mitigation approach will be similar to that used for collectors.

Design Considerations

Research basis: For new reinforced concrete collectors, see the discussion of tests by Bass, Carrasquillo, and Jirsa (1985) in Section 12.4.2.

For new steel collectors, Jiménez-Pacheco and Kreger (1993) tested single anchor connections between existing concrete and new steel members in order to examine shear transfer along the interface between these two elements. Results indicate that sandblasting the steel surface and applying a layer of epoxy at the interface between steel and concrete can substantially increase the force level at which the interface begins to slip. Also, the use of spring washers may reduce long-term anchor bolt relaxation, maintaining the first-slip force capacity over time. For applications where significant inelastic deformations are expected, a thick layer of nonshrink grout between the steel and concrete was found to increase deformation capacity, though it decreased ultimate strength slightly. Filling the annulus between the bolt and washer with epoxy resulted in greater connection stiffness than that exhibited by specimens with unfilled annuli or those filled with nonshrink grout.

Material selection - reinforced concrete or steel: In reinforced concrete buildings with some sort of concrete slab floor system, especially one with joists, waffle ribs or beams crossing the path of the collector, the most common material choice for the new collector is reinforced concrete. Often, this choice is made because concrete is aesthetically compatible with the surrounding structure, especially in a condition exposed to view. However, concrete is selected principally because it is compatible with the deformation characteristics of the diaphragm it is connected to. A concrete collector is bonded to, and is integral with, the concrete slab diaphragm and the strain deformations of the collector are the same as the deformations of the diaphragm system. At a steel plate collector, the elongation of the plate is not compatible with the diaphragm slab. As the collector load accumulates towards the connection to the new wall or brace, the elongation of the plate accumulates as well. The threaded rod anchors connecting the plate collector to the diaphragm in the zone of greatest elongation can become overloaded to failure by the plate bearing on the bolts. This can lead to a "zipper-like" failure mode as the adjacent anchors assume the load of the failed anchors and become overloaded in turn. This behavior can occur even at relatively short collectors if the elongation exceeds the available annulus gap around the anchor. To avoid this, special detailing is required as discussed in the *Detailing Considerations* section below.

Impact on architectural and M/E/P systems and components: A new collector often must extend one or more entire bays away from the new wall or brace in order to draw the necessary load from the existing concrete diaphragm. Installation of the new collector at the underside of the existing floor slab impacts any existing ceilings, partitions, ductwork, plumbing, lighting, etc., located along its entire length. As a result, the new collectors will often have a greater impact on the building's other systems than the new walls or braces themselves. Furthermore, consideration of these impacts will often affect placement of the new walls or braces. In many cases, the new walls and their associated collectors are located along the exterior edge of the building specifically to avoid or minimize these impacts on other building systems, especially in a case where building occupancy is maintained during the construction.

In some cases, it may be possible to locate the new collector at the top surface of the existing diaphragm. At roofs, new collectors can be placed on top of the roof diaphragm, provided that any conflicts with roof mounted equipment, pads or penthouses can be accommodated or avoided. More importantly, placement of collectors on top of the roof slab requires careful consideration of the impact on roof drainage and waterproofing systems. At floors, the opportunity exists if a new concrete topping or structural overlay is proposed. In this case, the reinforcement for the collector can be embedded in the topping. Also, if a new raised floor system is being installed, it may be possible to locate a new collector in the space beneath the new floor.

Weight of new collector: The gravity load capacity of the existing slab, waffle ribs, joists or adjacent beams must be adequate to support the additional weight of the new collector, especially for a new concrete collector that may represent a considerable load. In some cases, the new collector may need to be designed to support itself as it spans between existing girders or columns. In others, it may be required to adjust the location of the collector, and the new shear walls, if the existing floor or roof slab cannot support the new loads.

Detailing Considerations
Connection of a reinforced concrete collector to existing concrete diaphragms or collectors: A typical detail of the installation of a new reinforced concrete collector to the underside of an existing concrete slab diaphragm is shown in Figure 12.4.3-1. The primary considerations are to provide a good bond between the new and existing concrete and to provide adequate access ports for concrete placement and consolidation. The contact surface must be thoroughly cleaned and roughened for good shear transfer performance. It is best to place concrete from above through pour ports made in the diaphragm. The ports will need to be at least 4 inches in diameter. Care must be taken to locate existing diaphragm reinforcement before cutting the ports to avoid cutting any bars in what is likely to be a lightly reinforced slab.

THROUGHLY CLEAN AND
ROUGHEN EXISTING CONCRETE
SURFACE TO BE IN CONTACT
WITH NEW CONCRETE

POUR HOLES
*Use for consolidation access.
Do not cut any existing
diaphragm rebar.*

DRILLED DOWELS
*Size and number to meet
collector demand.*

EXISTING CONCRETE
SLAB/DIAPHRAGM

LONGITUDINAL
COLLECTOR
REINFORCING
BARS

NEW REINFORCED
CONCRETE
COLLECTOR

STANDARD HOOKS

CAPPED STIRRUP TIES

INTERMEDIATE HAIRPIN
CROSSTIES

SECTION

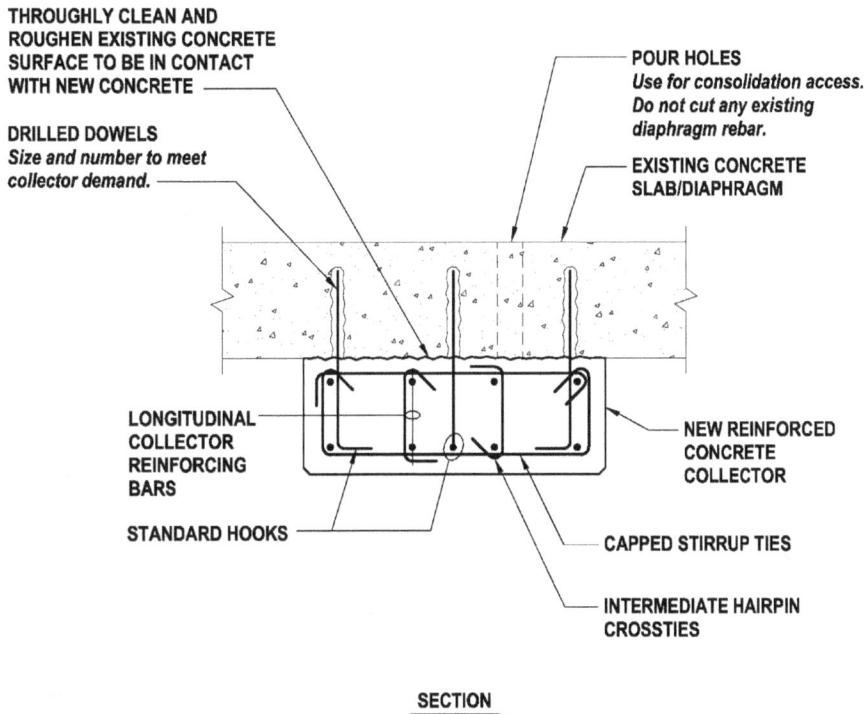

Figure 12.4.3-1: Concrete Collector at Concrete Slab

The required length of collector will be determined primarily by the existing diaphragm shear capacity. Lightly reinforced diaphragms can deliver only a limited load per foot, requiring long collectors. Also, for thin diaphragm slabs, the shear capacity of each drilled dowel will be limited, requiring more dowels. If the collector crosses any existing beam or girder, a splice must be made through the existing member. Horizontal holes can be drilled through the member and dowels installed to lap with the main collector reinforcing bars on each side. Care must be taken to avoid cutting any reinforcement, either main longitudinal bars or stirrups, in the existing beam.

If the existing floor or roof diaphragm is a waffle or pan joist system, the continuous collector will almost always be placed below the ribs, as shown in Figure 12.4.3-2, to avoid excessive drilling and rebar splicing. In this condition, the voids between the ribs, above the dropped collector, will be filled with reinforced concrete. Advantages of this condition are that the drilled dowels can be installed into the sides of the ribs instead of the relatively thin cover slab, and making pour ports through the slab is likely to be less problematic. Also, although the new collector may weigh more in this condition, the waffle or joist ribs are much more likely to have adequate strength to support the added weight.

CLEAN AND ROUGHEN EXISTING
CONCRETE SURFACES TO BE IN
CONTACT WITH NEW CONCRETE

4- ☐ EACH WAY AT EACH
WAFFLE DOME.

DRILLED DOWELS,
2 MIN., EACH SIDE OF
WAFFLE DOME

POUR HOLE. USE FOR
CONSOLIDATION ACCESS.
DO NOT CUT EXISTING
REINFORCING.

EXISTING WAFFLE
SLAB RIBS

CAPPED STIRRUP TIES

SUPPLEMENTAL COLLECTOR
BARS AT 12" MAX. SPACES
TOP AND BOTTOM

*Width of new concrete collector
as required or as best suited for forming*

MAIN COLLECTOR BARS
ALIGNED WITH FRAME
AND/OR NEW SHEAR
WALL BEYOND.
Alternate:
Use an embedded steel
plate with welded studs.

SECTION

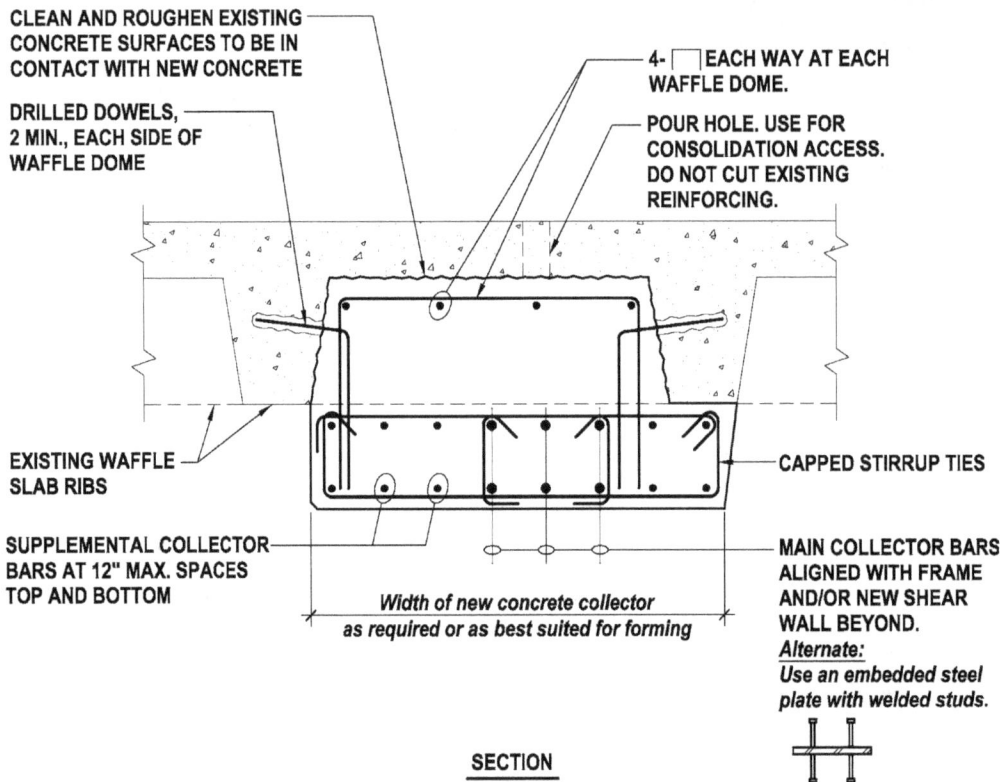

Figure 12.4.3-2: Concrete Collector at Waffle Slab

In many cases, the new collector will occur along an existing beam or girder line. Often, this will occur if the task is to strengthen a diaphragm edge chord or if the new walls occur at the building's exterior. Figure 12.4.3-3 shows two generic conditions that can be used in this case. In this condition, the dowels will always be placed into the beam, and the combined beam-collector member will easily be designed support the added weight. However, special care should be taken to avoid cutting any beam stirrups or slab diaphragm reinforcement, especially at an exterior edge condition, with the pour holes.

In any of these collector configurations, a significant portion of the main reinforcement can be provided by a steel plate instead of by bar reinforcement (refer to Figure 12.4.3-2). This option may be best for conditions of very high loads, where installation of a high strength steel plate may be preferred over placement of many large bars.

Connection of a steel collector to existing concrete diaphragms: Steel plate also may be used as the collector in lieu of a reinforced concrete member. A steel collector will have to be installed in

POUR HOLES
DO NOT CUT EXISTING
REINFORCEMENT.

NEW COLLECTOR BARS

NEW HAIRPIN TIES

CLEAN AND ROUGHEN
SURFACE OF EXISTING
CONCRETE IN CONTACT
WITH NEW CONCRETE

NEW CONCRETE
COLLECTOR BOTH
SIDES OF EXISTING
COLLECTOR/BEAM
AS REQUIRED.

DRILLED DOWELS

SIDE CONDITION   A

POUR HOLES
EACH SIDE

NEW CAPPED
STIRRUPS

±2" DIAMETER HOLES
THROUGH EXISTING BEAM.
FILL WITH FLOWABLE
GROUT.
Alternate:
Install drilled dowels
staggered each side,
similar to A .

DRILLED DOWELS

SIDE AND BOTTOM CONDITION   B

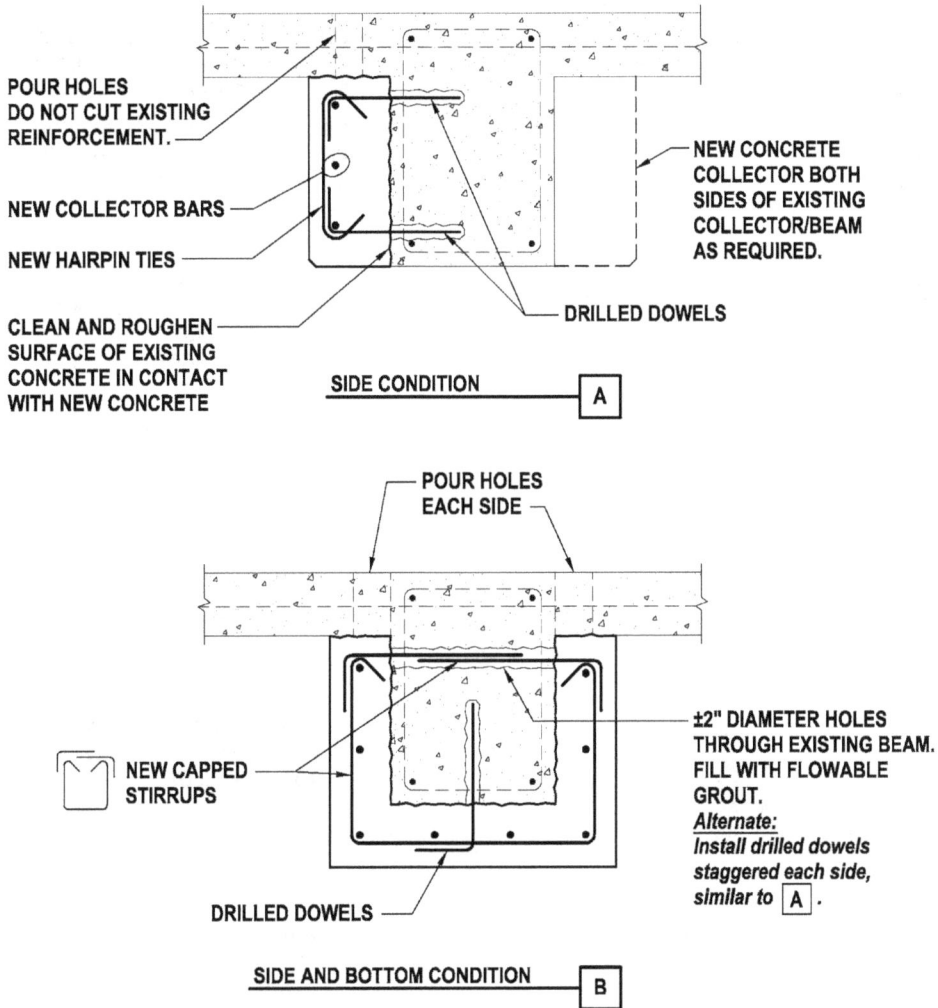

Figure 12.4.3-3: Concrete Collector at Existing Beam

manageable sections, generally about 10 to 20 feet in length, and will be connected to the
concrete diaphragm with drilled threaded rod anchors set in adhesive or epoxy. In almost all
cases, the steel plates will be installed at the top of the diaphragm as shown in Figure 12.4.3-4.
Although possible, it is extremely difficult to install heavy plate sections, connect the bolts and
make the necessary welded splices from below.

As discussed in the *Design Considerations* section above, the primary concern with a steel plate
collector is its lack of strain compatibility with the concrete diaphragm, unless the collector is

STEEL PLATE COLLECTOR
Size to control elongation.

*Provide additional slotted
holes to allow for rejected
anchors due to interference
with existing rebar.*

*Install collector plate in
direct contact with cleaned and
lightly roughened surface of
existing concrete to provide
maximum friction capacity.*

CP WELD
SPLICE TO
DEVELOP
SMALLER
PLATE
SECTION.

LONGITUDINALLY SLOTTED HOLES
*Size to allow for collector elongation
plus placement tolerance.*

DRILLED THREADED ROD ANCHOR
SET IN ADHESIVE OR GROUT
*Size to provide design clamping
force.*

STANDARD NUT

SPRING WASHER. PRETENSION
ANCHORS TO DESIGN CLAMPING
FORCE AFTER GROUT HAS CURED
AND TENSION PROOF TESTING IS
COMPLETED.

EXISTING CONCRETE
SLAB/DIAPHRAGM

DO NOT CUT EXISTING
DIAPHRAGM REINFORCEMENT.
WHERE INTERFERENCE OCCURS
PLACE ANCHOR IN NEAREST
ADJACENT SLOT.

TRANSVERSE SECTION

LONGITUDINAL SECTION

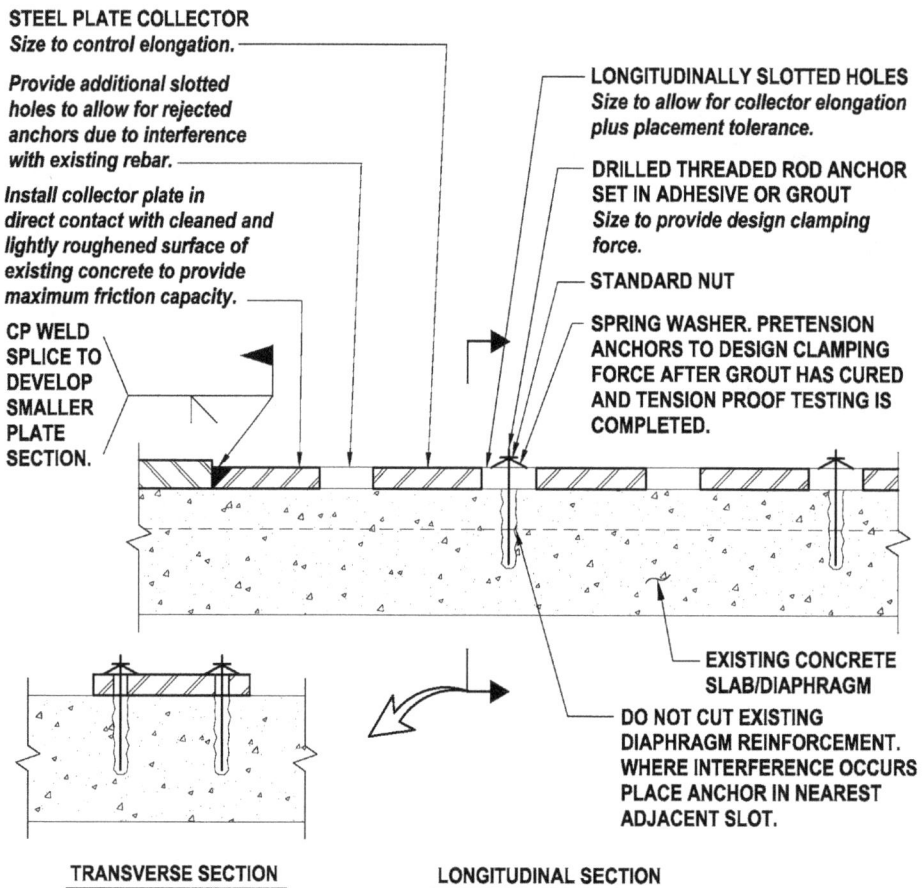

Figure 12.4.3-4: Steel Plate Collector

very short. The strain deformation of a steel collector will vary from zero at its free end to a
maximum at the connection to the wall or brace while the concrete diaphragm will not
experience similar deformations. In effect, the steel collector will stretch like a very stiff rubber
band relative to the concrete diaphragm. This relative deformation is difficult to accommodate,
especially in relatively long collectors. To do that, several conditions must be considered. First,
the various plate sections of the collector must be stepped in size so the strain is distributed
relatively equally along the length of the collector. Second the plates must be sized to limit the
maximum elongation to a reasonable amount of about one or two inches. Third, the threaded rod
anchors must be installed in slotted holes to allow the design elongation to occur without bearing
on and overloading the anchors. Fourth, to allow the slip to occur between the collector and
diaphragm, load transfer must be accomplished by friction using specially calibrated spring
washers to generate the appropriate clamping force in the anchors.

Cost/Disruption Considerations

Collectors have significant cost/disruption impact in a retrofit project primarily due to their length. They impact many building systems over a relatively large area compared to the impact associated with the walls themselves. This is especially true if general renovation of architectural and M/E/P systems is not included in the project. Thus, any available means of reducing collector length will probably be cost effective. A collector installed at the exterior edge of a diaphragm will generally be less costly than one installed in the interior and one installed above the diaphragm will be easier to install and, generally, less costly than one installed from below. However, installation of any collector can be very disruptive to any building occupants, due to the noise and vibration caused by drilling and coring through concrete, as well as the likely need to relocate various utilities and service distribution systems.

A comparison of the cost between reinforced concrete and steel plate collectors is very difficult to make. In general, the cost of either type of collector installed from below the diaphragm will likely be similar, because so much of the cost will be related to the impact on other systems. The cost of a steel plate collector may be less than one of reinforced concrete, but only for collectors installed from above the diaphragm, and particularly on a roof.

Construction Considerations

Existing concrete surfaces to be in contact with new concrete or steel plate collector should be cleaned of all finishes, paint, or other substances that could impair bond and shear transfer capability. Surfaces to receive new concrete should be roughened to provide ¼" amplitude aggregate interlock to prevent slip. However, since slip is expected to occur as a steel collector elongates, only light sandblasting may be required to assure development of the appropriate friction.

Installation of grouted anchors and/or dowels for steel plate collectors will require relative precision. They must be installed at the middle of the long slotted holes, with only a small tolerance, to allow the plate to elongate without bearing on the anchor. If existing rebar is encountered at a location, that location should be abandoned and the anchor installed in an available adjacent or nearby slot. Since it is reasonable to expect that this will occur with some frequency, a substantial amount of extra slotted holes must be available. For instance, if anchors are required at 12" on center, slotted holes should be provided at 6" on center.

The complete penetration welded splices of the relatively thick steel plate collectors are likely to be problematic. Making one-sided complete penetration welds in relatively thick plates will cause the plates to curl. To control this, the plate sections may need to be anchored down, with anchors placed in addition to the required shear anchors, and welded in place. Also, removal of backup bars will be difficult or impossible. Notches may be made into the concrete slab, and any remaining gap between the bottom of the steel plate collectors and the concrete diaphragm slab must be filled with grout to assure adequate friction at the concrete-steel interface.

Overhead construction of reinforced concrete collectors will require careful consideration and planning of how the reinforcing steel is placed and secured, prior to closing up the forms from below. Making the pour access ports and any sleeve holes for continuous rebar will require careful scanning of the slabs, waffle ribs, and beams to locate existing reinforcement to avoid cutting any existing rebar.

While the inspection, sampling, and testing required for reinforced concrete collectors is not particularly different from what is required for other seismic force-resisting reinforced concrete work, some special considerations do occur for steel plate collectors. The welded splices will require careful, non-destructive testing and thorough inspection. The shear anchors must be located at the middle of the slotted holes with some precision, and they must be extensively proof tested in tension. The holes with anchors must be free of any grout that could reduce the range of slip. Any gap between the bottom of the plate and the concrete slab must be grouted. The installation of the spring washers must be carefully inspected and tested to assure development of the design clamping force.

Proprietary Concerns

The basic materials are generic.

12.4.4   Enhance Column with Fiber-Reinforced Polymer Composite Overlay

Deficiencies Addressed by the Rehabilitation Technique

Inadequate shear capacity
Inadequate concrete compression strain and stress capacity due to lack of concrete confinement
Inadequate lap splice

Description of the Rehabilitation Technique

The use of a fiber-reinforced polymer (FRP) overlay with columns has proven to be an efficient rehabilitation technique in both the building and bridge construction industries. Columns are overlaid with unidirectional fibers in a horizontal orientation, thus providing shear strengthening and confinement similar to that provided by hoops and spirals used with circular columns, and stirrups and ties used with rectangular columns. The confinement enhances the concrete compression characteristics, provides a clamping action to improve lap splice connections, and provides lateral support for column longitudinal bars.

The preferred strength hierarchy for a building type structure is strong-column, weak-beam. Where the strength hierarchy results in weak-column, strong-beam (and is not considered acceptable by the designer due to, perhaps, concern for a soft story mechanism), the use of FRP overlay as flexural strengthening should not be used, unless there are extenuating circumstances and a very detailed analysis and design are performed. The uncertainty of strain compatibility between the FRP and column longitudinal bars and between the FRP and substrate, the lack of vertical strain capacity as a result of using FRP as longitudinal reinforcement, and the anchorage of the FRP at column ends and at points of contra-flexure deem this approach as undesirable. Other techniques presented in this document should be used in this situation.

See Section 13.4.1, "Enhance Shear Wall with Fiber-Reinforced Polymer Composite Overlay, Fiber-Reinforced Polymer Composite Overview," for background information.

Design Considerations

Research basis: Seible and Innamorato (1995) is one of the original ground-breaking research efforts of this topic and serves as an excellent source for understanding and design equations.

The primary deficiencies of a column are typically the lack of shear strength capacity and post-yield shear deformation capacity, as observed during many earthquake events and in laboratory testing. For shear assessment, two column locations need to be evaluated for shear: the end region within the plastic hinge zone (where the concrete shear strength degrades), and the region away from the flexural hinges, where there is no concrete shear strength degradation.

The FRP overlay provides confinement to enhance the concrete strain and stress capacity. Confinement is more effective for circular sections than rectangular sections. For circular sections, the passive radial pressure exerted by the FRP overlay on the gross concrete section, which is a result of the concrete lateral dilation, provides confinement. Dilation (similar to the concrete splitting action when performing a pure axial compression test on a concrete cylinder) occurs as result of the compression force, which is influenced by the level of axial load and flexural forces. For a rectangular section, dilation is only arrested at the corners of the section, thus relying on the concrete to arch between the corners, resulting in a reduced concrete core size. Due to the lack of effective confinement by the FRP, it is recommended to limit the rectangular section to a 1.5 depth-to-width aspect ratio and a width or depth dimension less than 36 inches, unless a special study is performed.

The confinement afforded by this technique does marginally increase the flexural strength and stiffness of the column, but not to the degree of concrete jacketing. The marginal increase is due to the higher concrete stress capacity of the cover and core concrete, hence reduced neutral axis depth, and is located within the what is called the *primary* hinge zone. This increase is over about half the depth of the column at each column end (where double-curvature occurs). Consequently, there is a greater moment demand just beyond this region, within the *secondary* hinge zone. Confinement enhancement, therefore, extends from the end of the column through the primary and secondary hinge zones, as shown in Figure 12.4.4-1A. Note that the categorization of *primary* and *secondary* hinge zones comes from Seible and Innamorato(1995).

The confinement pressure also serves to clamp the lap splice connection of the column longitudinal bars. The thickness (effective clamping pressure) needed for lap splices is derived differently from the confinement requirements, however, as test results indicate that, at a dilation strain of about 0.001, lap splice slippage is initiated. These results, combined with the empirically derived radial pressure requirement to prevent slippage, determine the FRP overlay thickness.

The FRP overlay thickness is determined for each of the three deficiencies; zones for these are shown in Figure 12.4.4-1A. As noted by Priestley, Seible, and Calvi (1996), the maximum of the three at any section should be provided; it should not be the sum, as reported in some other documents. This is because the lap-splice clamping action and compression concrete confinement occurs on opposite sides of the column, so the maximum requirement of the two will serve both well. Shear resistance of the FRP occurs along the column face parallel to shear load direction. The FRP anchors the concrete compression strut and is designed to maintain

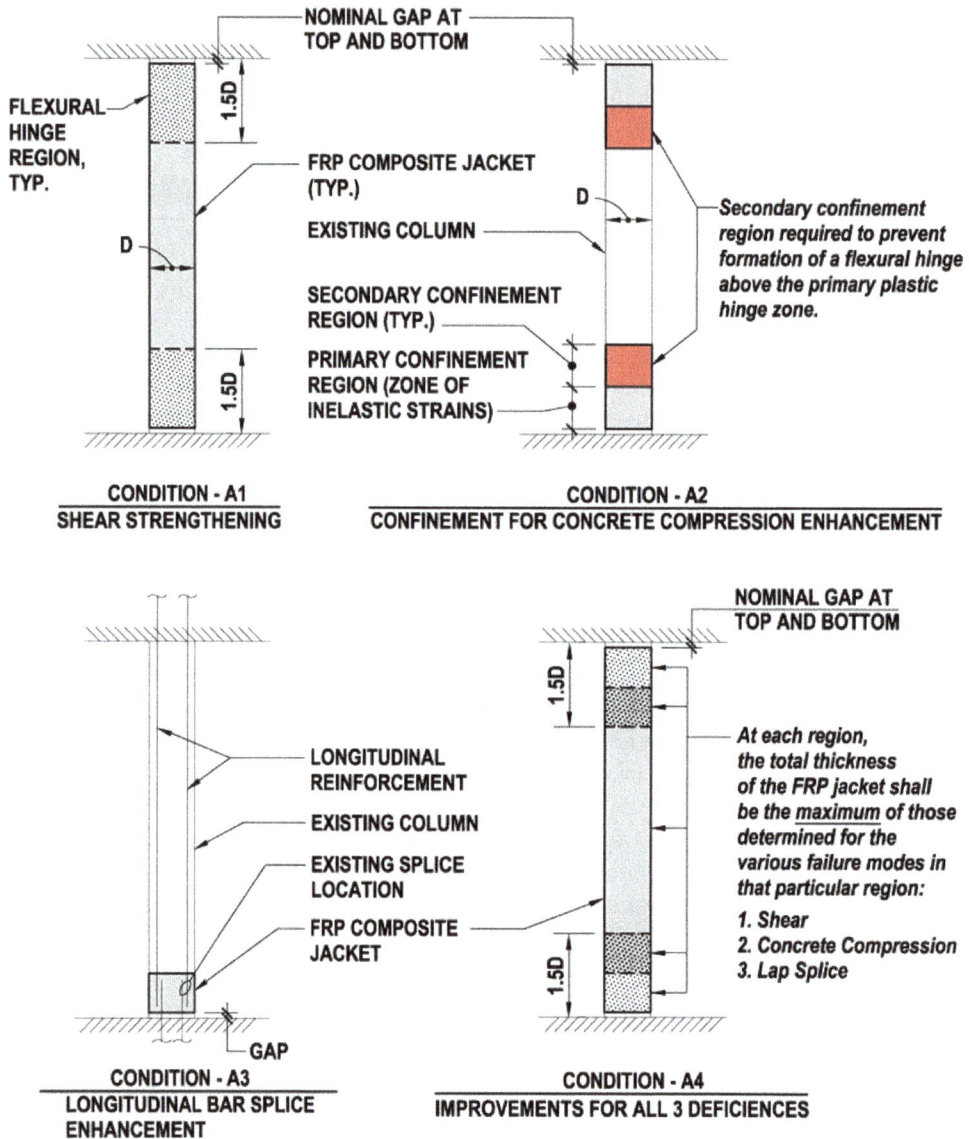

Figure 12.4.4-1A: Seismic Retrofit of Columns Using FRP Composites

ADDITIONAL FRP REINFORCEMENT AT OPENINGS

FRP OPENING FOR UTILITY PENETRATION

ROUND CORNER

FRP COMPOSITE OVERLAY

EXISTING CONCRETE COLUMN

NOTE: $\dfrac{D}{b} \leqslant 1.5$

$D\ \&\ b \leqslant 36"$

b

LAP

D

RECTANGULAR COLUMN   B1

LAP

FRP COMPOSITE OVERLAY

EXISTING CONCRETE COLUMN

CIRCULAR COLUMN

PLAN DETAIL   B2

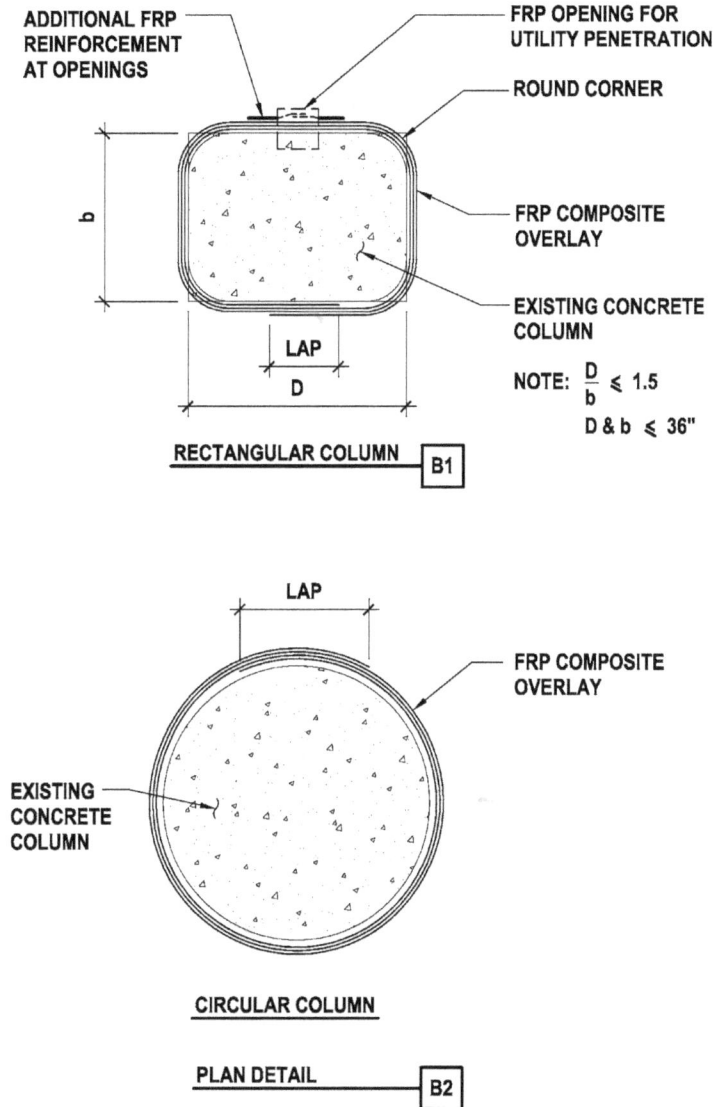

Figure 12.4.4-1B: Seismic Retrofit of Columns Using FRP Composites

aggregate interlock over the crack length. This contribution is a by-product of the hoop tension required for the confinement, so FRP composite thickness for shear need not be added to that required for confinement.

With successful mitigation of the three deficiencies, column flexural hinges can be developed and deformation capacity will be increased.

Detailing Considerations

The lap splices should be staggered, like that done with steel reinforcement, to mitigate a weak plane. As shown in Figure 12.4.4-1A, a nominal gap of about ½" between the FRP composite and the boundary elements (slab, beam or footing) is provided to prevent the overlay from bearing and, consequently, increasing the flexural capacity or stiffness of the column.

Cost/Disruption

To appropriately evaluate the cost of a retrofit scheme using and FRP overlay in comparison to traditional retrofit concepts (such as concrete or steel jacketing), one needs to consider the cost of the raw material, the level of specialization required by the contractor to install the system, the cost of labor and equipment, the cost of quality control and quality assurance, the temporary impact of disruption during construction, and the permanent impact to the building functions. Although FRP overlays are relatively expensive compared to steel and concrete, they can offer advantages when only limited access is available or minimal disruption of existing conditions is desired.

Construction Considerations

Access all around the column is often limited due to partition walls, ceilings and other architectural components, as well as structural elements of a building. Conditions at the base of the column in the lowest story must be carefully considered. The extent of FRP application normally will extend to the top of footing which may require local slab removal. The floor slab in the area may also interact with the column and affect strengthening requirements. This effect should be considered or a gap placed between the slab and column to prevent interaction.

Proprietary Concerns

See Section 13.4.1 for brief discussion of proprietary concerns.

12.4.5   Enhance Concrete Column with Concrete or Steel Overlay

Deficiencies Addressed by the Rehabilitation Technique

    Inadequate shear capacity
    Inadequate axial compression capacity
    Inadequate flexural plastic hinge confinement
    Inadequate lap splice

Description of the Rehabilitation Technique

Adding a fiber reinforced polymer (FRP) composite overlay to a concrete column is a recent approach to addressing seismic deficiencies, and is discussed in Section 12.4.4. Adding a concrete or steel jacket is a more traditional method of enhancing a deficient concrete column,

Figure 12.4.5-1 provides examples of concrete and steel jacketing for a rectangular column. Because FRP overlays have become more common, this section will focus on FRP overlays.

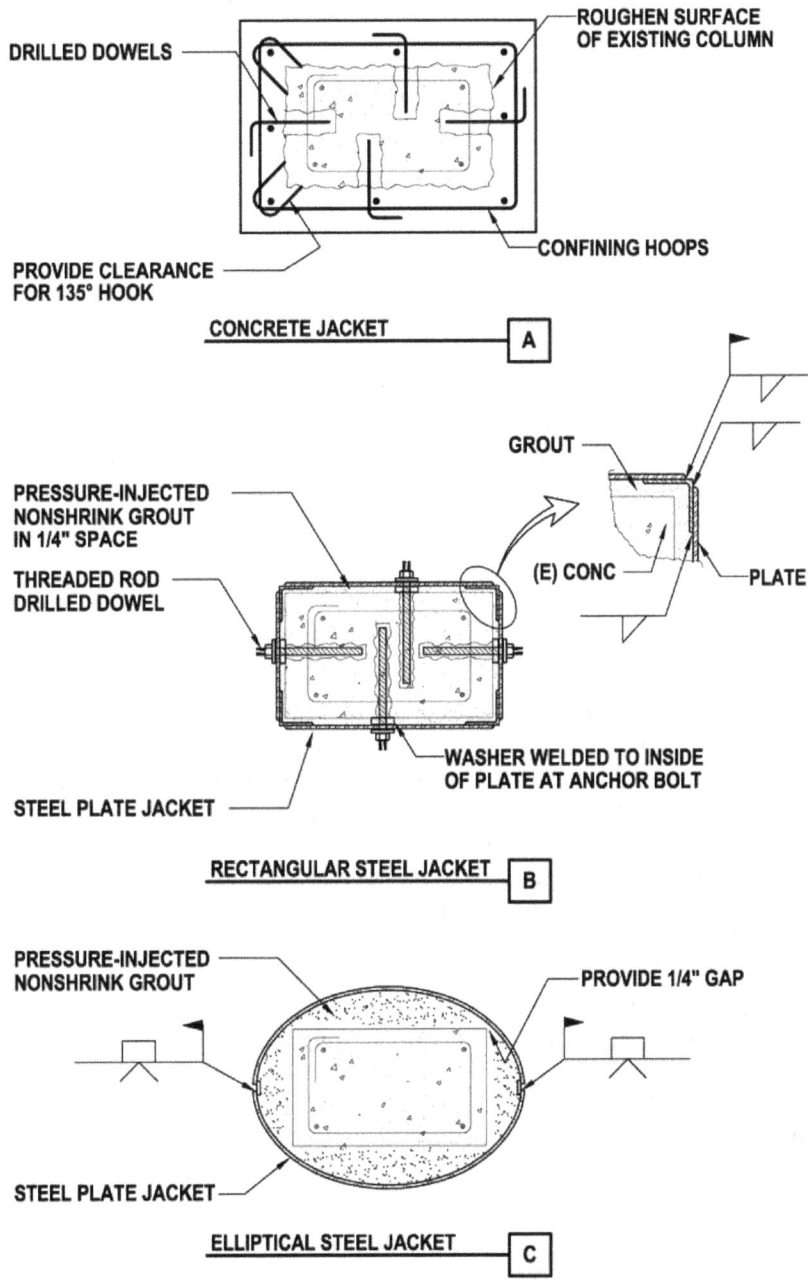

Figure 12.4.5-1: Concrete and Steel Overlays for Concrete Columns

Design Considerations

Research basis: Some research on concrete overlays is contained in FIB (2003); other related work is discussed in Section 12.4.5. Design of the overlay or jacket uses typical ACI 318 concrete design principles. Sufficient drilled dowels between the overlay and existing concrete should be provided to achieve composite action. Research on steel overlays includes Engelhardt, et al., (1994).

Detailing Considerations

Concrete overlays: Concrete jacketing will take up a larger cross section than either FRP overlays or steel overlays. The surface of the existing concrete must be roughened appropriately. Reinforcing steel will need to be in at least two pieces to get it around the existing column. 135 degree hooks are required for confining ties and may dictate the size of the overlay in order to provide enough room for the hook extension.

Steel jackets: Steel jackets require at least two pieces to get around the existing column and involve field welding. Like FRP overlays, when the aspect ratio of a rectangular column gets too large, the jacket becomes less effective. The California Department of Transportation (Caltrans) uses elliptical jackets in these situations. The corners of the existing column will need to be trimmed so the steel can pass by. There is a gap between the steel and the concrete of at least ¼" that is filled with grout. A gap is also provided at the ends of the column to permit rotation without engagement in bearing of the steel jacket.

Cost/Disruption

To appropriately evaluate the cost of a retrofit scheme using and FRP overlay in comparison to traditional retrofit concepts (such as concrete or steel jacketing), one needs to consider the cost of the raw material, the level of specialization required by the contractor to install the system, the cost of labor and equipment, the cost of quality control and quality assurance, the temporary impact of disruption during construction, and the permanent impact to the building functions. Although FRP overlays are relatively expensive compared to steel and concrete, they can offer advantages when only limited access is available or minimal disruption of existing conditions is desired.

Construction Considerations

Concrete overlays: Because of the difficulty of placing an overlay on all sides of an existing column, concrete overlays are typically done with cast-in-place concrete, rather than shotcrete. The need for formwork is a significant disadvantage for concrete overlays, compared to FRP and steel overlays. Placing the concrete and vibrating are also challenging due to access limitations at the top of the column where it runs into beams or slabs. Pour ports or holes in the diaphragm are needed.

Steel overlays: Steel overlays are typically 3/16" or ¼" thick and become quite heavy. Access and lifting issues in existing buildings can force the overlay to be broken down into pieces, increasing the amount of field welding necessary to join the pieces back together.

Proprietary Concerns

Unlike FRP overlays, no proprietary issues have been identified with using concrete of steel jackets.

12.4.6   Enhance Concrete Moment Frame

Deficiencies Addressed by the Rehabilitation Technique

Inadequate global shear capacity

Inadequate lateral displacement (global stiffness) capacity

Inadequate ductile detailing for shear strength, confinement, or strong column/weak beam

Description of the Rehabilitation Technique

An alternative method to either add strength and stiffness to an existing concrete moment frame or correct non-ductile detailing deficiencies of the frame members is by direct enhancement: increasing the size of the columns and beams of the frame with new reinforced concrete. This method entails adding a jacket of reinforced concrete around the existing columns and beams, an approach similar to jacketing by steel or fiber wrap. The new concrete may be either cast-in-place or shotcrete. This approach is relatively rarely employed in the U.S. because it is labor intensive, but when existing openings must be maintained and walls or braced frames cannot be installed, it may be rehabilitation technique worth considering.

Design Considerations

Research basis: Alcocer and Jirsa (1991) tested reinforced concrete beam-column connections subjected to unidirectional and bidirectional cyclic loading up to displacements equivalent to 4% interstory drift. Jacketing of columns resulted in strong column – weak beam behavior, with increases in peak strength of over four times the existing assembly. Jacketing of beams as well as columns improved joint confinement, decreased stresses on existing beam reinforcement, and provided some additional strength increases. The use of jacketing to repair damaged frames was also tested. When comparing the performance of a specimen in which a damaged column and joint were repaired by jacketing to that of a similar specimen in which the original column and joint were undamaged, the damaged and repaired assembly exhibited 65% of the strength at 2% drift and 50% of the stiffness at 0.5% drift relative to the undamaged retrofitted specimen.

Reinhorn, Bracci, and Mander (1993) performed shake table tests on a one-third scale model of a three-story, one-bay by three-bays, concrete moment frame. The tests compared the performance of an unretrofitted building, subjected to peak ground acceleration 0.30 g, to that of a similar structure retrofitted by jacketing interior columns, post-tensioning added longitudinal column reinforcement, and providing a reinforced concrete "fillet" infill around beam-column joints. These retrofit measures were intended to ensure strong column - weak beam behavior, enhance joint shear capacity, and improve anchorage for discontinuous beam reinforcement. Tests demonstrated that the retrofitted structure exhibited significantly reduced column damage, especially in the first story, and improved ductility in yielding beams; however, lateral displacements remained quite large, reaching a maximum inter-story drift of 2.1% at the first floor.

Impact on architectural and M/E/P systems and components: Enlarging columns, beams, or joints may impact existing ceilings, partitions, ductwork, plumbing, lighting, or usable floor space. The designer should keep in mind that the space required for access, formwork, and finishes will be greater than the final dimensions of the jacketed member.

System design: New transverse and longitudinal reinforcement should be designed such that members are flexurally-governed and beams yield before columns do. The jacketing of the frame will result in increases in system stiffness that may result in increased attracted load. Both new and existing concrete should be considered in developing composite properties for modeling and design. Joint shear can be the weak link in existing moment frames. A significant focus of the Alcocer and Jirsa (1991) research was to provide special angles at the joint region to improve joint confinement. Stiffness at 0.5% drift was well predicted when cracked properties of 0.5EI were used for the beams and 0.5EI were used for the columns. Stiffness decreased about 40% to 50% as drifts increased from 0.5% to 2.0%.

Detailing Considerations

Surface preparation: In order for a new reinforced concrete jacket to act compositely with an existing member, sufficient bond must exist between the new and existing concrete. The Alcocer and Jirsa (1991) tests made use of findings from Bass, Carrasquillo, and Jirsa (1985). In the Alcocer and Jirsa (1991) tests, they used a handheld electric chipping hammer to reveal some aggregate. The amplitude is not documented in the report, but the implication is that it was less than the traditional ¼" amplitude value. Dust was cleaned with a thick brush and vacuum cleaner. A bonding agent was not used, and the existing concrete surfaces were not saturated in all cases. Bonding agents in current practice are not common, but some engineers do recommend prewetting the existing concrete.

Column jacketing: Beams are typically the same plan dimension or narrower that the supporting column. New column bars can then pass by the existing beam bars through holes cut in the floor slab. Alcocer and Jirsa (1991) explored distributed longitudinal bars vs. bundled bars in the corners of the new jacket, but did not find significant differences. It is not possible to put a one-piece closed hoop around an existing column. One approach is to place a U-bar on three sides with a single leg closure piece, such as shown in Figure 12.4.4-2A. Alcocer and Jirsa (1991) used two overlapping L-shaped bars, each with 135-degree hooks at the ends. A minimum column jacket thickness of 4" was recommended.

Beam jacketing: When beams are jacketed, the bottom and sides can be increased in dimension, but typically the top of beam must remain at the existing top of slab level. In the Alcocer and Jirsa (1991) tests, slots were cut in the top of the slab just above the beam with holes at the ends of the slot through slab. Inverted U-bars were placed in the slot and through the holes. The bottom of the inverted U-bars overlapped around beam longitudinal steel with U bars surrounding the beam. A minimum beam jacket thickness of 3" is recommended. Care must be taken to investigate existing beam reinforcing, so that appropriate locations for new beam longitudinal bars can be located to miss existing reinforcing in orthogonal beams and in columns.

Joint enhancement: In the Alcocer and Jirsa tests (1991), vertical steel angles were placed in holes cut through the slab at the four corners of the column and then welded to horizontal steel bars above the slab and below the beam, to form a cubical cage. In the Reinhorn, Bracci, and Mander (1993) tests, joint strength was enhanced by infilling between orthogonal beams with concrete to create a diamond-shaped "fillet" in plan. Reinforcing at the outer ends of the fillet was placed through holes cut into the beam just under the slab and at the base of the beam.

Cost/Disruption

The cost and disruption associated with jacketing a concrete frame is significant because it potentially involves complicated formwork, reinforcement, and concrete pouring over much of the building, as opposed to more common shear wall approaches where the walls are only placed in localized areas. Beam jacketing is much more invasive and time consuming and is typically less important than the column jacketing. Alcocer and Jirsa (1991) reported that the construction time required for jacketing beams and columns was nine times that required for jacketing columns alone.

Construction Considerations

Noise and disruption: Removing existing column, beam, ceiling and floor finishes is disruptive. Chipping the cover concrete off of existing frame members is noisy and disruptive to occupants, as is cutting holes in the floor slab. As such this particular technique can be less desirable than many others.

Mix design: Due to the narrow thickness of the jacket and difficult working conditions, the mix design should emphasize ease of placement, by using small size aggregate and water-reducing admixtures. In the Alcocer and Jirsa (1991) tests, they used 3/8" maximum size aggregate and superplasticizer.

Concrete placement: Concrete should be placed from above through holes in the slab for both columns and slabs. Holes have to be big enough for both the concrete hose and the vibrator. Consideration can be given to leaving gaps at the top of the forms on the sides of the beams where they meet the slabs for air relief vents.

Proprietary Concerns

The basic materials are generic.

12.5   References

ACI 440.2R-02, 2002, "Guide for the Design and Construction of Externally Bonded FRP Systems for Strengthening Concrete Structures," American Concrete Institute, Chicago, IL.

Alcocer, S.M. and J.O. Jirsa, 1991, *Reinforced Concrete Frame Connections Rehabilitated by Jacketing*, Phil M. Ferguson Structural Engineering Laboratory Report 91-1, University of Texas at Austin, Austin, TX, July.

Bass, R.A., R.L. Carrasquillo, and J.O. Jirsa, 1985, *Interface Shear Capacity of Concrete Surfaces Used in Strengthening Structures*, Phil M. Ferguson Structural Engineering Laboratory Report 85-4, University of Texas at Austin, Austin, TX, December.

Engelhardt, M.D., et al., 1994, "Strengthening and Repair of Nonductile Reinforced Concrete Frames Using External Steel Jackets and Plates," *Repair and Rehabilitation Research for Seismic Resistance of Structures*, ed. James Jirsa, Report R/R 1994-1, June.

FIB (Federation Internationale de Beton), 2003, *Seismic Assessment and Retrofit of Reinforced Concrete Buildings*, State-of-Art Report, Bulletin 24.

Jiménez-Pacheco, J. and M.E. Kreger, 1993, *Behavior of Steel-to-Concrete Connections for Use in Repair and Rehabilitation of Reinforced Concrete Structures*, Phil M. Ferguson Structural Engineering Laboratory Report 93-2, University of Texas at Austin, Austin, TX, March.

Jones, E.A. and J.O. Jirsa, 1986, *Seismic Strengthening of a Reinforced Concrete Frame Using Structural Steel Bracing*, PMFSEL Report No. 86-5, Department of Civil Engineering/Bureau of Engineering Research, University of Texas, Austin, TX.

Priestley, N., Seible, F. and G.M. Calvi, 1996, *"Seismic Design and Retrofit of Bridges,"* John Wiley & Sons, New York, NY.

Reinhorn, A.M., J.M. Bracci, and J.B. Mander, 1993, "Seismic Retrofit of Gravity Load Designed Reinforced Concrete Buildings," *Proceedings of the 1993 National Earthquake Conference: Earthquake Hazard Reduction in the Central and Eastern United States: A Time for Examination and Action*, Memphis, TN, May, Vol. 2, pp. 245-254.

Seible, F. and D. Innamorato, 1995, *"Earthquake Retrofit of Bridge Columns with Continuous Carbon Fiber Jackets, Volume II, Design Guidelines,"* Report to Caltrans, Division of Structures, University of California, San Diego, La Jolla, CA.

Chapter 13 - Building Type C2b: Concrete Shear Walls (Bearing Wall Systems)

13.1   Description of the Model Building Type

Reinforced concrete walls in a building will act as shear walls whether designed for that purpose or not. Therefore, cast-in-place concrete buildings that contain any significant amount of concrete wall will fall into this category. However, there are two distinctly different types of concrete wall buildings, those that contain an essentially complete beam/slab and column gravity system, and those that use bearing walls to support gravity load and have only incidental beam and column framing. In this document, these building types have been separated and are designated **C2f** for the gravity frame system and **C2b** for the bearing wall. This section covers the bearing wall type. In this type of building, all walls usually act as both bearing and shear walls. The building type is similar and often used in the same occupancies as Building Type **RM2**, namely in mid- and low-rise hotels and motels. This system is also used in residential apartment/condominium type buildings.

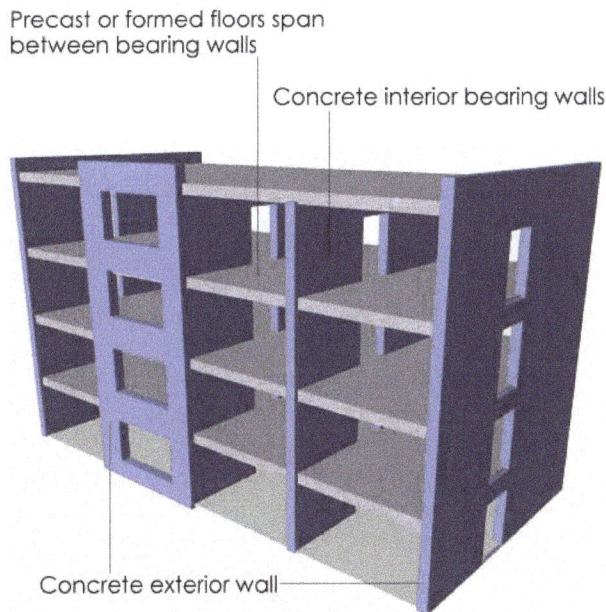

Figure 13.1-1: Building Type C2b: Concrete Shear Walls (Bearing Wall Systems)

Variations Within the Building Type

In order for this framing system to be efficient, a regular and repeating pattern of concrete walls are required to provide support points for the floor framing. In addition, since it is difficult and expensive to make significant changes in the plan during the life of the building, planning flexibility is not normally an important characteristic when this structural system is employed. The occupancy type that most often fit these characteristics are residential buildings, including dormitories, apartments, motels, and hotels. These buildings will often be configured with reinforced concrete bearing walls between rooms—also acting as shear walls in the transverse direction, and reinforced concrete walls on the interior corridor acting primarily as shear walls in the longitudinal direction. Sometimes the longitudinal lateral system includes the exterior wall system, although this wall is normally made as open as possible. In any case, the wide variation in structural layouts and occupancies that is included in Building Type **C2f** is not seen in Type **C2b**.

It is seldom possible to plan a building layout that provides complete gravity support with walls. Often, local areas are supported with isolated columns, and sometimes beams and girders are also necessary, but story heights in these buildings are usually small and added depth in the floor framing system is difficult to obtain. The extent of such beam and column framing often causes confusion between Building Types **C2b** and **C2f**, but buildings should have an essentially complete gravity frame system to be placed in **C2f**. If significant plan area is supported solely by walls, the structures are normally classified as **C2b**.

There are important variations in floor framing systems employed in this building type, and their adequacy to act as a diaphragm is an important characteristic of this building type as discussed below.

Floor and Roof Diaphragms

The parallel layouts of supporting walls and the need to minimize story heights normally leads to the use of one-way uniform-depth concrete floor systems. Cast-in-place and precast systems, both conventionally reinforced and prestressed, have been employed. The precast systems are often built up of narrow planks, which may not provide an adequate diaphragm unless a cast-in-place topping is provided. In addition, the precast systems may be placed with only a very narrow bearing area on the supporting walls, which may be inadequate to provide vertical support during seismic movements. The adequacy of the shear connection between slab and walls is also often an issue for both cast-in-place and precast systems.

Foundations

The bearing walls obviously require some kind of starter beam at grade for construction purposes and this often leads to a simple continuous grade beam system. In poor soils, piles or drilled piers may be added below the grade beam. A continuous mat foundation may also be employed due to the short spans and total length of bearing points in this building type.

13.2   Seismic Response Characteristics

Due to the extent of wall, bearing wall buildings will be quite stiff. Elastic and early post-elastic response will therefore be characterized with lower-than average drifts and higher-than-average floor accelerations. Damage in this range of response should be minimal.

Overall post-elastic response may often include rocking at the foundation level. If rocking does not occur, the height-to-length ratio of shear walls in these buildings may force shear yielding near the base, which may lead to strength and stiffness degradation.

Global stability may also be compromised by poor connections between floor slab construction and bearing walls.

Shear Wall Behavior

When subjected to ever increasing lateral load, individual shear walls or piers will first often force yielding in spandrels, slabs, or other horizontal components restricting their drift, and eventually either rock on their foundations, suffer shear cracking and yielding, or form a flexural hinge near the base. Shear and flexural behavior is quite different, and estimates of the controlling action are affected by the distribution of lateral loads over the height of the structure.

Yielding of spandrels, slabs, or other coupling beams can cause a significant loss of stiffness in the structure. Flexural yielding will tend to maintain the strength of the system, but shear yielding, unless well detailed, will degrade the strength of the coupling component and the individual shear wall or pier will begin to act as a cantilever from its base. In this building type, the coupling elements are often slabs, and their lack of bending stiffness may reduce or eliminate significant coupling action.

Rocking is often beneficial, limiting the response of the superstructure. However, the amplified drift in the superstructure from rocking must be considered. In addition, if varying wall lengths or different foundation conditions lead to isolated or sequencing rocking, the transfer of load from rocking walls must be investigated. In buildings with basements, the couple created from horizontal restraint at the ground floor diaphragm and the basement floor/foundation (sometimes called the "backstay" effect) may be stiffer and stronger than the rocking restraint at the foundation and should be considered in those configurations.

Shear cracking and yielding of the wall itself is generally considered undesirable, because the strength and stiffness will quickly degrade, increasing drifts in general, as well as potentially creating a soft story or torsional response. However, in accordance with FEMA 356, shear yielding walls or systems can be shown to be adequate for small target displacements. Type **C2b** buildings will often fall into this category.

Flexural hinging is considered ductile in FEMA 356 and will degrade the strength of the wall only for larger drifts. Similar to rocking, the global effect of the loss of stiffness of a hinging wall must be investigated.

13.3   Common Seismic Deficiencies and Applicable Rehabilitation Techniques

See Table 13.3-1 for deficiencies and potential rehabilitation techniques particular to this system. Selected deficiencies are further discussed below by category.

Global Strength

Due to the extensive use of walls, buildings of this type seldom have deficiencies in this category, unless significant degradation of strength occurs due to shear failures.

Global Stiffness

Similar to strength, global stiffness is seldom a problem in this building type. However, the effect of coupling slabs on initial stiffness and the potential change in stiffness due to yielding of these coupling slabs or wall-beams over doors should be investigated.

Configuration

The most common configuration deficiencies in this building type are weak or soft stories created by walls that change configuration or are eliminated at the lower floors. It is difficult to provide the needed ductility at the weak story, and often strength must be added. Completely discontinuous walls also create a load transfer deficiency for both overturning and shear. In such cases, collectors are often needed in the floor diaphragm, and supporting columns need axial strengthening.

Load Path

A common deficiency in this building is weakness in the load path from floor to walls, either collector weaknesses or shear transfer weakness immediately at the floor wall interface. Local transfer can be strengthened by adding concrete or steel corbel elements, dowels, or combinations of these components. As indicated above discontinuous walls also often create load path deficiencies.

Component Detailing

The most common detailing problem in this building type is an imbalance of shear and flexural strength in the walls, leading to pre-emptive shear failure. This deficiency may be shown to be not critical with small displacement demands, walls can be strengthened in shear with overlays of concrete, steel, or FRP.

The layout of walls often forces coupling between walls through the slab system or across headers of vertically aligned doors. These coupling components are seldom designed for the coupling distortions that they will undergo, particularly in older buildings. Short lengths of slabs between adjacent walls receive damage by coupling action that could compromise the gravity capacity. It is difficult to add strength or ductility to these slab areas, but vertical support at support points can be supplemented by corbels of steel or concrete. Damage to headers over doors often does not contribute to deterioration of overall response and can sometimes be acceptable. Local areas of wall can also be strengthened by overlays of concrete, steel, or FRP.

Table 13.3-1: Seismic Deficiencies and Potential Rehabilitation Techniques for C2b Buildings

Deficiency		Rehabilitation Technique				
Category	Deficiency	Add New Elements	Enhance Existing Elements	Improve Connections Between Elements	Reduce Demand	Remove Selected Components
Global Strength	Insufficient in-plane wall shear strength	Concrete/masonry shear wall [12.4.2]	Concrete wall overlay [21.4.5] Fiber composite wall overlay [13.4.1] Steel overlay		Seismic isolation [24.3] Reduce flexural capacity [13.4.4]	
	Insufficient flexural capacity	Concrete/masonry shear wall [12.4.2]	Add chords [12.4.3]			
	Inadequate capacity of coupling beams	Concrete/masonry shear wall [12.4.2]	Strengthen beams [12.4.2] Improve ductility of beams [12.4.2]			Remove beams
Global Stiffness	Excess drift (normally near the top of the building)	Concrete/masonry shear wall [12.4.2]	Fiber composite wrap of columns to improve lateral displacement capability [12.4.4] Provide detailing of all other elements to accept drifts Concrete wall overlay [21.4.5]		Supplemental damping [24.4]	
Configuration	Discontinuous walls	Add wall or adequate columns beneath [12.4.2]	Fiber composite wrap of supporting columns [12.4.4] Concrete/steel jacket of supporting columns [12.4.5]	Improve connection to diaphragm [13.4.3]		Remove wall
	Soft story or weak story	Add strength or stiffness in story to match balance of floors				

13-5

Table 13.3-1: Seismic Deficiencies and Potential Rehabilitation Techniques for C2b Buildings

Deficiency		Rehabilitation Technique				
Category	Deficiency	Add New Elements	Enhance Existing Elements	Improve Connections Between Elements	Reduce Demand	Remove Selected Components
Configuration (continued)	Re-entrant corner	Add floor area to minimize effect of corner		Provide chords in diaphragm [12.4.3]		
	Torsional layout	Add balancing walls [12.4.2]				
Load Path	Inadequate collector	Add steel or concrete collector [12.4.3]				
	Inadequate slab bearing on walls			Add diagonal dowels [13.4.3] Add steel ledger [13.4.3]		
Component Detailing	Wall inadequate for out-of-plane bending	Add strongbacks [21.4.3]	Concrete wall overlay [21.4.5]			
	Wall shear critical		Concrete wall overlay [21.4.5] Fiber composite wall overlay [13.4.1]		Reduce flexural capacity [13.4.4]	
Diaphragms	Precast components without topping		Improve interconnection [22.2.11] Add topping			
	Inadequate in-plane shear capacity		R/C topping slab overlay Fiber composite overlay [22.2.5]			
	Inadequate shear transfer to walls		Add diagonal drilled dowels [13.4.3] Add steel angle ledger [13.4.3]			

Table 13.3-1: Seismic Deficiencies and Potential Rehabilitation Techniques for C2b Buildings						
Deficiency		**Rehabilitation Technique**				
Category	**Deficiency**	**Add New Elements**	**Enhance Existing Elements**	**Improve Connections Between Elements**	**Reduce Demand**	**Remove Selected Components**
Diaphragms (continued)	Inadequate chord capacity	New concrete or steel chord member [12.4.3]				
	Excessive stresses at openings and irregularities	Add chords [12.4.3]				Infill openings [22.2.4]
Foundations	See Chapter 23					
[ ] Numbers noted in brackets refer to sections containing detailed descriptions of rehabilitation techniques.						

Diaphragm Deficiencies

Precast floor systems used in this building type often provide inadequate diaphragm behavior that could lead to bearing failures at the floor wall interface, particularly when no topping slab is present. Some topping slabs used primarily for leveling and smoothing the floor are inadequately tied to the precast elements or the walls, and are too thin or poorly reinforced to act as diaphragms on their own. See Chapter 22.

Foundation Deficiencies

This building type often places large demands on the foundation system. If rocking is shown to be a controlling displacement fuse for the building, the foundations must be investigated to assure that these displacements can safely occur. See Chapter 23.

13.4 Detailed Description of Techniques Primarily Associated with This Building Type

13.4.1 Enhance Shear Wall with Fiber-Reinforced Polymer Composite Overlay

Fiber-Reinforced Polymer Composite Overview

In addition to this section, seismic rehabilitation techniques using fiber-reinforced polymers are described in several other sections in this document, including Section 12.4.4, "Enhance Column with Fiber-reinforced Polymer Composite Overlay," and Section 22.2.5, "Enhance Slab with Fiber Reinforced Polymer Composite Overlay." This section provides a general overview of FRP characteristics.

Composite makeup and application: The construction industry's term *Fiber-Reinforced Polymer* (FRP) refers to a composite material made up of carbon, fiberglass or aramid (Kevlar) fibers that are bound together by either a resin or ester polymer. These are commonly referred to as glass fiber-reinforced polymer (GFRP) composite, carbon fiber-reinforced polymer (CFRP) composite, or aramid fiber-reinforced polymer (AFRP) composite, respectively. The raw fibers (synonymous to filaments) can be woven to form mesh (uni- or bi-directional with the orientation of the warp to the weft at 45 or 90 degrees), collected together to form a carbon tow winding (an untwisted bundle of continuous filaments) or sheet, or pultruded to form a prefabricated plate or other shape.

Usually, at least two layers of FRP are applied to the exterior concrete (substrate) surface. For beam and column applications, one layer should be considered sacrificial, due to the possibility of abrasion and the fact that lap splices are used and potentially compromise a layer's effectiveness.

Mechanical properties: The mechanical property in the direction of the fibers of GFRP, CFRP and AFRP is an essentially linear-elastic response followed by sudden rupture. The three material systems have different rupture strains and moduli. This variation becomes an important consideration when selecting the fiber type. While the rupture strain of each material is different, they are all significantly less (by approximately an order of magnitude) than that of conventional concrete reinforcement, which leads to compatibility issues.

Creep (increase in strain), stress rupture (reduced tensile capacity) and stress corrosion (corrosion that is dependent on presence of stress to occur) are phenomena that occur when FRP is subjected to sustained loads, such as flexural strengthening of slabs for long-term gravity loads. Each type of composite responds differently to loading regime, environmental conditions, and matrix and fiber make-up: AFRP is prone to creep under sustained load and moderately so to stress corrosion, GFRP is prone to stress rupture (as low as 20% of the ultimate), but CFRP is robust to sustained loads. Page 115 of FIB (2001) reference provides more information. This document addresses seismic loading, which is of short duration; hence, FRP is not affected by these characteristics. Accidental sustained stress loading for seismic applications may occur and should be given special study. An example is the FRP diaphragm overlays where the fiber composite transfer on top of a floor to a wall can act as negative slab reinforcement and resist subsequent live loads. Gravity load enhancement is not addressed in this document, but it should be noted that it is affected by this characteristic.

For the same FRP composite, mechanical properties can vary between manufacturers and sometimes within the same manufacturer over the course of material production for a project. It is recommended that the contract documents clearly specify the performance requirements (such as force per unit width for each application) and the minimum ultimate rupture strain.

Requirements at the FRP-to-substrate interface: There are "contact-critical" and "bond-critical" applications between the composite material and the substrate surface.

Contact-critical applications are mostly limited to beam and column shear and confinement enhancement techniques, and it is preferred that the composite be wrapped around all sides of the element (i.e., made continuous). These applications do not require shear flow capacity between the composite to the substrate, so paint and other smooth finish materials may be left in place.

Bond-critical applications, such as discontinuous applications and wall applications, require shear flow capacity between the FRP overlay and the substrate. (Note that flexural strengthening for gravity load enhancement of slabs and beams is a bond-critical application, but it is not included in this document.) The surface preparation is important and a concrete surface profile of 3 (CSP 3) is required (as described in ICRI, 2003), which calls for all loose laitance, the weaker outer cement paste layer, and any unsound concrete to be removed. Light sand or water blasting readily achieves the desired result. The surface should be dust free at the time of applying the first resin layer. Resin putty is used to fill the voids, provide a smooth surface, and create a chamfered corner. Paint and other surface finishes must be checked for hazardous materials before removal.

All of the fiber composites have the inherent tendency to rupture prematurely at stress concentrations. Such concentrations are formed by burrs, sharp edges, protrusions, etc., in the substrate. Any sharp edge, or protrusion of any kind, must be removed during the preparation of the substrate to ensure a smooth surface free of dirt, grease, oil and finishes. Existing elements damaged by cracks or corrosion should be repaired prior to applying this rehabilitation technique.

The substrate surface should be essentially flat, so that the fibers are straight when positioned. An uneven surface prevents the fibers from lying straight, and upon loading, will tend to straighten, which will compromise the bond capacity. An out-of-plane angle of 1-2% will compromise the bond strength by initiating peeling of the FRP overlay from the substrate.

Typically, the weak link in bond-critical applications is the tensile rupture of the substrate (although the resin should also be checked). This weak link obtains peak strength at about 1/32-inch displacement, followed by complete loss of bond strength at about 1/3-inch. Further, test results and observed behavior show a finite development length of about a few inches and beyond that no additional load is developed in the composite material, meaning that the fiber material does not necessarily develop its full strength (see Teng et al. (2002) Figures 2.3 and 2.7). This behavior is unlike rebar reinforcement, where the bar development length is sized, and the bond area is provided to attain full rebar strength. The design, therefore, for FRP composite requires focus on bond strength and an awareness of crack patterns and the locations of inelastic deformations.

Fiber and mechanical anchors: Significant research on new means of anchoring FRP composite to a substrate is occurring at this time. The most common anchor consists of inserting carbon fiber or a mixture of carbon and glass fibers into a drilled hole in the substrate. The hole is then filled with epoxy, and the protruding strands are splayed to a cone or fan shape that is used to lap with the composite overlay. The splay is located between the overlay layers to enhance the splice connection. There are many variations of this approach being studied. Anchor spacing, edge distance and other issues that influence the performance are also being investigated. The design of the anchor and FRP composite overlay system, therefore, should be case specific and based on the most recent research.

Durability: Composite materials, if manufactured correctly and with the appropriate finish applied, can provide corrosion resistance, ultraviolet light resistance, fire resistance, and tolerance to variations in temperature – making them generally suitable for most environments. Specific consideration should be given to galvanic corrosion of CFRP, particularly where there is an electrolyte present and potential for mixing of metals. Although CFRP is otherwise corrosion resistant, it is a conductor and will change properties when heated. Locations where lightning strikes may occur, such as garage rooftops or exterior of buildings, will warrant the use of other FRP materials or grounding of metal grid to protect the CFRP. See FIB (2001), Section 9.11. For each environment application, the experience of manufacturers and researchers should be considered, and the manufacturer's warranties should be carefully considered. The designer must consider the myriad of environmental factors and develop an FRP solution that is appropriate for those conditions, and, more importantly ensure that the appropriate fiber and polymer material, surface protection, and finish is specified. As an example, where high humidity and/or high alkalinity are present, carbon fiber is the preferred choice. Where ultraviolet light is present, surface protection of the composite can be achieved with an acrylic or polyurethane based paint applied when the polymer is still tacky. For more durability information, refer to FIB (2001), Chapter 9.

Constructibility: In addition to the costs of material and installation of FRP composite, concrete surface preparation and final appearance requirements must also be considered. The surface finish requirements, as described above in the *Interface Requirements* section, need to be included and made clear in the contract documents. The woven texture of the mat applications does read through to the finish surface. If additional surface finishing is required, such as paint or a cementitious appearance, this can add significant cost, and should be coordinated with the architect and owner. For seismic rehabilitation applications, the FRP overlay is similar to structural steel braces, in that they are not typically required to support gravity loads and thus do not require fireproofing. However, some local jurisdictions do require that a fire-protecting surface be provided, such as intumescent paint.

The resin's shelf-life, pot-life, ambient temperature, ventilation, substrate moisture state and other environmental factors need to be carefully considered by the design team. For example, due to issues with offgassing during and following installation for some time, not only will ventilation be needed, but occupants may need to be temporarily displaced. Quality control procedures, which should be required by the contract documents, need to address these issues and be verified as acceptable by the engineer of the project and the inspector of record.

Research basis: Teng, et al. (2002) provides a summary of recent research projects and design equations suggested by researchers. Design equations and construction quality assurance requirements are also present in model codes, such as ACI 440.2R-02 (2002) and FIB 14 (2001). For a summary of the mechanical properties and design procedures, see Priestley, et al. (1996). For detailed material mechanical properties and chemical background information, see Kaw (1997).

Deficiency Addressed by the Rehabilitation Technique
Inadequate shear capacity in a concrete shear wall

Description of the Rehabilitation Technique
An FRP overlay is a technique that is used to enhance the in-plane shear capacity of a reinforced concrete shear wall. The overlay can be applied to one or both sides of a wall, and where possible, should wrap around the ends of the wall to aid in anchoring the overlay. The rehabilitation technique is bond-critical, regardless of whether or not the material wraps around the end of the walls. Uni-directional (horizontal-oriented) fibers are used to enhance the shear capacity, creating a predominately flexural post-yield response. Vertically-oriented fibers in bi-directional layouts will limit the vertical strains to that of the FRP composite, inhibiting the ductile behavior. Therefore, the rehabilitation technique for walls is limited to horizontally oriented fibers, unless there are extenuating circumstances. The shear enhancement may change the wall's response from a shear-dominated behavior to a flexural, sliding shear, or rocking behavior. At coupling beams, vertically-oriented fibers are typically used. See Figure 13.4.1.1A for examples of wall and coupling beam layouts.

Design Considerations
Research basis: For wall-specific criteria, there has been an increasing amount of research on strengthening of unreinforced masonry walls, but less information on reinforced concrete walls. Ghobarah (2004) is one source; and Laursen, Seible, and Hegemier (1995), though based on CMU specimens, can be extended to some degree to reinforced concrete walls.

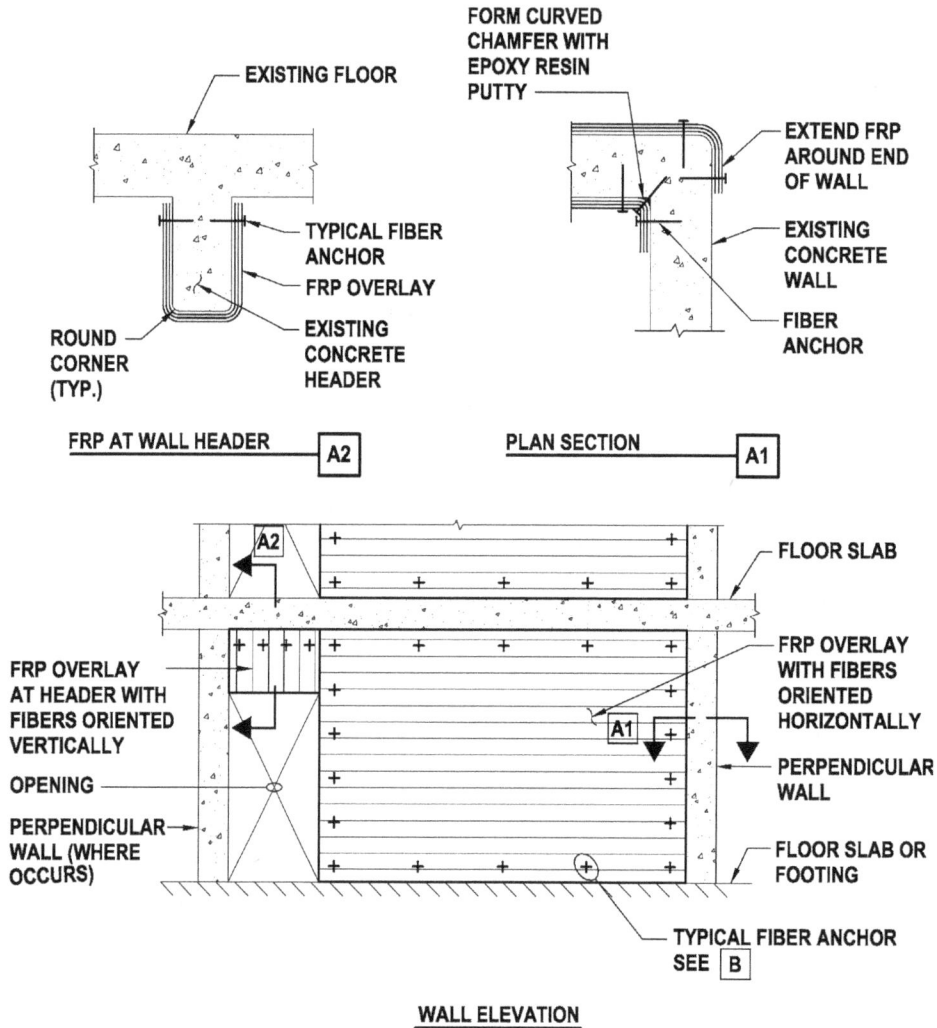

Figure 13.4.1-1A: Shear Strengthening of Concrete Shear Walls Using FRP Composite

The shear resistance contribution from the FRP is obtained in a similar manner to that used for wall reinforcement; the horizontal bars and FRP resist the horizontal shear (see Section 5.1 in Laursen, Seible, and Hegemier, 1995). The effective fiber area per unit width, and its contribution to shear resistance, is limited to the bond and anchorage strength capacity; providing additional fiber area will not provide additional shear capacity. Testing has been limited to mostly single and double layers of FRP per side of wall. Wrapping the material around the ends of the wall and/or providing fiber anchors will enhance the effectiveness of the overlay, particularly where cracks form, by increasing the anchorage and bond of the FRP to the substrate

EDGE DISTANCE

SPLAY OUT FIBER
ROVINGS AFTER
ANCHORAGE W/ EPOXY

EMBEDMENT

INITIAL LAYER(S) OF
FIBER OVERLAY

FINAL LAYER(S)
OF FIBER OVERLAY

SPLAY

EXISTING
CONCRETE
SLAB

INSTALL FIBER ANCHOR
BETWEEN INITIAL LAYER
AND SECOND LAYER OF
FIBER OVERLAY.

ROUND
CORNER

TYPICAL FIBER ANCHOR DETAIL   B

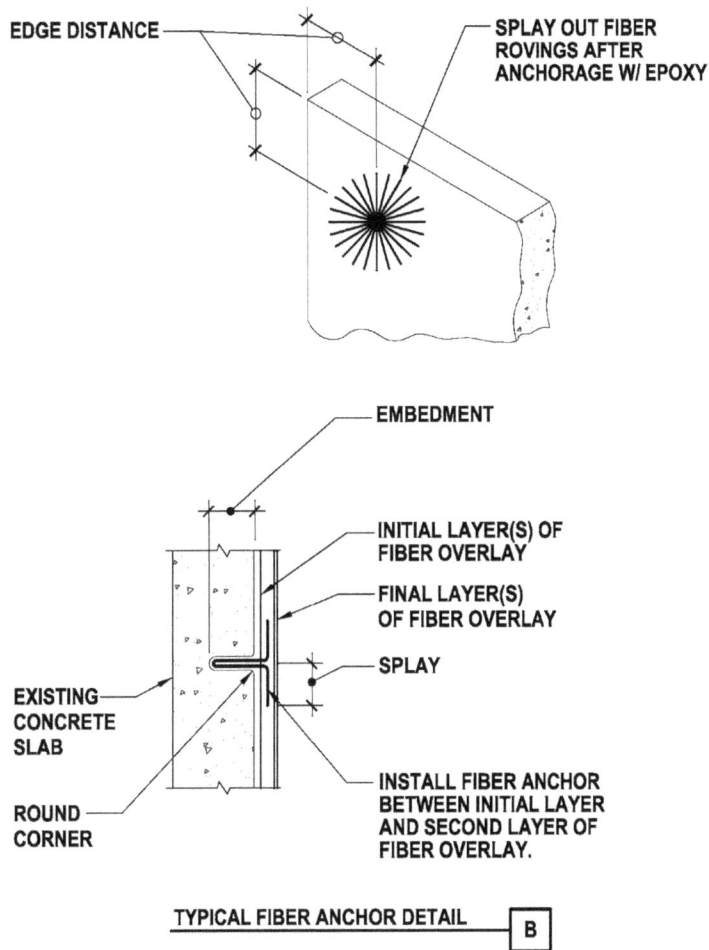

Figure 13.4.1-1B: Fiber Anchor Details

In terms of wall shear strength, the FRP overlay can be additive to the wall's concrete and reinforcement contribution. The ductility, however, is dependent on the type of governing mechanism. For shear-dominated walls (where the post-yield deformations require slippage at crack locations), the FRP has limited ability to accommodate such deformation; hence, there is likely very limited ductility. Conversely, a wall dominated by flexural yielding is able to accommodate the plastic deformations by the yielding of the wall vertical reinforcement, provided that the wall vertical reinforcement is developed. As a result, the typical goal in adding an FRP overlay is to make a wall flexurally-critical. For the case where flexural yielding occurs first, and is then followed by shear yielding as a result of the reduced concrete shear contribution (due to the reduced aggregate interlock effectiveness), ductility is significantly less than that with a flexural yielding response. Moreover, there is essentially no confining pressure afforded by the use of FRP overlay, except locally, where a fiber anchor is used. Therefore, lap splice performance is not enhanced with this technique and, depending on the size of bar, cover, and lap length, limited ductility may result.

Although testing with bi-directional (at plus and minus 45 degrees) fibers in wall overlays shows enhanced shear and flexural strengths and moderately enhanced displacement ductility capacity, the vertical force component of the fibers contribution to the flexural strength may have adverse affects. This may reduce the distribution of inelastic straining and/or change the strength hierarchy to that of shear controlled. Until further research is performed, horizontally-oriented FRP strengthening is recommended.

Detailing Considerations

Given the high dependence on the bond strength of the FRP overlay to the substrate, in-situ bond testing should be included as a requirement in the contract documents. A testing program will verify the design assumptions and assist in providing quality assurance. If pilasters are present, either within the wall length or at wall ends, installing fiber anchors or removing portions of the pilaster should be considered to enhance the anchoring of the FRP overlay.

Construction Considerations

As discussed in the *FRP Composite Overview* section earlier, the engineer should inspect the surface of the wall elements to be rehabilitated and note in the contract documents the surface condition and wall configuration (e.g., wall corner profile and wall-to-slab configuration). To aid in developing a sound bid price, the contract documents need to record the as-built condition, including surface anomalies and configuration, and the surface preparation requirements. If the surface has been board formed with wood planks, for example, calling this out in the construction documents will enable a more accurate bid. Bid documents should also require the contractor to view existing conditions before bidding.

Proprietary Concerns

Although the basic materials are generic, the fabric that will be supplied is proprietary. Strand orientation and density, epoxy overlays, preparation requirements, and installation procedures may be different between suppliers. Some suppliers may not have experienced applicators in the area of the project. These variations must be considered to achieve an adequate specification, particularly if competition between suppliers is desirable.

13.4.2   Enhance Deficient Coupling Beam or Slab

Deficiencies Addressed by the Rehabilitation Technique

  Inadequate shear and bending capacity of coupling beams or slabs
  Inadequate ductile detailing for shear strength

Description of the Rehabilitation Technique

Coupling beam deficiencies can be encountered in any type of building structure that has shear walls for their primary lateral force-resisting system. In a bearing wall structure, however, the coupling beam is often not a beam at all but only the relatively thin concrete slab linking the adjacent walls across a corridor. In some cases, particularly near the top of taller buildings, the restraint of a slab or beam in-line with the flexural deformed shape of a narrow shear wall can create coupling beam type issues at the tip of a single wall. In many cases, the slab consists of precast (and prestressed) hollow core planks, with or without a topping, with limited reinforcing steel at the link beam-wall joint instead of cast-in-place concrete. Furthermore, because bearing wall buildings are often residential buildings with short story heights, there is often no ability to create a dropped beam, and it is very difficult or impossible to increase the strength or ductility of the linking slab-beam itself. In these cases, the mitigation approach is to install a steel or concrete corbel to provide supplemental vertical support.

For those cases where there actually is a beam or deeper header linking the walls, the most common mitigation method to strengthen it or to correct its non-ductile detailing deficiencies is by direct enhancement with new added reinforced concrete. The new concrete is generally added to one face of the coupling beam, in concert with concrete strengthening of the adjacent shear walls as well. In some cases, new reinforced concrete can be added to both faces and extended along the walls for development. In addition, or as an alternative, it may be possible to install a limited number of diagonal dowels to enhance the shear capacity of the coupling beam to wall joint. In a case where there is enough beam or header depth, two groups of tied bars could be placed in an X-configuration, embedded within the new beam reinforcement cage. The new concrete may be either cast-in-place or shotcrete.

Design Considerations

If the restraint of the slabs and beams described above is not needed for overall structural stiffness, the element should be evaluated for gravity support in a damaged state. No mitigation may be required, particularly in cases where cast-in-place concrete slabs exist. For cases where gravity support is inadequate, the approach may be the installation of a supplemental support in the form of new corbels placed under locations of damage. If there is no ceiling, care should be taken to minimize the visual impact of the corbel and connectors. If existing M/E/P system components conflict with installation of these supplemental supports, they should be relocated as required to make this minimum mitigation effort.

For cases where an actual coupling beam or header is being augmented, the contribution of both the existing and new portions of the now composite member should be considered. Also, since it will be very difficult to provide sufficient strength in the augmented beam, the primary goal should be to provide adequate ductility to survive the expected rotations and continue to provide gravity load support and shear transfer capability. If the new concrete work is exposed to view, consideration should be given to the nature of the forming and surface finishes desired in relation to the surrounding existing elements. For cases where a deeper beam or header exists, the new ductile member should be designed to provide all required strength and ductility.

Detailing Considerations

Connection of new corbel to existing concrete elements: A typical detail showing the installation of a new steel corbel is shown in Figure 13.4.2-1. The steel angle provides supplemental support for the coupling slab at the wall joint where the most damage is expected to occur. For a precast plank slab condition (as shown in the detail), especially without a cast-in-place topping, consideration should be given to extending the supplemental steel supports along the full length of the link between the coupled walls. Installation of drilled anchors into the bottom of the

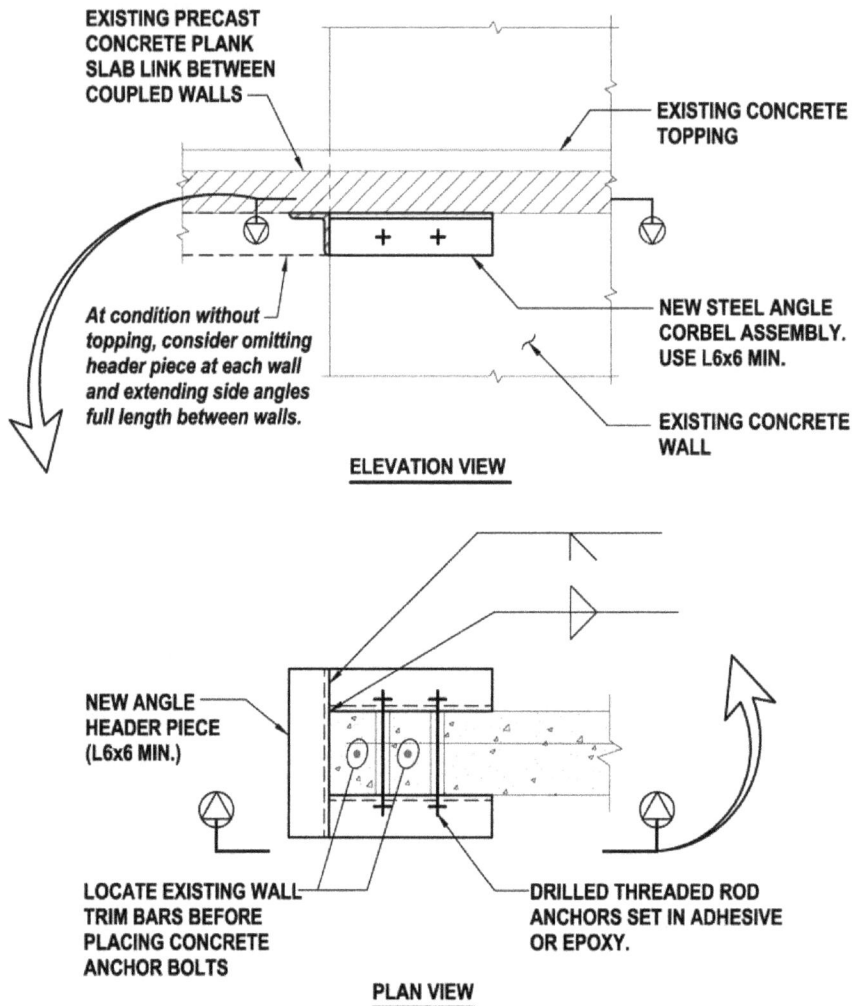

EXISTING PRECAST CONCRETE PLANK SLAB LINK BETWEEN COUPLED WALLS

EXISTING CONCRETE TOPPING

At condition without topping, consider omitting header piece at each wall and extending side angles full length between walls.

NEW STEEL ANGLE CORBEL ASSEMBLY. USE L6x6 MIN.

EXISTING CONCRETE WALL

ELEVATION VIEW

NEW ANGLE HEADER PIECE (L6x6 MIN.)

LOCATE EXISTING WALL TRIM BARS BEFORE PLACING CONCRETE ANCHOR BOLTS

DRILLED THREADED ROD ANCHORS SET IN ADHESIVE OR EPOXY.

PLAN VIEW

Figure 13.4.2-1: Typical Corbel at Linking Slab

planks may be considered. However, it may be difficult to avoid both prestressing tendons and hollow core voids. Consider use of screen-tube anchors (see Chapter 21) if voids cannot be avoided.

Connection of new concrete to the existing concrete beam and adjacent walls: Figure 13.4.2-2 shows an elevation and section view of a typical augmentation of a relatively shallow coupling beam or header. The surface of the existing concrete should be thoroughly cleaned and roughened to provide a good bond and interaction between the existing and new portions of the

NEW CONCRETE ON FACE OF
EXISTING SHALLOW COUPLING
BEAM OR HEADER

DOWELED TIES
AT 4" O.C. MAX.

2d   2d

EXISTING TOPPING

EXISTING PRECAST
PLANK SLAB

LENGTH AS
REQUIRED TO
DEVELOP ADDED
LONGITUDINAL
BARS

EXISTING CONCRETE
WALL

*In some cases wall will be
augmented with concrete or
shotcrete.*

ELEVATION VIEW

ACCESS HOLES. DO NOT
CUT EXISTING PRESTRESSING
TENDONS IN PRECAST PLANKS.

d

'C'-SHAPED DOWEL TIES
@ d/2 O.C., MAX.

2-#6 MIN., TOP
AND BOTTOM

8" MIN.

NEW CONCRETE OR
SHOTCRETE

EXISTING CONCRETE
COUPLING BEAM

SECTON VIEW

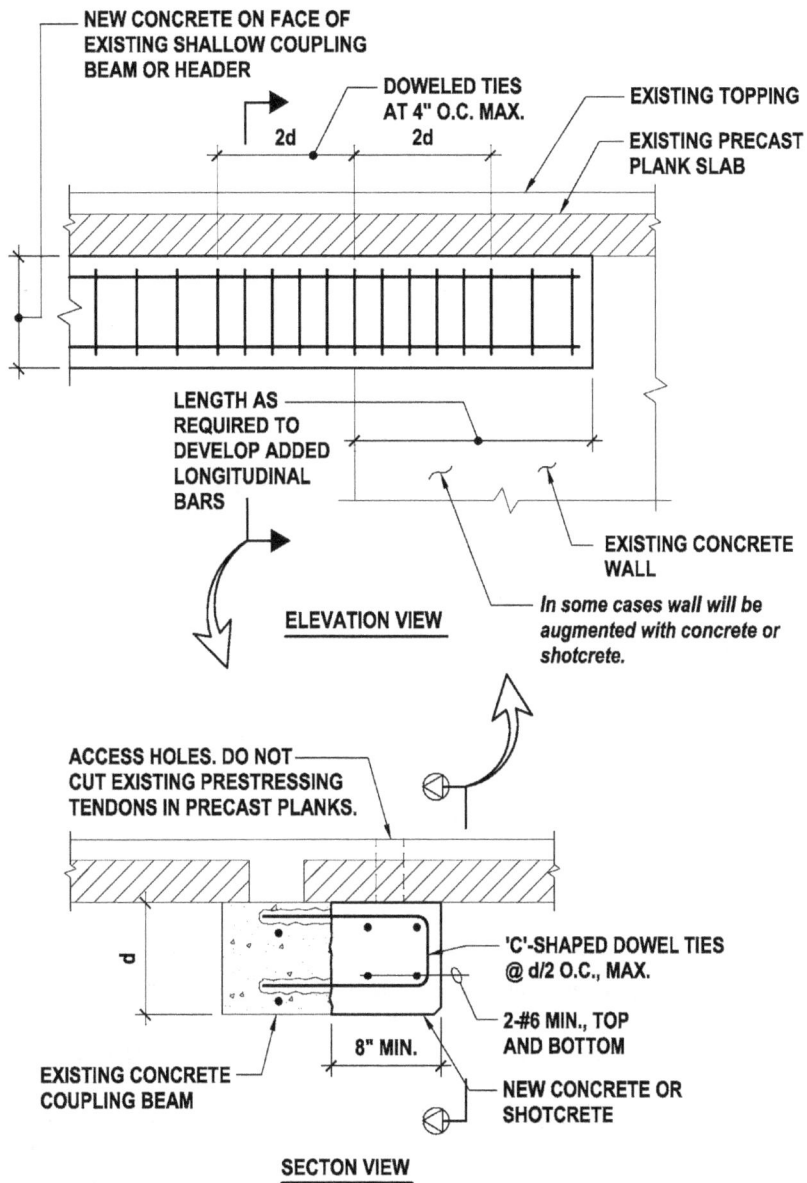

Figure 13.4.2-2: Typical Strengthening of Shallow Coupling Beam

composite member. The new concrete should be extended along the walls far enough to develop the longitudinal bars. If cast-in-place concrete is used, access holes will be required through the slab for placement and consolidation. Care must be taken to avoid damage to any prestressing tendons in the precast planks. Figure 13.4.2-3 shows typical details for installation of X-shaped rebar cages at deeper coupling beams or headers. The new concrete should be thick enough to allow placement of the crossing longitudinal reinforcement cages inside of the confining beam ties.

Cost/Disruption Considerations

Installation of steel corbel supplemental supports is an inexpensive, minimal mitigation approach. However, costs could increase substantially if relocations of M/E/P systems are required. The disruption associated with a single installation of this work is relatively local, but similar work will likely be required at all or most of the walls throughout the building. At all cases where new concrete or shotcrete is added to existing coupling beams, the cost and level of disruption will increase substantially. Refer to similar discussion in Sections 12.4.2 and 12.4.3.

Construction Considerations

For installation of a steel corbel, care should be taken to locate existing reinforcing steel in the wall prior to drilling or coring. Also, the pattern of prestressing tendons should be located before any drilling into precast planks is begun. For augmentation of coupling beams, refer to similar discussions at Sections 12.4.2 and 12.4.3.

Proprietary Concerns

The basic materials are generic.

13.4.3   Enhance Connection Between Slab and Walls

Deficiencies Addressed by the Rehabilitation Technique

Inadequate bearing for precast slab planks at wall
Inadequate shear transfer capacity from precast plank diaphragms to shear wall

Description of the Rehabilitation Technique

For most cases involving inadequate bearing support for precast floor planks at the walls, the typical mitigation approach will be to install steel (or concrete) ledgers secured to the wall under the slab-wall joint. This approach can also be used to increase joint shear capacity by including concrete anchors drilled up into the slab. An alternative approach to increasing joint shear capacity is to install diagonal dowels, or shear pins, through the joint. Occasionally, if adequate story height is available and gravity load issues can be addressed adequately, a new reinforced concrete topping slab could be placed over the existing diaphragm.

Design Considerations

For cases where the concern is focused only on providing adequate bearing, installation of a steel ledger angle anchored to the wall is a very simple and direct technique. Provision of about six inches of additional bearing will be enough to assure against loss of vertical support, so anchors drilled into the slab or precast concrete planks may be omitted. Use of a concrete ledger is

℄ COUPLING BM

4-BARS (MIN.) TYP.
WITH TIES AT
4" O.C. MAX.

EXTEND NEW CONCRETE
AT WALL TO DEVELOP
ALL NEW MAIN BEAM
REBAR (AT A MIN.)

NEW CONCRETE ON
FACE OF EXISTING
DEEP COUPLING
BEAM OR HEADER

6"

l_d

EXISTING CONCRETE WALL
(Wall may often be augmented
with concrete or shotcrete as well)

ELEVATION VIEW

ACCESS HOLES. DO NOT CUT
(E) SLAB REBAR OR
PRESTRESSED TENDONS

DRILLED REBAR
DOWELS

BEAM STIRRUPS

X-SHAPED TIED CAGES

CLEAN AND ROUGHEN
FACE OF EXISTING
CONCRETE TO
RECEIVE NEW
CONCRETE

12" MIN.

SECTION VIEW

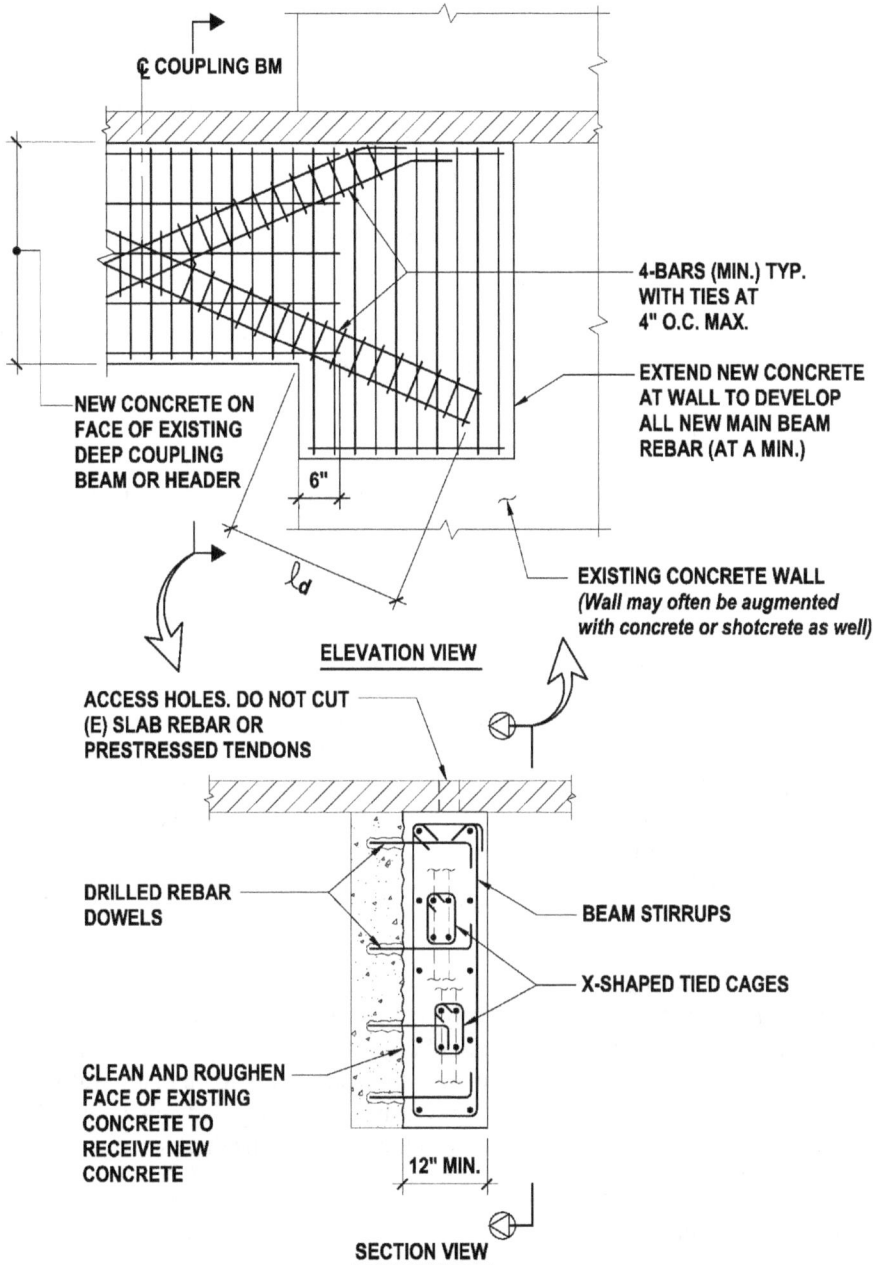

Figure 13.4.2-3: Strengthening of Deep Coupling Beam

certainly possible, but would seem appropriate only if concrete collectors were also being installed. However, a steel angle or channel section ledger could still be used, linking the end of a concrete collector to the wall.

For cases where diaphragm-wall shear transfer is to be augmented, either vertical slab anchors or diagonal dowels, drilled and placed from above, may be added. However, the added shear capacity that can be provided by these techniques is limited by the shear capacity of the diaphragm slab immediately adjacent to the wall. If the local diaphragm capacity is inadequate, then a new collector will be required to engage a greater extent of the diaphragm (refer to Section 12.4.3).

If a new concrete topping slab is to be placed, sufficient reinforcement probably can be included to serve as the collector. However, new diagonal dowels or perhaps a new ledger will likely be needed to insure transfer of the collected diaphragm shear demand down through the lightly reinforced slab-wall construction joints into the wall below.

Consideration should be given to the treatment of any architectural finishes and M/E/P system components mounted on the wall in the affected areas.

Detailing Considerations

Typical details for installation of new steel ledgers and drilled diagonal dowels are shown in Figures 13.4.3-1 and 13.4.3-2, respectively. For these techniques, little or no prior cleaning or preparation of the existing concrete surfaces will be required. Alternatively, refer to Section 12.4.3 for discussion related to installation of a concrete ledger. Where drilled concrete anchors or drilled diagonal dowels (acting as shear pins through the slab-wall joint region) are to be installed through precast concrete planks, the prestressed tendons must be avoided. Also, consideration should be given to means of dealing with any hollow core voids in the planks that may be encountered. Screen tubes similar to those used in brick masonry anchorage details could be used, or the voids could be filled, at least locally, with grout.

Cost/Disruption

Installation of steel section ledgers and drilled concrete anchors and dowels are simple and well known basic techniques that should be relatively inexpensive. There will be noise and vibration associated with the drilling, but the work is essentially "dry" and not particularly messy. However, given the nature of this type of structure, there are likely to be many walls distributed throughout the building, thus the work will be pervasive. If concrete ledgers are installed, the level of cost and disruption will be considerably higher (refer to the discussions regarding installation of concrete collectors at Section 12.4.3).

Construction Considerations

Installation of the drilled concrete anchors or diagonal dowels may require some precision to avoid prestressing tendons or hollow core voids. Once the pattern of these items is defined, a steel template, perhaps the steel ledger itself, can be used. If some flexibility in the exact location of the anchors is required, consider using oversized holes and welded plate washers, and provide extra holes.

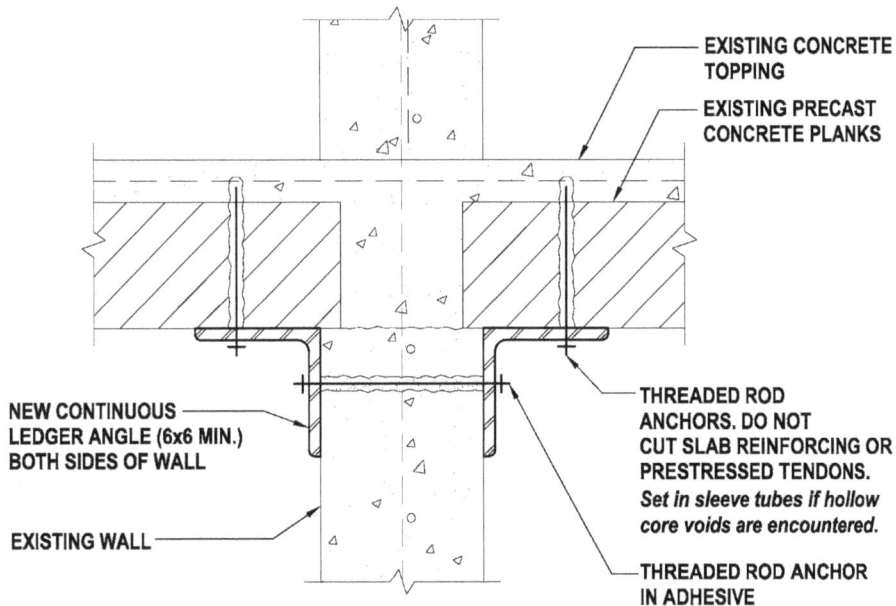

Figure 13.4.3-1: Added Support and Shear Strength at Slab-Wall Joint

The drilled anchors and dowels will require testing, which should be performed by persons experienced in torque and/or tension testing of diagonal dowels and overhead installations. For the drilled shear pin dowels that do not project out of the slab, additional similar dowels can be installed for testing purposes only.

Proprietary Concerns

The basic materials are generic.

13.4.4   Reduce Flexural Capacity of Shear Walls to Reduce Shear Demand

Deficiency Addressed by the Rehabilitation Technique

Degradation of shear-critical walls

Description of the Rehabilitation Technique

In buildings with many walls and high strength, many of the walls may be shear-critical and prone to strength degradation and/or possible reduction of gravity support capacity. A sudden

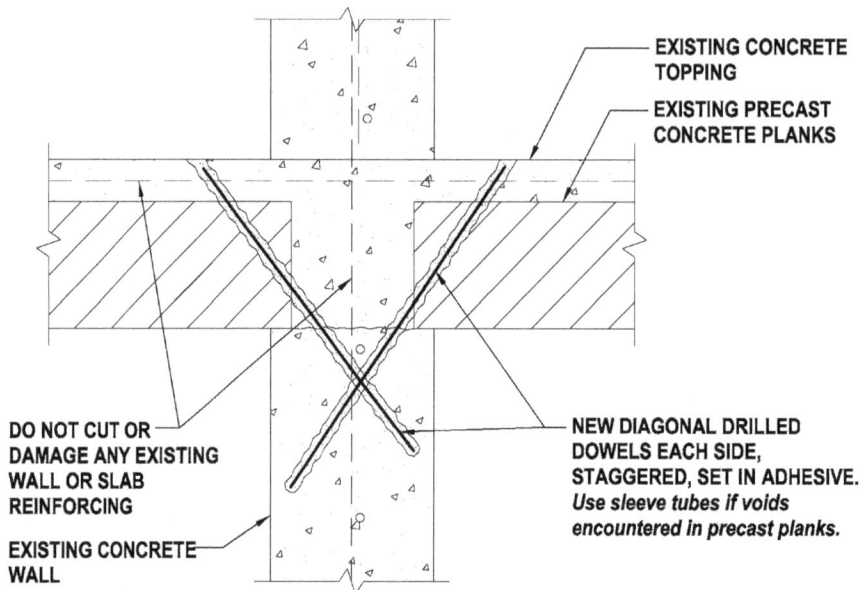

Figure 13.4.3-2: Added Shear Capacity at Slab-Wall Joint

loss of lateral strength and stiffness, particularly all at one floor, should be avoided. To avoid this condition, an alternative to shear enhancement is to reduce the level of possible shear demand that the wall can experience by reducing flexural capacity. This approach may be applied to one or more walls, in a system with adequate global strength, to change the expected post-yield behavior from a brittle shear failure to a more ductile flexural yielding. This approach also may be applied in other circumstances to an individual wall that cannot be strengthened for some reason, but must be protected from serious damage, even as other walls are strengthened or new walls are added in other locations within the system. This technique will generally involve cutting a number of the existing wall chord bars. However, an alternative approach that may be applied to a long wall would be to make one or more vertical cuts in the wall, creating a series of shorter panels with reduced flexural capacity.

Design Considerations

Application of this technique to a particular shear wall is intended to replace a brittle shear failure mechanism with a more ductile flexural hinging mechanism. It is not recommended that this technique be applied in conditions that may result in creating a brittle tension failure mechanism. Therefore, the number and location of cut bars must be determined with careful analysis of the detail of the expected mechanism (strain compatibility, flexural hinge length, etc.). Bar cuts should be sufficiently staggered and a major reduction in flexural capacity should not be attempted with this method. The impact of this technique on architectural and M/E/P systems and components is likely to be relatively limited. Only fixtures in the immediate vicinity of the chord bar cuts or vertical wall slices are likely to be affected. However, client perception of this reductive technique may be decidedly negative.

Detailing Considerations

Cutting the selected existing wall chord bars must be done with great care and precision to avoid cutting adjacent chord bars or the confining transverse reinforcement. If the wall reinforcement can be adequately mapped by using metal detector, x-ray or ground penetrating radar techniques, then the selected bars could be cut by coring through the concrete cover. However, congested chord reinforcement and/or closely spaced transverse ties may require that the concrete cover be chipped away to expose the bars before cutting with either a core drill or a torch.

In the case of reducing the overall length of a long wall with vertical cuts, the cuts can be made with a circular concrete saw. In order to extend the cuts as close to the floor slabs as possible the cuts may be made from each side of the wall, to keep the depth of the cut and the radius of the saw blade to about half the wall thickness.

Cost/Disruption

This technique will be less costly than alternative methods to increase the shear capacity of a wall. The extent of the work is very localized and the impact on nonstructural components will be limited. However, there will always be noise and vibration associated with any concrete chipping, coring or sawcutting, and the latter two operations can be very wet and messy.

Construction Considerations

Access must be available to the locations where chord bars are to be cut or the wall is to be sawcut. However, this will certainly be less of an imposition than any other alternative to increase the shear capacity of the wall.

Employment of this technique will most likely require special scheduling and/or sequencing considerations relative to the other strengthening work, to avoid creating a weakened structure during the course of the overall retrofit project.

Proprietary Concerns

The basic materials are generic.

13.5   References

ACI 440.2R-02, 2002, "Guide for the Design and Construction of Externally Bonded FRP Systems for Strengthening Concrete Structures," American Concrete Institute, ACI Committee 440, Chicago, IL

ICRI (International Concrete Repair Institute), 2003, "Guide for Surface Preparation for Repair of Deteriorated Concrete Resulting from Reinforcing Steel Corrosion," Publication 03730, International Concrete Repair Institute.

FIB (Federation Internationale de Beton), 2001, *Externally Bonded FRP Reinforcement for RC Structures*, Technical Report, Task Group 9.3 FRP Reinforcement for Concrete Structures Bulletin 14, July.

Ghobarah, A., 2004, "Seismic Retrofit of RC Walls Using Fibre Composites," McMaster University, Hamilton, Onterio, Canada.

Kaw, A.K., 1997, "Mechanics of Composite Materials," CRC Press LLC, FL.

Laursen, P.T., Seible, F., and G.A. Hegemier, 1995, *Seismic Retrofit and Repair of Reinforced Concrete with Ovelays*, University of California, San Diego, CA.

Priestley, N., Seible, F. and G.M. Calvi, 1996, *Seismic Design and Retrofit of Bridges,* John Wiley & Sons, New York, NY.

Teng, J.G., J.F. Chen, S.T. Smith, and L. Lam, 2002, *FRP Strengthened RC Structures,* John Wiley & Sons, New York, NY.

Chapter 14 - Building Type C2f: Concrete Shear Walls (Gravity Frame Systems)

14.1   Description of the Model Building Type

Reinforced concrete walls in a building will act as shear walls whether designed for that purpose or not. Therefore, concrete buildings that contain any significant amount of concrete wall will fall into this category. However, there are two distinctly different types of concrete wall buildings: those that contain an essentially complete beam/slab and column gravity system, and those that use bearing walls to support gravity load and have only incidental beam and column framing. In this document, these building types have been separated and are designated **C2f** for the gravity frame system and **C2b** for the bearing wall. This section covers the building with gravity framing system. Although it is typically assumed that the gravity framing is not part of the lateral force-resisting system, the framing could add stiffness to the building, particularly near the top of taller buildings. This building type is very common and has been used in a wide variety of occupancies and in all sizes.

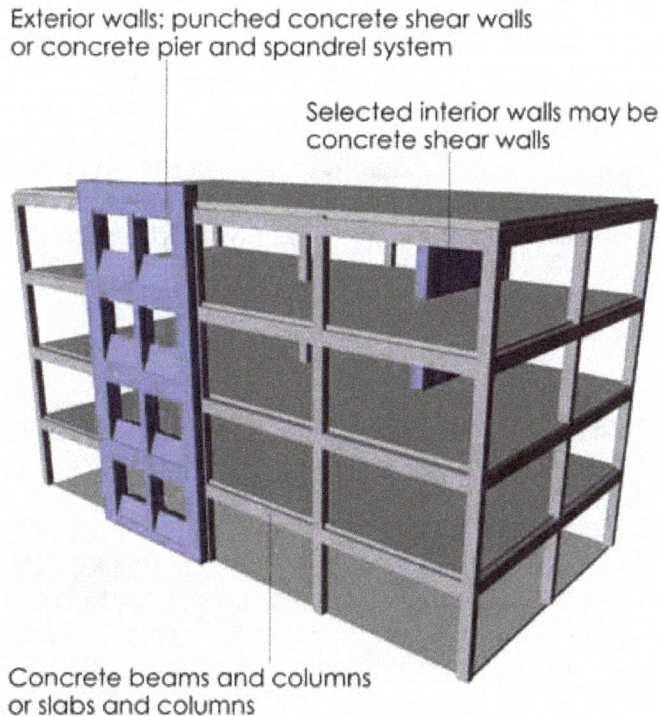

Exterior walls: punched concrete shear walls
or concrete pier and spandrel system

Selected interior walls may be
concrete shear walls

Concrete beams and columns
or slabs and columns

**Figure 14.1-1: Building Type C2f: Concrete Shear Walls
(Gravity Frame Systems)**

Variations in Framing Systems

There are wide overall variations within this building type due to the possible configuration and extent of the concrete walls, the many types of vertical framing systems used, and the lateral stiffness interaction between the two. In buildings with incidental concrete walls and a substantial beam-column gravity frame system, this building type merges with Building Type **C1**. If the building type is unclear, reference should be made to both Chapter 12 and this chapter.

Gravity frame systems in this building type include cast-in place concrete beam and slab, one-way joists, two-way waffles, and two-way or flat slabs.

In older buildings that are seismically deficient, the walls were often intended for fire protection of vertical shafts, as exterior closure walls, or as bearing walls. However, buildings built in regions of high seismicity in the 1950s, 1960s or early 1970s often were designed with a shear wall lateral force-resisting system, but they are now found deficient due to low global strength, a highly torsional plan layout or detailing that leads to premature shear failure

In buildings designed with shear walls, the walls are either strategically placed around the plan, or at the perimeter. Shear walls systems placed around the entire perimeter almost always contain windows and other perimeter openings and are often called punched shear walls. Older buildings will have concrete walls somewhat arbitrarily placed in plan.

Floor and Roof Diaphragms

The floor and roof diaphragms in this building type are essentially the same as the bearing wall system and are almost always cast-in-place concrete. The diaphragms are stiff and strong in shear because the horizontal slab portion of the gravity system is either thick or frequently braced with joists. However, one way joist systems could be inadequate in shear in the direction parallel to the joists. Collectors are seldom in place and transfer of load from diaphragm to shear wall must be carefully considered.

Foundations

There is no typical foundation for this building category. Foundations could be found of every type depending on the height of the building, the span of the gravity system and the site soil.

14.2   Seismic Response Characteristics

Shear wall buildings, unless configured with only incidental or minimal walls, will typically be quite stiff. Elastic and early post-elastic response will therefore be characterized with lower-than-average drifts and higher-than-average floor accelerations. Damage in this range of response should be minimal.

Overall post-elastic response is highly dependent on the specific characteristics of the shear walls and the gravity frame components.

Shear Wall Behavior

The walls must first be evaluated to determine if they contribute sufficient strength or stiffness to be considered significant. Often walls around vertical shafts are thin and lightly reinforced and will have little effect on the overall building response. Although retrofit techniques are similar such a building may be classified as Type **C1.**

When subjected to ever increasing lateral load, individual shear walls or piers will first often force yielding in spandrels or other horizontal components restricting their drift, and eventually either rock on their foundations, suffer shear cracking and yielding, or form a flexural hinge near the base. Shear and flexural behavior is quite different, and estimates of the controlling action are affected by the distribution of lateral loads over the height of the structure.

Yielding of spandrels or other coupling beams can cause a significant loss of stiffness in the structure. Flexural yielding will tend to maintain the strength of the system, but shear yielding, unless well detailed, will degrade the strength of the coupling component and the individual shear wall or pier will begin to act as a cantilever from its base.

Rocking is often beneficial, limiting the response of the superstructure. However, the amplified drift in the superstructure from rocking must be considered. In addition, if varying wall lengths or different foundation conditions lead to isolated or sequencing rocking, the transfer of load from rocking walls must be investigated. In buildings with basements, the couple created from horizontal restraint at the ground floor diaphragm and the basement floor/foundation (often termed the "backstay" effect) may be stiffer and stronger than the rocking restraint at the foundation and should be considered in those configurations.

Shear cracking and yielding of the wall itself is generally considered undesirable, because the strength and stiffness will quickly degrade, increasing drifts in general, as well as potentially creating a soft story or torsional response. However, in accordance with FEMA 356 (FEMA 2000), shear yielding walls or systems can be shown to be adequate for small target displacements.

Flexural hinging is considered ductile in FEMA 356 and will degrade the strength of the wall only for larger drifts. Similar to rocking, the global effect of the loss of stiffness of a hinging wall must be investigated.

Gravity Frame Behavior

The lateral strength and stiffness of gravity frames will vary considerably among buildings in this type. In some configurations of this building type, the gravity frame will not significantly participate in the response. However, it is not uncommon in these buildings for a stiff and brittle gravity system to dominate both response and the extent of damage. For example, if concrete spandrels or sills on the perimeter of the building restrain the gravity columns (the "short column"), the column must take the full story drift over a short height, potentially causing shear failure and loss of gravity load capacity. Other gravity systems, such as flat slab or heavy beam and column systems, can also be sensitive to drifts, particularly to the increased drifts near the top of buildings with walls of a height-to-width ratio over 3. The frame action of the gravity system of these buildings may be beneficial or could form a deficiency, but in any case the interaction with the shear walls should be considered.

14.3 Common Seismic Deficiencies and Applicable Rehabilitation Techniques

See Table 14.3-1 for deficiencies and potential rehabilitation techniques particular to this system. Selected deficiencies are further discussed below by category.

Global Strength

Older buildings placed in this category may have minimal shear walls and seismic displacements will likely put excessive demand on the walls regardless of the shear or flexural behavior. The most common method of mitigating this deficiency is to add more shear walls, although for smaller buildings steel braced frames have been used. The nonlinear behavior of steel braced frames must be carefully studied for compatibility with the other shear walls and the gravity frame. Additional bending resistance of existing walls can be obtained by enhancing existing chords, although such walls should not be strengthened to become shear critical. If the majority of walls are shear-critical and strength degradation is the primary concern, shear strength can be added to the existing walls with concrete or FRP overlays.

Global Stiffness

The most important issue in a building of this type that exhibits large interstory drifts is the ability of the gravity system to accept the drifts while sustaining their loads. Excess drifts could be caused by inadequate length of wall, by rocking at the foundations, or, at the upper stories, by a deformed shape characterized by bending deformations. In most buildings, strength and stiffness are closely related, and inadequate stiffness is mitigated by adding new elements or stiffening existing ones, which generally will also increase strength. Damping can also be added to reduce drifts but care must be taken to achieve the desired damping with small displacement expected in a shear wall building.

Configuration

The two most common configuration deficiencies in this building type are 1) severe torsion caused by eccentrically placed shafts or towers, and 2) shear walls or full stories that are weak or soft from openings in the walls or from discontinuous walls that may not run through the ground floor or basement floor. Completely discontinuous walls also create a load transfer deficiency for both overturning and shear.

Load Path

Common load path deficiencies include discontinuous shear walls, as discussed above, and collectors for the shear walls. Collectors can be added with steel members or new reinforcing and concrete.

Component Detailing

Shear walls in most older buildings meet none of the current detailing requirements covering minimum shear reinforcement, for confinement of chords, and the walls are commonly shear critical. FEMA 356 allows these deficiencies at controlled displacement levels. A common

Table 14.3-1: Seismic Deficiencies and Potential Rehabilitation Techniques for C2f Buildings						
Deficiency		Rehabilitation Technique				
Category	Deficiency	Add New Elements	Enhance Existing Elements	Improve Connections Between Elements	Reduce Demand	Remove Selected Components
Global Strength	Insufficient in-plane wall shear strength	Concrete/masonry shear wall [12.4.2] Steel braced frame [12.4.1] Steel plate shear wall	Concrete wall overlay [21.4.5] Fiber composite wall overlay [13.4.1] Steel wall overlay		Seismic isolation [24.3] Reduce flexural capacity [13.4.4]	
	Insufficient flexural capacity	Concrete/masonry shear wall [12.4.2] Steel braced frame [12.4.1]	Add or enhance chords [12.4.3]			
	Inadequate capacity of coupling beams	Concrete/masonry shear wall [12.4.2] Steel Braced frame [12.4.1]	Strengthen beams Improve ductility of beams [13.4.2]			Remove beams
Global Stiffness	Excess drift (normally near the top of the building)	Concrete/masonry shear wall [12.4.2] Steel braced frame [12.4.1]	Fiber composite column wrap [12.4.4] Concrete/steel column jacket [12.4.5] Provide detailing of all other elements to accept drifts Thicken walls		Supplemental damping [24.4]	
Configuration	Discontinuous walls	Concrete/masonry shear wall [12.4.2]	Enhance existing column for overturning loads	Improve connection to diaphragm [13.4.3]		Remove wall
	Soft story or weak story	Add strength or stiffness in story to match balance of floors				

14-5

Table 14.3-1: Seismic Deficiencies and Potential Rehabilitation Techniques for C2f Buildings						
Deficiency		Rehabilitation Technique				
Category	Deficiency	Add New Elements	Enhance Existing Elements	Improve Connections Between Elements	Reduce Demand	Remove Selected Components
Configuration (continued)	Re-entrant corner	Add floor area to minimize effect of corner		Provide chords in diaphragm [12.4.3]		
	Torsional layout	Add balancing walls, braced frames, or moment frames				
Load Path	Inadequate collector	Add steel collector [12.4.3] Add concrete collector [12.4.3]	Strengthen existing beam or slab Enhance splices or connections of existing beams			
	Discontinuous Walls	Provide new wall support components to resist the maximum expected overturning moment	Strengthen the existing support columns for the maximum expected overturning moment Provide elements to distribute the shear into the diaphragm at the level of discontinuity			
Component Detailing	Wall inadequate for out-of-plane bending	Add strongbacks	Concrete wall overlay [21.4.8]			
	Wall shear critical		Concrete wall overlay [21.4.8] Fiber composite wall overlay [13.4.1]		Reduce flexural capacity of wall [13.4.4]	
	Inadequate displacement capacity of gravity columns		Enhance ductility (see also global stiffness)			

14-6

Table 14.3-1: Seismic Deficiencies and Potential Rehabilitation Techniques for C2f Buildings

Deficiency		Rehabilitation Technique				
Category	Deficiency	Add New Elements	Enhance Existing Elements	Improve Connections Between Elements	Reduce Demand	Remove Selected Components
Diaphragms	Inadequate in-plane shear capacity		Concrete topping slab overlay Fiber composite overlays [22.2.5]			
	Inadequate chord capacity	New concrete or steel chord member [12.4.3]				
	Excessive stresses at openings and irregularities	Add chords [12.4.3]				Infill openings [22.2.4]
Foundations	See Chapter 23					
[ ] Numbers noted in brackets refer to sections containing detailed descriptions of rehabilitation techniques.						

improvement to these walls is to enhance shear strength to be equal or greater than the maximum that can be developed in the wall, based on bending strength.

Diaphragm Deficiencies

The most common diaphragm deficiency in this building type is a lack of adequate collectors. The addition of effective collectors in an existing diaphragm is difficult and disruptive. Existing strength to deliver loads to the shear walls should be studied carefully before adding new collectors.

Foundation Deficiencies

See Chapter 23.

14.4   Detailed Description of Techniques Primarily Associated with This Building Type

The most relevant recommendations listed in Table 14.3-1 are similar to techniques also associated with other concrete building types such as **C1** and **C2b** or general techniques applied to concrete diaphragms. Details concerning these techniques can be found in other chapters.

14.5   References

FEMA, 2000, *Prestandard and Commentary for the Seismic Rehabilitation of Buildings*, FEMA 356, Federal Emergency Management Agency, Washington, D.C.

Chapter 15 - Building Types C3/C3A: Concrete Frames with Infill Masonry Shear Walls

15.1   Description of the Model Building Type

Building Type **C3** is normally an older building that consists of an essentially complete gravity frame assembly of concrete columns and floor systems. The floors consist of a variety of concrete systems including flat plates, two-way slabs, and beam and slab. Exterior walls, and possibly some interior walls, are constructed of unreinforced masonry, tightly infilling the space between columns horizontally and between floor structural elements vertically, such that the infill interacts with the frame to form a lateral force-resisting element. Windows and doors may be present in the infill walls. The buildings intended to fall into this category often feature exposed clay brick masonry on the exterior. Figure 15.1-1 shows an example of this building type.

It is important to note that similar buildings with exterior masonry infill sills below windows that extend column to column do not behave with strut action and should be classified as Building Type **C1** or **C2f**. In fact, such infill sills often create "short columns" that must absorb the entire story drift over the unrestrained height of the window, which can often be an extreme deficiency in poorly reinforced columns. The **C3A** building type is similar but has floors and roof that act as flexible diaphragms such as wood, or untopped metal deck.

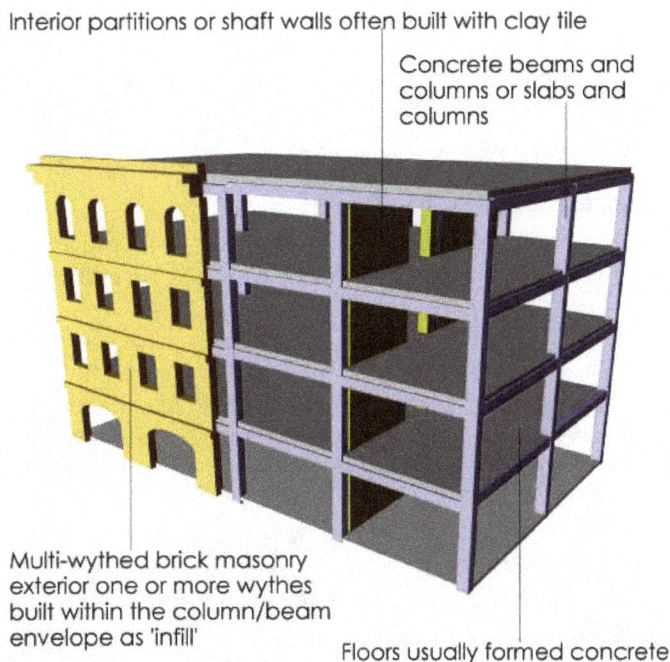

Figure 15.1-1: Building Type C3: Concrete Frames with Infill Masonry Shear Walls

Variations Within the Building Type

The building type was identified primarily to capture the issues of interaction between unreinforced masonry and concrete gravity framing. The archetypical building has solid clay brick at the exterior with one wythe of brick running continuously past the plane of the column and beam and two or more wythes infilled within the plane of the column and beam. The exterior wythe of clay brick forms the finish of the building although patterns of terra cotta, stone, or precast concrete may be embedded into the brick. However, there can be many variations to this pattern depending on the number and arrangement of finished planes on the exterior of the building. For example, the full width of the infill wall may be located with the plane of the column and beam with a pilaster built out and around the column and a horizontal band of brick or other material covering the beam; the beam may also be slightly offset from the centerline of the column to accommodate the pattern of exterior finishes.

Hollow clay tile masonry may also be used as an exterior infill material. Although this material often has a high compression strength, the net section of material available to form the compression strut within the frame will normally contribute a lateral strength of only a small percentage of the building weight. The material being brittle and the wall being highly voided, these walls may also lose complete compressive strength quite suddenly. Therefore, walls of hollow clay tile infill will probably not contribute a significant portion of required lateral resistance except in areas of low seismicity and/or when walls are arranged as infill on both the exterior and interior of the building.

More recent buildings may have unreinforced concrete block masonry configured as an exterior infill wall, with a variety of finish materials attached to the outside face of the concrete block. Similar to hollow clay tile walls, these walls may exhibit moderate to low compressive strength and brittle behavior that marginalizes their usefulness as lateral elements. In addition, hollow concrete block exterior walls often will not be installed tight to the surrounding framing, eliminating infill compression strut behavior.

Floor and Roof Diaphragms

Floors are often flat plates or two-way slabs. Beam and slab or beam and joist systems will also be found in this building type. Typically, these slabs provide adequate diaphragms.

Building Type **C3A** will have heavy timber floors with one or more layers of sheathing forming a diaphragm. The flexibility of such diaphragms will often form a seismic deficiency because, assuming no interior shear elements, the large drift at the diaphragm mid-span will damage perpendicular walls and gravity framing. Specific strengthening techniques for this building type are not covered here. For generalized strengthening of diaphragms, see Chapter 22.

Foundations

There is no typical foundation for this building type. Foundations can be found of every type depending on the height of the building, the span of the gravity system and the site soil. The exterior walls are exceptionally heavy and typically will be supported by a continuous concrete footing or often a continuous concrete wall forming a basement space below.

15.2   Seismic Response Characteristics

Both in terms of stiffness and strength, the exterior infill walls will form the effective lateral system for this building type. The effectiveness of the system depends on the size and extent of openings and articulation of the plane of the wall. With solid or nearly solid infill panels, strut action will be stiff and strong. As openings in panels increase in size, struts or combinations of struts cannot effectively form around the opening, and the concrete columns and beams may begin to work as a moment frame, with "fixity" at the beam-column joint provided by the masonry. For low and moderate intensity shaking, the exterior walls may provide adequate strength to satisfy the specified performance objective. As the shaking demand increases, the masonry will tend to crack and spall, losing stiffness and potentially creating a falling hazard. The complete concrete gravity system, characteristic of this building type, will provide additional stability, but may quickly degrade in strength and stiffness due to inadequate column reinforcing. However, in wall configurations with large height-to-width ratios, end or corner columns could experience large compression or tension loads from overturning, leading to rapid degradation of column lateral and gravity capacity.

This building type is often characterized by a commercial store-front first floor with little or no infill at that level on one or more faces of the building. This condition can cause a soft story condition or a severe torsional response if open on one or two sides only. Such conditions can lead to concentration of seismic deformation at the open level, degradation of the columns, and possible P-delta failure.

15.3   Common Seismic Deficiencies and Applicable Rehabilitation Techniques

See Table 15.3-1 for deficiencies and potential rehabilitation techniques particular to this system. Deficiencies related to specifically to concrete moment frames and to masonry shear walls are shown in Table 12.3-1 and Table 18.3-1, respectively. Selected deficiencies are further discussed below by category.

Global Strength

The overall strength provided by the exterior walls may be insufficient to prevent serious degradation and resulting amplified displacements that can lead to irreparable damage or even instability. The strength may be limited by inadequate number of panels of infill, excessive openings, or masonry weak in compressive strength. The standard approach to such deficiencies will be to add new, relatively stiff lateral force-resisting elements such as concrete shear walls or steel braced frames often located on the interior between existing columns. The infill itself could also be strengthened by adding a layer of reinforced concrete on the interior surface, although whether such a system is acting as infill or as a shear wall must be checked by analysis. Unless all infill has been backed by a support system, the damage state of the infill wall must be estimated for the expected drifts of the combined system to determine if the desired performance has been achieved.

Global Stiffness

The most important issue in a building of this type that exhibits large interstory drifts is the ability of the gravity system to accept the drifts while sustaining their loads. In most buildings,

Table 15.3-1: Seismic Deficiencies and Potential Rehabilitation Techniques for C3/C3A Buildings					
Deficiency		Rehabilitation Technique			
Category	Deficiency	Add New Elements	Enhance Existing Elements	Improve Connections Between Elements	Reduce Demand	Remove Selected Components
Global Strength	Inadequate length of exterior wall	Interior concrete walls [12.4.2] Interior steel braced frames [12.4.1]	Concrete wall overlay [21.4.5] Fiber composite wall overlay [21.4.6]			
	Excessive sized openings in infill panels	Interior concrete walls [12.4.2] Interior steel braced frames [12.4.1]	Infill selected openings [21.4.7] Fiber composite wall overlay [21.4.6]			
	Inadequate columns for overturning forces		Add confinement Add tensile capacity on outside surface of column.			
	Weak or deteriorated masonry	Interior concrete walls [12.4.2] Interior steel braced frames [12.4.1]	Point outside and/or inside wythes of masonry Inject wall with cementitious grout Fiber composite overlay [21.4.6]			
Global Stiffness	See Global Strength					
Configuration	Soft or weak story	Interior concrete walls [12.4.2] Interior steel braced frames [12.4.1]				
	Torsion from one or more solid walls	Balance with Interior concrete walls Balance interior steel braced frames				Remove selected infill panels on solid walls

Table 15.3-1: Seismic Deficiencies and Potential Rehabilitation Techniques for C3/C3A Buildings						
Deficiency		**Rehabilitation Technique**				
Category	**Deficiency**	**Add New Elements**	**Enhance Existing Elements**	**Improve Connections Between Elements**	**Reduce Demand**	**Remove Selected Components**
Configuration (continued)	Irregular Plan Shape	Balance with Interior concrete walls Balance with Interior steel braced frames				
Load Path	Out-of-plane failure of infill due to loss of anchorage or slenderness of infill		Provide surface wall supports [13.4.3] Shotcrete overlays [21.4.5] Fiber composite overlay [21.4.6]			Remove infill
	Inadequate connection of finish wythe to backing		Add connections			
	Inadequate collectors	Add steel collector or concrete collector [12.4.3]				
Component Detailing	Inadequate columns splice for tension due to uplift force induced by infill			Add splice plates Provide splice through added reinforced concrete encasement		
	Inadequate beam column connection to resist compression thrust			Strengthen connection in shear with steel or fiber composite		

Table 15.3-1: Seismic Deficiencies and Potential Rehabilitation Techniques for C3/C3A Buildings						
Deficiency		**Rehabilitation Technique**				
Category	**Deficiency**	**Add New Elements**	**Enhance Existing Elements**	**Improve Connections Between Elements**	**Reduce Demand**	**Remove Selected Components**
Component Detailing (continued)	Weak or incompletely filled joint between masonry and surrounding steel components			Create appropriate clean void and repack with masonry and/or mortar. Inject voids with cementitious grout or epoxy		
Diaphragms	See Chapter 22					
Foundations	See Chapter 23					
[ ] Numbers noted in brackets refer to sections containing detailed descriptions of rehabilitation techniques.						

strength and stiffness are closely related, and inadequate stiffness is mitigated by adding new elements or stiffening existing ones, which generally will also increase strength. Damping can also be added to reduce drifts, but care must be taken to achieve the desired damping with the small displacement expected in a shear wall building.

Configuration

Two configurational deficiencies are common in this building type. The first is a soft and weak story at the street level created by commercial occupancies with exterior bays with little or no infill. This deficiency can be corrected by adding selected bays of infill or by adding shear walls or braced frames at this level. The second common issue is a plan torsional irregularity created by solid masonry walls on property lines coupled with walls with many openings on street fronts. If shown by analysis to be necessary, torsional response can be minimized by stiffening the more flexible side of the building with more infill or by the addition of lateral elements. In rare cases, the solid walls can be balanced with the open side by selected removal of panels or disengagement of the infill strut action.

Load Path

The primary load path issue with this building type is to assure that the mass of the exterior walls will not become disengaged from the frame which will both prevent infill strut action as well as to create a significant falling hazard on the street below.

In-plane, the articulation of the exterior walls may result in offsets of the wall plane between floors. The presence of a complete load path and maintenance of confinement for strut formation must be reviewed in such instances.

If new lateral load-resisting elements are added, existing slab and/or beam construction may need to be strengthened to provide adequate collectors.

Component Detailing

In order to qualify as an infill lateral force-resisting element, the infill must be installed tight to the surrounding concrete elements. Loose or incomplete infill can be mitigated with local patching of the masonry or by injection of cemetitious or epoxy grout.

As previously noted, the infill adjacent to columns must be of sufficient stiffness to provide a floor to floor diagonal strut. With narrow piers surrounding columns, the jamb and header masonry may restrain the column such that the entire story drift must be absorbed in the central length of the column, often creating dangerous shear failures.

The detailing of the concrete frame forming the confinement for the masonry is important to achieve infill strut behavior. The connection of beam to column must be capable of resisting the strut compression forces from the masonry. The shear capacity of the beam-to-column connection is often critical. Strengthening of these connections may require removal of considerable masonry to obtain adequate access. Some of the techniques developed to strengthen concrete moment frames may be applicable. In addition, splices in vertical reinforcing of column splices may be inadequate to transfer the tensile overturning forces created by strut action. These areas can be confined to reduce the required splice length or augmented with additional tensile strength from the surface.

Diaphragm Deficiencies

Concrete slab diaphragms often will be adequate. The connection of slab to exterior wall should be reviewed.

See Chapter 21 on URM construction for discussion of wood diaphragms in this type of building.

Foundation Deficiencies

No systematic deficiency in foundations should be expected solely due to the characteristics of this building type.

15.4 Detailed Description of Techniques Primarily Associated with This Building Type

Most significant recommendations listed in Table 15.3-1 are similar to techniques more commonly associated with other building types such as various concrete buildings (**C1**, **C2b**, or **C2f**), unreinforced masonry bearing wall buildings (**URM**), or general techniques applied to concrete diaphragms. Details concerning these techniques can be found in other chapters.

Chapter 16 - Building Type PC1: Tilt-up Concrete Shear Walls

16.1   Description of the Model Building Type

Building Type **PC1** is constructed with concrete walls, cast on site and tilted up to form the exterior of the building. **PC1** buildings are used for many occupancy types including warehouse, light industrial, wholesale and retail stores, and office. The majority of these buildings are one story; however, tilt-up buildings of up to three and four stories are common, and a limited number with more stories exist. For many years, tilt-up buildings have been primarily large box-type buildings with the tilt-up walls at the building perimeter; this is by far the largest group of **PC1** buildings in the U.S. inventory. In recent years, tilt-up construction has expanded to more varied uses and building configurations. Figure 16.1-1 illustrates one example of a **PC1** building.

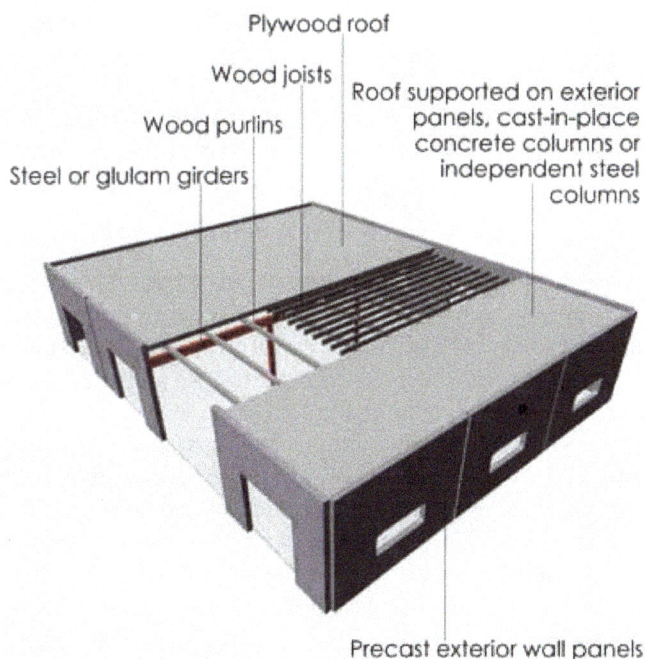

Plywood roof

Wood joists

Wood purlins

Steel or glulam girders

Roof supported on exterior panels, cast-in-place concrete columns or independent steel columns

Precast exterior wall panels

Figure 16.1-1: Building Type PC1: Tilt-Up Concrete Shear Walls

Key to the **PC1** building type as addressed by this chapter is the combination of a flexible roof diaphragm and rigid walls. Lateral forces in **PC1** buildings are resisted by flexible wood sheathed or steel deck roof diaphragms, wood, composite steel deck, or precast floor diaphragms, and tilt-up concrete shear walls. In some local areas, walls are concrete panels or T-beams cast off-site, rather than on-site; **PC1** also applies to this variation.

One-story **PC1** buildings with flexible roof diaphragms are the primary focus of this chapter; however, discussion of wall-to-flexible diaphragm anchorage is equally applicable to the roof of multistory **PC1** buildings. For other rehabilitation issues in multistory buildings, refer to the **C1** Building Type. For variations with rigid diaphragms at floors and roof, see Building Type **PC2**.

Guidelines for Seismic Evaluation and Rehabilitation of Tilt-Up Buildings and Other Rigid Wall/Flexible Diaphragm Structures or *SEAONC Guidelines* (SEAONC, 2001) provides a substantial collection of information on West Coast **PC1** building configurations, experience with earthquake performance, rehabilitation priorities, and techniques for rehabilitation. This chapter highlights the major rehabilitation considerations from the *SEAONC Guidelines* document, and provides suggested adaptations for construction variations. Provisions addressing rehabilitation can also be found in *International Existing Building Code (IEBC) Appendix Chapter A2* (ICC, 2003b), *City of Los Angeles Building Code Chapter 91* (City of Los Angeles, 2002) and *Guidelines for Seismic Retrofit of Existing Buildings* (GSREB) (ICBO, 2001). These provisions focus on wall anchors, diaphragm cross-ties, and collectors, with the goal of hazard reduction. This chapter and the *SEAONC Guidelines* address a broader range of rehabilitation issues and techniques.

The *Tilt-up Construction and Engineering Manual* (Tilt-Up Concrete Association, 2004), now in its sixth edition, serves as a primary resource for design and detailing practice for new tilt-up construction.

Walls

Tilt-up exterior walls are the primary vertical elements in the lateral force-resisting system. Tilt-up buildings with large plan areas may have interior tilt-up walls or braced steel frames providing additional lateral resistance. Large-scale construction of tilt-up buildings began in the 1950s and 1960s with primarily solid wall panels for warehouse and light-industrial use. As tilt-up construction expanded to commercial use in the 1970s and 1980s, the wall panels changed to include large window and door openings and multistory construction. The publications *Recommended Tilt-up Wall Design* (SEAOSC, 1979) and *Test Report on Slender Walls* (ACI-SEASC, 1982) document the change from use of code-prescribed height to thickness (h/t) limits of 25 for bearing walls and 36 for other walls to use of much higher h/t ratios, in combination with rational analysis of slenderness effects. Other code changes of interest have occurred for wall pier reinforcing requirements and wall panel connection to the foundation or slab-on-grade.

Most tilt-up panels are a single piece from the foundation to the top of the building; however, some tilt-up systems use separate wall panels at upper stories, or lintel (spandrel) panels that are supported on other tilt-up wall panels. Welded connections of these panels can be damaged when they restrict panel movement under earthquake load. Prior to start of rehabilitation design, it is important to identify the tilt-up panel joint and support locations and connection condition. Surface treatments (such as exposed aggregate) and reveal joints are commonly used to visually enhance tilt-up wall panels. These treatments effect the location and dimensions of critical sections for design.

Connections between adjacent tilt-up panels are often relied on to provide continuity for diaphragm chords and collectors. Connection types have varied over the years. Early tilt-up walls were often joined by cast-in-place pilasters. Later, welded connections between cast-in embedments or between horizontal reinforcing steel were commonly used. Both of these connection systems experience some problems with fractures at welds due to panel shrinkage and temperature movement. As a result, the codes placed carbon equivalent requirements on rebar to be welded, and detailing to minimize restraint of panel movement was pursued. In *Recommended Tilt-up Wall Design* (SEAOSC, 1979), minimal interconnection between the panels and special detailing of chord and collector connections are suggested to allow movement. Detailing suggestions included 1) breaking the bond between the chord reinforcing and concrete for one fourth the panel width in from each panel end and 2) use of steel angle chord/collector members connected in the center portion of the panel, but slotted near the panel ends. The panel connection detailing should be considered in evaluating the behavior of the chords and collectors and in the distribution of shear to the tilt-up panels.

Gravity Load Support at the Building Perimeter

It is most common for roof and floor framing members to be supported on the tilt-up panels at the building perimeter; however, some contractors have found it advantageous to provide steel gravity columns at the inside face of the tilt-up walls. This separation of the gravity and lateral systems makes construction tolerances for items embedded in the tilt-up panels and construction sequence less critical. The use of columns requires some modification of wall anchorage details at girders, but has little or no effect on the typical wall to diaphragm attachments, or fundamental building behavior.

Floor and Roof Diaphragms

The **PC1** roof system will generally be of either wood or steel construction. In the western states (primarily California, Oregon, Washington), roof systems are almost exclusively wood structural panel sheathed. Subpurlins (joists), purlins, and girders are most often wood; however, open web trusses with wood chords or nailers are also used. Wood girders are most often supported on steel columns at the building interior. Sheathing fastening and therefore unit shear capacity in wood structural panel diaphragms generally varies by nailing zone or area. In older West Coast **PC1** buildings, roofs can be lumber sheathed, and roof framing can include bow-string trusses.

Outside of the western states, roof systems are almost exclusively sheathed with steel decking with rigid foam or nonstructural concrete insulation. The roof framing system is most commonly of steel open web trusses (bar joists) used in combination with truss girders or hot-rolled steel beams.

Interior Additions

Mezzanines and interior second stories constructed within tilt-up buildings are common. The interior addition may be seismically separated and braced independently of the building shell (exterior walls and roof), or may be attached to and supported by the shell. Attachment to the building shell raises two potential issues for the tilt-up: 1) whether the seismic load to the tilt-up system is significantly increased beyond that considered in initial design, and 2) whether the interior addition restrains seismic deflections of the building and creates an unintended load path. Both of these issues should be addressed in evaluation and rehabilitation.

Foundations

Tilt-up buildings are often constructed on continuous or isolated spread footings; however, drilled pier and grade beam foundations are common in some regions. Often tilt-up wall panels are set on top of shims and grout pads on top of the foundation. Where continuous foundations are provided, grout may then be placed under the full length of the panel to provide continuous bearing. Often, no direct connection is provided between the wall panel and foundation. While not common practice in older tilt-up construction, in many buildings constructed in the 1980s and later, wall panel connection to the slab-on-grade were provided, allowing transfer of horizontal seismic forces to the slab. The most common connection uses rebar dowels, cast into the wall panel and a slab closure strip. Other approaches include welded or bolted connections between cast-in connection plates or threaded inserts. See Section 16.4.5 for further discussion.

Due to the stiffness of the wall system, tilt-up buildings are best located on sites with very stable soils; however, they are often relegated to poor soil sites. Where tilt-up buildings are located on sites with soils subject to expansion, consolidation, or liquefaction, the effects of any damage to the foundation and wall connections due to soil movement should be considered in building evaluation.

16.2   Seismic Response Characteristics

The seismic response of the classic large box-type one-story **PC1** building is characterized by rigid wall and flexible diaphragm behavior. In this type of building, the concrete walls will have a very short period, while the diaphragm has long-period behavior. Amplification of seismic forces near the center of wood sheathed diaphragms loaded in the transverse direction can be significant. This creates high demand on the diaphragm and out-of-plane wall anchorage near the center of the diaphragm, and it can also generate very high shear demand at the diaphragm connection to the end walls. This behavior has been replicated in instrumented buildings and in laboratory testing (Fonseca, Wood and Hawkins, 1996). Roof diaphragm amplification in the longitudinal direction can be lower, depending on the diaphragm aspect ratio; however, significant overstrength can result in large anchorage forces in this direction also.

Observation of the earthquake performance of these buildings has identified as important 1) understanding the magnitude of the wall anchorage force and 2) detailing to eliminate weak links in the wall anchorage connection. With each subsequent earthquake since 1971, building code requirements have been revised to reflect changing understanding of the magnitude of forces generated in wall anchors and requirements for proper detailing; simply providing positive connections from the walls to the diaphragm has not resulted in adequate performance. A variety of weaknesses in the connections have kept anchorages from developing adequate capacity. Included are cross-grain tension of wood ledgers, rotation of non-symmetric connectors, low-cycle fatigue of straps that buckle under compression loads, net section fracture of straps with punches bolt holes, etc. The current code wall anchorage force and detailing requirements reflect a history of knowledge gained for wood structural panel diaphragms in large box-type buildings.

Recent construction of **PC1** buildings has moved away from this classic large box-type building with wood structural diaphragms. In most regions of the United States, **PC1** buildings are now constructed with steel deck diaphragms. Many **PC1** box-type buildings are being constructed with very tall walls, in which out-of-plane behavior could potentially modify building seismic response. Tilt-up construction has expanded to include a wide variety of building uses with different building characteristics, including smaller and less regular plans, less regular diaphragm configurations, and some with interior as well as exterior tilt-up walls. The seismic response characteristics of these building configurations are not known. Their behavior is of much interest, however, because current building code provisions may not adequately address these variations and materials. The Tilt-Up Concrete Association has just initiated the TCA Seismic Performance Initiative, with the intent of identifying and developing strategic plans to resolve issues of building performance and code requirements for design and detailing. Initial work will focus on design models developed from instrumented building behavior (Freeman, Searer and Gilmartin, 2002). Research on tilt-up building performance is also in progress at Canterbury University, New Zealand.

16.3   Common Seismic Deficiencies and Applicable Rehabilitation Techniques

PC1 buildings with wood sheathed roof diaphragms have experienced structural damage and partial building collapse in a number of California earthquakes, as well as the 1964 Alaska earthquake. Partial collapse is almost exclusively associated with inadequate connection of the walls to the flexible roof diaphragm for out-of-plane loading. A variety of other types of damage to the wall panels, connections and roof diaphragms have been observed, including some interior diaphragm failures observed in the Northridge earthquake. These observations have been almost exclusively of buildings with wood diaphragms.

Currently used design provisions for new buildings [IBC (ICC, 2003a), ASCE 7 (ASCE, 2005), and NEHRP (FEMA, 2004)] contain one set of requirements addressing wall anchors and cross-ties for concrete and masonry walls used with flexible diaphragms. These requirements do not do not differentiate between wood and steel deck diaphragms in application these requirements, nor do they differentiate between different possible tilt-up building configurations and behaviors. Opportunities to observe earthquake performance of steel deck diaphragms and newer building configurations have been limited to date. Until performance information is available from research or earthquake experience, the vulnerable behavior seen in wood diaphragm **PC1** buildings needs to be considered a possibility for all **PC1** buildings. The fundamental rehabilitation concept is positive anchorage of tilt-up walls to supporting diaphragms, with anchorage load paths adequate for forces both away from and towards the diaphragm.

The *SEAONC Guidelines* document provides a detailed list of rehabilitation measures and relative priorities based on both potential hazard and the level of design of the existing construction. The major categories of deficiencies and rehabilitation (not prioritized) are:

> Out-of-plane wall anchors to walls and pilasters
> Diaphragm cross-ties
> Collectors
> Diaphragm strength, stiffness and openings
> Wall in-plane shear connections
> Wall in-plane capacity
> Wall in-plane base anchorage
> Wall out-of-plane bending

These deficiencies and rehabilitation measures are included in Table 16.3-1 and the general discussion below; however, the compilation of information in the *SEAONC Guidelines* is recommended as a useful resource.

Global Strength and Stiffness

Global strength and stiffness of the tilt-up walls have not been seen as a significant source of damage in **PC1** buildings to date. This is likely due to the length of solid wall provided in older buildings, minimum reinforcing ratios, and the tendency to design at low stress levels so that a single layer of reinforcing without special detailing can be used. Strength and stiffness deficiencies are most likely to occur at wall lines with significant number of large penetrations and other locations where loads are carried by a limited number of panels. Rehabilitation measures include addition of new vertical elements, enhancing existing walls, and infilling openings in existing walls.

Configuration

Poor distribution of shear walls can result in torsionally irregular behavior of **PC1** buildings. Common occurrences include an entire line of highly perforated tilt-up panels such as at loading dock walls in distribution and storage facilities and street front walls in commercial buildings. Concrete cracking and spalling have been seen in perforated wall panels that act as frames. The most direct approach to rehabilitation of this condition is the addition of strength and stiffness in line with the perforated wall. This can be accomplished through addition of new shear walls, enhancing of existing shear walls, or addition of steel braced frames.

Re-entrant corners are reasonably common in large box-type **PC1** buildings, either due to in-set panels (Figure 16.3-1), or an L-shaped building plan. The in-set walls create a hard spot in the diaphragm, restraining it from the deflection required to transmit load to the end wall. In most cases, the in-set walls will have to be considered shear walls supporting the diaphragm. Where chords and collectors have not been provided at the re-entrant corner, the diaphragm has been seen to pull away from the wall, damaging gravity and lateral load connections. Rehabilitation at re-entrant corners requires the provision of adequate chords and collectors, shear transfer to the in-set wall panels, and possibly the strengthening of the diaphragm, wall panels, and connections to the foundation. In some cases, high earthquake loads in existing elements may make it necessary to add new vertical elements in line with the re-entrant corner. The *SEAONC Guidelines* suggest that there may be diaphragm continuity over this interior diaphragm support, increasing the diaphragm reaction to the in-set wall line. Alternately, it may be possible to allow the wall at the re-entrant corner to rock, or to separate the wall from the diaphragm allowing the diaphragm to span to the exterior wall. These approaches require complex detailing, however.

Some very large tilt-up buildings are constructed in configurations that resemble multiple buildings alongside each other, as seen in Figure 16.3-2. This figure illustrates a single large tilt-up building constructed in three sections, separated by roof expansion joints. Thermal movements in the steel deck diaphragms make the expansion joints necessary. If each of the three building sections were provided a complete lateral force-resisting system and separated by adequate seismic joints, this building configuration would be of little concern. As shown in Figure 16.3-2, however, building sections are often laterally braced off of adjacent building sections, using shear transfer through the expansion joint. This configuration raises concerns of

Table 16.3-1: Seismic Deficiencies and Potential Rehabilitation Techniques for PC1 Buildings						
Deficiency		Rehabilitation Technique				
Category	Deficiency	Add New Elements	Enhance Existing Elements	Improve Connections Between Elements	Reduce Demand	Remove Selected Components
Global Strength	Insufficient in-plane strength of shear walls or frames	Steel braced frame [7.4.1] Concrete/masonry shear wall [21.4.8]	Fiber composite wall overlay [13.4.1] Concrete wall overlay [21.4.5] Infill openings			
Global Stiffness	Insufficient in-plane stiffness of shear walls or frames	Steel braced frame [7.4.1] Concrete/masonry shear wall [21.4.8]	Fiber composite wall overlay [13.4.1] Concrete wall overlay [21.4.5] Infill openings			
Configuration	Torsionally irregular plans (highly perforated wall line)	Steel braced frame [7.4.1] Concrete/masonry shear wall [21.4.8]	Enhance existing collector [7.4.2] Fiber composite wall overlay [13.4.1] Concrete wall overlay [21.4.5] Infill openings			
	Re-entrant corners	Steel braced frame [7.4.1] Collector [7.4.2] Concrete/masonry shear wall [21.4.8]	Enhance existing collector [7.4.2] Fiber composite wall overlay [13.4.1] Concrete wall overlay [21.4.5] Infill openings			
	Incidental bracing					Isolate component from lateral force-resisting system
Load Path	Inadequate or missing wall-to-diaphragm tie for out-of-plane load – exterior and interior walls			Wall-to-diaphragm anchorage [16.4.1] plus diaphragm cross-ties [22.2.3]		

16-7

Table 16.3-1: Seismic Deficiencies and Potential Rehabilitation Techniques for PC1 Buildings					
Deficiency		**Rehabilitation Technique**			
Category	**Deficiency**	**Add New Elements**	**Enhance Existing Elements**	**Improve Connections Between Elements**	**Reduce Demand**	**Remove Selected Components**
Load Path (continued)	Inadequate anchorage to diaphragms for in-plane forces – exterior and interior walls			Wall-to-diaphragm shear anchors [21.4.8]		
	Beam or girder connection to tilt-up wall inadequate for wall out-of-plane loads		Enhance beam or girder connection [16.4.2]			
	Inadequate connection at base of tilt-up panel			Wall-to-foundation connections [16.4.3]		
	Inadequate collectors	Add collector [7.4.2]	Enhance existing collector [7.4.2]			
Component Detailing	Wall inadequate for out-of-plane bending		Wall strongback or pilaster [21.4.3]			
	Inadequate detailing of narrow wall piers	Steel braced frame [7.4.1] Concrete/masonry shear wall [21.4.7]	Supplement component to provide adequate load path [13.4.1], [21.4.5] Add backup vertical supports where bearing might be lost [21.4.11]			
Diaphragms	Inadequate in-plane strength and/or stiffness	Steel braced frame [7.4.1] Concrete or masonry wall [21.4.8]	Enhance existing diaphragm [22.2.1] Horizontal braced frame			
	Inadequate chord capacity	Add chord [22.2.2]	Enhance existing chord			
	Excessive stresses at openings and irregularities		Enhance diaphragm detailing			
	Excessive stresses at openings and irregularities		Enhance diaphragm detailing			
Foundations	See Chapter 23					
[ ] Numbers in brackets refer to sections containing detailed descriptions of rehabilitation techniques.						

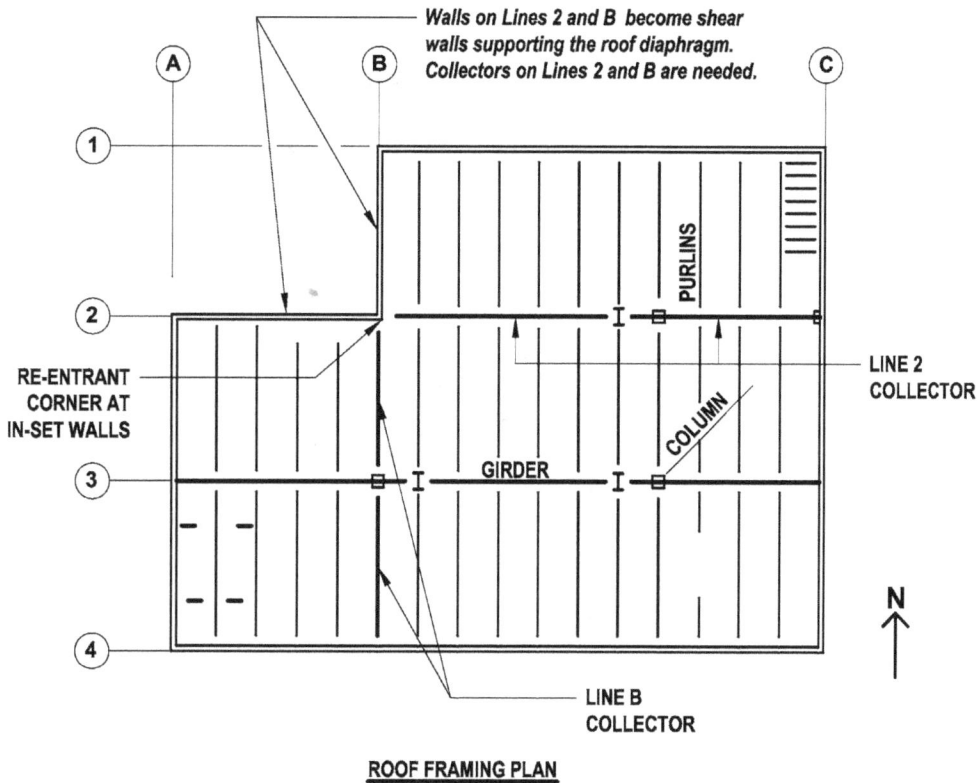

Figure 16.3-1: Plan of PC1 Building with Re-Entrant Corner at In-Set Panels

deformation compatibility between the building sections under earthquake loading. Incompatible deformations are likely to significantly compromise shear transfer through the expansion joint. In addition, common details used for the expansion joint may not perform adequately under expected seismic forces. Details often involve significant eccentricities. The eccentricities may create forces in roof framing members that were not envisioned in the framing member design. The most direct approach to rehabilitation of this building type is the addition of vertical elements such that each building section is independently braced. Alternatively, the expansion joint connection can be improved such that it can reliably transfer anticipated earthquake forces while accommodating anticipated building movement.

Another configuration concern is the occurrence of components (such as mezzanines) that act as incidental bracing, creating an unintended load path. For the **PC1** building seismic resisting system to work as intended, the diaphragm must be able to deflect, and the walls must be able to deflect out-of-plane to follow the diaphragm. Ideally rehabilitation of incidental bracing would involve isolation from the building shell. Alternately, the incidental bracing could be analyzed as part of the structural bracing system.

ROOF PLAN

SECTION   A

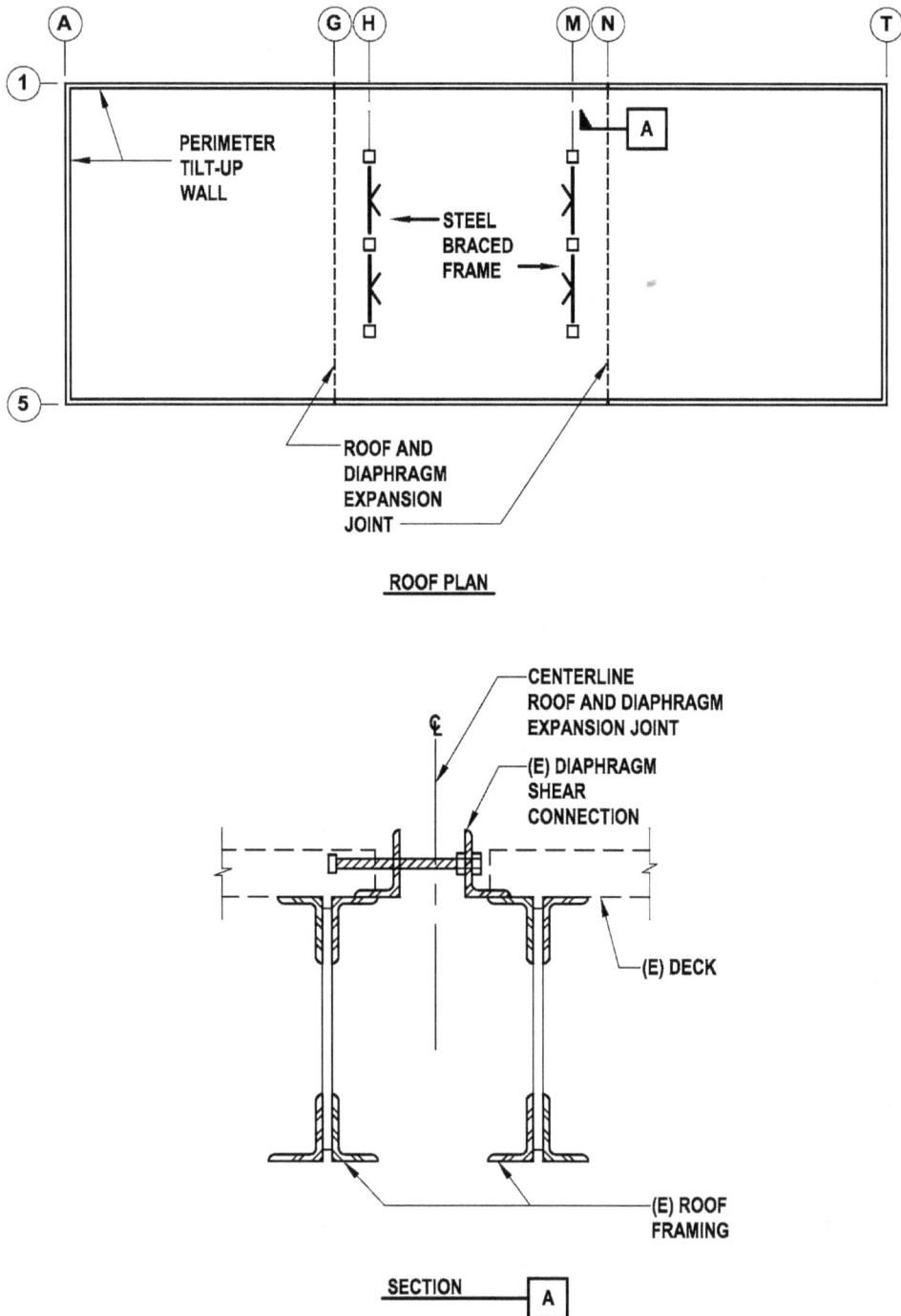

Figure 16.3-2: Plan and Detail of Large PC1 Building Constructed in Three Sections

Load Path

As previously mentioned, out-of-plane anchorages between the tilt-up walls and the diaphragm have been the primary source of damage and focus of rehabilitation in wood diaphragm **PC1** buildings. Conceptually, the approach for both new construction and rehabilitation has been to create continuous diaphragm cross-ties between exterior walls on opposite sides of the building. Rehabilitation of wall out-of-plane anchorage is discussed in this chapter. Diaphragm cross-ties are discussed in Chapter 22. Wall anchorage for in-plane shear commonly uses different connectors than for out-of-plane loads. In-plane shear connection is discussed as part of the diaphragm chord discussion in Chapter 22.

Girder gravity load connections to tilt-up walls provide diaphragm-to-wall anchorage and will therefore have to resist wall out-of-plane anchorage forces. These anchorage forces may or may not have been considered in initial connection design. Even if considered, the connection may not be adequate for currently required loads. Rehabilitation of these connections is often required.

The addition of or enhancement of existing collectors may be required in order to transmit diaphragm forces to the resisting shear walls. This is particularly of concern when a limited number of solid panels are intended to carry a significant portion of the building shear. Although not as common, there is also significant concern when vertical offsets in the roof diaphragm result in incomplete chords or collectors. Any breaks or offsets in chords or collectors need to be carefully evaluated. In the 1993 Guam earthquake (EERI, 1995), a high bay portion of a forklift repair shop was braced off of lower bays on each side. Incomplete collectors from the high-bay diaphragm to low bay shear walls resulted in damage.

The anchorage at the tilt-up panel at the wall base may also be deficient. In some older tilt-ups, no positive connections were made from the wall to the foundation or slab; friction was relied on for force transfer both in-plane and out-of-plane. Most tilt-up panels in recently constructed buildings will have a base connection, either to the foundation or more likely to the adjacent slab-on-grade; however, it is possible for the connection to be inadequate. Rehabilitation most often involves the addition of new wall to slab connections.

Component Detailing

Component detailing deficiencies include inadequate out-of-plane wall capacity. This deficiency may occur due to increases in design seismic forces or inadequate consideration of panels with openings. It is seldom practical to address wall capacity by adding reinforcing and concrete thickness to individual wall sections, so addition of wall pilasters or strongbacks is common. Where pilasters are added to tilt-up walls, the pilasters stiffen the wall for out-of-plane forces, allowing two-way spanning of the wall and attracting high out-of-plane forces to the pilaster and pilaster-to-diaphragm connection. A pilaster-to-roof diaphragm anchorage must be provided to accommodate the concentrated wall out-of-plane force.

Tilt-up panels with large openings generally have narrow wall panels on each side. These panels and the spandrel wall over the opening act as a concrete frame. In loading dock and storage facilities, entire walls can be made up of frames. Because the narrow wall piers do not meet the ACI 318 (ACI, 2005) definition of a column, many have been constructed with standard wall detailing. The 1994 and later editions on the UBC (ICBO, 1994) defined and created reinforcing requirements for wall piers, which have been brought into the IBC (ICC, 2003a), but are not in ACI 318. Where the wall piers are required for resistance to gravity and lateral loads, rehabilitation may be required. Rehabilitation for lateral loads may involve the additional of new vertical elements, enhancing of existing elements, or filling in some of the openings. In addition, loss of vertical support for the roof adjacent to the panel may be of concern. Rehabilitation for vertical loads may be approached by either enhancing the wall element, or providing back-up vertical supports.

Diaphragm Deficiencies

Due to changes in building code requirements, it is very common for diaphragms in areas of high seismic hazard to have inadequate in-plane shear capacity. Regardless, the *SEAONC Guidelines* indicates that diaphragm overstresses have rarely been associated with significant earthquake damage. In areas of moderate seismic hazard, this may or may not be a significant deficiency. Diaphragm strength and stiffness deficiencies are most often rehabilitated by enhancing the existing diaphragm. The addition of new vertical elements to reduce diaphragm span and therefore diaphragm shears is also an effective approach, but such an approach is not as commonly used. Diaphragm enhancement is addressed in Chapter 22.

Many California tilt-up buildings have been constructed with solid-sawn roof purlins with sizes ranging from 4x12 to 4x16. These framing members may have calculated overstresses under existing dead and live loads due to reductions in allowable stresses from in-grade values introduced in the 1991 NDS (AF&PA, 1991). When this is the case, removal and replacement of roof sheathing is a preferred alternative to overlays in order to keep additional dead load a minimum. Diaphragm enhancement using staples may also provide required strength. Sometimes, lighter roofing finish materials can be installed where existing roofing is removed, to further reduce overstresses.

Other diaphragm deficiencies include inadequate chord capacity and stress concentrations at large diaphragm openings and re-entrant corners. Rehabilitation at re-entrant corners primarily involves the provision of adequate chords and collectors. The same is true at large diaphragm openings.

16.4   Detailed Description of Techniques Primarily Associated with This Building Type

16.4.1   Enhance Wall-to-Diaphragm Anchorage

Deficiency Addressed by Rehabilitation Technique

This rehabilitation technique addresses enhanced anchorage of walls into wood sheathed or steel deck diaphragms for out-of-plane loads. Cross-ties are a required continuation of the wall anchorage system. See Section 22.2.3.

Wall anchors and cross-ties should be the highest priority for rehabilitation of wood diaphragm **PC1** buildings. This section illustrates the basic rehabilitation requirements for wall anchorage of wood and steel diaphragm buildings. Refer to the *SEAONC Guidelines* for more exhaustive treatment of detailing.

The *SEAONC Guidelines* provides details of code requirements and observed damage over many years. Earthquake damage to wall anchorage has been observed not only in older buildings, but also in recently constructed buildings, where the type and installation of wall anchors was critical to their ability to perform. Just providing anchors is not enough; attention to detailing and field installation is critical.

Description of the Rehabilitation Technique

Figure 16.4.1-1A illustrates a roof plan for a **PC1** building with a wood diaphragm. Figures 16.4.4-1B through 16.4.4-1D, located as shown on the roof plan, illustrate common rehabilitation measures for wall-to-diaphragm anchorage. Also illustrated in Figure 16.4.1-1A are subdiaphragms used as part of the diaphragm cross-tie system, described in detail in Chapter 22. While subdiaphragms are always used in wood structural panel diaphragms, they are only occasionally used in steel deck diaphragms.

Figures 16.4.1-1B and 16.4.1-1C illustrate anchorage from the west wall to a subdiaphragm extending between Lines A and B. Similarly, the east wall is anchored to a subdiaphragm between Lines G and H. The primary objective is to create an adequate load path for out-of-plane loads acting both into the diaphragm (compression) and away from the diaphragm (tension). This type of anchorage is generally provided every four to eight feet on center. The load path into the diaphragm includes both wall to framing member anchorage and sheathing nailing to transfer loads from the framing member to the sheathing. Figure 16.4.1-1B relies on existing sheathing nailing, while Figure 16.4.1-1C adds sheathing nailing.

Where framing member ends are tight against the wood ledger and the wood ledger is tight against the exterior wall, the compression load path can be carried by the framing members. Experience has shown that gaps often occur between member ends and the ledger. The movement required to close these gaps has been enough to damage the roof sheathing and damage devices used for tension anchorage.

Tie-down devices can provide both the tension and compression load path. In order to do this, the device must be rated for both tension and compression loading by the manufacturer, and it must be installed in accordance with the manufacturer's instructions for compression load. This generally involves limiting the unsupported length of the tie-down rod and providing additional nuts and washers at the tie-down seat. Very few tie-down devices are currently available that are rated for both tension and compression loads. Building codes such as the now require that forces and stresses induced by eccentricities in the connection be addressed, and the *SEAONC Guidelines* encourages use of tie-downs placed symmetrically on the purlin or sub-purlin in order to minimize eccentric beam loading and encourages use of stiff tie-downs in order to minimize deformation demand and possible diaphragm damage resulting from deformation. For girders, symmetrical tie-downs are recommended but not as critical as with smaller framing members.

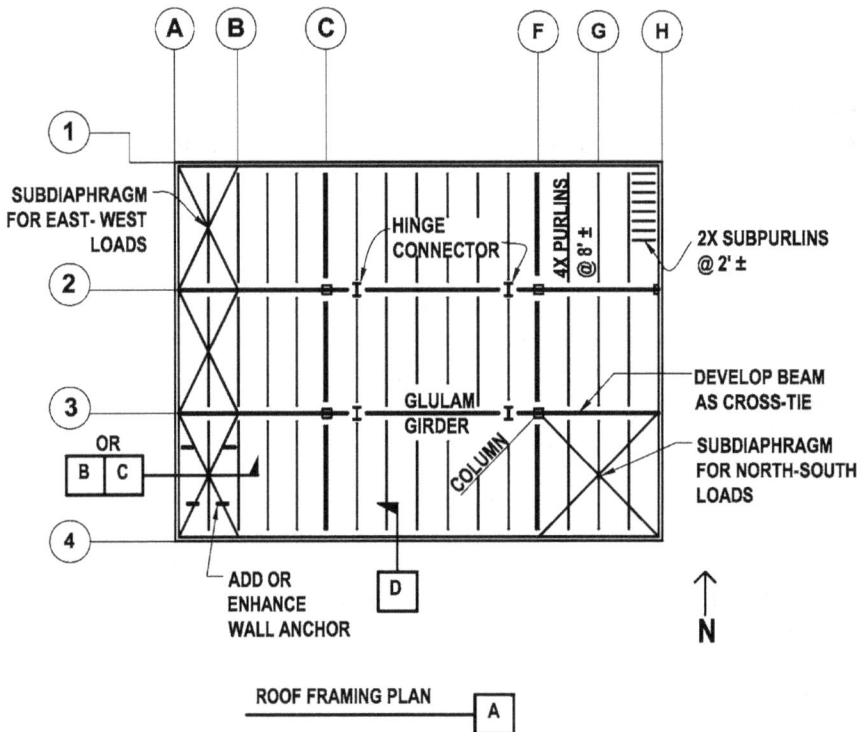

**Figure 16.4.1-1A: Roof Plan with Wall Out-of-Plane Anchorage
for Flexible Wood Diaphragm**

Figures 16.4.1-1A and 16.4.1-1C do not show wall anchors attached to single existing 2x subpurlins. It is recommended that anchorage be to a 3x or wider member, or multiple 2x's as shown in Figure 16.4.1-1B. While it may be possible for a single 2x4 or 2x6 to be shown adequate in calculation, adequate installation and performance are extremely difficult to achieve.

The wall anchorage system needs to extend across the width of the subdiaphragm. In Figure 16.4.1-1C, this involves providing extra pairs of tie-down devices between subpurlins across the first purlin. Again, both tension and compression load paths are needed. The required subdiaphragm depth is determined from the number of nails required to transfer subdiaphragm forces into the main diaphragm. Depending on subdiaphragm requirements, additional tie-down pairs could be required across more purlins. Subdiaphragm requirements are discussed in Chapter 22.

Details 16.4.1-1B and 16.4.1-1C show work from top of the diaphragm. See Chapter 22 for alternates for working from below. Location of access needs to be decided early on in the design process and will drive both calculations and detailing of the rehabilitation work.

SUBDIAPHRAGM

B1

(E) ROOFING AND
SHEATHING NOT MODIFIED

(E) FIELD NAILING
ASSUMED

(E) EDGE NAILING
ASSUMED - VERIFY

(E)
CONCRETE
WALL

CHECK PURLIN
CAPACITY WITH
HOLE FOR TIE-ROD

4X BLOCK DRY
(MC ≤ 19%)

(E) PURLIN

(E) SUBPURLIN

SECTION   B

VERIFY TIGHT
BEARING OR SHIM
TO PROVIDE
COMPRESSION
LOAD PATH

(E)
CONCRETE
WALL

SPECIFY
MIN.
EMBED
LENGTH
FOR
ADHESIVE
ANCHOR

HANGER
EACH END

(E) SUBPURLIN

(E) PURLIN

STEEL PLATE
WASHER OR
MALLEABLE IRON
WASHER
UNDER NUT

PLAN   B1

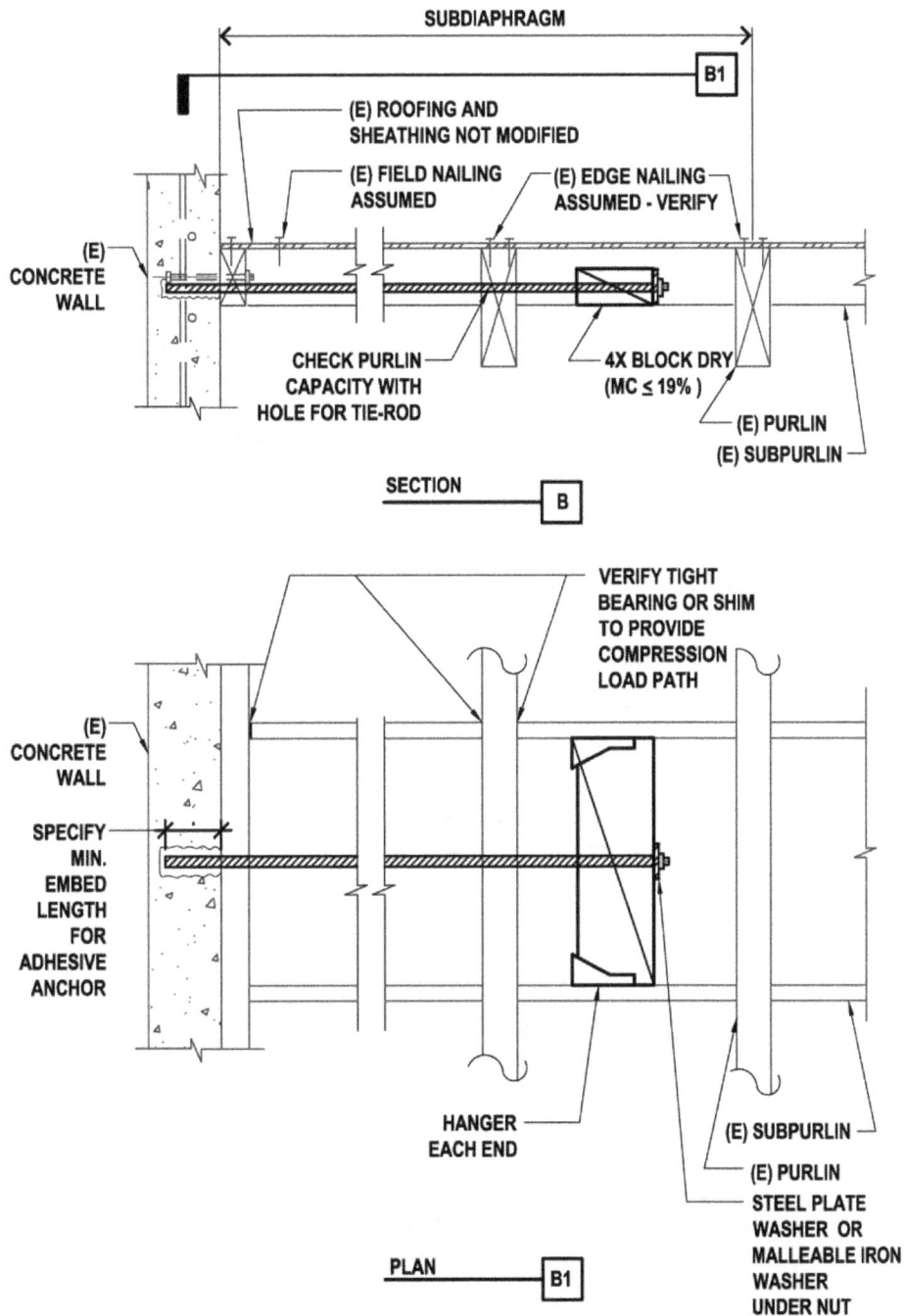

**Figure 16.4.1-1B: Wall Out-of-Plane Anchorage for Flexible
Wood Diaphragm at Subpurlins – Roofing Not Removed**

SUBDIAPHRAGM

C1

(E) ROOFING REMOVED
(E) SHEATHING NAILING
MODIFIED

EDGE NAILING

(E)
CONCRETE
WALL

TIE-DOWN EACH
SIDE 3x OR 4x

3x OR 4x BLOCK
DRY (MC ≤ 19%)

(E)
SUBPURLIN

(E) PURLIN

SECTION   C

(E)
CONCRETE
WALL

SPECIFY
MIN.
EMBED
LENGTH
FOR
ADHESIVE
ANCHOR

3x MINIMUM
FRAMING
MEMBER

(E)
SUBPURLIN

(E) PURLIN

PLAN   C1

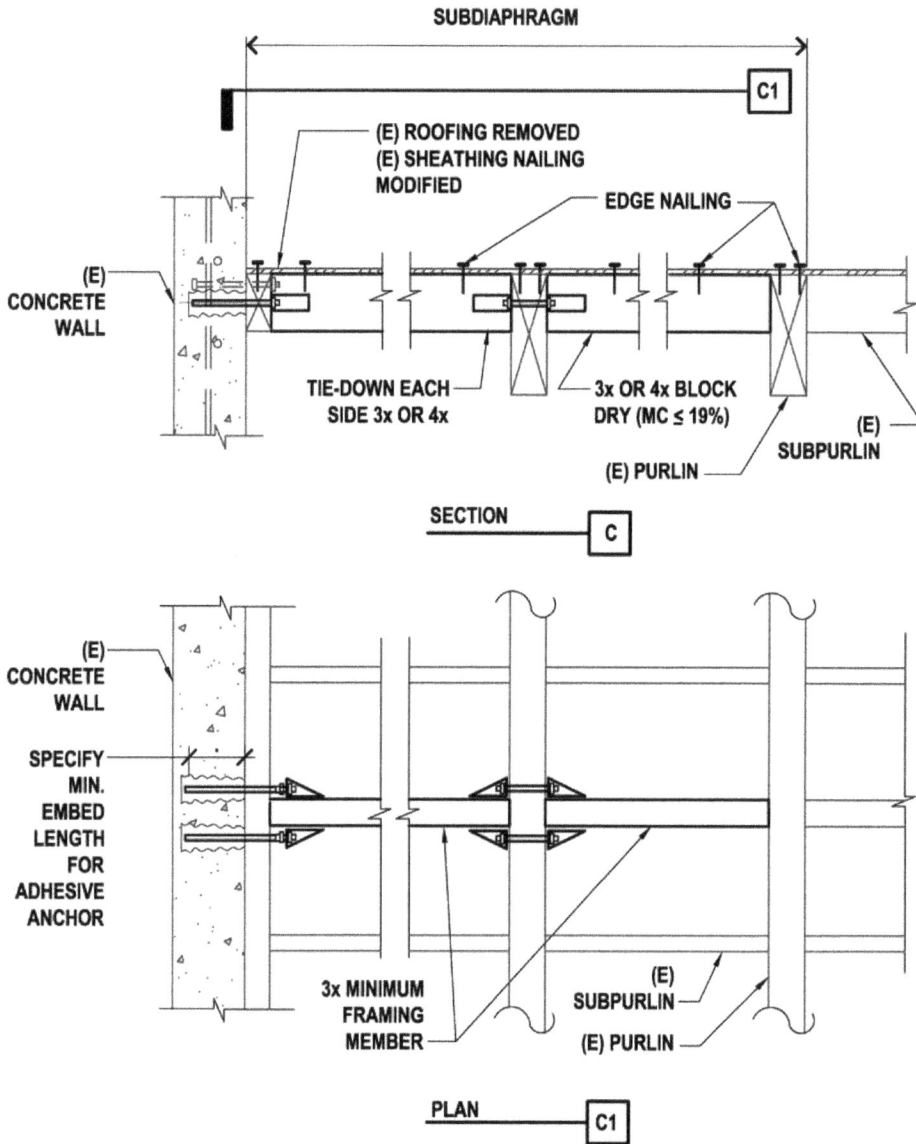

**Figure 16.4.1-1C: Wall Out-of-Plane Anchorage for Flexible
Wood Diaphragm at Subpurlins – Roofing Removed**

Figure 16.4.1-1D illustrates anchorage of the north and south walls to a purlin. In this case, the purlin is long enough to extend across the subdiaphragm width (extending between Lines 1-2 and 3-4), so additional pairs of tie-downs are not needed. As is previous details, both tension and compression load paths must be maintained.

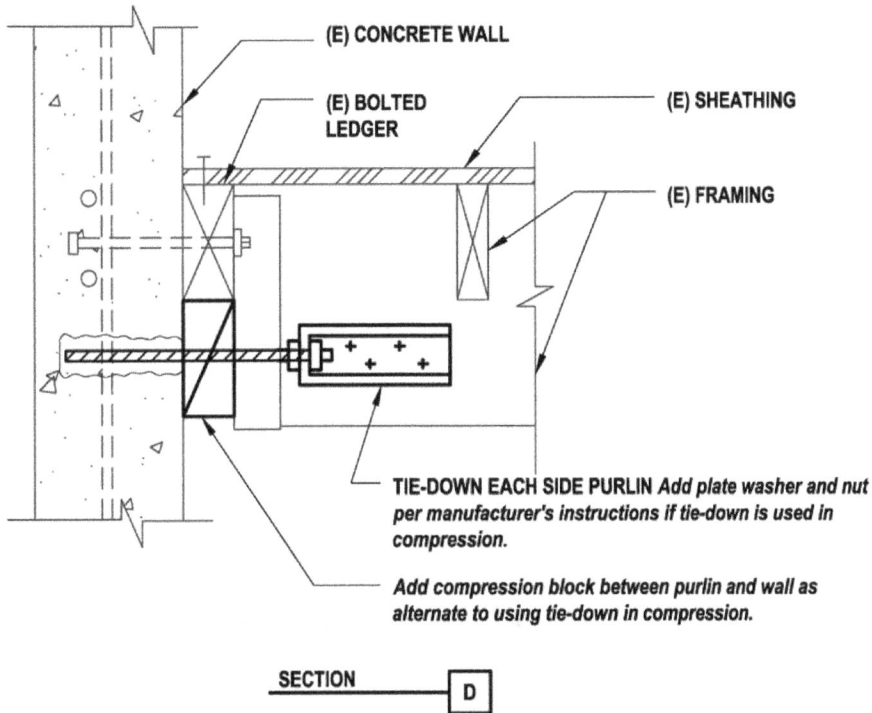

**Figure 16.4.1-1D: Wall Out-of-Plane Anchorage
for Flexible Wood Diaphragm at Purlins**

Figure 16.4.1-2A illustrates a similar roof plan for a **PC1** building with a steel deck diaphragm. It is important to note in this figure that subdiaphragms (as shown in Figure 16.4.1-1A) are not used. Instead, the steel deck provides a continuous cross-tie in the east-west direction, while in the north-south direction open web joists provide direct cross-ties across the entire diaphragm width at each wall anchor location. This is the primary approach used in new steel deck diaphragm construction. Subdiaphragm concepts can be applied to steel deck construction, but are not common.

Figures 16.4.1-2B and 16.4.1-2C provide wall to diaphragm anchorage details. In Figure 16.4.1-2B, wall anchorage forces are transmitted to the steel deck. The deck section, deck edge fastening, and deck end splices need to be checked for wall anchorage tension and compression forces. Justification of the capacity may be by calculation or testing. The balance of the load path also needs to be checked and enhanced as required. In the detail shown, supplemental adhesive

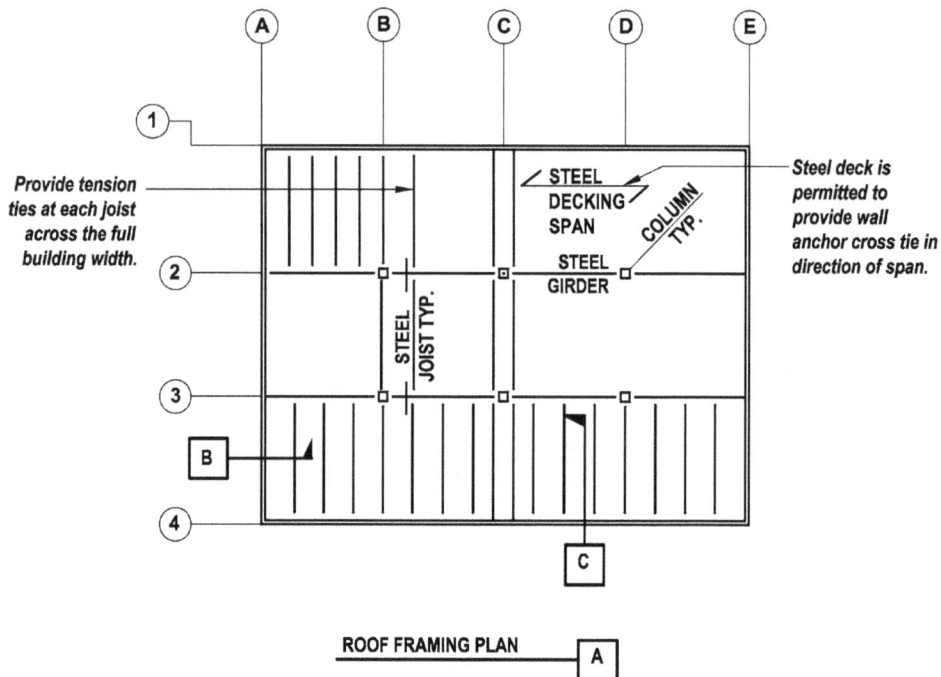

**Figure 16.4.1-2A: Roof Plan with Wall Out-of-Plane Anchorage
for Flexible Steel Diaphragm**

anchors to the wall and a second ledger angle are provided. Prying action in the steel ledger angle needs to be considered in determining wall anchor forces.

Figure 16.4.1-2C shows wall anchorage to a steel open web joist. The wall anchorage connection may be through the joist seat in new construction; however, a supplemental anchor to the joist is likely needed in rehabilitation. The joist must be checked for wall anchorage forces and any applicable eccentricities. Details in Chapter 22 address joist to joist connections to complete the cross-tie.

Compression forces can be carried in tie-down devices, if rated for compression by the manufacturer. Unsupported tie-down rod lengths must be kept short, and additional nuts and washers are needed to transfer compression. Again, wall anchorage loads need to be transferred into the decking. In this case existing welding or screwing of the decking is relied upon. If this fastening is not adequate, the roofing will need to be removed to allow additional fastening.

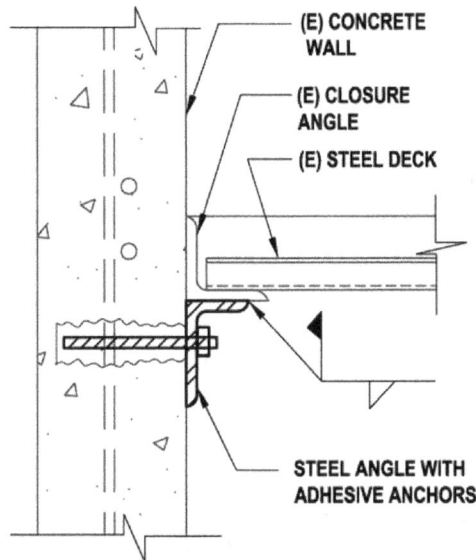

Note: Steel deck is permitted to provide wall anchor and diaphragm
cross-tie in direction of span. Added steel angle may be needed to
enhance wall to deck anchorage. See Detail C for perpendicular direction.

SECTION   B

**Figure 16.4.1-2B: Wall Out-of-Plane Anchorage for Flexible Steel Diaphragm
– to Decking for Load Parallel to Flutes**

Design and Detailing Considerations

Research basis: No research applicable to the performance or adequacy of enhanced anchorage methods has been identified; however, the demands created in flexible diaphragms have been studied by Fonseca, Wood and Hawkins (1996); Hamburger and McCormick (1994); Ghosh and Dowty (2000); and Freeman, Searer, and Gilmartin (2002).

As discussed in Section 16.1, even wall anchorages constructed or rehabilitated in the 1980s and early 1990s were observed to have been damaged in the 1994 Northridge earthquake. The reader is referred to the extensive discussion in the *SEAONC Guidelines* for design and detailing considerations and lessons learned.

Anchor type and installation: A variety of proprietary anchors are available for anchorage to existing concrete walls. Both manufacturer literature and ICC Evaluation Service reports should be consulted for information on conditions of use, allowable loads, and installation and inspection requirements. It is important to make sure that the anchor type is appropriate for the material to which it will be connected and is approved for seismic loads. The diameter of drilled holes is specified in installation requirements for each anchor type; variation from this size often

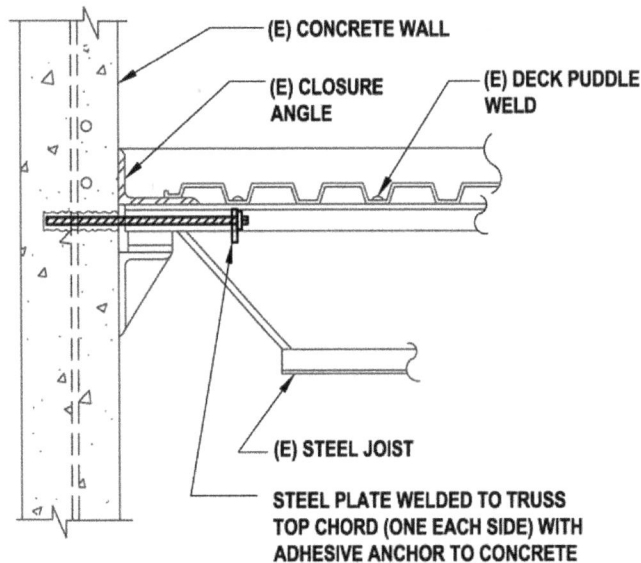

Note:
1. *Verify that steel joist top chord is adequate for combined gravity and wall anchorage loads.*
2. *Provide double washers and nuts where required for compression loads.*
3. *Verify that steel deck connection to joist is adequate for wall anchorage loads.*

SECTION ──── C

Figure 16.4.1-2C: Steel Open Web Joist Anchorage to Exterior Wall

leads to inadequate anchor capacity. Most manufacturers have caulking gun-like devices that make field placement of epoxy fairly simple and automatically mix two-part adhesives. Generally, these types of adhesives provide more than adequate strength, and there is no need to use more complicated high-strength adhesive types. The cleaning of holes prior to placing adhesive anchors is paramount for anchor capacity. When not well cleaned, the anchors can pull out at a small fraction of the design load. It is common to pull-test a portion of the adhesive anchors to verify adequate installation. The pull test load is usually in the range of one to two times the tabulated allowable stress design tension load. The bridge used for testing generally makes a concrete cone pull-out failure unlikely. The test load should not be near yield load for bolts or adhesive pull-out failure loads.

Anchors added in rehabilitation will often have to work in combination with existing cast-in anchors. In order to allow load sharing, anchorage of similar stiffness is desirable. This is often best achieved with adhesive anchors. In addition, anchorage provisions for new buildings have moved towards having the attachment to concrete be capable of developing the yield capacity of the steel anchor in order to promote ductile connection behavior. Again, this is best achieved with adhesive anchors. Anchor types other than adhesives should be carefully evaluated.

Cost, Disruption and Construction Considerations

When rehabilitation work is undertaken on the roof diaphragm of a **PC1** building, it is important that the cost and the preferred location for work (from the underside or top of roof) take into account the combination of work, rather than considering one piece at a time. If several diaphragm measures will be undertaken, it will quickly become cost-effective to remove the roof and allow work from the top. This is particularly true if a steel deck requires several rehabilitation measures.

Proprietary Concerns

There are no proprietary concerns with this rehabilitation technique, other than the use of proprietary connectors and adhesive anchors as part of the assemblage.

16.4.2   Enhance Beam and Girder Connection to Supporting Elements

Deficiency Addressed by Rehabilitation Technique

This rehabilitation technique addresses enhancement of girder gravity load connections to tilt-up walls. While primarily intended to carry gravity loads, these connectors should also be adequate to resist wall out-of-plane loads.

Description of the Rehabilitation Technique

Tilt-up walls may have pilasters supporting girder loads. In older tilt-up buildings, these are often cast-in-place pilasters, while in newer buildings pilasters are cast as part of the tilt-up panel. The pilaster acts as a wall stiffener, allowing the wall to span both horizontally and vertically under out-of-plane earthquake loading. This attracts more load to the pilaster, and the top of pilaster reaction for out-of-plane loading will be higher than at a typical wall anchor. The gravity load connection generally also serves as the girder-to-pilaster connection for out-of-plane loads. Two deficiencies are common with this connection: 1) inadequate confinement around anchor bolts embedded in the pilaster top and 2) inadequate connection to the girder. Tension loads on the connector have led to splitting of the column top, pulling away the wedge of concrete in front of the anchor bolts. In recent codes, the placement of three closely spaced ties at anchor bolts has been required to reinforce across the anticipated concrete crack. Where added ties have not been provided, a collar around the pilaster top can provide external reinforcement (Figure 16.4.2-1). The collar should be relatively stiff to minimize splitting of the concrete before load is taken up. The second issue is inadequate connection to the wood girder, including bolt capacity and placement. Where the girder seat connection to concrete is adequate or can be enhanced, inadequate bolt capacity can be mitigated with addition of bolt tabs (Figure 16.4.2-2). Alternately, it is possible to use new wall anchors from the girder to the wall, bypassing the girder seat (Figure 16.4.2-3). Again, this connection must be stiff.

(E) WALL PANEL

(E) GLULAM GIRDER

STEEL BENT PLATE COLLAR
AROUND PILASTER TOP WITH
ADHESIVE ANCHORS

(E) PILASTER WITH WIDELY
SPACED TIES

SECTION

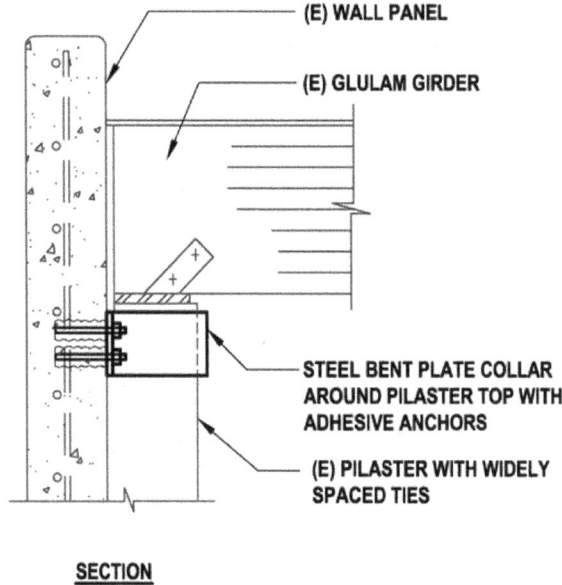

Figure 16.4.2-1: Enhanced Girder Connection – Collar at Pilaster
Adapted From SEAONC (2001)

Girders that are supported directly on a flat wall panel using a steel U-bracket bolted or welded
to the panel (Figure 16.4.2-3) will also attract wall out-of-plane forces. As is true with wall
pilasters, a girder and U-bracket are likely to provide a stiffer load path for wall out-of-plane
loads than adjacent anchors. For this reason, use of a wall anchorage force greater than used for
adjacent anchors is encouraged. The girder connection should have the ability to resist wall
anchorage loads in combination with gravity loads. Anchorage of the bracket to the panel will
often be adequate for both gravity and lateral loads; however, the bracket attachment to a wood
girder will often not have the quantity or placement of bolts required for tension loads. Addition
of steel tabs and bolts will add capacity and place bolts where end distances are adequate for
tension loads. Where the steel connection to the concrete is not adequate, the out-of-plane
anchor might bypass the existing connection and connect the girder directly to the wall. Figure
16.4.2-2 shows two approaches, one with a tie-down on each side of the girder and a second with
a tie-down on the girder bottom. The out-of-plane wall anchor should be as stiff as possible to
minimize damage to the gravity connection.

Design and Detailing Considerations

Research basis: No research applicable to this rehabilitation measure has been identified.

See Section 16.4.1 wall anchorage and the *SEAONC Guidelines* for additional detailed
discussion.

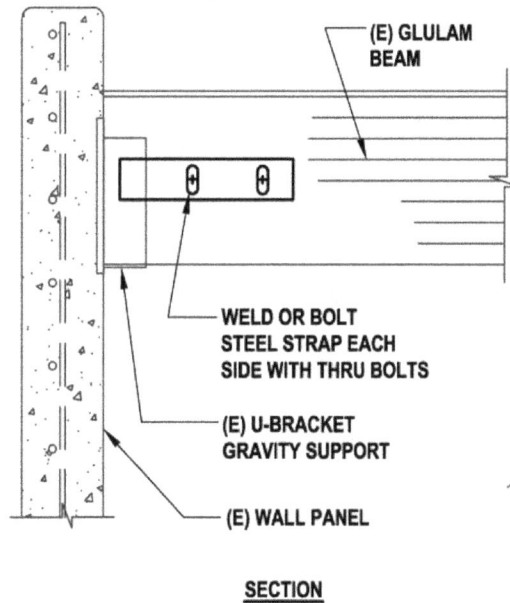

SECTION

Figure 16.4.2-2: Enhanced Girder Connection at U-hanger
Adapted From SEAONC (2001)

Proprietary Concerns

There are no proprietary concerns with this rehabilitation technique, other than the use of
proprietary connectors as part of the assemblage.

16.4.3   Enhance Anchorage at Base of Tilt-Up Panels

Deficiency Addressed by Rehabilitation Technique

This rehabilitation technique addresses the addition or enhancement of connections between the
tilt-up wall panel and foundation or adjacent slab-on-grade to resist in-plane shear and
overturning forces and out-of-plane wall anchorage forces.

Description of the Rehabilitation Technique

Rehabilitation of base connections in **PC1** buildings for in-plane and out-of-plane shear loads is
most commonly accomplished by addition of steel angles and adhesive anchors between the wall
panel and adjacent slab-on-grade. This is illustrated in Figure 16.4.3-1. In some instances, the
slab-on-grade may not have been thickened adjacent to the tilt-up panel. When this is the case, it
may be necessary to remove and recast a thicker pour strip in order to get adequate anchorage.
The connection shown would flex if the wall were to uplift. Where uplift connection capacity is
required, a direct tension connection of the wall to the footing below is recommended.

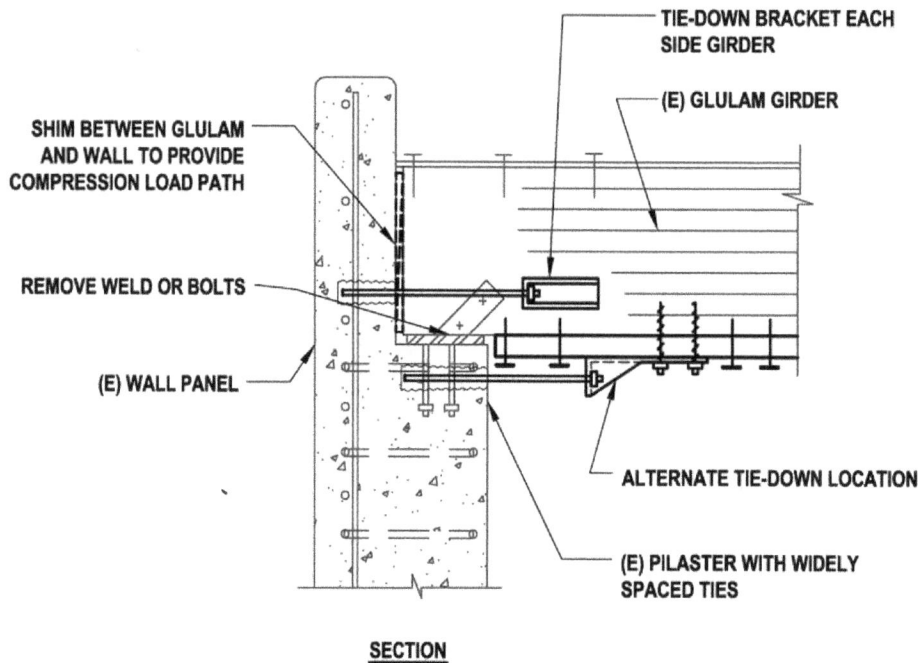

Figure 16.4.2-3: Enhanced Girder Connection at Pilaster
Adapted From SEAONC (2001)

Variations in base conditions include 1) older **PC1** buildings that may not have any doweling because friction was relied on to resist forces at the base of the panel and 2) welded connections between cast-in embeds in the wall panel and slab, similar to PC2 wall panel connections.

Design and Detailing Considerations

Research basis: No research applicable to this rehabilitation technique has been identified.

The *SEAONC Guidelines* provide discussion of a variety of possible existing conditions, changes in code requirements, and implications for retrofit.

Proprietary Concerns

There are no proprietary concerns with this rehabilitation technique, other than the use of proprietary adhesive anchors as part of the assemblage.

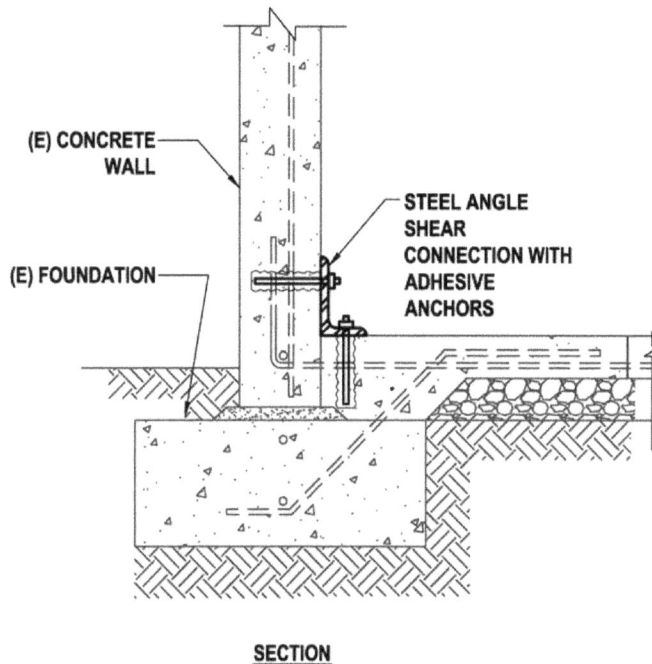

Figure 16.4.3-1: Enhancement of Tilt-up Panel Base Connection

16.5   References

Extensive reference and bibliography listings are included in SEAONC (2001). The following only includes references specifically cited in this chapter.

ACI, 2005, *Building Code Requirements for Structural Concrete*, American Concrete Institute, Farmington Hills, MI.

ACI-SEASC Task Group on Slender Walls, 1982, *Report of the Task Group on Slender Walls* (Green Book), American Concrete Institute, Southern California Chapter and the Structural Engineers Association of Southern California, Los Angeles, CA.

AF&PA, 1991, *ANSI/NFoPA NDS-1991 National Design Specification for Wood Construction*, American Forest and Paper Association, Washington, D.C.

City of Los Angeles, 2002, "Chapter 91," *City of Los Angeles Building Code*, Los Angeles, CA.

EERI, 1996, *Supplement C to Volume 11, Northridge Earthquake of January 17, 1994, Reconnaissance Report, Earthquake Spectra*, Volume 2, EERI, Oakland, CA.

EERI, April 1995, *Guam Earthquake of August 8, 1993 Reconnaissance Report,* Earthquake Spectra, Supplement B to Volume 11, Earthquake Engineering Research Institute, Oakland, Ca.

FEMA, 2004, *NEHRP Recommended Provisions for Seismic Provisions for New Buildings and Other Structures*, FEMA 450, Federal Emergency Management Agency, Washington, D.C.

Freeman, S., S. Searer and U. Gilmartin, 2002, "Proposed Seismic Design Provisions For Rigid Shear Wall / Flexible Diaphragm Structures," *Proceedings of the Seventh U.S. National Conference on Earthquake Engineering*, EERI, Oakland, CA.

Fonseca, F., S. Wood and N. Hawkins, 1996, "Measured Response of Roof Diaphragms and Wall Panels in Tilt-Up Systems Subject to Cyclic Loading," *Earthquake Spectra,* Volume 12, Number 4, Earthquake engineering Research Institute, Oakland, CA.

Ghosh, S.K. and S. Dowty, 2000, *Anchorage of Concrete or Masonry Walls to Diaphragms Providing Lateral Support*, Draft 2000, Not Published.

Hamburger, R.O. and D. McCormick, 1994, "Implications of the January 17, 1994 Northridge Earthquake on Tilt-up Wall and Masonry Wall Buildings with Wood Roofs," *Proceedings of the 1994 Convention of the Structural Engineers Association of California*, Sacramento, CA.

ICBO, 1994, *Uniform Building Code*, International Conference of Building Officials, Whittier, California.

ICBO, 2001, *Guidelines for Seismic Retrofit of Existing Buildings* (GSREB), International Conference of Building Officials, Whittier, CA.

ICC, 2003a, *International Building Code*, International Code Council, Country Club Hills, IL.

ICC, 2003b, *International Existing Building Code*, International Code Council, Country Club Hills, IL.

SEAONC (Structural Engineers Association of Northern California), 2001, *Guidelines for Seismic Evaluation and Rehabilitation of Tilt-Up Buildings and Other Rigid Wall/Flexible Diaphragm Structures*, SEAONC, San Francisco, California.

SEAOSC (Structural Engineers Association of Southern California), 1979, *Recommended Tilt-up Wall Design* (Yellow Book), SEAOSC, Los Angeles, CA.

Tilt-up Concrete Association, 2004, *Tilt-up Construction and Engineering Manual*, Sixth Edition, Tilt-up Concrete Association, Mount Vernon, IA.

Chapter 17 - Building Type PC2: Precast Concrete Frames with Shear Walls

17.1   Description of the Model Building Type

Buildings designated as **PC2** include wide ranging combinations of precast and cast-in-place concrete elements. Precast members may be limited to a floor system of hollow core or T-beam construction, or may include all elements of the gravity and lateral load systems. For this chapter, Building Type **PC2** includes concrete wall or frame buildings in which any of the horizontal or vertical elements of the lateral load system are of precast concrete, except for flexible diaphragm buildings which are addressed as Building Type **PC1** in Chapter 16.

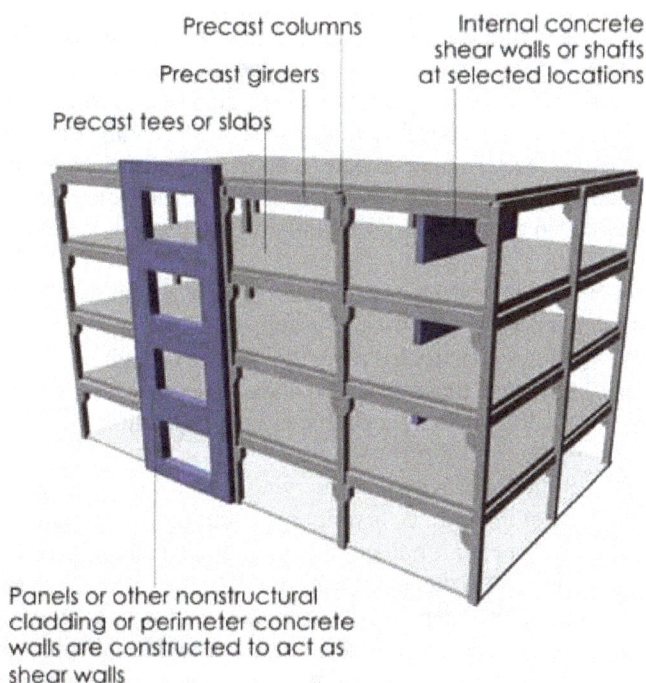

Precast columns

Internal concrete
shear walls or shafts
at selected locations

Precast girders

Precast tees or slabs

Panels or other nonstructural
cladding or perimeter concrete
walls are constructed to act as
shear walls

Figure 17.1-1: Building Type PC2: Precast Concrete Gravity Frames with Shear Walls

Extensive use of hollow core floor systems in buildings with concrete and masonry walls in southern regions of the United States makes this the single largest group of **PC2** buildings. Parking garages (used exclusively for parking rather than mixed use) represent the next largest group of **PC2** buildings and a substantial portion of the current **PC2** building inventory in the U.S. The **PC2** building type has also been used for a variety of other occupancy types in the U.S. and internationally, including mid-rise office, hotel, and residential buildings, low-rise residential, commercial, and prison buildings.

Over the past decade, significant effort has been devoted to development and testing of precast ductile moment frame systems through the PRESSS (Priestley et al., 1999). These systems are not addressed in this chapter due to their very recent development and state-of-the-art detailing.

Gravity-Carrying Load Systems

Special attention is needed to **PC2** buildings in which concrete frames (beams, girders, and columns or moment frames) resist gravity load, or a combination of gravity and seismic load. Very important to the performance of all concrete buildings with frames, including **PC2** buildings, is the lack of ductile detailing in concrete columns not designated as part of the seismic force-resisting system. These columns in many instances do not have confining steel adequate to accommodate the drift imposed by the seismic force-resisting system and as a result fail through longitudinal bar buckling and concrete crushing. Requirements for estimation of building drift have changed over time, and understanding of potential building deflection has improved with each observation of earthquake performance. As a result, it is important to revisit the ability of non-ductile columns to accommodate estimated drifts, even if they were checked when initially designed. In some precast buildings, the division of initial design responsibility between one engineering firm for the gravity load system and a second firm for the seismic force-resisting system may have contributed to inability to accommodate estimated building deflections. In earthquake performance to date, diaphragm deflections have been a large contributor to deflection of non-ductile gravity systems. Vertical elements, and most particularly moment frames, could also contribute significantly to gravity system deflection.

Following the 1994 Northridge earthquake, column detailing requirements and methods of estimating building deflection for purposes of gravity frame design were modified in codes and standards. Concrete buildings constructed recently in areas of high seismic hazard should perform significantly better than those designed under older provisions. Research continues to develop a better understanding of sources of diaphragm deflection.

When considering the ability of gravity load-carrying systems to accommodate building deflection, a related issue of importance is proper accounting for column stiffness and restraint in analysis. This, again, is a concern for all concrete buildings including **PC2** buildings. Short columns will attract higher forces due to increased stiffness and have been seen to fail as a result. Short columns can be created accidentally due to inadequate separation of the column from nonstructural components such as guardrails. In addition, systematic problems with short columns can occur at parking garage ramps. Analysis models need to pay special attention to these and other sources of shortened columns or columns with increased end fixity.

PC2 buildings with gravity and lateral loads supported exclusively by structural walls do not have the same issues of deflection of non-ductile columns. Connections tend to be the primary issue of importance to these systems for both gravity and seismic load systems.

Shear Walls and Frames

Building Type **PC2** may have a lateral force-resisting system of concrete shear walls or moment frames, cast-in-place or precast. In **PC2** buildings, critical behavior of shear walls is generally governed by connections including: diaphragm to shear wall, shear wall above to shear wall or foundation below, and interconnection of shear walls within a story. In **PC2** buildings with precast frames, field connections within the frame are critical to performance, as is ductile detailing. Connection practice has varied widely over time and by geographic region.

Floor and Roof Diaphragms

In California, precast floor T-beams or hollow core planks are covered by a cast-in-place topping slab, reinforced to provide diaphragm action. These toppings need to be clearly differentiated from topping slabs plant-applied to individual precast members, which do not serve the same function of structurally interconnecting adjacent members. Welded connections between embedded inserts or plates may also be used to aid in alignment of members during erection, but are generally not relied on for diaphragm action. Reinforcing bars are often added in the cast-in-place topping slab to act as diaphragm chords and collectors, and welded wire fabric is used to provide shear reinforcement.

In other areas of the United States, common methods of joining floor sections include use of grouted hollow core joints (grout placed in the joint between two adjacent panels, relying on adhesion and/or friction for shear transfer) or welded insert plates. Cast-in-place topping slabs are not commonly used. In some areas outside the U.S., hollow core planks are installed with no connection or grouting between adjacent planks.

As per the discussion of gravity load systems, deflection of the diaphragm system has been seen as a significant contributor to building deflection is past earthquakes. Discussion of diaphragm behavior and rehabilitation can be found in this chapter, Chapter 20 for masonry wall buildings, and Chapter 22 for detailed discussion of diaphragm rehabilitation.

Parking Structure Issues

Parking structure **PC2** buildings have unique characteristics that deserve specific discussion, some applicable to parking structures regardless of structural system and others specific to precast construction. These issues include the following.

> Many parking structures have large plan areas, and considerations of security and restraint against temperature, creep, and shrinkage movement lead to concentration of the shear walls at the building perimeter near the center of each side (Figure 17.2-1). This configuration leads to long diaphragm spans with significant shear, moment, and collector demands. With these high demands, it is possible for the diaphragm, rather than the vertical elements to control building dynamic behavior. This is of concern in all systems, but particular in precast systems due to the lower level of inherent diaphragm continuity.
>
> Compared to other building uses, parking garages have greatly minimized finish and cladding systems, resulting in low levels of nonstructural damping and energy dissipation.
>
> Ramps in parking structures may act as tension and compression struts between floors, resulting in demands not anticipated during design. This behavior can be avoided by inclusion of seismic joints at one end of each ramp; however, seismic joint detailing is difficult to accommodate in precast concrete construction, making use of a fixed connection more likely. Unless the effect of the ramp is specifically considered in analysis, force transfer through the ramp can result in seismic forces bypassing the

intended resisting system and significant redistribution of forces in the diaphragms and vertical elements. An analytical study of ramp effects discussed in Lyons, Bligh, Purlinton, and Beaudoin (2003) suggests that while the effect of ramps is significant in moment frame buildings, it is less significant and can often be managed in design of shear wall buildings. The effect of ramps should be considered in evaluation and rehabilitation of parking structure buildings.

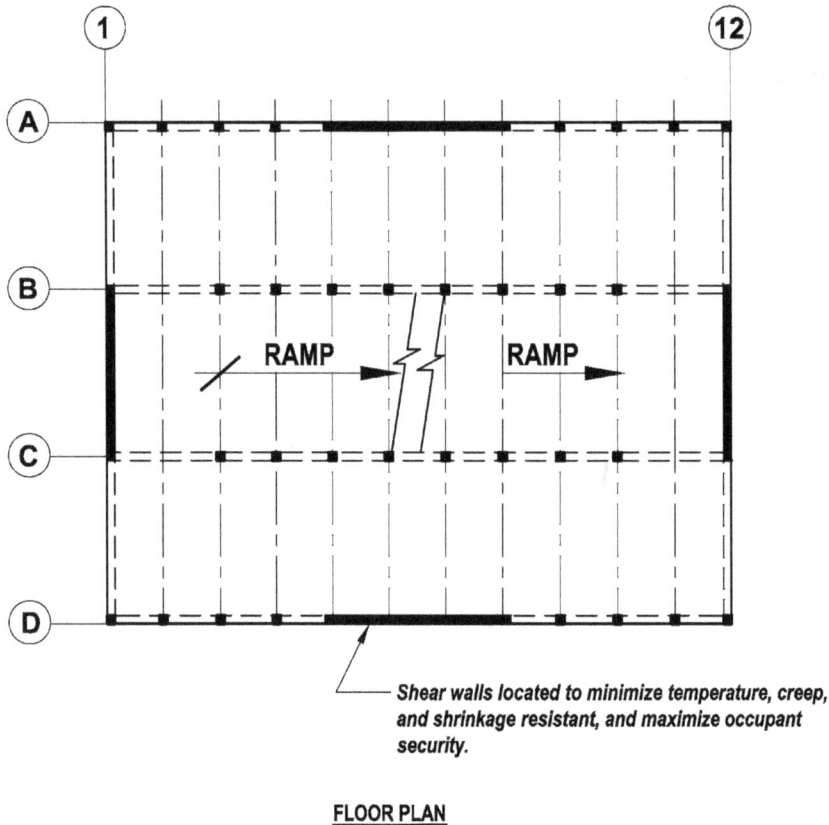

Shear walls located to minimize temperature, creep, and shrinkage resistant, and maximize occupant security.

FLOOR PLAN

Figure 17.2-1: Plan of Common Parking Structure Configuration

17.2   Seismic Response Characteristics

PC2 buildings occur with a wide range of vertical element types. In most cases, the vertical element type will dictate the building seismic response: shear wall buildings will have short period response, while frame buildings will have a longer period. In **PC2** buildings, stiff diaphragm behavior will generally be intended. Parking structure **PC2** buildings with long diaphragm spans, however, have been observed to have inelastic behavior concentrated in the diaphragms rather than vertical shear wall or frame elements. To date, this has been brittle behavior resulting in premature diaphragm failure; however, with development of proper detailing it may be possible to achieve stable long-period diaphragm behavior.

17.3   Common Seismic Deficiencies and Applicable Rehabilitation Techniques

Construction of **PC2** buildings in areas of high seismic hazard in the U.S. has been of limited quantity and recent compared to most other building types, resulting in limited opportunities to observe earthquake performance; the City of Los Angeles and SEAOSC (1994) and EERI (1996) reported on performance of one group of **PC2** buildings following the 1994 Northridge earthquake. Out of an estimated 100 parking garages (precast and cast-in-place) in heavily shaken areas in the Northridge earthquake, eight had partial collapses, and an additional 20 had at least 25% damage (City of Los Angeles and SEAOSC, 1994). The task force looked at approximately 30 structures; of 26 structures with damage, approximately half contained some precast elements (Mooradian, 2005).

Within limited experience to date, life-safety performance of other **PC2** buildings in the U.S. has been good; however, performance in other countries has identified concerns that could be applicable to U.S. construction. See below for general discussion and Table 17.3-1 for a detailed compilation of common seismic deficiencies and rehabilitation techniques for the **PC2** building type.

Global Strength and Stiffness

Insufficient in-plane shear wall strength and stiffness are possible seismic deficiencies in **PC2** buildings and particularly in parking garages where shear wall length is generally limited. Rehabilitation to address inadequate shear wall strength and stiffness can include addition of new vertical elements or strengthening of existing elements, as summarized in Table 17.3-1. Addition of and enhancement to elements in **PC2** buildings is very similar to other concrete building types; however, several additional cautions are in order. First, the configuration of existing precast members, including cast-in voids and prestressing tendons must be carefully studied to allow connection of new elements to existing construction. Second, precast and post-tensioned systems are configured to minimize damaging effects of movement due to temperature variation, shrinkage and creep; these effects should be considered in the addition of new vertical elements.

Insufficient in-plane moment frame strength is a possible seismic deficiency in **PC2** buildings and particularly of concern where the frame might not have been initially designed for seismic loads. Where strength is a concern, it is likely that stiffness, connections, and ductile detailing will also be inadequate and that major addition or enhancement of vertical elements is required.

Configuration

Torsional irregularities can lead to possible seismic deficiencies in **PC2** buildings, as in any other building type, increasing deformation demand in local portions of the structure. One of the parking garages investigated by the City of Los Angeles and SEAOSC Task Force (City of Los Angeles and SEAOSC, 1994) had shear walls on three sides and an open front of the fourth side. The report speculates that excessive deflection at the open-front allowed girders to move sideways off of supporting columns, resulting in total collapse. The torsionally irregular building configuration appears to have contributed to collapse, along with inadequate diaphragm stiffness and component connections. Torsional deficiencies are most directly addressed through the addition of new vertical elements, as indicated in Table 17.3-1. Design to accommodate the concentrations of force and deformation demand may be an alternative.

Table 17.3-1: Seismic Deficiencies and Potential Rehabilitation Techniques for PC2 Buildings						
Deficiency		Rehabilitation Technique				
Category	Deficiency	Add New Elements	Enhance Existing Elements	Improve Connections Between Elements	Reduce Demand	Remove Selected Components
Global Strength	Insufficient in-plane strength of shear walls or frames	Steel braced frame [7.4.1] Concrete/masonry shear wall [12.4.2]	Fiber composite wall overlay [13.4.1] Concrete wall overlay [21.4.5] Infill openings			
Global Stiffness	Inadequate stiffness of shear walls or frames	Steel braced frame [7.4.1] Concrete/masonry shear wall [12.4.2]	Fiber composite wall overlay [13.4.1] Concrete wall overlay [21.4.5] Infill wall openings			
Configuration	Torsional irregularity	Steel braced frame [7.4.1] Concrete/masonry shear wall [12.4.2]				
	Incompatible deformation of building sections					Provide seismic separation of portions with different behavior. See general discussion of seismic separation.
	Distance between shear walls too large	Steel braced frame [7.4.1] Concrete/masonry shear wall [12.4.2]				
Load Path	Inadequate force transfer, diaphragm to shear wall, shear wall above to shear wall below, shear wall to foundation			Enhance anchorage [17.4.2]		

Table 17.3-1: Seismic Deficiencies and Potential Rehabilitation Techniques for PC2 Buildings					
Deficiency		**Rehabilitation Technique**				
Category	**Deficiency**	**Add New Elements**	**Enhance Existing Elements**	**Improve Connections Between Elements**	**Reduce Demand**	**Remove Selected Components**
Load Path (continued)	Inadequate connection of beam or girders to supporting elements	Supplemental vertical supports [21.4.11]		Enhance anchorage		
	Inadequate collectors	Add collector [17.4.1]	Enhance existing collector [17.4.1]			
Component Detailing	Gravity columns inadequate to accommodate drift	Reduce building drift to level acceptable for gravity elements – see global stiffness	Enhance column ductility with jacketing [12.4.4]			Remove or reconfigure portions of structure creating short columns [17.4.3]
	Inadequate wall strength	See Global Strength	See Global Strength			
	Inadequate frame connection detailing	Steel braced frame [7.4.1] Concrete/masonry shear wall [12.4.2]	Enhance existing frame connections			
Diaphragms	Inadequate strength and/or stiffness	Steel braced frame [7.4.1] Concrete/masonry shear wall [12.4.2]	Enhance shear transfer within diaphragm [22.2.11]			
	Inadequate shear transfer to walls		See load path			
	Inadequate chord capacity	Add chord [17.4.1]	Enhance existing chord [17.4.1]			
	Excessive stresses at openings and irregularities		Enhance diaphragm detailing			
	Re-entrant corners		Enhance diaphragm detailing			
Foundations	See Chapter 23					
[ ] Numbers noted in brackets refer to sections containing detailed descriptions of rehabilitation measures.						

One of the observations from the Northridge earthquake was that performance of parking garages decreased as the distance between supporting shear walls increased. Long diaphragm spans resulted in high shear and flexure demands on the diaphragms, causing yielding and fracture of diaphragm reinforcing (City of Los Angeles and SEAOSC, 1994; and Wood, Stanton & Hawkins, 2000). This can be considered either a configuration deficiency or a diaphragm deficiency. The addition of vertical elements will contribute significantly to reduction of demand and deflection. See global strength and stiffness discussion. Diaphragm shear capacity, chords, and collectors can also be enhanced, as discussed in the diaphragm deficiency section.

Where **PC2** buildings are comprised of several sections separated by seismic joints, movement can be incompatible and separation inadequate.

Load Path

Deficiencies in the load path connections between diaphragms and vertical elements, story to story and wall to foundation are of significant concern in every type of **PC2** building. Connection and load path detailing in the existing **PC2** building stock is thought to range from systems with no positive connections at all, to potentially brittle welded connections, to recently developed connections and systems that may allow better performance than cast-in-place buildings. Good performance was reported for limited examples of low-rise **PC2** shear wall buildings in the Northridge earthquake (Iverson and Hawkins, 1994) and the Kobe earthquake (Ghosh, 1995). In the Guam earthquake (EERI, 1995) damage to the shear wall to foundation connections in a mid-rise hotel building caused extensive local spalling, apparently related to eccentricities within the connections. In the Armenia earthquake (EERI, 1989) low to mid-rise large panel precast buildings performed well. The good performance was attributed to floor panels that spanned to bearing walls on all four sides, and to the redundancy of these systems. In contrast, the connections between precast column members in frame buildings were very vulnerable, likely due to eccentricities introduced in site modifications to longitudinal bar splices and poor column confinement. Overall load path connections are important to the performance of **PC2** buildings, and attention to detail and eccentricities is important. Rehabilitation of load path connections in shear wall type buildings will generally involve external mechanical connections.

While not as obvious an issue in **PC2** buildings as in **PC1**, as part of load path considerations concrete wall panels must be adequately connected to resist out-of-plane forces. The connections that are transferring diaphragm forces to the shear walls are generally also used to resist out-of-plane forces, so rehabilitation of these connections should consider both demands.

The connection between girders and supporting columns and other similar connections may require rehabilitation in order to provide a continuous load path. The movement of a girder off of the supporting column due to building deflection is the most obvious concern. It has additionally been postulated that, in the Northridge earthquake, some gravity members may have been pulled free of supports due to high vertical accelerations or vibrations. At one time, it was common to have fairly heavy connections between girders and columns at the point of bearing. Connection design would likely have controlled by calculated forces, and welded connection behavior could be brittle. It is now more common to minimize or eliminate connection between the column and girder at the bearing point and resulting restraint of support movement and to rely instead on the column and girder each being doweled into the diaphragm system. This approach minimizes unintended restraint and resulting damage. As in new construction, rehabilitation of this connection would best be accomplished by connection of each member into the diaphragm system. This approach is best suited to the intended behavior of the building system. Alternate approaches could include use of restraint cables, as is common in bridge rehabilitation or use of secondary vertical supports, as is common in rehabilitation of unreinforced masonry buildings (Section 21.4.10). There is no broad consensus on the contribution of vertical accelerations or vibrations to damage of **PC2** buildings in the Northridge earthquake. In locations where vertical thrust is a concern, rehabilitation measures could specifically take vertical demands into consideration.

In some precast systems, design of connections between elements has been based on the concept that the precast system emulates performance of an equivalent cast-in-place system. In frame members, the splicing of reinforcing steel is key to this performance. Over the years, the understanding of demands on splices and adequacy of splice technologies under earthquake loading has changed, as well as understanding of desirable locations of splices and controlling behavior modes within concrete frame systems, both precast and cast-in-place. Rehabilitation of connections between members in precast frames is difficult, and addition of vertical elements to limit frame drift may be a more practical solution.

Component Detailing

The inability of columns in gravity load systems to accommodate building drift has been pointed out as a significant deficiency in earlier discussion. Rehabilitation approaches include either adding additional vertical elements to reduce drift or enhancing columns with fiber-reinforced polymer wraps or similar systems to allow ductile behavior. These rehabilitation measures are discussed further in Chapter 12. The City of Los Angeles/SEAOSC report (1994) identified column wrapping as the only rehabilitation measure for parking garages that could be recommended as both practical and economical. Rehabilitation for the related issue of short columns can sometimes be as simple as creating an adequate joint between the column and the incidental restraint. Where this is not possible, reducing drift or enhancing column ductility are recommended rehabilitation approaches.

Diaphragm Deficiencies

Insufficient in-plane strength and stiffness of diaphragms are of significant concern over a range of precast systems.

In parking garages with long diaphragm spans, insufficient shear strength has been identified as a likely contributor to poor performance of **PC2** parking garages in the 1994 Northridge earthquake (Wood, Stanton & Hawkins, 2000), where pre-earthquake cracking of cast-in-place topping slabs occurred along the joists between T-beams. Analytical studies identified two strength-related issues that had not been considered previously. First, the pre-earthquake cracking of the topping slab along T-beam joints meant that the topping slab concrete was not contributing shear strength, leaving the reinforcing behavior acting in a shear-friction mode. Second, the limited strain capacity of the welded wire fabric reinforcing was being exceeded in commonly observed pre-earthquake crack widths, leaving the reinforcing vulnerable to brittle fracture during earthquake loading. Further, inadequate performance of chord and collector reinforcing in topping slabs has also been identified as a deficiency contributing to damage. In the Northridge earthquake, reinforcing bars serving as chords and collectors in cast-in-place topping slabs appear to have yielded and subsequently buckled (City of Los Angeles/SEAOSC, 1994). This behavior is a combination of strength and stiffness issues. In order to limit diaphragm deflection, maintain the integrity of the chord and collector members, and have the vertical elements control building dynamic behavior, it would be desirable to not have the chord and collector reinforcing yield. Since the Northridge earthquake, ACI 318 (ACI, 2005) and the building codes have taken initial steps towards addressing observed problems. Nakaki (1998) observed that the prescriptive steps taken by ACI 318 did not always result in improved behavior. Nakaki proposed simplified approaches to estimation of force and deformation demand for the purpose of ensuring elastic diaphragm design. This work has been incorporated into an appendix chapter of the *NEHRP Provisions* (FEMA, 2004) for untopped diaphragms.

In diaphragms without cast-in-place topping slabs, connections between adjacent planks or T-beams often use embedded steel plates with field-welded connections, or grout connections. Questions arise as to the pre-earthquake adequacy of these connections. Welded connections are often used to correct differences in camber between adjacent members during initial erection and are often stressed by moving vehicle point loads and shrinkage and creep movement of the building. Observations of connections suggest that reduced capacity prior to earthquake loading may be common. This combines with a changing understanding of earthquake demands on the connection and the interaction of shear demands and deformation due to flexural or tension loading. The complete lack of connection between hollow core floor planks within diaphragms appears to have been a primary contributor to collapse of nine-story residential precast concrete frame buildings in the 1988 Armenia earthquake (EERI, 1989).

A significant integrated analytical and experimental research program is currently underway to develop a comprehensive design methodology for precast concrete diaphragm systems. The project intends to address the discrepancy between current design practice, based on inelastic behavior concentrating in vertical elements and observed performance in which substantial inelastic behavior has occurred in diaphragms (Wan et al., 2004; and Naito and Cao, 2004). The project proposes to determine force and deformation demands required for design, connection details to support the performance, and address deformation relative to the gravity load-carrying system. This information will be invaluable for both new design and rehabilitation. Testing will include individual connections, joints, and half-size components. Analytical modeling of full buildings is being used to identify critical demands. Of particular interest is the simultaneous occurrence of shear and tension or compression on connections normally considered to carry only shear. Published information to date (Naito and Cao, 2004) provides a database of connector properties from existing literature and suggests a simplified analysis model based on initial finite element testing. Additional information should be available over the next several years.

Rehabilitation of diaphragm chord and collector members is reasonably practical due to the focused locations of work. Where possible, it is easiest to add reinforcing steel collectors on the top of the floor system in new cast concrete curbs. Where chords and collectors need to occur at building interior locations where traffic must cross, more complex solutions are required. Unstressed post-tensioning tendons may be a desirable alternative to rebar in some locations; however, it must be kept in mind that stresses must be kept low in order to minimize deformation, so the high strength does not give particular advantage. In undertaking rehabilitation of the chord and collector members, it must be acknowledged that there is lack of consensus on the interaction of shear and flexure demand in these buildings. Care must be taken not to induce brittle shear failure of the floor diaphragm as a result of flexural strengthening.

Rehabilitation of inadequate shear capacity is significantly more difficult. Most precast floor systems will have little capacity to support vertical load from additional topping slab thickness, and removal of existing toppings over large areas is not practical. Research conducted on connections between precast wall panels can be applied to connections between precast diaphragm members. Research by Pantelides, Volnyy, Gergeley, and Reaveley (2003), discussed in relation to load path connection in Section 17.4.2, could be applied to precast diaphragms. In addition a research program is currently being conducted by the Precast/Prestressed Concrete Institute investigating development of ductile panel-to-panel connections in precast diaphragms. See Section 22.2.11 for further discussion.

Rehabilitation for large openings and re-entrant corners in **PC2** buildings involves providing adequate chord and collector members in the vicinity of the opening or corner. Rehabilitation methods discussed in Section 17.4.1 are applicable. Framing bays at ramps in parking garages may need to be treated as openings for purposes of diaphragm design.

17.4   Detailed Description of Techniques Primarily Associated with This Building Type

To date, very little rehabilitation of **PC2** buildings has occurred in the U.S. As a result, the following discussion of rehabilitation measures draws from limited available research, suggested details for new **PC2** construction, and application of rehabilitation techniques for concrete buildings to the specific configurations of precast elements.

17.4.1   Add or Enhance Collector or Chord in Existing Precast Diaphragm

Deficiencies Addressed by Rehabilitation Technique

This rehabilitation technique addresses deficient diaphragm boundary members – chords and collectors at diaphragm boundaries and at interior openings and re-entrant corners.

Description of the Rehabilitation Technique

In diaphragms with cast-in-place topping slabs, existing chords and collectors are likely to be reinforcing bars at the edge of the cast-in-place slab and in line with shear walls. The most practical rehabilitation technique is addition of structural steel sections or reinforcing bars at the diaphragm boundary locations. Where boundary members occur at the perimeter of parking structures, it may be possible to encase the steel sections or reinforcing in new concrete curbs on top of the existing deck, as shown conceptually in Figure 17.4.1-1. Where boundary members extend across the floor plate where foot or vehicle traffic will occur, alternate chord and collector locations are required, as shown conceptually in Figures 17.4.1-2 and 17.4.1-3. For both chords and collectors, shear transfer capability between the boundary member and the structural diaphragm needs to be provided. Adhesive anchors or reinforcing dowels are the most likely

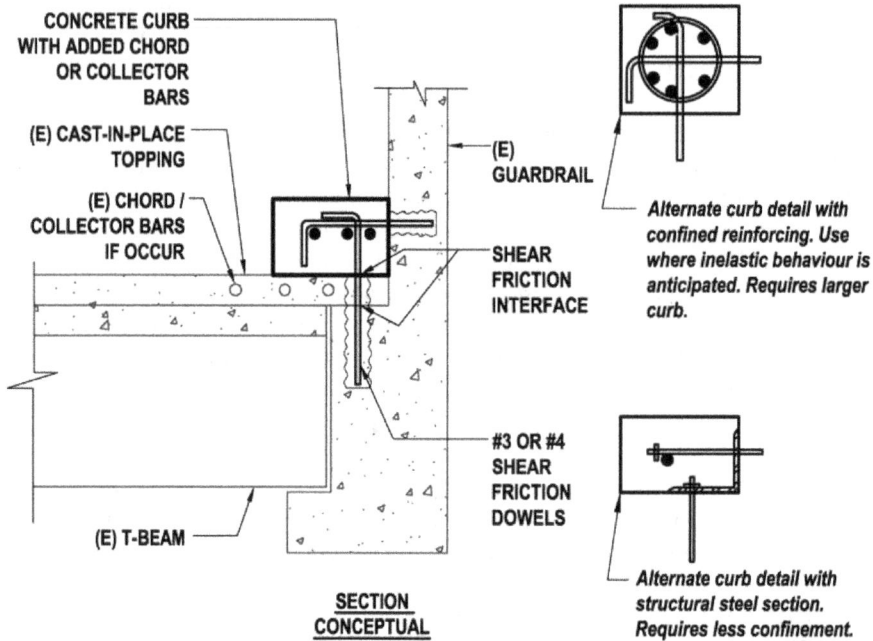

**Figure 17.4.1-1: Added or Enhanced Chord or Collector at Floor Perimeter
with Cast-In-Place Topping Slab**

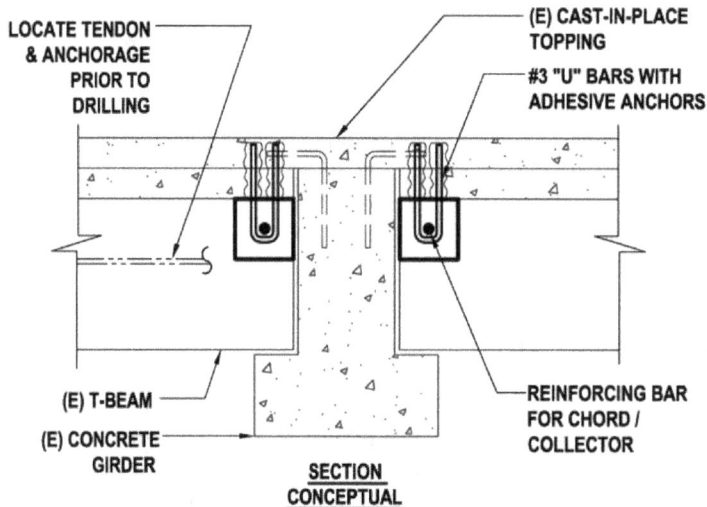

**Figure 17.4.1-2: Added or Enhanced Chord or Collector at Floor Interior
with Cast-In-Place Topping Slab**

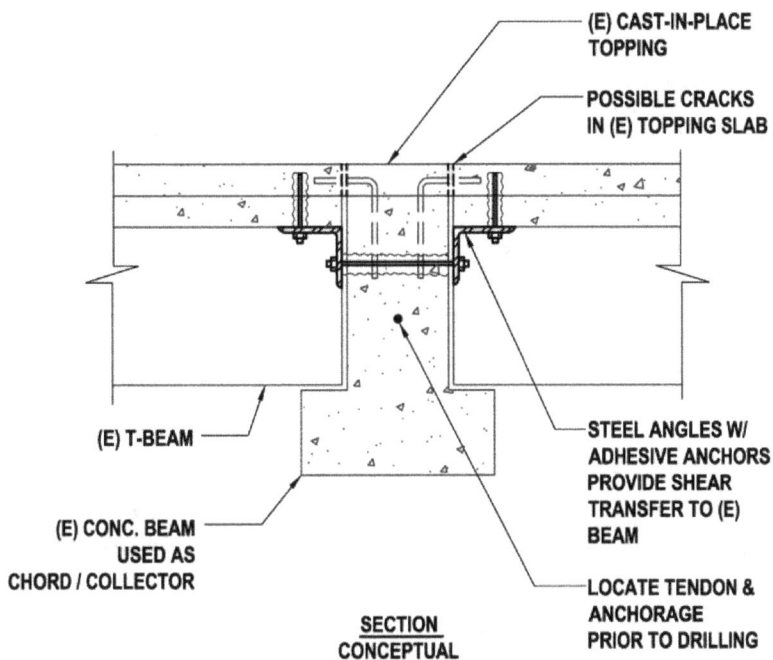

Figure 17.4.1-3: Added or Enhanced Chord or Collector at Floor Interior with Cast-In-Place Topping Slab

methods of attachment. Chord and collector splices must also be detailed. For collectors, transfer of the collector force to the vertical elements of the lateral force-resisting systems is required. Figure 17.4.1-4 shows a conceptual approach where the chord/collector curb runs by the face of the shear wall and dowels in over the full shear wall length. Figure 17.4.1-5 illustrates considerations when collector steel is to be doweled into the end of an existing shear wall, which may not be advisable.

In diaphragms without cast-in-place topping slabs, reinforcing in wall panels, floor panels, or beams will likely serve as chord and collector reinforcing. Where these floor panels or beams are precast, the connection between members is likely to be the weak link in chord and collector capacity. Figures 17.4.1-6 and 17.4.1-7 show the concept of added steel angles used as chords and collectors. The steel angle also serves as the connection between the wall and diaphragm for in-plane and out-of-plane forces.

Design Considerations

Research basis: No research applicable to this rehabilitation measure has been identified.

The objective of this rehabilitation method is to enhance the ability of the diaphragm to perform adequately and to deliver forces to the vertical elements of the lateral force-resisting system. In

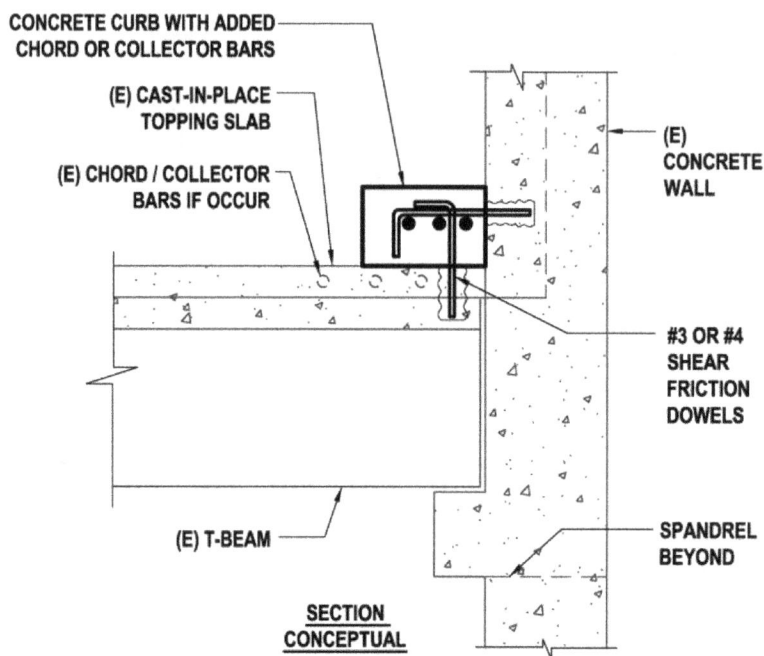

**Figure 17.4.1-4: Added Collector Anchorage to Shear Wall with
Cast-In-Place Topping Slab**

order to achieve this, careful consideration of the existing diaphragm configuration and shear
strength is needed in order to determine what behavior can be achieved and what detailing
approach is best. In long-span diaphragms, yielding of a chord member may be a preferred
behavior in order to protect against in-plane shear failure. Where this is true, it should be
anticipated that the chord member is likely to elongate in one or more locations under tension
loading, opening up gaps between adjacent precast members. When loading reverses,
compression will be carried by the chord member until gaps close. In order to prevent local
buckling failure of the chord, it is advisable to either use a structural steel section that can be
adequately braced against buckling with a reasonable adhesive anchor spacing, or provide
confinement of reinforcing bars, as would be provided for a concrete column. These alternatives
are shown at the right hand side of Figure 17.4.1-1. Both of these alternatives will be more costly
and difficult to install. In diaphragm configurations where yielding of the chord and collector
members can be avoided, providing additional reinforcing may be less costly than detailing for
buckling restraint. Where this approach is taken, a careful evaluation of anticipated forces is
needed. In general, it is the philosophy of ASCE 7 seismic provisions for inelastic behavior to be
focused in vertical elements rather than collectors, allowing the vertical elements of the seismic
force-resisting system to control building dynamic behavior. This may not be achievable in
diaphragms where the collector serves as a chord for loading in the perpendicular direction, and
avoiding shear failure is paramount. In this case, use of a detail that restrains buckling is
recommended.

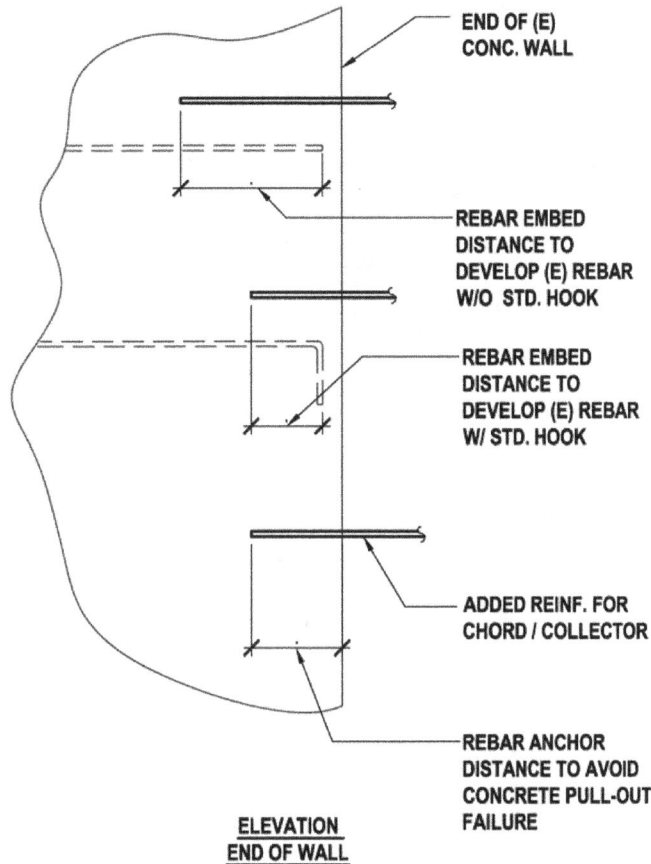

**END OF (E)
CONC. WALL**

**REBAR EMBED
DISTANCE TO
DEVELOP (E) REBAR
W/O STD. HOOK**

**REBAR EMBED
DISTANCE TO
DEVELOP (E) REBAR
W/ STD. HOOK**

**ADDED REINF. FOR
CHORD / COLLECTOR**

**REBAR ANCHOR
DISTANCE TO AVOID
CONCRETE PULL-OUT
FAILURE**

ELEVATION
END OF WALL

**Figure 17.4.1-5: Rebar Embedment Considerations in Collector Anchorage
to Existing Shear Wall**

For the chord configurations shown in Figure 17.4.1-6, the existing original connection (if any)
likely consists of intermittent cast-in embed plates and a welded plate connection. The capacity
of anchors embedded in the concrete and welds would have been sized to meet load requirements
(wind or earthquake) applicable at time of construction. Even if earthquake loading were
considered, the need to allow for forces in excess of design levels, ductility, and energy
dissipation likely would not have been considered. Because inelastic behavior in the precast
walls and diaphragms is very unlikely, it must be anticipated that inelastic behavior will
concentrate in the connections between members. With many existing connections, evaluation
would likely identify failure of the anchors embedded in concrete as the weak link. This is an
undesirable weak link due to lack of ductility. Unless extreme overstrength has been provided, to
allow the connection to remain elastic, rehabilitation of the connections is needed in order to
avoid this weak link. ACI 318 Appendix D (ACI, 2005) provisions for anchorage to concrete
require that that design be governed by tensile or shear strength of a ductile steel element rather

the concrete capacity. This requirement is particularly appropriate for rehabilitation of connections in precast wall buildings.

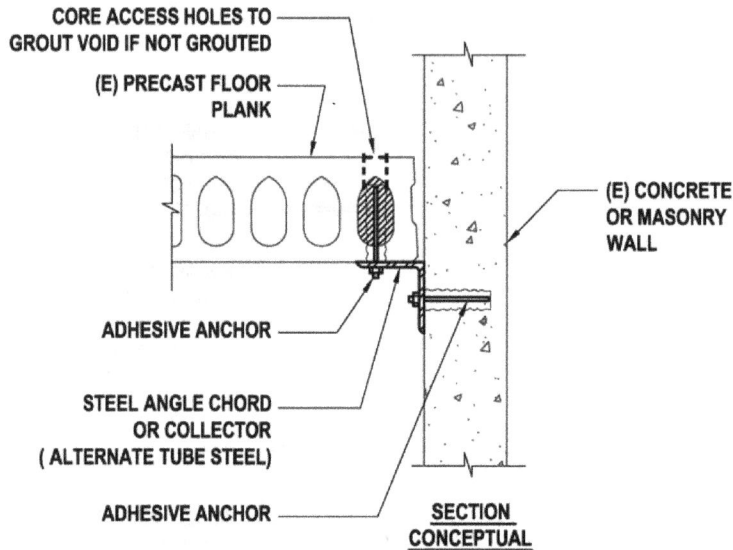

**Figure 17.4.1-6: Steel Chord or Collector at Floor Perimeter
without Cast-In-Place Topping**

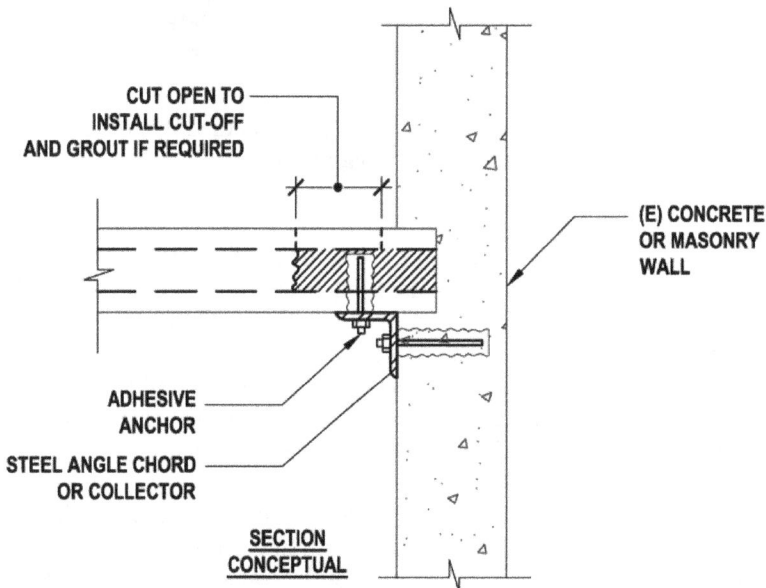

**Figure 17.4.1-7: Steel Chord or Collector at Floor Perimeter
without Cast-In-Place Topping**

The deformation compatibility of the existing and added connections needs to be carefully considered. It may be desirable to design new connections to carry all forces, neglecting the contribution of the existing connections. In some cases it might even be desirable to cut the weld on the original connection to ensure that it will not carry load.

Detailing Considerations

Use of reinforcing dowels to anchor new chord or collector reinforcing to existing concrete, as shown in Figures 17.4.1-1 and 17.4.1-2, relies on shear friction. Roughening of the existing concrete to the amplitude required by ACI 318 requires a very heavy sandblasting or chipping with a jackhammer, in order to expose the aggregate. This is expensive and messy. It is often preferable to use a reduced μ coefficient of less than 1.0 and add more dowels, rather than roughening the surface. Removal of finishes and cleaning of the concrete surface is still required. Use of shear friction also requires that the yield capacity of the dowel be developed on both sides of the joint. This forces use of smaller dowels and a curb dimension adequate to develop standard hooks per ACI 318 requirements. Adhesive anchor embedment requirements for development of the bar yield are generally available from the manufacturer.

Figure 17.4.1-2 places new chord or collector reinforcing at the underside of the floor so that vehicle or foot traffic above is not disrupted. Where chord or collector reinforcing runs parallel to the T-beams, use of reinforcing bars is likely feasible. Where the chord or collector runs perpendicular, reinforcing bars would have to be installed in short lengths threaded thru the T-beam webs. An unstressed tendon that can be more easily placed might be a preferable alternative.

Figure 17.4.1-3 uses the existing girder as the chord or collector member and adds angles for shear transfer between the slab and collector. This must be used in combination with enhanced connections at the girder ends. The figure illustrates the possible pre-earthquake cracks in the topping slab described by Wood, Stanton and Hawkins (2000). The shear transfer angles bypass this potentially weak location. Where steel splice angles are used, splice details will be required. A field welded splice would be common in concrete rehabilitation. Where yielding of the steel chord/collector member can be anticipated, it may be desirable to proportion the splice plates and welds to develop the anticipated strength of the chord/collector member.

Figure 17.4.1-5 illustrates considerations when collector steel is to be doweled into the end of an existing shear wall. This is only possible when there is adequate reinforcing within the wall in the vicinity of the collector anchorage to distribute the collector force over the wall length. Anchoring two No. 9 bars to the end of walls with only two No. 4 bars should not occur. Further, it is important that the collector lap adequately with reinforcing in the shear wall that can distribute the collector force over the wall length, as shown in the upper two figures. If embedment were limited to the requirements of the anchor manufacturer, as shown in the bottom figure, failure of the collector may occur due to inability of the wall reinforcement to develop.

Cost/Disruption and Construction Considerations

The work required to add or enhancement chords and collectors is generally spread out over considerably over the building area. This distribution of work is reasonably easy in parking garages because there are generally no finishes to remove and replace, all areas are reasonably accessible for materials and equipment, and user relocation does not involve a big effort. In contrast, the distribution of work will create significantly greater cost and coordination in other building types such as commercial or residential. In these building types, it may well be easier and more cost effective to add vertical elements, rather than enhancing chords and collectors, because this work may be located in one or more local areas.

Proprietary Concerns

There are no proprietary concerns with this rehabilitation technique other than the use of proprietary anchors as part of the assemblage.

17.4.2   Enhance Connections Between Existing Precast Diaphragm, Shear Wall and Foundation Elements

Deficiency Addressed by Rehabilitation Technique

This rehabilitation technique addresses rehabilitation of inadequate connections between precast concrete elements including: diaphragm to shear wall, shear wall above to shear wall below, shear wall to foundation, and shear wall panel to panel connections.

Description of the Rehabilitation Technique

Existing connections of precast wall and diaphragm elements are generally either welded connections between cast-in embed plates, or connections involving mechanical, grouted or welded rebar splices. Where precast elements are combined with cast-in-place elements, Figures 17.4.2-1 and 17.4.2-2 show connections between precast wall panels and foundation and diaphragms. The connections depict one possible configuration for existing welded connections and rehabilitation approaches for enhancing connection shear capacity. The rehabilitation measures involve adding new steel angles and adhesive anchors as required to carry seismic forces. Figures 17.4.2-1B shows added anchorage to a hollow core wall section using grout, which provides a high connection capacity. Alternately an anchor specifically designed for attachment to hollow masonry could be used, resulting in a lower connection capacity.

Design Considerations

Research basis: FRP composite connections between wall panels have been tested at the University of Utah (Pantelides, Volnyy, Gergeley, and Reaveley, 2003).

Existing precast shear wall connections most often use cast-in embed plates and welded plate connections. The capacity of anchors embedded in the concrete and welds would have been sized to meet load requirements (wind or earthquake) applicable at time of construction. Even if earthquake loading were considered, the need to allow for forces in excess of design levels, ductility and energy dissipation likely would not likely have been considered. Because inelastic behavior in the precast walls and diaphragms is very unlikely, it must be anticipated that inelastic behavior will concentrate in the connections between members. With many existing connections, evaluation would likely identify failure of the anchors embedded in concrete as the weak link. This is an undesirable weak link due to lack of ductility. Unless extreme overstrength has been provided, allowing the connection to remain elastic, rehabilitation of the connections is needed in order to avoid this weak link. ACI 318 Appendix D (ACI, 2005) provisions for anchorage to concrete require that the design be governed by tensile or shear strength of a ductile steel element rather the concrete capacity. This requirement is particularly appropriate for rehabilitation of connections in precast wall buildings. The deformation compatibility of the existing and added connections needs to be carefully considered. It may be desirable to design new connections to carry all forces, neglecting the contribution of the existing connections. In some cases, it might even be desirable to cut the weld on the original connection to ensure that it will not carry load.

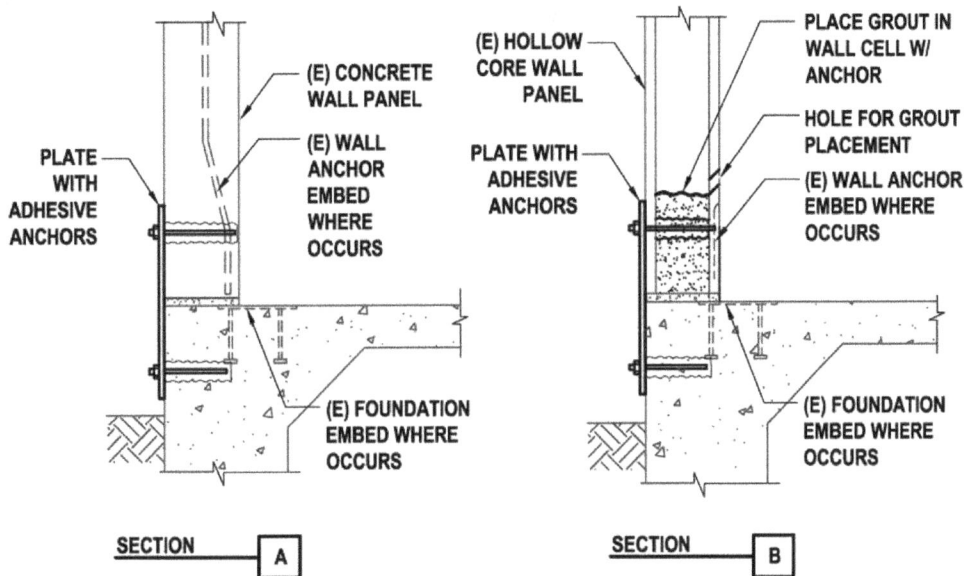

Figure 17.4.2-1: Added or Enhanced Precast Wall-to-Foundation Connection

Each precast wall panel tends to have two existing welded connections at the wall top and two at the wall bottom, and connections are generally not provided panel to panel. As a result, the two connections at the wall bottom must resist both shear and overturning forces from in-plane loads as well as out-of-plane loads. This is illustrated in Figure 17.4.2-3A. Where multiple wall panels are in line, as shown in Figure 17.4.2-3B, it may be possible to add panel to panel connections to resist overturning, reducing the demand on the bottom of panel connections. Overturning capacity will still be needed at each end of the panel group. The perpendicular wall panel at the right side of Figure 17.4.2-3 should not be counted on to resist the uplift, since it may also be overturning.

Figure 17.4.2-4 illustrates an FRP composite connection between wall panels developed and tested at the University of Utah (Pantelides, Volnyy, Gergeley, and Reaveley, 2003) that would be applicable to this use. While various surface preparations and FRP applications were investigated, the researchers settled on use of a high-pressure water jet preparation of the concrete surface to expose aggregate, a bonding agent and dry lay-up of carbon fibers, and saturation with an epoxy resin. A lay up of six layers of 12k (12,000 threads per tow) carbon fiber reinforcing in a 16 inch by 48 inch rectangle on one face of the wall panels provided failure loads on the order of 40 kips. The failures were sudden and brittle, indicating that design of this connection type should be considered force-controlled, and ductility should be provided in other connections or members. See Section 13.4.1 for general information on FRP composite overlays.

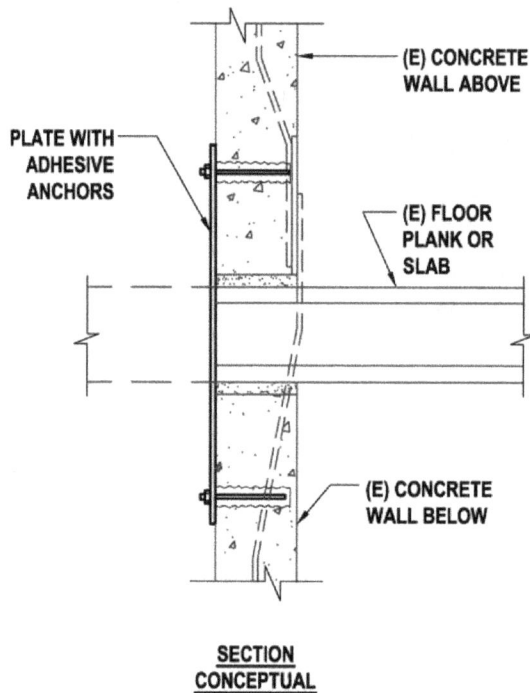

Figure 17.4.2-2: Precast Wall Connection

The new connections shown have both vertical and horizontal eccentricities that must be considered in design of the angle section and the anchors.

Detailing Considerations

Anchorage to hollow core wall panels is more difficult than solid precast panels because the voids do not leave solid concrete sections large enough to meet embedment and edge distance requirements for adhesive anchors. For most anchorage, drilling grout access and inspection holes and filling the lower portion of the void with grout can provide a solid cell for anchor placement. Similar grouting of floor planks is sometimes used in new construction. For very light anchorage loads, it may be possible to use screen-tube adhesive anchors specifically designed for attachment to hollow masonry units. Along with low capacity, failure of this anchorage should be expected to have brittle behavior.

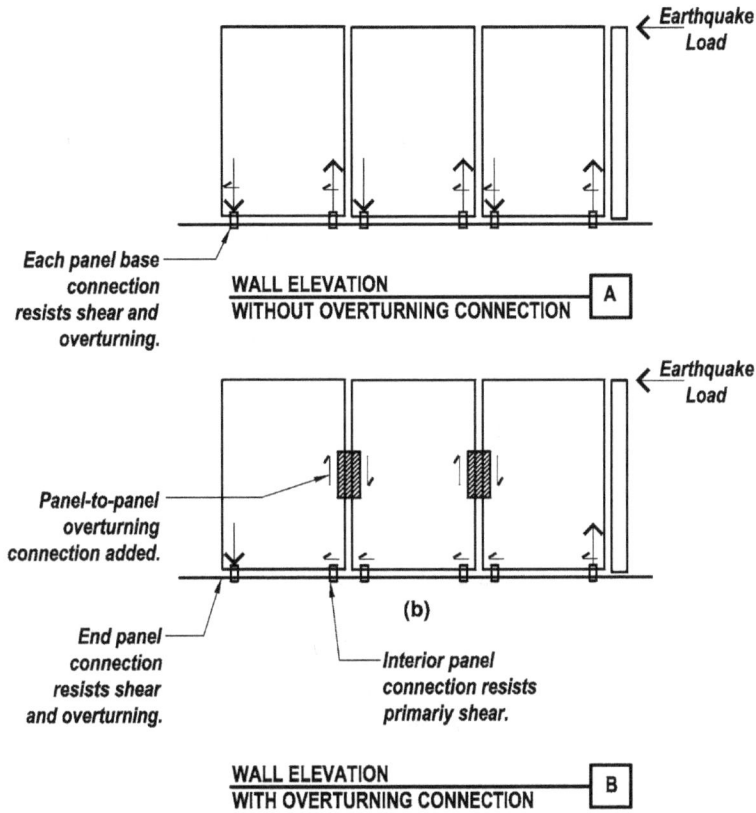

Figure 17.4.2-3: Modification of Demand on Anchors
Through Use of Panel-to-Panel Connections

Cost, Disruption and Construction Considerations

The cost of retrofitting connections in areas covered by finishes can be very expensive due to finish removal and replacement. The use of high-pressure water jets for surface preparation is not practical for an occupied building with finishes. It may be possible in an open building such as a parking garage.

Proprietary Concerns

FRP systems are proprietary. Manufacturers should be contacted for appropriate uses and limitations. Proprietary adhesives are used as part of connection details.

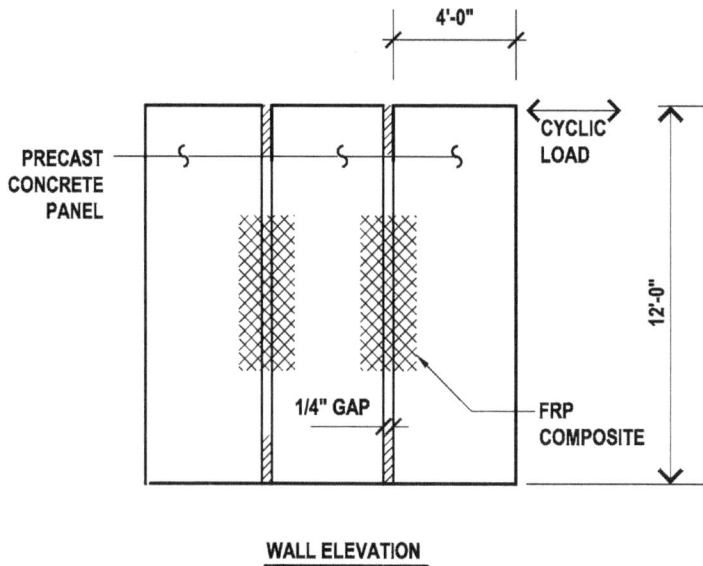

WALL ELEVATION

Figure 17.4.2-4: Panel Setup for Testing of FRP Panel-to-Panel Connections

17.4.3   Mitigate Configurations Creating Short Columns

Deficiency Addressed by Rehabilitation Technique

This technique addresses mitigation of unintended shortening and fixity of concrete columns.

Description of the Rehabilitation Technique

Figures 17.4.3-1, 17.4.3-2 and 17.4.3-3 illustrate unintended shortening and stiffening of columns, common in parking garages. The condition in Figure 17.4.3-1 is very easily mitigated by sawcutting the required gap between the column and the guardrail. The connection at the base of the guardrail may need to be improved as a result. The condition in Figure 17.4.3-2 can be improved by creating a hinge or joint in the column at the top of slab level. This will require shoring to take the load off of the column during modification, and the development of a shear and tension connection to the pedestal and footing. The condition in Figure 17.4.3-3 is difficult to address. Columns with this configuration are not easily confined. A solution that can minimize column shortening for new construction is to provide separate columns for the ramp and level floors. This approach could also be attempted for a retrofit, but would be costly.

Design Considerations

Research basis: No research applicable to this rehabilitation measure has been identified.

Proprietary Concerns

There are no proprietary concerns with this rehabilitation technique.

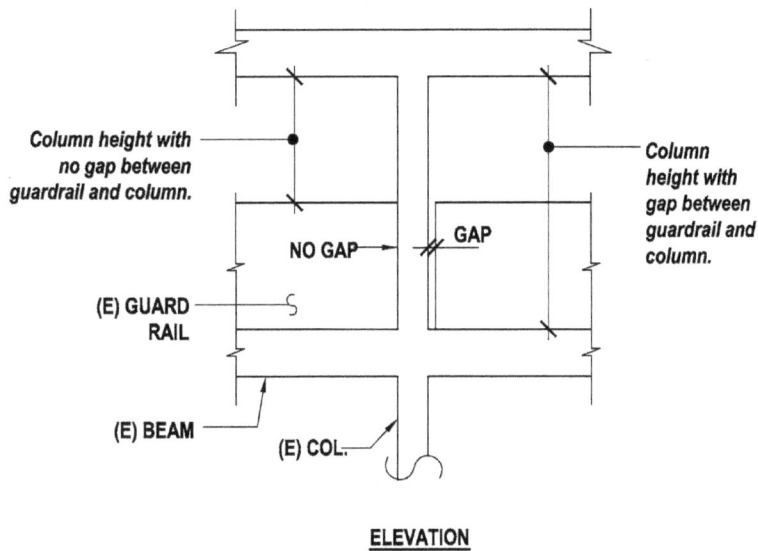

Figure 17.4.3-1: Accidental Reduction of Column Height Due to Guardrail

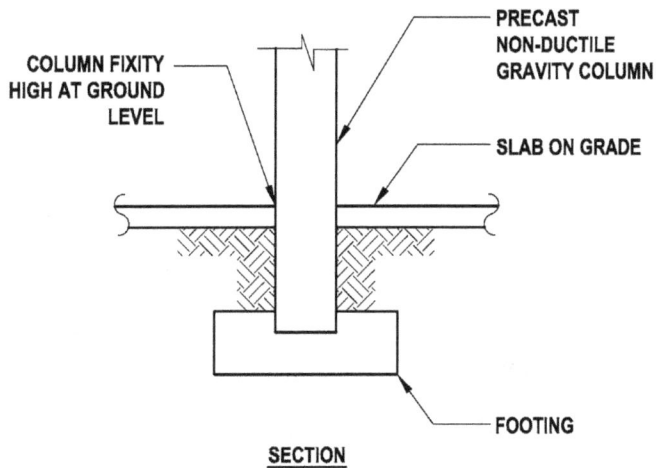

Figure 17.4.3-2: Accidental Increase in Column Fixity
Due to Embedment into Grade

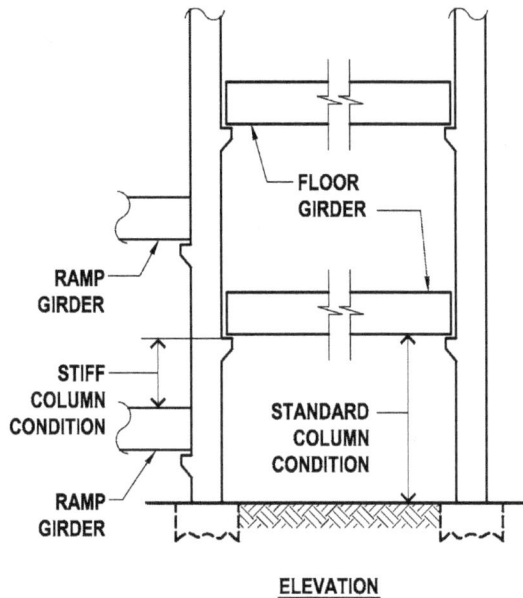

ELEVATION

Figure 17.4.3-3: Accidental Reduction of Column Height at Ramp

17.5   References

ACI, 2005, *Building Code Requirements for Structural Concrete* (ACI 318), American Concrete Institute, Farmington Hills, MI.

City of Los Angeles and SEAOSC (Structural Engineers Association of Southern California), *Findings & Recommendations of the City of Los Angeles/SEAOSC Task Force on the Northridge Earthquake*, 1994, City of Los Angeles, CA.

EERI, 1989, *Armenia Earthquake Reconnaissance Report*, Earthquake Spectra Special Edition, Earthquake Engineering Research Institute, Oakland, CA, August.

EERI, 1995, *Guam Earthquake of August 8, 1993 Reconnaissance Report*, *Earthquake Spectra*, Supplement B to Volume 11, Earthquake Engineering Research Institute, Oakland, CA, April.

EERI, January 1996, *Northridge Earthquake of January 17, 1994, Reconnaissance Report, Volume 2*, Earthquake Spectra, Supplement C to Volume 11, Earthquake Engineering Research Institute, Oakland, CA.

FEMA, 2004, *NEHRP Recommended Provisions for Seismic Provisions for New Buildings and Other Structures*, FEMA 450, Federal Emergency Management Agency, Washington, D.C.

Ghosh, S.K., "Observations on the Performance of Structures in the Kobe Earthquake of January 17, 1995," *PCI Journal*, Vol. 40, No. 2, March/April 1995, Precast/Prestressed Concrete Institute, Chicago, IL.

Iverson, J.K. and N.M. Hawkins, 1994, "Performance of Precast/ Prestressed Concrete Building Structures During the Northridge Earthquake," *PCI Journal*, Vol. 39, No. 2, Precast / Prestressed Concrete Institute, Chicago, IL.

Lyons, S., R. Bligh, J. Purlinton, and D. Beaudoin, 2003, "What about the Ramp? Seismic Analysis and Design of Ramped Parking Garages," *Proceedings of the 2003 SEAOC Convention*, Structural Engineers Association of California, Sacramento, CA.

Mooradian, D., 2005, personal communication.

Naito, C. and L. Cao, 2004, "Precast Diaphragm Panel Joint Connector Performance," *Proceedings of the 13th World Conference on Earthquake Engineering*, Vancouver, B.C. Canada.

Nakaki, S., 1998, *Design Guidelines: Precast and Cast-in-Place Concrete Diaphragms*, Earthquake Engineeering Research Institute, Oakland, CA.

Pantelides, C. P., V. A. Volnyy, J. Gergeley, and L.D. Reavely, 2003, "Seismic Retrofit of Precast Concrete Panel Connections with Carbon Fiber Reinforced Polymer Composites," *PCI Journal*, Vol. 48, No. 1, January/February, Precast/Prestressed Concrete Institute, Chicago, IL.

Priestley, M.J.N., S. Sritharan, JR. Conley, and S. Pampanin, 1999, "Preliminary Results and Conclusions From the PRESSS Five-Story Precast Concrete Test Building," *PCI Journal*, Vol. 44, No. 6, November/December, Precast/Prestressed Concrete Institute, Chicago, IL.

Wan, G., R. Fleischman, J. Restrepo, R. Sause, C. Naito, S.K. Ghosh, L. Cao, and M. Schoettler, 2004, "Integrated Analytical and Experimental Research Program to Develop a Seismic Design Methodology for Precast Diaphragms," *Proceedings of the 2004 SEAOC Convention*, Structural Engineers Association of California, Sacramento, CA.

Wood, S. L., J.F. Stanton, and N.M. Hawkins, 2000, "New Seismic Design Provisions for Diaphragms in Precast Concrete Parking Structures," *PCI Journal*, Vol. 45, No. 1, January/February 2000, Precast/Prestressed Concrete Institute, Chicago, IL.

Chapter 18 - Building Type RM1t: Reinforced Masonry Bearing Walls (Similar to Tilt-up Concrete Shear Walls)

18.1 Description of the Model Building Type

Building Type **RM1** is constructed with reinforced masonry (brick cavity wall or concrete masonry unit) perimeter walls with a wood or metal deck flexible diaphragm. For this document, Building Type **RM1** is separated into two categories. Chapter 19 describes **RM1u**, which is multistory, and typically has interior CMU walls and shorter diaphragm spans. This chapter covers **RM1t**, the large, typically one-story buildings with relatively open interiors that are similar to concrete tilt-ups. The exterior walls are commonly bearing, with an interior post and beam system of steel or wood. Older buildings of this type are generally small and used for a wide variety of occupancies. Recently, the building type has become commonly used for one-story warehouse and wholesale/retail occupancies similar to tilt-up (Building Type **PC1**) buildings. Figure 18.1-1 illustrates one example of the **RM1t** Building Type.

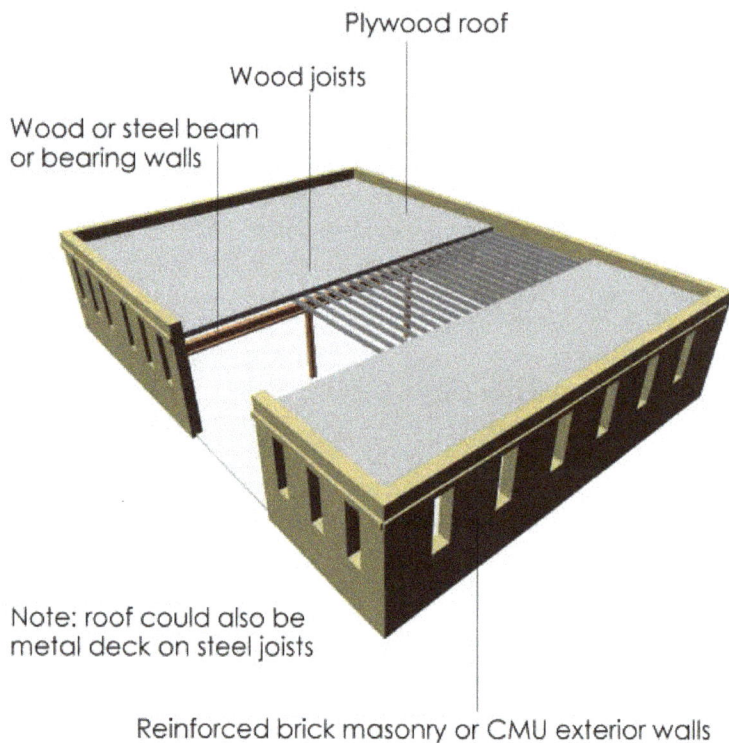

Figure 18.1-1: Building Type RM1t: Reinforced Masonry Bearing Walls

Guidelines for Seismic Evaluation and Rehabilitation of Tilt-Up Buildings and Other Rigid Wall/Flexible Diaphragm Structures or *SEAONC Guidelines* (SEAONC, 2001) provides a substantial collection of information on flexible diaphragm / rigid wall building issues, the West Coast experience with earthquake performance, rehabilitation priorities, and techniques for rehabilitation. This document was a primary source of information for Chapter 16 (**PC1** buildings), and it is also recommended for **RM1t** buildings.

Walls

Exterior reinforced masonry walls are the primary vertical elements in the lateral load-resisting system. Buildings with large plan areas may have interior walls providing additional lateral resistance; however, this is not common. Like tilt-up construction, masonry walls have recently transitioned from use of code-prescribed height to thickness (h/t) limits to use of much higher h/t ratios, in combination with rational analysis of slenderness effects. Most existing construction, however, will have been designed using prescriptive ratios and allowable stress design methods.

Reinforced masonry walls in **RM1t** buildings share many issues with **PC1** buildings, discussed in Chapter 16. Two distinctive aspects of reinforced masonry walls require discussion: movement control joints and partially grouted masonry.

The masonry industry recommends providing vertical control joints in new masonry wall construction to accommodate masonry wall shrinkage and thermal movement. The currently recommended maximum spacing of control joints is the lesser of 1.5 times the wall height or 25 feet (CMACN, 2003). The inclusion of and spacing of control joints varies significantly, however, with the age and region of the building. Wall-to-diaphragm anchorage remains a primary concern irrespective of whether control joints are provided. Where vertical control joints occur, diaphragm chords and collectors will be provided by either horizontal reinforcing that continues across the control joint (typically provided at the diaphragm level only), or a steel angle or similar member on the face of the masonry. Rehabilitation of chords, collectors, and shear transfer for **RM1t** building will be much the same as **PC1** buildings. Masonry wall construction without control joints will not have to rely as much on discreet chords and collectors; adequacy of shear transfer may still be a focus of rehabilitation.

In areas of high seismic hazard, it is most common for masonry walls to be fully grouted. In many other areas, however, grout is only provided at required reinforcing. A typical reinforced masonry wall in a **RM1t** building might have vertical reinforcing and grout alongside window and door openings, and at between four and ten feet on center horizontally and vertically in the piers and spandrels. Partial grouting has a significant effect on the weight of the wall and calculated seismic forces, as well as wall strength for both in-plan and out-of-plane forces. Where partial grouting has been provided, grout locations need to be known in order to design wall to diaphragm anchorage.

Gravity Load-Carrying Support at the Building Perimeter

It is most common for wood girders to be supported on the building exterior walls in **RM1t** buildings. As in **PC1** buildings, girder connections to the exterior walls are required to resist wall out-of-plane loading in addition to gravity loads. Connections of girders to the exterior walls require evaluation and possible rehabilitation. Section 16.4.2 discusses applicable rehabilitation measures. Where the existing masonry is partially grouted, it may be necessary to open up the masonry face and grout at new anchorage locations. Where this is done, cast-in anchors can be provided in lieu of adhesive anchors.

Roof Diaphragms

Like the **PC1** building, the roof system will generally be either of wood or steel construction. In the western states, roof systems are almost exclusively wood sheathed. Outside of the western states, roof systems are almost exclusively sheathed with steel decking topped with rigid insulation or vermiculite concrete. For both the wood and the steel roof systems, the roof diaphragm in the **RM1t** building is almost always flexible compared to the walls. See Chapter 16 for additional discussion.

Wall-to-Diaphragm Connections

Like **PC1** buildings, wall-to-diaphragm connections are thought to be the aspect of **RM1t** buildings most vulnerable to earthquake damage, due to the significant force and deformation demands imposed on this connection. The wall-to-diaphragm connections are therefore recommended as the first focus of rehabilitation measures for one-story **RM1t** buildings. See Chapter 16 for a discussion of past performance of these connections in tilt-up buildings. Similar to tilt-up buildings, even in **RM1t** buildings constructed or upgraded recently, it should be assumed that these connections require review and possible rehabilitation.

Interior Additions

Mezzanines and interior second stories, commonly constructed within large box-like **RM1t** buildings, can restrain building movement under earthquake loading, resulting in unintended load paths and damage. See Chapter 16 for discussion.

Foundations

RM1t buildings are generally constructed on continuous perimeter footings, with dowels to the reinforced masonry walls at vertical reinforcing locations.

18.2   Seismic Response Characteristics

RM1t buildings, like **PC1** buildings, are distinguished by rigid shear walls and flexible diaphragms. Like **PC1** buildings, amplification of seismic forces near the center of the diaphragm is of concern for both diaphragm capacity and wall anchorage capacity. The in-place construction of the reinforced masonry walls may provide wall continuity that makes **RM1t** buildings somewhat less vulnerable to partial collapse than **PC1** buildings; however, the potential for significant performance problems exists.

18.3   Common Seismic Deficiencies and Applicable Rehabilitation Techniques

Unlike **PC1** buildings, damage to **RM1t** buildings has not been noted as significant or wide spread. In the 1994 Northridge earthquake, little in the way of damage to **RM1t** buildings was reported (EERI, 1996; and Klingner, 1994). One item of interest was damage to masonry walls at building corners near the roof line, attributed to interaction between the masonry wall and flexible wood diaphragm. See below for general discussion and Table 18.3-1 for a detailed

Table 18.3-1: Seismic Deficiencies and Potential Rehabilitation Techniques for RM1t Buildings						
Deficiency		**Rehabilitation Technique**				
Category	**Deficiency**	**Add New Elements**	**Enhance Existing Elements**	**Improve Connections Between Elements**	**Reduce Demand**	**Remove Selected Components**
Global Strength		Steel braced frame [7.4.1] Concrete/masonry shear wall [21.4.8]	Concrete wall overlay [21.4.5] Infill openings			
Global Stiffness		Steel braced frame [7.4.1] Concrete/masonry shear wall [21.4.2]	Concrete wall overlay [21.4.5] Infill openings			
Configuration	Torsionally irregular plans	Steel braced frame [7.4.1] Concrete/masonry shear wall [21.4.8]	Concrete wall overlay [21.4.5] Infill openings			
	Re-entrant corners	Steel braced frame [7.4.1] Concrete/masonry shear wall [21.4.8] Collector [7.4.2]	Enhance existing collector Concrete wall overlay [21.4.5] Infill openings	Collector [7.4.2]		
	Incidental bracing					Separate component from incidental bracing
Load Path	Inadequate or missing wall-to-diaphragm tie for out-of-plane load			Wall-to-diaphragm tension anchors plus subdiaphragms and cross-tie [16.4.1]		
	Inadequate anchorage to diaphragms for in-plane forces			Wall-to-diaphragm shear anchors [21.4.2]		
	Inadequate collectors	Add collector [7.4.2]	Improve collector member and connections			

Table 18.3-1: Seismic Deficiencies and Potential Rehabilitation Techniques for RM1t Buildings						
Deficiency		**Rehabilitation Technique**				
Category	**Deficiency**	**Add New Elements**	**Enhance Existing Elements**	**Improve Connections Between Elements**	**Reduce Demand**	**Remove Selected Components**
Component Detailing	Wall inadequate for out-of-plane bending		Wall strongback or pilaster [21.4.3]			
	Inadequate detailing of slender walls	Steel braced frame [7.4.1] Concrete/masonry shear wall [21.4.8]	Concrete wall overlay [21.4.5] Infill openings Add backup vertical supports where bearing might be lost [21.4.11]			
Diaphragms	Inadequate in-plane strength and/or stiffness	Steel braced frame [7.4.1] Concrete/masonry shear wall [21.4.8]	Enhance existing diaphragm [22.2.1] Horizontal braced frame [21.2.10]			
	Inadequate chord capacity	Enhance chord [22.2.2]	Enhance chord [22.2.2]			
	Excessive stresses at openings and irregularities		Enhance diaphragm detailing			
	Re-entrant corners		Enhance diaphragm detailing			
Foundations	See Chapter 23					
[ ] Numbers noted in brackets refer to sections containing detailed descriptions of rehabilitation measures.						

compilation of common seismic deficiencies and rehabilitation techniques for Building Type
RM1t. See also Chapter 16 for similar issues in **PC1** buildings.

Global Strength and Stiffness

Global strength and stiffness are rarely a concern for large box-like **RM1t** buildings, but can be
for smaller buildings that have very short walls along a street-front side. Rehabilitation of global
strength and stiffness deficiencies will commonly involve adding new vertical elements,
enhancing existing elements, or infilling openings in existing walls.

Configuration

Poor distribution of shear walls can result in torsionally irregular behavior of **RM1t** buildings.
Common occurrences include street-front walls in commercial buildings. The most direct
approach to rehabilitation of this condition is the addition of strength and stiffness in line with
the perforated wall. This can be accomplished through addition of new shear walls, enhancing
existing shear walls, or addition of steel braced frames.

Rehabilitation at re-entrant corners requires the provision of adequate chords and collectors,
shear transfer to the in-set wall panels, and possibly the strengthening of the wall panels and
connections to the foundation. The *SEAONC Guidelines* suggest that there may be diaphragm
continuity over this interior diaphragm support, increasing the diaphragm reaction to the in-set
wall line. See Chapter 16 for illustration and additional discussion.

Load Path

As previously mentioned, load path connections between the masonry walls and the flexible
diaphragm are suggested as the first focus of rehabilitation in **RM1t** buildings. Diaphragm cross-
ties, as addressed in Chapter 22, are a required continuation of the wall anchorage system.
Section 16.4.1 discusses applicable rehabilitation measures. Where the existing masonry is
partially grouted, it may be necessary to open up the masonry face and grout at new anchorage
locations. Where this is done, cast-in anchors can be provided in lieu of adhesive anchors.
Connection between the wall and diaphragm may also be inadequate for in-plane shear loads.

The addition of or enhancement of existing collectors may be required in order to transmit
diaphragm forces to the resisting shear walls. This is particularly of concern when a limited
length of shear wall intended to carry a significant portion of the building shear. Although not as
common, there is also significant concern when vertical offsets in the roof diaphragm result in
incomplete chords or collectors. Any breaks or offsets in chords or collectors need to be carefully
evaluated.

Component Detailing

Component detailing deficiencies include inadequate out-of-plane wall capacity. Where existing
walls are partially grouted concrete masonry, it may be possible to place additional vertical
reinforcing and grout in ungrouted cells, accessed by cutting open face shells. Where this
approach is taken, doweling of the vertically reinforcing at the top and bottom of the grouted
cells would commonly be provided. In addition, the increase in wall weight should be considered
in building seismic forces. Where existing masonry wall construction is solid, it is seldom
practical to address wall capacity by adding reinforcing and concrete thickness to individual wall

sections, so addition of wall pilasters or strongbacks is common. Where pilasters are added to masonry walls, pilaster-to-roof diaphragm anchorage must be provided to accommodate the concentration of wall out-of-plane force.

Diaphragm Deficiencies

Due to changes in building code requirements, it is very common for diaphragms in areas of high seismic hazard to have inadequate in-plane shear capacity. These diaphragms may also have inadequate in-plane stiffness due to high unit shear stresses. Regardless of this, the *SEAONC Guidelines* indicate that diaphragm overstresses have rarely been associated with significant earthquake damage. Diaphragm strength and stiffness deficiencies are most often rehabilitated by enhancing the existing diaphragm.

Other diaphragm deficiencies include inadequate chord capacity and stress concentrations at large diaphragm openings and re-entrant corners. Rehabilitation at re-entrant corners primarily involves the provision of adequate chords and collectors. The same is true at large diaphragm openings.

18.4   Detailed Description of Techniques Primarily Associated with This Building Type

No techniques have been developed for this building type. See other chapters for detailed descriptions of relevant rehabilitation techniques.

18.5   References

CMACN, 2003, "Movement Control Joints," *Masonry Chronicles Winter 02-03*, Concrete Masonry Association of California and Nevada, Citrus Heights, CA.

EERI, January 1996, *Northridge Earthquake of January 17, 1994, Reconnaissance Report, Volume 2*, *Earthquake Spectra*, Supplement C to Volume 11, Earthquake Engineering Research Institute, Oakland, CA.

Klingner, R., 1994, *Performance of Masonry Structures in the Northridge California Earthquake of January 17, 1994* (Technical Report 301-94), The Masonry Society, Boulder, CO.

SEAONC (Structural Engineers Association of Northern California), 2001, *Guidelines for Seismic Evaluation and Rehabilitation of Tilt-Up Buildings and Other Rigid Wall/Flexible Diaphragm Structures*, Structural Engineers Association of Northern California, San Francisco, CA.

Chapter 19 - Building Type RM1u: Reinforced Masonry Bearing Walls (Similar to Unreinforced Masonry Bearing Walls)

19.1   Description of the Model Building Type

Building Type **RM1** takes a variety of configurations, but they are characterized by reinforced masonry walls with flexible diaphragms such as wood or metal deck. The walls are commonly bearing, but the gravity system also contains post and beam construction of wood or steel in interior or some façade locations. For this document, Building Type **RM1** is separated into two categories. Chapter 18 describes **RM1t**, the large, typically one-story buildings with relatively open interiors that are similar to concrete tilt-ups. This chapter covers **RM1u**, which is multistory, and typically has interior CMU walls and shorter diaphragm spans. It is similar to Building Type **URM** (Chapter 21) and has many of the same deficiencies. Figure 19.1-1 shows an example of this building type.

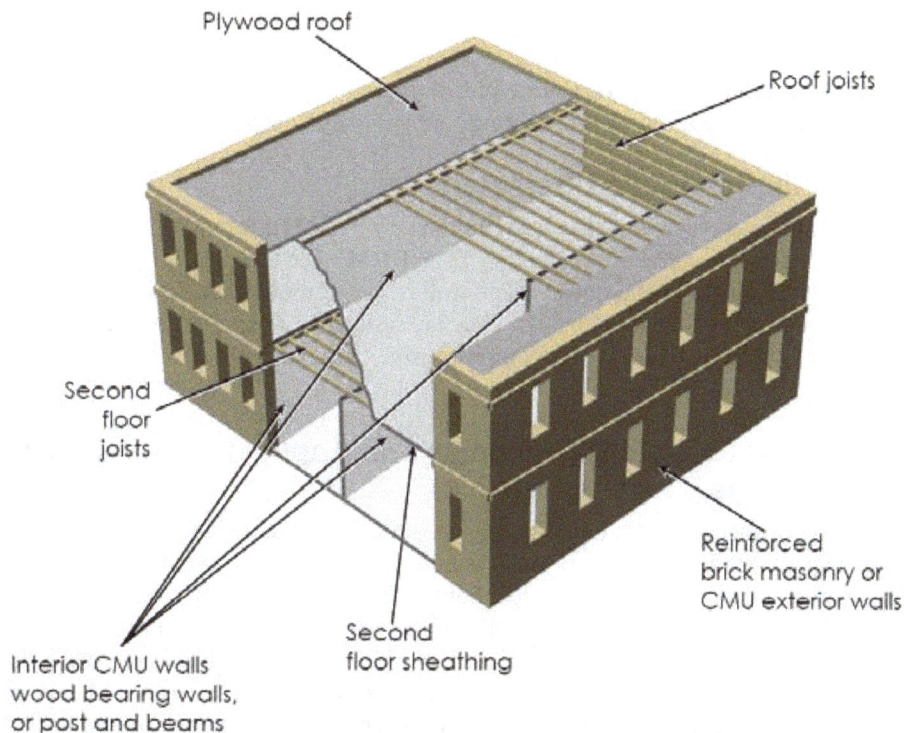

Figure 19.1-1: RM1u Building Type: Reinforced Masonry Bearing Walls

Masonry Wall Materials

FEMA 306 (FEMA, 1999) identifies several common reinforced masonry wall types. These are:

> Fully-grouted hollow concrete block
> Partially-grouted hollow concrete block
> Fully-grouted hollow clay brick
> Partially-grouted hollow clay crick
> Grouted-cavity wall masonry (two wythes of clay brick or hollow units with a reinforced grouted cavity)

Brick veneer facing may be placed on the exterior façade with the above walls used as backing walls.

Floor and Roof Diaphragm

Floor and roof diaphragm construction is similar to those of Building Type **URM**, although unfilled metal deck diaphragms can be found at the roof and occasionally at floors. See Chapter 21.

Foundations

Foundations for Building Type **RM1u** are typically spread footings at interior columns and strip footings under masonry bearing walls. Footings are typically concrete.

19.2   Seismic Response Characteristics

As a flexible diaphragm, stiff wall structure Building Type **RM1u** is expected to have dynamic behavior similar to that described for Building Type **URM**. See Chapter 21. Since the walls are reinforced, however, in-plane and out-of-plane wall behavior modes are substantially different from those of unreinforced masonry walls. They are instead more similar to those of reinforced concrete. FEMA 306 provides the most comprehensive categorization of reinforced masonry wall behavior modes.

19.3   Common Seismic Deficiencies and Applicable Rehabilitation Techniques

RM1u buildings, while similar to **URM** buildings, are generally considered to be less hazardous. In-plane damage to reinforced masonry walls is much less likely to reach levels compromising life safety. Parapets can still be overstressed, but the risk to life safety is much less than those of unreinforced parapets, and out-of-plane failures of the walls spanning between diaphragms are relatively unlikely. The most significant risk to loss of life is due to inadequate connections between the walls and diaphragms. See below for general discussion and Table 19.3-1 for a detailed compilation of common seismic deficiencies and rehabilitation techniques for Building Type **RM1u**.

Global Strength

As shear wall buildings, global strength in **RM1u** buildings is dependent on the in-plane shear capacity of the walls. Relatively large seismic forces are needed to lead to life safety concerns

Table 19.3-1: Seismic Deficiencies and Potential Rehabilitation Techniques for RM1u Buildings						
Deficiency		**Rehabilitation Technique**				
Category	**Deficiency**	**Add New Elements**	**Enhance Existing Elements**	**Improve Connections Between Elements**	**Reduce Demand**	**Remove Selected Components**
Global Strength	Insufficient in-plane wall strength	Wood structural panel shear wall [5.4.1], [6.4.2] Concrete/masonry shear wall [21.4.8] Steel braced frame [7.4.1] Steel moment frame [21.4.9]	Concrete wall overlay [21.4.5] Fiber composite wall overlay [21.4.6] Grouting Infill openings [21.4.7]		Seismic isolation [24.3]	
Global Stiffness						
Configuration	Soft story, weak story, excessive torsion	Wood structural panel shear wall [6.4.2] Concrete/masonry shear wall [21.4.8] Steel braced frame [7.4.1] Steel moment frame [21.4.9]				
Load Path	Inadequate or missing wall-to-diaphragm tie			Tension anchors [16.4.1] Shear anchors [21.4.2] Subdiaphragms and cross-ties [22.2.3]		
	Missing collector	Add collector [7.4.2]				
	Inadequate girder-to-column connection			Improve connection Supplemental vertical supports [21.4.11]		
	Inadequate wall-foundation dowels			Wall-to-foundation improvements		

Table 19.3-1: Seismic Deficiencies and Potential Rehabilitation Techniques for RM1u Buildings					
Deficiency		Rehabilitation Technique			
Category	Deficiency	Add New Elements	Enhance Existing Elements	Improve Connections Between Elements	Reduce Demand	Remove Selected Components
Component Detailing	Wall inadequate for out-of-plane		Exposed interfloor wall supports [21.4.3] Reinforced cores [21.4.4] Concrete wall overlays [21.4.5] Fiber composite overlays [21.4.6]			
	Poorly anchored veneer or appendages		Add ties			Remove veneer or appendages
Diaphragms	Inadequate in-plane strength and/or stiffness	Wood structural panel shear wall [6.4.2] Steel braced frame [7.4.1] Concrete/masonry shear wall [21.4.8] Add wood structural panel or moment frame crosswall [21.4.10] Horizontal braced frame [22.2.9]	Enhance existing diaphragm [22.2.1] Enhance crosswall [21.4.10]			
	Inadequate chord capacity	Add steel strap or angle				
	Excessive stresses at openings and irregularities	Add wood or steel strap reinforcement				
	Re-entrant corner	Wood structural panel shear wall [6.4.2] Concrete/masonry shear wall [21.4.8] Steel braced frame [7.4.1] Steel moment frame [21.4.9]		Collector [7.4.2]		
Foundations	See Chapter 23					
[ ] Numbers noted in brackets refer to sections containing detailed descriptions of rehabilitation techniques.						

with in-plane wall behavior, though cracking damage will occur in relatively moderate events. When walls are found to be deficient, new vertical lateral force-resisting elements can be added at interior locations or existing walls can be enhanced.

Global Stiffness

Reinforced masonry walls are generally quite rigid, even if punctured with window openings, so global stiffness deficiencies are relatively uncommon.

Configuration

Many commercial **RM1u** buildings will have a fairly open street façade at the ground level, leading to a weak and soft first story and torsional irregularities. This is usually addressed by the addition of a moment frame at the façade or another vertical lateral force-resisting element at some distance back from the façade.

Load Path

As noted above, it is the lack of adequate ties between the walls and diaphragms that is the single most significant deficiency in **RM1u** buildings. Rehabilitation measures include tension ties for out-of-plane forces and shear ties for in-plane forces along the typical wall-diaphragm interface. In many reinforced masonry walls, pilasters are formed by thickening the wall in order to provide support of key girder lines. They may not have adequate bearing length for the girder seat or sufficient reinforcement at the top of the pilaster. Anchor bolts from the ledger to the wall are likely to be present, but cross-grain bending under out-of-plane tension loading will be a common deficiency. Rehabilitation measures include supplemental column supports and connection enhancements.

Component Detailing

When brick veneer is present, it may not be adequately anchored back to the backing masonry, creating a falling hazard. Veneer ties can be added.

Diaphragm Deficiencies

Wood diaphragms may lack both strength and stiffness in **RM1u** buildings. This can be addressed by adding new interior elements to cut the diaphragm span or by enhancing the diaphragm itself with wood structural panel overlays.

Foundation Deficiencies

Foundation deficiencies for existing elements are relatively uncommon in **RM1u** buildings. Foundation rehabilitation work usually is focused on the support for new lateral force-resisting elements that are added to the superstructure.

19.4   Detailed Description of Techniques Primarily Associated with This Building Type

Rehabilitation techniques for Building Type **RM1u** are typically similar to those used in **URM** buildings. See Chapter 21 for detailed descriptions of techniques. When metal deck floors are present, techniques in Chapter 16 provide examples of rehabilitation methods for connecting the metal deck to the tilt-up concrete walls; details for connecting to reinforced masonry walls are similar.

19.5   References

FEMA, 1999, *Evaluation of Earthquake Damaged Concrete and Masonry Wall Buildings: Basic Procedures Manual*, FEMA 306, May.

Chapter 20 - Building Type RM2: Reinforced Masonry Bearing Walls (Similar to Concrete Shear Walls with Bearing Walls)

20.1   Description of the Model Building Type

This building consists of reinforced masonry walls and concrete slab floors that may be either cast-in-place or precast. In this type of building, all walls usually act as both bearing and shear walls. The building type is similar and often used in the same occupancies as Building Type **C2b**, namely in mid- and low-rise hotels and motels. This system is also commonly used in residential apartment/condominium type buildings.

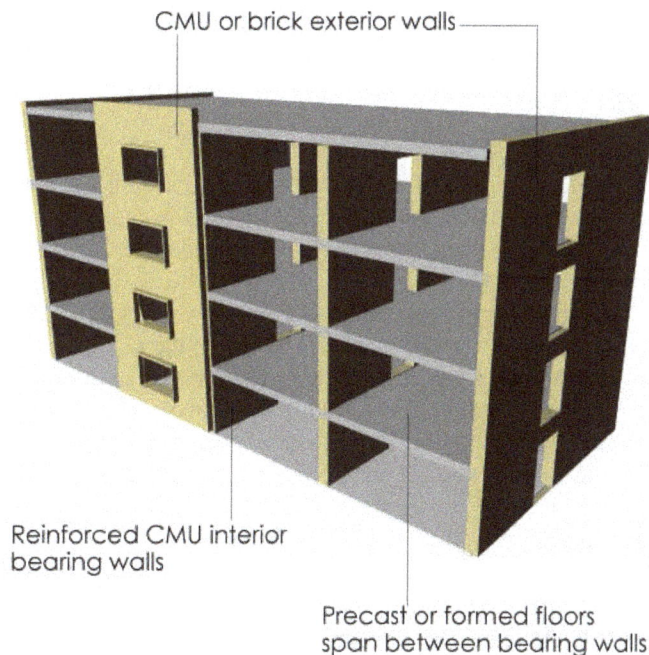

CMU or brick exterior walls

Reinforced CMU interior
bearing walls

Precast or formed floors
span between bearing walls

Figure 20.1-1: RM2 Building Type: Reinforced Masonry Bearing Walls

Variations Within the Building Type

In order for this framing system to be efficient, a regular and repeating pattern of bearing walls are required to provide support points for the floor framing. In addition, since it is difficult and expensive to make significant changes in the plan during the life of the building, planning flexibility is not normally an important characteristic when this structural system is employed. The occupancy types that most often fit these characteristics are residential buildings, including dormitories, apartments, motels, and hotels. These buildings will often be configured with reinforced masonry bearing walls between rooms which act as shear walls in the transverse direction, and reinforced masonry walls on the interior corridor which act as shear walls in the longitudinal direction. Sometimes the longitudinal lateral system includes the exterior wall system, although this wall is normally made as open as possible. In any case, the wide variation in structural layouts and occupancies that is included in other shear wall buildings such as **C2f** is not seen in **RM2**.

It is seldom possible to plan a building layout that provides complete gravity support with walls, and often local areas are supported with isolated columns, sometimes with beams and girders. However, story heights in these buildings are usually small, and added depth in the floor framing system is difficult to obtain. The extent of such beam and column framing often causes confusion as to the classification of the structure as a bearing wall system. However, if significant plan area is supported solely by walls, the structures are normally classified as **RM2**.

There are important variations in floor framing systems employed in this building type, and their adequacy to act as a diaphragm is an important characteristic of this building type as discussed below.

Floor and Roof Diaphragms

The parallel layouts of supporting walls and the need to minimize story heights normally leads to the use of one-way uniform-depth concrete floor systems. Cast-in-place and precast systems, both conventionally reinforced and prestressed, have been employed. The precast systems are often built up of narrow planks, which may not provide an adequate diaphragm unless a cast-in-place topping is provided. In addition, the precast systems may be placed with only a very narrow bearing area on the supporting walls, almost always on the outer masonry wythe or the CMU shell. When prestressed, the planks may be connected into the wall system only with the tail of the stressing tendon, and this connection may be inadequate to provide vertical support during seismic movements. The adequacy of the shear connection between slab and walls is also often an issue for both cast-in-place and precast systems.

Foundations

The bearing walls obviously require some kind of starter beam at grade for construction purposes, and this often leads to a simple continuous grade beam system. In poor soils, piles or drilled piers may be added below the grade beam. A continuous mat foundation may also be employed due to the short spans and total length of bearing points in this building type.

20.2   Seismic Response Characteristics

Due to the extent of wall, bearing wall buildings will be quite stiff. Elastic and early post-elastic response will therefore be characterized with lower-than average drifts and higher-than-average floor accelerations. Damage in this range of response should be minimal.

Overall post-elastic response may often include rocking at the foundation level. If rocking does not occur, the height-to-length ratio of shear walls in these buildings may force shear yielding near the base, which may lead to strength and stiffness degradation.

Global stability may also be compromised by poor connections between floor slab construction and bearing walls.

Shear Wall Behavior

When subjected to every increasing lateral load, individual shear walls or piers will first often force yielding in spandrels, slabs, or other horizontal components restricting their drift, and eventually walls and piers either rock on their foundations, suffer shear cracking and yielding, or form a flexural hinge near the base. Shear and flexural behavior are quite different, and estimates of the controlling action are affected by the distribution of lateral loads over the height of the structure.

Yielding of spandrels, slabs, or other coupling beams can cause a significant loss of stiffness in the structure. Flexural yielding will tend to maintain the strength of the system, but shear yielding, unless well detailed, will degrade the strength of the coupling component and the individual shear wall or pier will begin to act as a cantilever from its base. In this building type, the coupling elements are often slabs, and their lack of bending stiffness may reduce or eliminate significant coupling action.

Rocking is often beneficial, limiting the response of the superstructure. However, the amplified drift in the superstructure from rocking must be considered. In addition, if varying wall lengths or different foundation conditions lead to isolated or sequencing rocking, the transfer of load from rocking walls must be investigated. In buildings with basements, the couple created from horizontal restraint at the ground floor diaphragm and the basement floor/foundation (often termed the "backstay" effect) may be stiffer and stronger than the rocking restraint at the foundation and should be considered in those configurations.

Shear cracking and yielding of the wall itself are generally considered undesirable, because the strength and stiffness will quickly degrade, increasing drifts in general, as well as potentially creating a soft story or torsional response. However, in accordance with FEMA 356 (FEMA, 2000), shear yielding walls or systems can be shown to be adequate for small target displacements. Type **RM2** buildings will often fall into this category.

Flexural hinging is considered ductile in FEMA 356 and will degrade the strength of the wall only for larger drifts. Similar to rocking, the global effect of the loss of stiffness of a hinging wall must be investigated.

20.3   Common Seismic Deficiencies and Applicable Rehabilitation Techniques

See Table 20.3-1 for deficiencies and potential rehabilitation techniques particular to this system. Selected deficiencies are further discussed below by category.

Global Strength

Due to the extensive use of walls, buildings of this type seldom have deficiencies in this category, unless significant degradation of strength occurs due to shear failures.

Table 20.3-1: Seismic Deficiencies and Potential Rehabilitation Techniques for RM2 Buildings						
Deficiency		**Rehabilitation Technique**				
Category	**Deficiency**	**Add New Elements**	**Enhance Existing Elements**	**Improve Connections Between Elements**	**Reduce Demand**	**Remove Selected Components**
Global Strength	Insufficient in-plane wall shear strength	Concrete/masonry shear wall [12.4.2]	Concrete wall overlay [21.4.5] Fiber composite wall overlay [13.4.1] Steel wall overlay		Seismic isolation [24.3] Reduce flexural capacity [13.4.4]	
	Insufficient flexural capacity	Concrete/masonry shear wall [12.4.2]	Add chords [12.4.3]			
	Inadequate capacity of coupling beams	Concrete/masonry shear wall [12.4.2]	Strengthen beams [13.4.2] Improve ductility of beams [13.4.2]			Remove beams
Global Stiffness	Excess drift (normally near the top of the building)	Concrete/masonry shear wall [12.4.2]	Concrete/steel column jackets [12.4.5] Provide detailing of all other elements to accept drifts Concrete wall overlay [21.4.5]		Supplemental damping [24.4]	
Configuration	Discontinuous walls	Add wall or adequate columns beneath [12.4.2]	Fiber composite wrap of supporting columns [12.4.4] Concrete/steel jacket of supporting columns [12.4.5]	Improve connection to diaphragm [13.4.3]		Remove wall
	Soft story or weak story	Add strength or stiffness in story to match balance of floors				
	Re-entrant corner	Add floor area to minimize effect of corner		Provide chords in diaphragm [12.4.3]		

Table 20.3-1: Seismic Deficiencies and Potential Rehabilitation Techniques for RM2 Buildings

Deficiency		Rehabilitation Technique				
Category	**Deficiency**	**Add New Elements**	**Enhance Existing Elements**	**Improve Connections Between Elements**	**Reduce Demand**	**Remove Selected Components**
Configuration (continued)	Torsional layout	Add balancing walls [12.4.2]				
Load Path	Inadequate collector	Add steel or concrete collector [12.4.3]				
	Inadequate slab bearing on walls			Add diagonal dowels [13.4.3] Add steel ledger [13.4.3]		
Component Detailing	Wall inadequate for out-of-plane bending	Add strongbacks [21.4.3]	Concrete wall overlay [21.4.5]			
	Wall shear critical		Concrete wall overlay [21.4.5] Fiber composite wall overlay [13.4.1]		Reduce flexural capacity [13.4.4]	
Diaphragms	Precast components without topping		Improve interconnection [22.2.11] Add topping			
	Inadequate in-plane shear capacity		Concrete slab overlay Fiber composite overlays [22.2.5]			
	Inadequate shear transfer to walls		Add diagonal drilled dowels [13.4.3] Add steel angle ledger [13.4.3]			
	Inadequate chord capacity	New concrete or steel chord member [12.4.3]				
	Excessive stresses at openings and irregularities	Add chords [12.4.3]				Infill openings [22.2.4]
Foundation	See Chapter 23					
[ ] Numbers noted in brackets refer to sections containing detailed descriptions of rehabilitation techniques.						

Global Stiffness

Similar to strength, global stiffness is seldom a problem in this building type. However the effect of coupling slabs on initial stiffness and the potential change in stiffness due to yielding of these coupling slabs or wall-beams over doors should be investigated.

Configuration

The most common configuration deficiencies in this building type are weak or soft stories created by walls that change configuration or are eliminated at the lower floors. It is difficult to provide the needed ductility at the weak story and often strength must be added. Completely discontinuous walls also create a load transfer deficiency for both overturning and shear. In such cases, collectors are often needed in the floor diaphragm, and supporting columns need axial strengthening.

Load Path

A common deficiency in this building is weakness in the load path from floor to walls, either collector weaknesses or shear transfer weakness immediately at the floor wall interface. Local transfer can be strengthened by adding concrete or steel corbel elements, dowels, or combinations of these components. As indicated above, discontinuous walls also often create load path deficiencies from the wall into the diaphragm at the discontinuity.

Component Detailing

The most common detailing problem in this building type is an imbalance of shear and flexural strength in the walls, leading to pre-emptive shear failure. This condition may be shown to be acceptable with small displacement demands. Walls can be strengthened in shear with overlays of concrete, steel, or FRP.

The layout of walls often forces coupling between walls through the slab system or across headers of vertically aligned doors. These coupling components are seldom designed for the coupling distortions that they will undergo, particularly in older buildings. Short lengths of slabs between adjacent walls receive damage by coupling action that could compromise the gravity capacity. It is difficult to add strength or ductility to these slab areas, but vertical support at support points can be supplemented by corbels of steel or concrete. Damage to headers over doors often does not contribute to deterioration of overall response and can sometimes be acceptable. Local areas of wall can also be strengthened by overlays of concrete, steel, or FRP.

Diaphragm Deficiencies

Precast floor systems used in this building type often provide inadequate diaphragm behavior that could lead to bearing failures at the floor wall interface, particularly when no topping slab is present. Some topping slabs used primarily for leveling and smoothing the floor are inadequately tied to the precast elements or the walls, and are too thin or poorly reinforced to act as diaphragms on their own. See Chapter 22.

Foundation Deficiencies

This building type often places large demands on the foundation system. If rocking is shown to be a controlling displacement fuse for the building, the foundations must be investigated to assure that these displacements can safely occur. See Chapter 23.

20.4 Detailed Description of Techniques Primarily Associated with This Building Type

Most significant recommendations listed in Table 20.3-1 are the same as techniques used in the similar building type, **C2b**, Concrete Shear Walls (Bearing Wall Systems) or general techniques applied to concrete diaphragms. Details concerning these techniques can be found in other chapters.

20.5 References

FEMA, 2000, *Prestandard and Commentary for the Seismic Rehabilitation of* Buildings, FEMA 356, Federal Emergency Management Agency, Washington, D.C., November.

Chapter 21 - Building Type URM: Unreinforced Masonry Bearing Walls

21.1   Description of the Model Building Type

Building Type **URM** consists of unreinforced masonry bearing walls, usually at the perimeter and usually brick masonry. The floors are typically of wood joists and wood sheathing supported on the walls and on interior post and beam construction. This building type is common throughout the United States and was built for a wide variety of uses, from one-story commercial or industrial occupancies to multistory warehouses to mid-rise hotels. It has consistently performed poorly in earthquakes. The most common failure is an outward collapse of the exterior walls caused by loss of lateral support due to separation of the walls from the floor and roof diaphragms. Figure 21.1-1 shows an example of this building type.

Building Type **URMA** is similar to the Building Type **URM**, but the floors and roof are constructed of materials that form a rigid diaphragm, usually concrete slabs or steel joists with flat-arched unreinforced masonry spanning between the joists. Building Type **URMA** is not covered by this document.

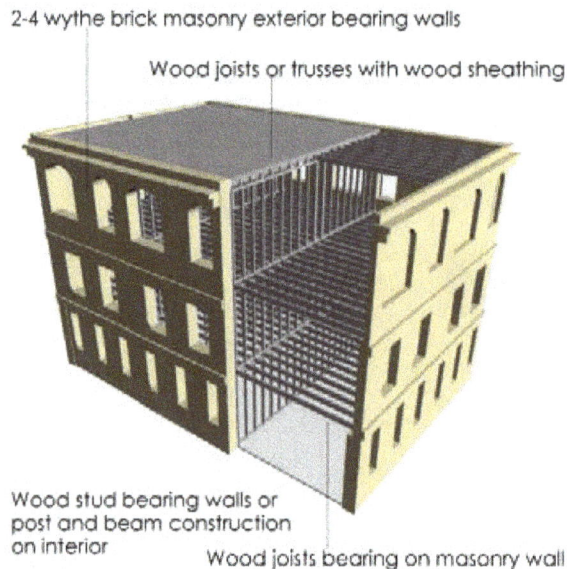

2-4 wythe brick masonry exterior bearing walls

Wood joists or trusses with wood sheathing

Wood stud bearing walls or post and beam construction on interior

Wood joists bearing on masonry wall

Figure 21.1-1: Building Type URM: Unreinforced Masonry Bearing Walls

Masonry Wall Materials

FEMA 306 (FEMA, 1999a) provides an overview of masonry wall material variables. It is paraphrased here. Unreinforced masonry is one of the oldest and most diverse building materials. Important material variables include masonry unit type, wall construction type, and material properties of various constituents.

Solid clay-brick unit masonry is the most common type of masonry unit, but there are a number of other common types, such as hollow clay brick, structural clay tile, concrete masonry, stone masonry, and adobe. Hollow clay tile (HCT) is a more common term for some types of structural clay tile. Concrete masonry units (CMU) can be ungrouted, partially grouted, or fully grouted. Stone masonry can be made from any type of stone, but sandstone, limestone, and granite are common. Other stones common in a local area are used as well. Sometimes materials are combined, such as brick facing over CMU backing, or stone facing over a brick backing.

Wall construction patterns also vary widely, with bond patterns ranging from common running bond in brick to random ashlar patterns in stone masonry to stacked bond in CMU buildings. The variety of solid brick bond patterns is extensive. Key differences include the extent of header courses, whether collar joints are filled, whether cavity-wall construction was used, and the nature of the ties between the facing and backing wythes. In the United States, for example, typical running-bond brick masonry includes header courses interspersed by about five to six stretcher courses. Header courses help tie the wall together and allow it to behave in a more monolithic fashion for both in-plane and out-of-plane demands. The 1997 UCBC (ICBO, 1997) and 2003 IEBC (ICC, 2003) have specific prescriptive requirements on the percentage, spacing, and depth of headers. Facing wythes not meeting these requirements must be considered as veneer and are therefore not used to determine the effective thickness of the wall. Veneer wythes must be tied back to the backing to help prevent out-of-plane separation and falling hazards. Although bed and head joints are routinely filled with mortar, the extent of collar-joint fill varies widely. Completely filled collar joints with metal ties between wythes help the wall to behave in a more monolithic fashion for out-of-plane demands. One form of construction where interior vertical joints are deliberately not filled is cavity-wall construction. Used in many northeastern United States buildings, the cavity helps provide an insulating layer and a means of dissipating moisture. The cavity, however, reduces the out-of-plane seismic capacity of the wall.

Material properties—such as compressive, tensile and shear strengths and compressive, and tensile and shear moduli—vary widely among masonry units, brick and mortar. An important issue for in-plane capacity is the relative strength of masonry and mortar. Older mortars typically used a lime/sand mix and are usually weaker than the masonry units. With time, cement was added to the mix and mortars became stronger. When mortars are stronger than the masonry, strength may be enhanced, but brittle cracking through the masonry units may be more likely to occur, resulting in lower deformation capacity.

Given the wide range of masonry units, construction and material properties, developing comprehensive mitigation techniques for all permutations is not practical. The rehabilitation measures in this document are most directly relevant to solid clay brick masonry laid in running bond with a typical spacing of header courses.

For additional general background on URM materials, see ABK (1981a), FEMA 274 (FEMA, 1997b), FEMA 307 (FEMA, 1999b), and Rutherford and Chekene (1997).

Floor and Roof Diaphragms

Building Type **URM**, by definition, is built with diaphragms that are considered, relative to the masonry walls, to be flexible. Typically, wood sheathing is attached to wood joists. Several types of wood diaphragm construction are common, including:

> Roofs with straight sheathing and roofing applied directly to the sheathing
> Roofs with diagonal sheathing and roofing applied directly to the sheathing
> Floors with straight tongue-and-groove flooring
> Floors with straight tongue-and-groove flooring over straight sheathing
> Floors with finished flooring over diagonal sheathing

Wood structural panel overlays may have been added as part of past renovation work, or there may be additional layers of sheathing materials. In some buildings with heavy live loads, like warehouses, 2x or 3x decking may have been used to span between joists.

Foundations

Foundations for URM buildings typically are spread footings at interior columns and strip footings under masonry bearing walls. Footings are typically either brick or concrete, though stone might be found under older walls, particularly if stone masonry was used in the walls.

21.2   Seismic Response Characteristics

In many building types, the horizontal diaphragms are more rigid than the vertical elements of the lateral force-resisting system. Such buildings are often thought of as lumped mass systems with the weight tributary to each diaphragm level lumped along a vertical cantilever with dynamic properties dependent on the stiffness of the vertical lateral force-resisting elements. Ground motion input at the base is dynamically amplified up the cantilever, increasing at each floor level. Each point within a floor has a similar acceleration.

Building Type **URM**, by contrast, has flexible diaphragms and stiff walls. Beginning with the ABK research program in the 1980s (see ABK, 1981a,b,c; and ABK, 1984), a different dynamic model was formulated for **URM** buildings. The ABK model assumes that there is relatively little dynamic amplification between the base and the top of the URM walls in the direction parallel to input motion. Significant amplification instead occurs at the midspan of the flexible diaphragms as they are driven by in-plane motion of the end walls. This generates large out-of-plane forces on the connections between the diaphragm and the coupled masonry walls. In some cases, the diaphragm may yield, limiting the forces that can be transmitted to the in-plane walls. If interior partitions are connected to the partitions, the deformation, cracking damage and resulting energy dissipation can help limit the movement of the diaphragms. Such existing partitions or newly added partitions or moment frames are termed "crosswalls".

The ABK program identified two modes of behavior for in-plane loading on the piers in the unreinforced masonry walls: shear-critical behavior and rocking-critical behavior. Each of these modes of behavior could be found acceptable if demands were below the capacity. These two modes are included in the UCBC as well. FEMA 273 (FEMA, 1997a) and then FEMA 356 (FEMA, 2000) expanded the characterization of in-plane behavioral modes into the more brittle force-controlled modes (toe crushing and diagonal tension) and the more ductile deformation-controlled modes (bed joint sliding and rocking). FEMA 306 and FEMA 307 identify a number of other in-plane modes and sequences of modes.

The ABK program also established criteria for determining the acceptability of out-of-plane resistance of the unreinforced masonry walls. Important variables are the height-to-thickness (h/t) ratio, presence of crosswalls, and overburden (axial compression) pressure on the walls. In short, a stocky lower story wall with overburden pressure from floors and walls above that is driven by a diaphragm damped by crosswalls is less likely to buckle and fail out-of-plane.

It is important to recognize that **URM** buildings vary substantially in structural layout and characteristics, and this can have a significant effect on seismic response. Fairly rectangular multistory residential, office, and commercial buildings often have relatively low story heights and many partitions that can serve as crosswalls. Walls adjacent to other buildings will usually be relatively solid. These buildings typically perform much better than structures like churches, which can have irregular plans, re-entrant corners, tall story heights, heavy walls, offset roofs, few partitions, and many windows. Churches can also be some of the most expensive structures to rehabilitate.

21.3   Common Seismic Deficiencies and Applicable Rehabilitation Techniques

URM buildings are generally considered to be one of the most hazardous building types. Significant property damage and loss of life have occurred in **URM** buildings during earthquakes around the world and in the United States. The primary deficiencies are due to unbraced parapets which can fall on adjacent pedestrian thoroughfares and poorly connected walls and diaphragms which can lead wall failure and loss of vertical support for diaphragms. See below for general discussion and Table 21.3-1 for a more detailed compilation of common seismic deficiencies and rehabilitation techniques for Building Type **URM**.

Global Strength

As shear wall buildings, global strength in **URM** buildings is dependent on the in-plane shear capacity of the walls. Relatively large seismic forces are needed to lead to life safety concerns with in-plane wall behavior, though cracking damage will occur in relatively moderate events. When walls are found to be deficient, new vertical lateral force-resisting elements can be added at interior locations or existing walls can be enhanced. At interior locations, new elements include wood structural panel shear walls, concrete shear walls, reinforced masonry shear walls, braced frames, and moment frames. At exterior locations, care must be taken to address relative rigidity concerns. Typically, concrete or shotcrete overlays are used to enhance the URM wall capacity. When the wall is highly punctured, braced frames or moment frames may be a viable option. The use of wood structural panel shear walls in buildings with masonry walls is permitted in new construction only in limited situations, such as one-story or two-story buildings with low story heights and no use of diaphragm rotation to resist loads, due to concerns about wood flexibility. Rehabilitation standards such as the 1997 UCBC and 2003 IEBC relax these restrictions significantly, though they do not permit the wood shear walls to resist lateral forces with other materials along the same line of resistance or when there are rigid diaphragms. Use of wood structural panel shear walls in rehabilitating masonry buildings should be carefully considered.

Table 21.3-1: Seismic Deficiencies and Potential Rehabilitation Techniques for URM Bearing Wall Buildings						
Deficiency		**Rehabilitation Technique**				
Category	**Deficiency**	**Add New Elements**	**Enhance Existing Elements**	**Improve Connections Between Elements**	**Reduce Demand**	**Remove Selected Components**
Global Strength	Insufficient in-plane wall strength	Wood structural panel shear wall [6.4.2], [5.4.1] Concrete/masonry shear wall [21.4.8] Steel braced frame [7.4.1] Steel moment frame [21.4.9]	Concrete wall overlay [21.4.5] Fiber composite wall overlay [21.4.6] Grouting Infill openings [21.4.7]		Seismic isolation [24.3]	
Global Stiffness						
Configuration	Soft story, weak story, excessive torsion	Wood structural panel shear wall [6.4.2] Concrete/masonry shear wall [21.4.8] Steel braced frame [7.4.1] Steel moment frame [21.4.9]				
Load Path	Inadequate or missing wall-to-diaphragm tie			Tension anchors [21.4.2] Shear anchors [21.4.2] Cross-ties and subdiaphragms [22.2.3] Supplemental vertical supports [21.4.11]		
	Missing collector	Add collector [7.4.2]				

Table 21.3-1: Seismic Deficiencies and Potential Rehabilitation Techniques for URM Bearing Wall Buildings					
Deficiency		**Rehabilitation Technique**			
Category	Deficiency	Add New Elements	Enhance Existing Elements	Improve Connections Between Elements	Reduce Demand	Remove Selected Components
Component Detailing	Wall inadequate for out-of-plane bending		Exposed interfloor wall supports [21.4.3] Reinforced cores [21.4.4] Concrete wall overlay [21.4.5] Fiber composite wall overlay [21.4.6]			
	Undesirable wall in-plane behavior mode		Sawcutting to change shear mode to rocking mode			
	Unbraced parapet		Brace parapet [21.4.1]			Remove parapet and improve roof-to-wall tie [21.4.1]
	Unbraced chimney		Brace chimney [5.4.6] Infill chimney [5.4.6]		Reduce chimney height [5.4.6]	Remove chimney [5.4.6]
	Poorly anchored veneer or appendages		Add ties [21.4.12]			Remove veneer or appendages
Diaphragms	Inadequate in-plane strength and/or stiffness	Add horizontal braced frame [22.2.9] Wood structural panel shear wall [6.4.2] Concrete/masonry shear wall [21.4.8] Steel braced frame [7.4.1] Steel moment frame [21.4.9] Wood structural panel or steel moment frame crosswall [21.4.10]	Enhance existing diaphragm [22.2.1] Enhance woodframe crosswall [21.4.10]			

Table 21.3-1: Seismic Deficiencies and Potential Rehabilitation Techniques for URM Bearing Wall Buildings

Deficiency		Rehabilitation Technique				
Category	Deficiency	Add New Elements	Enhance Existing Elements	Improve Connections Between Elements	Reduce Demand	Remove Selected Components
	Inadequate chord capacity	Add steel strap or angle				
Diaphragms (continued)	Excessive stresses at openings and irregularities	Add wood or steel strap reinforcement				
	Re-entrant corner	Wood structural panel shear wall [6.4.2] Concrete/masonry shear wall [21.4.8] Steel braced frame [7.4.1] Steel moment frame [21.4.9]		Collector [7.4.2]		
Foundation	See Chapter 23					
[ ] Numbers noted in brackets refer to sections containing detailed descriptions of rehabilitation techniques.						

Global Stiffness

URM bearing walls are generally quite rigid. When walls are solid or lightly punctured with window openings, global stiffness deficiencies are typically not an issue. In some buildings, though, facades facing the street can be highly punctured with relatively narrow piers between openings. In addition to lacking adequate strength, these wall lines may also be too flexible as well.

Configuration

Many commercial **URM** buildings will have a fairly open street façade at the ground level, leading to a weak and soft first story and torsional irregularities. This is usually addressed by the addition of a moment frame at the façade or another vertical lateral force-resisting element at some distance back from the façade.

Load Path

As noted above, it is the lack of adequate ties between the walls and diaphragms that is the single most significant deficiency in **URM** buildings. Rehabilitation measures include tension ties for out-of-plane forces and shear ties for in-plane forces. Bond beams are often employed for connections where the roof runs over the top of the walls. As a back-up vertical support system, supplemental vertical supports are added under trusses or girders where large gravity loads are concentrated on the wall in case the masonry is damaged locally.

Component Detailing

Since the masonry elements in Building Type **URM** are unreinforced by definition, they do not comply with modern ductile detailing requirements. Walls deemed susceptible to out-of-plane bending failures can be strengthened by strongbacks placed against them either on the outside or more commonly on the interior face. When preservation of finishes is critical, reinforced cores can be drilled and installed within the wall. Parapet bracing and chimney bracing are common. In some buildings, the exterior brick wythe will not be anchored back to the backing wall with mechanical ties or sufficient headers, and veneer ties are installed.

Diaphragm Deficiencies

Wood diaphragms may lack both strength and stiffness in **URM** buildings. This can be addressed by adding new interior elements to cut the diaphragm span or by enhancing the diaphragm itself with wood structural panel overlays. In some cases, such as sloped roofs, new horizontal braced frame diaphragms are added, in lieu of strengthening the existing diaphragm.

Foundation Deficiencies

Foundation deficiencies for existing elements are relatively uncommon in **URM** buildings. Foundation rehabilitation work usually is focused on the support for new lateral force-resisting elements that are added to the superstructure.

21.4   Detailed Description of Techniques Primarily Associated with This Building Type

21.4.1   Brace or Remove URM Parapet

Deficiency Addressed by Rehabilitation Technique

Past earthquakes have consistently shown that unreinforced masonry chimneys and parapets are the first elements to fail in earthquakes due to inadequate bending strength and ductility. Parapets tend to have greater damage at midspan of diaphragms due to higher accelerations and displacements from the oscillating diaphragm.

Description of the Rehabilitation Technique

URM parapets can be braced or removed to minimize the falling hazard risk. Bracing is usually done with a steel angle brace. The brace is anchored near the top of the parapet and to the roof. The existing roof framing may need localized strengthening to take the reaction from the brace. Roof-to-wall tension anchors are typically part of parapet bracing. See Figure 21.4.1-1 for an example of parapet bracing. If the top of the parapet is removed, the vertical compressive stress on roof-to-wall anchors is reduced, so removing the parapet is often combined with adding a concrete cap or bond beam as part of the roof-to-wall anchorage. See Figure 21.4.1-2 for an example of parapet removal and addition of a concrete cap beam.   See Section 21.4.2 for more details on wall-to-diaphragm anchorage.

Design Considerations

Research basis: No references directly addressing testing of parapet bracing have been identified.

Parapet height: Codes such as the 2003 IEBC and 1997 UCBC provide maximum allowable height-to-thickness (h/t) ratios for parapets. The height is taken from the lower of the either the tension anchors or the roof sheathing. Requirements are more stringent in higher seismic zones. With the 1997 UCBC, for example, the h/t ratio in Seismic Zone 4 is 1.5, so for a typical 13" thick, three-wythe wall, parapets taller than 19.5" above the roof-to-wall anchors require removal or bracing.   In the 2003 IEBC, locations with $S_{D1} > 0.4g$ have the same requirements.

Fire protection: The original purpose of extending the masonry wall up to form a parapet was to help limit the spread of fire between the wood roofs of adjacent buildings. Removing a parapet must be coordinated with local building code fire safety requirements.

Guardrails: Parapets often serve as a guardrail around a roof. Removing a parapet must be coordinated with local building code life safety requirements.

Load path: When parapet bracing/roof-to-wall tension anchorage is the only rehabilitation technique, the out-of-plane load path can be incomplete, particularly when the roof joists are perpendicular to the brace. A more complete rehabilitation strategy includes developing the parapet and tension anchorage forces back into the roof diaphragm through the use of subdiaphragms and cross-ties.

REMOVE (E) SHEATHING
BOARD AND REPLACE
WITH PLYWOOD OF
EQUIVALENT THICKNESS

ANGLE WITH BOLT EACH END

GUSSET PLATE

CONTINUOUS ANGLE

ANGLE WITH LAG SCREWS
INTO BLOCKING

WORK POINT

*Pitch pocket. Waterproofing
details typically provided
by the architect.*

22.5°

1'-0"

THROUGH BOLT
WITH BEARING
PLATE

1
2 MIN.

(E) JOISTS

JOIST HANGER

(E) URM WALL

4x BLKG

VARIES
VERIFY IN FIELD

Drilled dowel alternate

SECTION

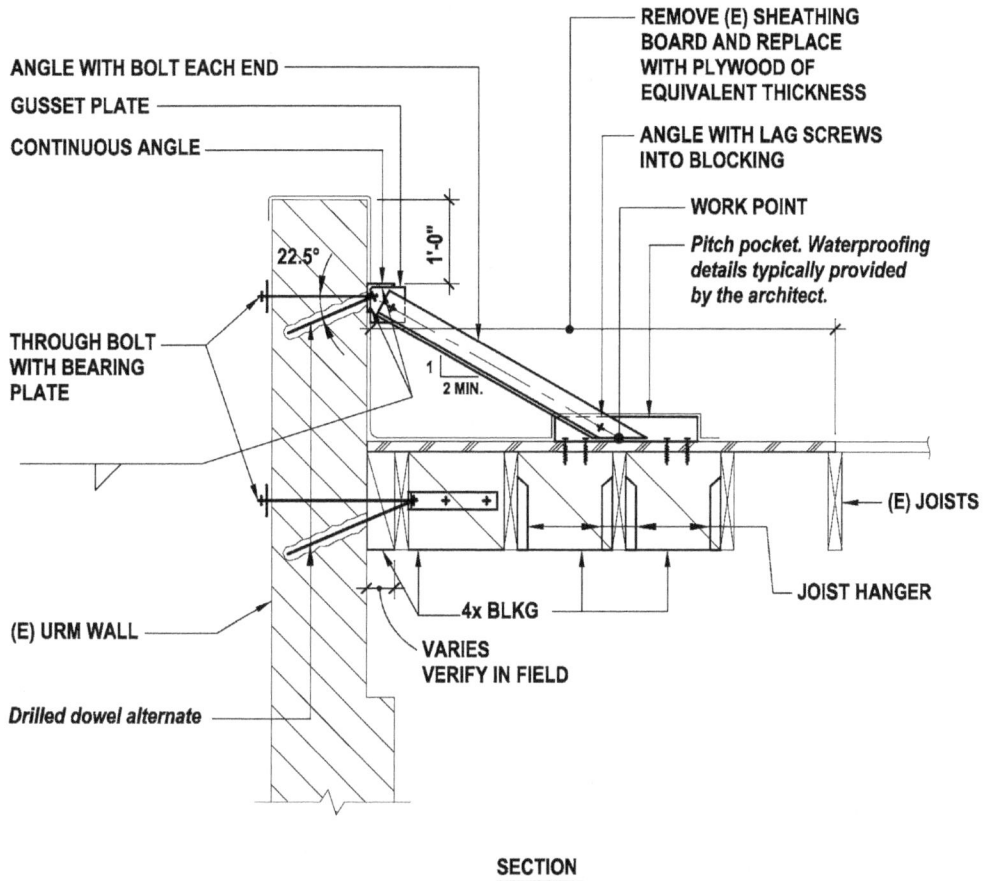

Figure 21.4.1-1: Parapet Bracing

Detailing and Construction Considerations

Parapet anchorage types: Drilled dowels connecting the top of the bracing to the masonry can
be with through bolts or adhesive anchors. See Section 21.4.2 for detailed discussion of drilled
dowels.

Top angle: Figure 21.4.1-1 shows a continuous angle running between braces in the roof. This
angle can be used to span between braces to reduce the number of bracing points. It also
increases redundancy over a localized connection of the brace to the parapet.

Load in the roof framing: The vertical reactions at the base of the brace are typically resisted by
roof framing. In Figure 21.4.1-1 the added blocking beneath the base of brace workpoint helps
to engage three joists in resisting vertical loads. Tall parapets can generate substantial brace
forces that existing wood roof joists may not be able to resist. Additional joists can be added, or
more braces can be used to distribute the load. Horizontal loads from the brace are distributed by
the blocking and new wood structural panel.

(E) ROOF MEMBRANE

(E) STRAIGHT SHEATHING

(E) URM WALL
AND PARAPET

(E) ROOF JOISTS

EXISTING CONDITION   A

CONCRETE CAP BEAM

STEEL TENSION TIE STRAP
AND ANCHOR BOLT

ROOFING TO COVER
TENSION TIE

REMOVE (E)
SHEATHING AND ADD
WOOD STRUCTURAL
PANEL SHEATHING
DIAPHRAGM

REMOVE (E)
PARAPET

(E) SHEATHING

B.N.

BLOCKING

SHEAR TIE

*Blocking/strap depth
depends on subdiaphragm
requirements.*

DRILLED DOWEL

REHABILITATED CONDITION   B

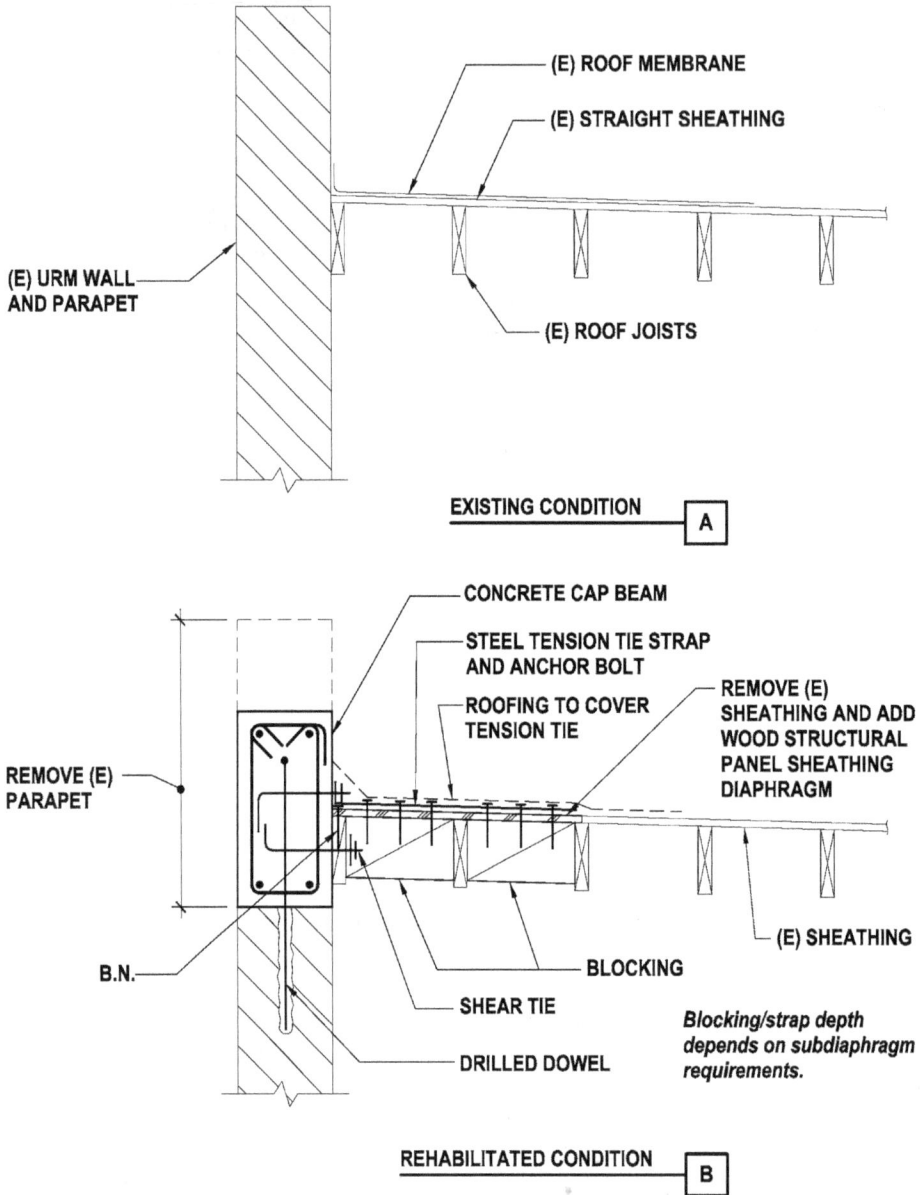

Figure 21.4.1-2: Parapet Removal and Concrete Cap Beam

Waterproofing at the roof: The brace anchor at the roof needs to attach to the structural framing members, so a penetration through the roof membrane will occur that needs proper waterproofing design by a qualified contractor or design professional. Often, parapet bracing and roof-to-wall ties and even roof diaphragm sheathing rehabilitation activities are combined with roofing replacement given the cost effectiveness of combining the work.

Cost and Disruption Considerations

Adding parapet bracing and roof-to-wall tension anchors provide some of the most effective seismic rehabilitation for reducing life safety risks. As a result, some communities—such as San Francisco—passed parapet safety ordinances requiring mandatory mitigation many years ago. Disruption is typically relatively low since occupants can remain in place. Combining parapet bracing and roof-to-wall ties and even roof diaphragm sheathing rehabilitation activities with roofing replacement can significantly reduce the total cost of the work.   Disruption can increase noticeably if the roof has to be removed for installation.

Proprietary Issues

There are no proprietary concerns with parapet bracing, other than use of proprietary anchors as part of the assemblage. See Section 21.4.2.

21.4.2   Add Wall-to-Diaphragm Ties

Deficiency Addressed by Rehabilitation Technique

Inadequate or missing shear and tension connections between the unreinforced masonry bearing wall and the wood floor or roof.

Description of the Rehabilitation Technique

The most significant deficiency in URM bearing wall buildings is the lack of an adequate positive (i.e. mechanical) tie between the masonry walls and the floor and roof diaphragms. Ties are usually separated into two categories: tension ties and shear ties. Tension ties transfer out-of-plane inertial loads perpendicular to the face of the masonry back into the diaphragm. Shear ties transfer loads from the diaphragm into the wall where they are resisted by in-plane action of the wall. Tension ties help keep the walls from falling away from the diaphragms; shear ties help keep the diaphragm from sliding along parallel to the wall. Ties are assemblages that consist of both the anchorage to the wall (shown in detail in Figures 21.4.2-1 and 21.4.2-2) and the anchorage back into the diaphragm (shown in the subsequent figures).

Design Considerations

Research basis: The focus of wall-to-diaphragm testing to date has been on the anchorage to the masonry and has been done primarily by manufacturers. Paquette, Bruneau and Brzev (2003) tested a specimen of a small full-scale one-story building with roof-to-wall ties, but the focus of the work was on wall and diaphragm response.

Anchor types and capacities: The 1997 UCBC and 2003 IEBC provide prescriptive values for tension and shear bolts meeting certain requirements. These are for a 2-1/2" diameter hole filled with nonshrink grout approach that is typically no longer used. The ICBO and now ICC evaluation report process has standardized procedures for vendors supplying adhesive ties for use in brick masonry. Three installations are included in most vendors' ICC Evaluation Service reports, and they have standardized installation techniques and capacities. Adapted versions of these installations are shown in Figures 21.4.2-1 and 21.4.2-2. Figure 21.4.2-1A shows a

CORE DRILL HOLE
Typically 1" diameter
x 8" deep

SCREEN TUBE
Typically 15/16"
diameter

THREADED ROD
Typically 3/4" diameter

SEE [B]

DRILLED DOWEL FOR SHEAR ONLY TIE [A]

BLOCKING OR
LEDGER

CORE DRILL HOLE
Typically 1" diameter

SCREEN TUBE
Typically 15/16"
diameter

See other figures

WASHERS

1"

22.5°

Hole in block
can be oversized
to place screen
tube. Fill annulus
in wood with
adhesive.

PREBENT THREADED ROD
Typically 3/4" diameter

DRILLED DOWEL FOR USE AS
TENSION AND/OR SHEAR TIE [B]

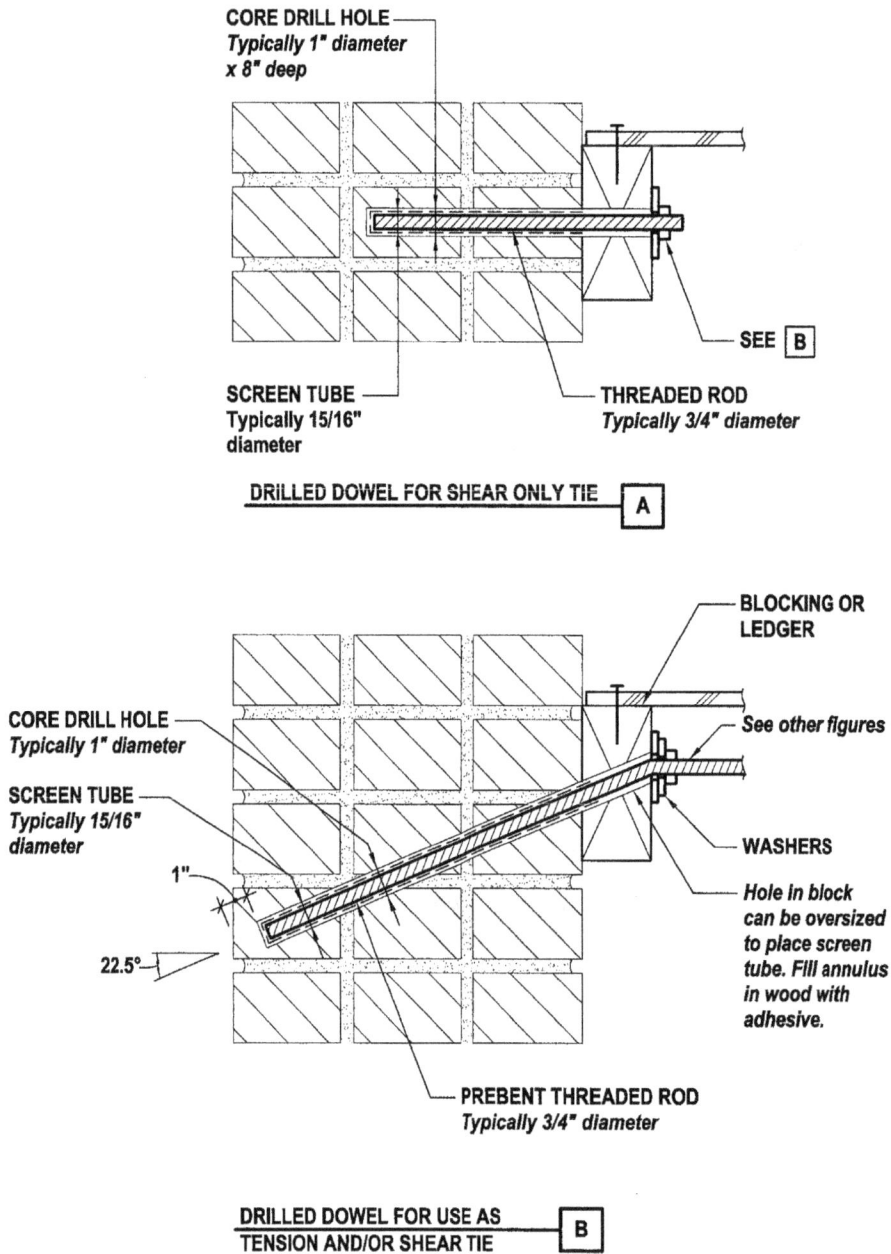

Figure 21.4.2-1: Drilled Dowels

**Anchor plates can be
decorative shapes and
castings.**

See other figures

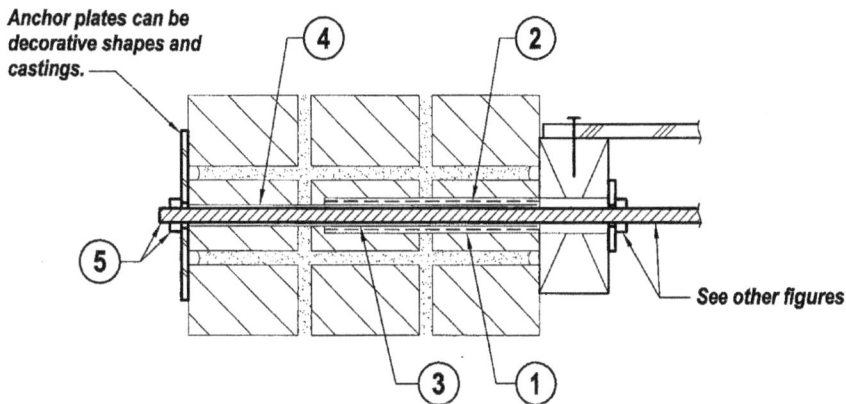

SEQUENCE OF INSTALLATION

1. **CORE DRILL HOLE.**
   Typically 1" diameter x 8" deep.

2. **PLACE SCREEN TUBE WITH ADHESIVE.**
   Typically 15/16" diameter x 8" deep with plug at end.

3. **INSERT STEEL SLEEVE.**
   Typically 13/16" outside diameter.

4. **AFTER CURING, DRILL HOLE THROUGH PLUG AND
   REMAINING MASONRY.**

5. **PLACE THREADED ROD AND ANCHOR PLATE.**
   Typically 5/8" diameter and 6"x6"x3/8" respectively.

Figure 21.4.2-2: Through Bolt Anchor

"combination" drilled dowel that can be used for resisting both tension and shear forces. It is
drilled into the wall at a 22.5 degree angle from horizontal at least 13" into the wall. The angle
allows the dowel to engage more courses of brick, theoretically improving the reliability.   At the
allowable stress design (ASD) force level, it is good for 1200 lbs in tension and 1000 lbs in
shear. Figure 21.4.2-1B shows a drilled dowel used only for resisting shear forces. It goes in 8"
deep into the masonry and is good for 1000 lbs at the ASD level.   Figure 21.4.2-2 shows a
special through bolt anchor using a steel sleeve in the first 8" that can take tension and shear and
has the same values as the combination anchor. These ICC capacities are typically used in
design; they come with a number of restrictions and requirements such as quality of masonry.
When higher values are needed, proof testing can be undertaken. In the ICC standards for both
shear and tension testing (ICC-ES, 2005) of adhesive anchors that manufacturers must use to
obtain ICC qualification, allowable stress design capacities are the lower of prescriptive values
and the average ultimate test value divided by a safety factors of 5.

It should be appreciated that the prescriptive values in the UCBC, the IEBC, and ICC Evaluation
Service reports are based on tests of the drilled dowel itself, not the full elements of the detail.

Capacities for nails, wood structural panels, bolts in wood and straps come from typical code provisions.

Detailing and Construction Considerations

There are many issues to consider in detailing for tension and shear ties. These include the following:

Aesthetics: Anchors that go all the way through the wall have a visible bearing plate on the exterior face, such as shown in Figure 21.4.2-2. There are simple circular or octagonal plates that can be purchased or fabricated. Some manufacturers make plates with a countersunk hole and use flathead bolt heads to reduce the surface projection. When the exterior face is stucco, a plate with a countersunk hole can be recessed into the stucco or just into the masonry and refinished with stucco so it is hidden. Special cast anchors can be made if there is a desire to match an historic exposed cast iron anchor. When the anchor plate approach cannot be used, drilled dowels are used such as those shown in Figure 21.4.2-1.

Nonshrink grout vs. chemical adhesive: Early ties used cementitious nonshrink grout. They required larger diameter holes (such as 2-1/2") to be cored in the masonry to place the grout. A number of vendors have now created special chemical adhesives and tools that have optimized the process. Standard details use ¾" diameter threaded rods in 1" diameter holes, though other sizes can be used, depending on manufacturer requirements. The typical installation approach is to drill the hole; clean it with a brush and compressed air; fill a nylon, carbon, or stainless steel screen tube (which looks like a test tube made out of wire mesh) with adhesive; place the screen tube into the hole; and then push the rod into the screen tube forcing the adhesive out of the tube into the annulus between the tube and the masonry. Figures 21.4.2-1 and 21.4.2-2 show the anchorage using chemical adhesives and screen tubes.

Chemical adhesive types: There are many different types of chemical adhesives, though most are epoxy. Epoxy products have the longest track record. Some vendors have begun to produce other types of chemicals. Key issues when considering an adhesive are the length of time the adhesive has been in use, the extent and quality of the testing, the ability to bond to damp or water filled surfaces, setting time, cost, the heat deflection temperature (an ASTM test method for quantifying the loss of strength as ambient temperature rises), and the capacities shown by test results. Most modern adhesives use two-component pre-packaged assemblies, rather than bulk products used in the past. This reduces the risk of improper mixing and not developing the adhesive to its proper strength. When adhesives are curing, the off-gassing can be unpleasant, and proper ventilation procedures are necessary.

Dowel material type: Threaded rod is commonly specified as ATSM A36 all-thread rod. It is a relatively ductile material, with a minimum yield strength of 36 ksi and ultimate strength of 60 ksi. When higher strength material is needed (which is rare), ASTM A193, Grade B7 threaded rod can be used with a minimum yield strength of 105 ksi and ultimate strength of 125 ksi. Rebar can be used as well, but this is not typically done in ties that connect to the wood diaphragms since the threaded connection is needed. Threaded rod is sometimes supplied with oil on it. This must be solvent cleaned, so that proper bonding with adhesives can occur.

Access: Installation of ties can be done either from below the diaphragm or above. Figure 21.4.2-3 shows installation of floor-to-wall tension ties from below. Figure 21.4.2-4 shows installation occurring from above the floor. Figure 21.4.2-5 shows installation of floor-to-wall shear dowels from above. Similar details are contained in Rutherford & Chekene (1990), SEAOSC (1982) and SEAOSC (1986). The choice of whether to install from above or below depends on whether there are finishes that need to be avoided, whether diaphragm strengthening is being done, and what type of diaphragm strengthening is planned. If there is a special plaster ceiling to be avoided, then access and installation would proceed from above. If there is no plaster ceiling and the floor or roof diaphragm is not being modified or is being enhanced by adding a wood structural panel overlay from above, then access and installation for wall-diaphragm ties would be from below. Angled dowels (see section below) installed from below can be angled upwards rather than the typical downward angle, provided non-sag adhesives are used.

Joist direction: Framing in most *buildings* is orthogonal so that joists or rafters are either perpendicular or parallel to the in-plane direction of the wall. Installations where the joists are perpendicular to the wall are easier to make; installations where the joists are parallel involve blocking and more complicated details. Figures 21.4.2-3 to 21.4.2-5 show variations for joist orientation.

Special issues at the top of the wall: In most **URM** buildings, the wall continues up past the roof forming a parapet that provides fire protection and serves as a guardrail during roof maintenance, as described in Section 24.4.1. In some buildings, though, the roof continues over the top of the wall. In these situations, the roof might be relatively flat or sloped. As a result, special issues arise. First, there is reduced overburden pressure at the top of the wall, reducing the reliability of drilled dowels. Second, eccentricities become more significant, such as the vertical eccentricity between the roof diaphragm and the top of the masonry. Making reliable connections between walls in these situations can be particularly challenging and is usually dependent on the specific geometry and characteristics of the existing details. A common strategy is to employ a concrete bond beam at the top of the wall. This ties the wall together, serves as a collector and chord, increases redundancy and often simplifies details. Figure 21.4.2-6 shows a bond beam placed on top of an existing wall under the roof framing. This is possible when the wall is wide, and there is sufficient distance between the masonry and rafter. Figure 21.4.2-7 shows an alternative when there is insufficient clearance between the rafter and top of wall that involves removing the top two courses of masonry to gain room for the bond beam.

Eccentricity: It is desirable to minimize the eccentricities in a connection. Figure 21.4.2-8 illustrates the issue and some alternate approaches with floor-to-wall tension ties. Figure 21.4.2-8A shows a common tension tie detail in plan view where a tie-down anchor is connected to the side of an existing joist. The plan offset between the drilled dowel at the center of the tie-down where load is applied and the center of the joist where it is resisted times the force is a moment that must be resisted by the joist in weak way bending. This stress can be quite significant. Figure 21.4.2-8B shows an alternative where two tie-downs are used to make a connection that is

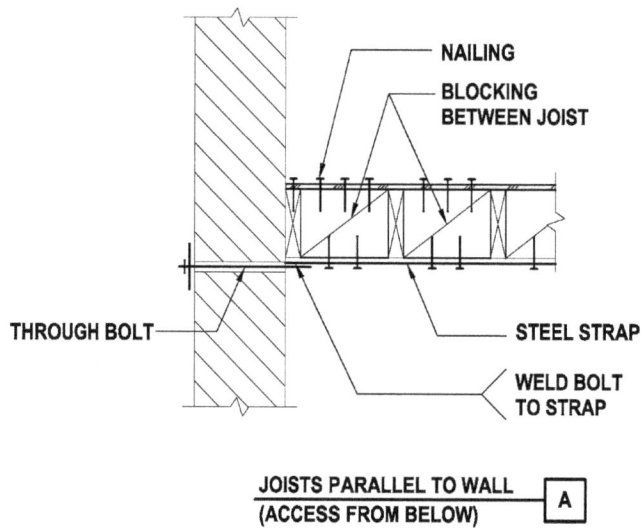

NAILING

**BLOCKING
BETWEEN JOIST**

THROUGH BOLT

STEEL STRAP

**WELD BOLT
TO STRAP**

JOISTS PARALLEL TO WALL
(ACCESS FROM BELOW)   A

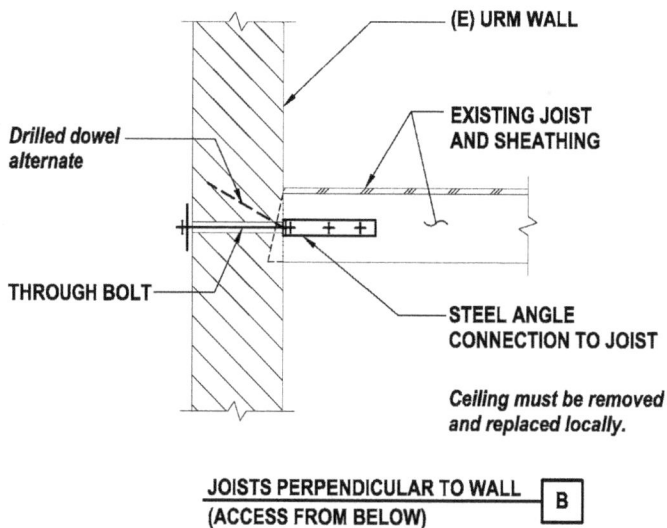

(E) URM WALL

*Drilled dowel
alternate*

**EXISTING JOIST
AND SHEATHING**

THROUGH BOLT

**STEEL ANGLE
CONNECTION TO JOIST**

*Ceiling must be removed
and replaced locally.*

JOISTS PERPENDICULAR TO WALL
(ACCESS FROM BELOW)   B

Figure 21.4.2-3: Tension Anchors Installed from Below the Floor

REMOVE & REPLACE EXISTING
FLOORING WITH WOOD STRUCTURAL
PANEL SHEATHING AND UNDERLAYMENT

EXISTING JOIST AND
SHEATHING

ADD 4x6 BLKG FLAT
WITH NAILING

Drilled dowel
alternate

ADD 2x4
CONTINUOUS LEDGER

ANCHOR STRAP
BAR WITH LAG BOLTS
CENTER ON NEW 4x6 BLKG.

WELD BOLT TO ANCHOR STRAP

JOISTS PARALLEL TO WALL
(ACCESS FROM TOP OF FLOOR)   **A**

LAG BOLTS
PREBORED ON
CENTER LINE OF
EXISTING JOISTS

STRAP

FINISH FLOORING

EXISTING JOISTS
AND SHEATHING

THROUGH BOLT

(E) URM WALL

*Structural elements can also be located
beneath sheathing similar to* **A** .

JOISTS PERPENDICULAR TO WALL
(ACCESS FROM TOP OF FLOOR)   **B**

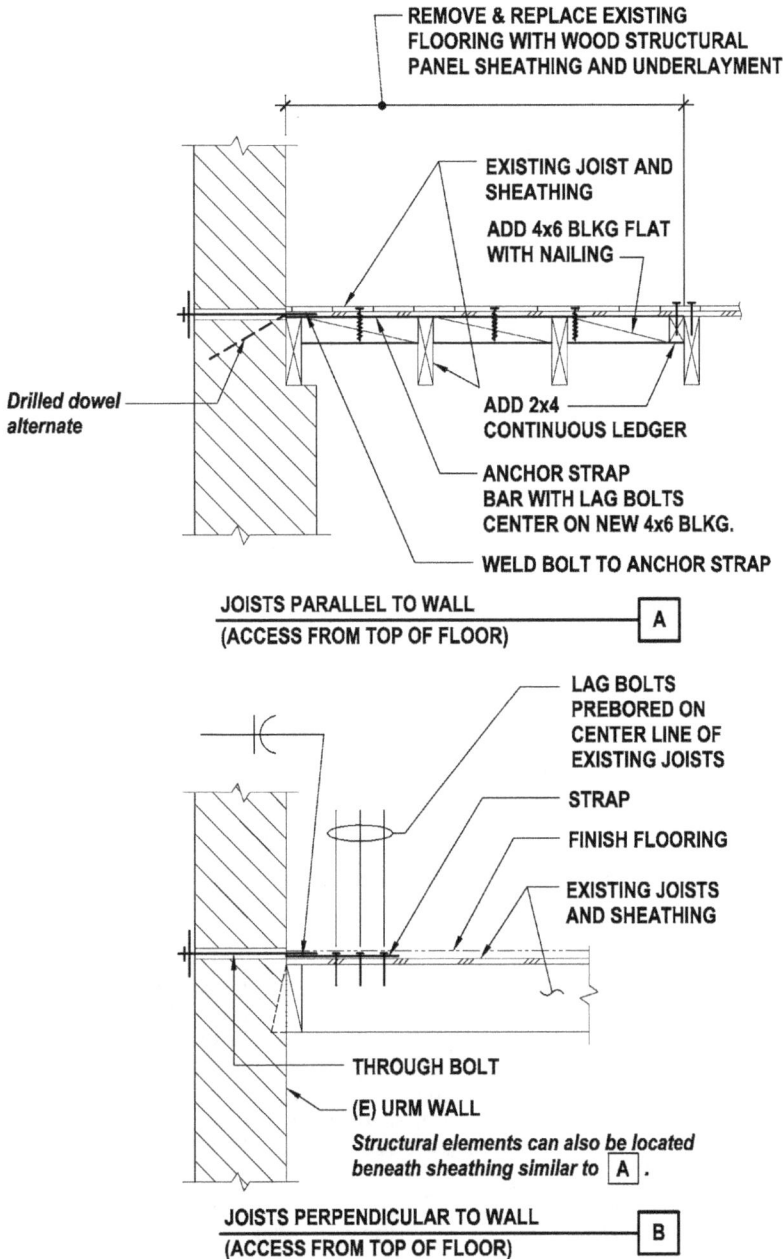

Figure 21.4.2-4: Tension Anchors Installed from Above the Floor

Remove and replace existing
flooring with wood structural
panel sheathing and underlayment
if there is no access from below.

NEW NAILS

DRILLED
DOWEL

ADD PLATE

NEW 2x4
CONTINUOUS

EXISTING
FLOOR JOIST

EXISTING
DIAGONAL
SHEATHING

New continuous nailing
may be required.

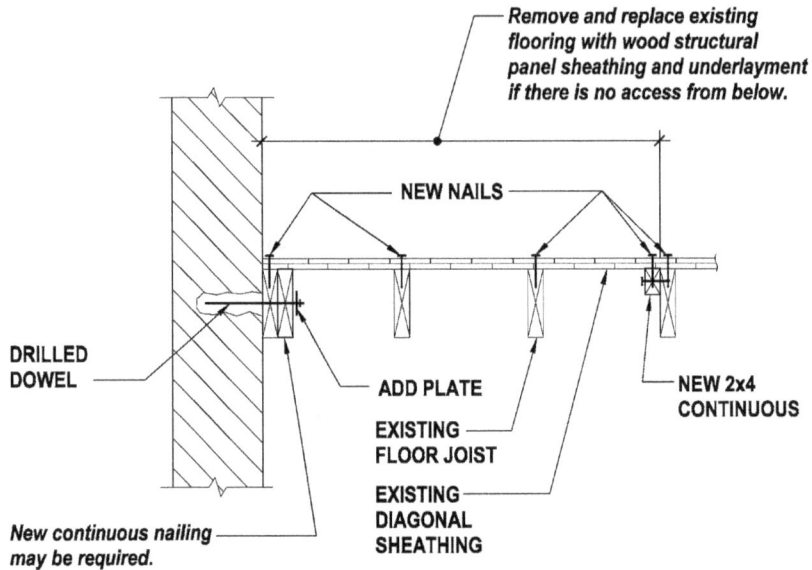

JOISTS PARALLEL TO WALL   A

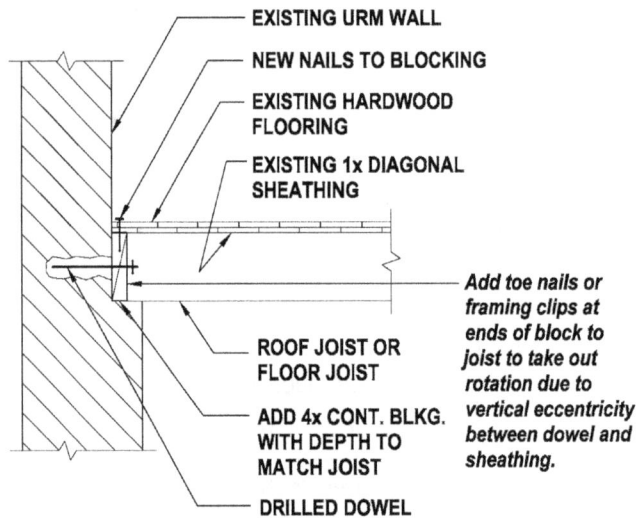

EXISTING URM WALL

NEW NAILS TO BLOCKING

EXISTING HARDWOOD
FLOORING

EXISTING 1x DIAGONAL
SHEATHING

Add toe nails or
framing clips at
ends of block to
joist to take out
rotation due to
vertical eccentricity
between dowel and
sheathing.

ROOF JOIST OR
FLOOR JOIST

ADD 4x CONT. BLKG.
WITH DEPTH TO
MATCH JOIST

DRILLED DOWEL

Note: See other details for tension tie requirements.

JOISTS PERPENDICULAR TO WALL   B

Figure 21.4.2-5: Floor-to-Wall Shear Anchors

RAFTER TAIL
FAR SIDE

B.N.

(N) BLKG. TO FIT
WOOD STRUCTURAL
PANEL SHEATHING

(E) RAFTER

E.N., WOOD STRUCTURAL PANEL
SHEATHING TO (E) JOIST, TYP.
AT STRAP

FRAMING CLIP

ANCHOR BOLT

(E) PURLIN

*Provide building
paper at existing
purlin for moisture
protection.*

DRILLED DOWEL
HOOK REBAR,
HEADED REBAR,
OR THREADED
ROD WITH PLATE
WASHER

STEEL STRAP
WITH LAG
SCREWS.
STAGGER
EACH SIDE.

STEEL PLATE
WITH HEADED
ANCHOR BOLT.

SHEAR TIE SECTION   A   TENSION TIE SECTION   B

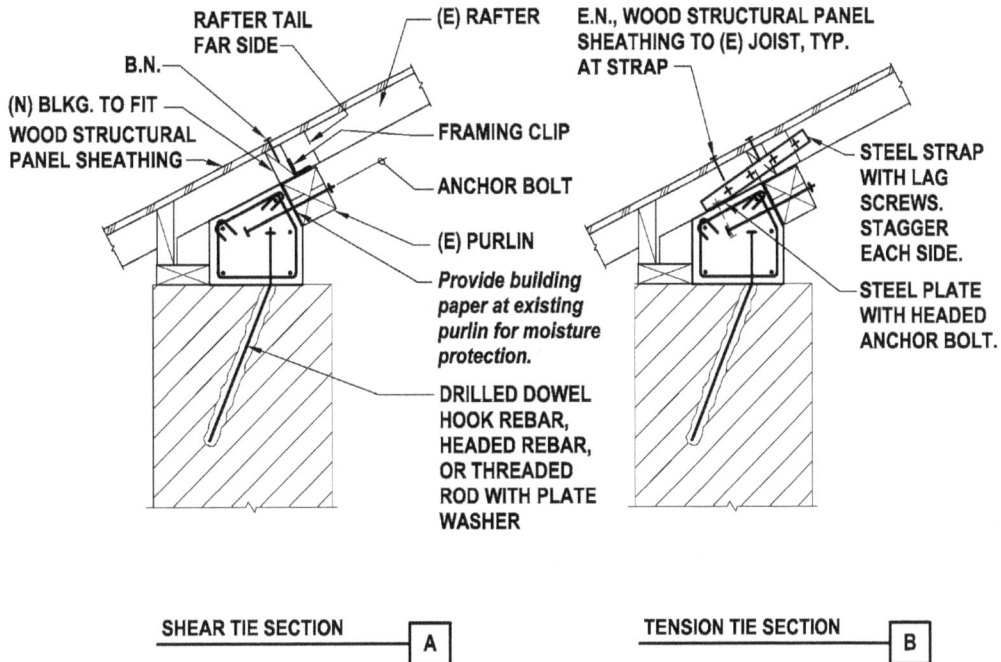

Figure 21.4.2-6: Bond Beam at a Sloping Roof

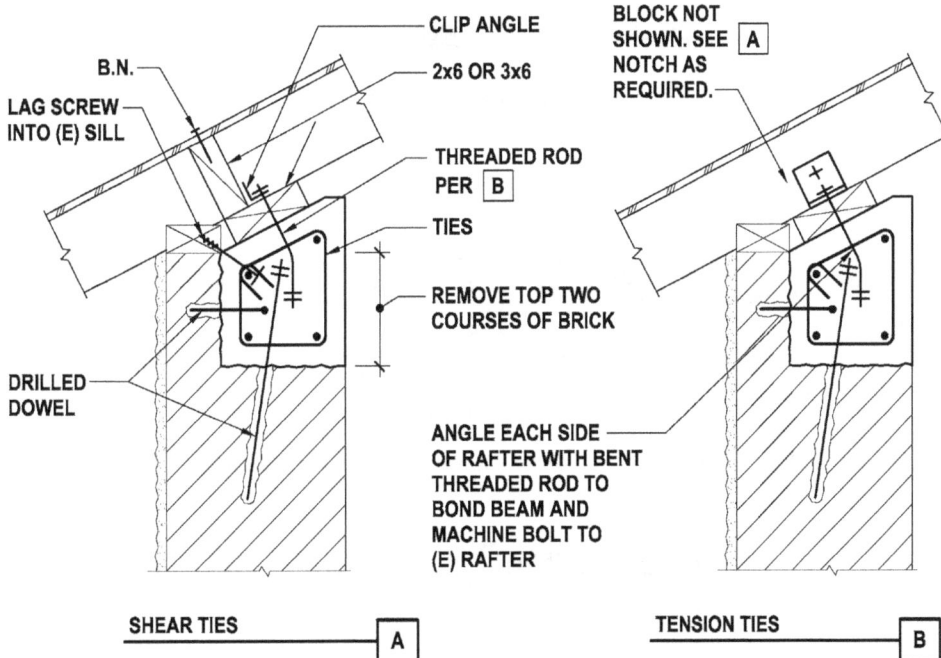

B.N.

LAG SCREW
INTO (E) SILL

DRILLED
DOWEL

CLIP ANGLE

2x6 OR 3x6

THREADED ROD
PER B

TIES

REMOVE TOP TWO
COURSES OF BRICK

ANGLE EACH SIDE
OF RAFTER WITH BENT
THREADED ROD TO
BOND BEAM AND
MACHINE BOLT TO
(E) RAFTER

BLOCK NOT
SHOWN. SEE A
NOTCH AS
REQUIRED.

SHEAR TIES   A   TENSION TIES   B

Figure 21.4.2-7: Bond Beam at a Sloping Roof with Limited Clearance

(E) JOIST POCKET **(E) JOIST**

F

e → **F**

*Moment from
eccentricity
is M = Fe.*

TIE-DOWN

BLOCKING

DRILLED DOWEL

ECCENTRIC TIE-DOWN PLAN DETAIL   **A**

*Offset holdowns so screws
do not conflict.*

SISTER (E) JOIST

CONCENTRIC TIE-DOWN PLAN DETAIL   **B**

*Drilled dowel misses existing
joist pocket.*

**PLATE
WASHER**

**STEEL
STRAP**

BLOCKING
*Takes reaction from
diagonal strap.*

**PLATE
WASHER**

*Drilled dowel is offset from end of blocking
so it can also serve as a shear tie.*

V-STRAP PLAN DETAIL   **C**

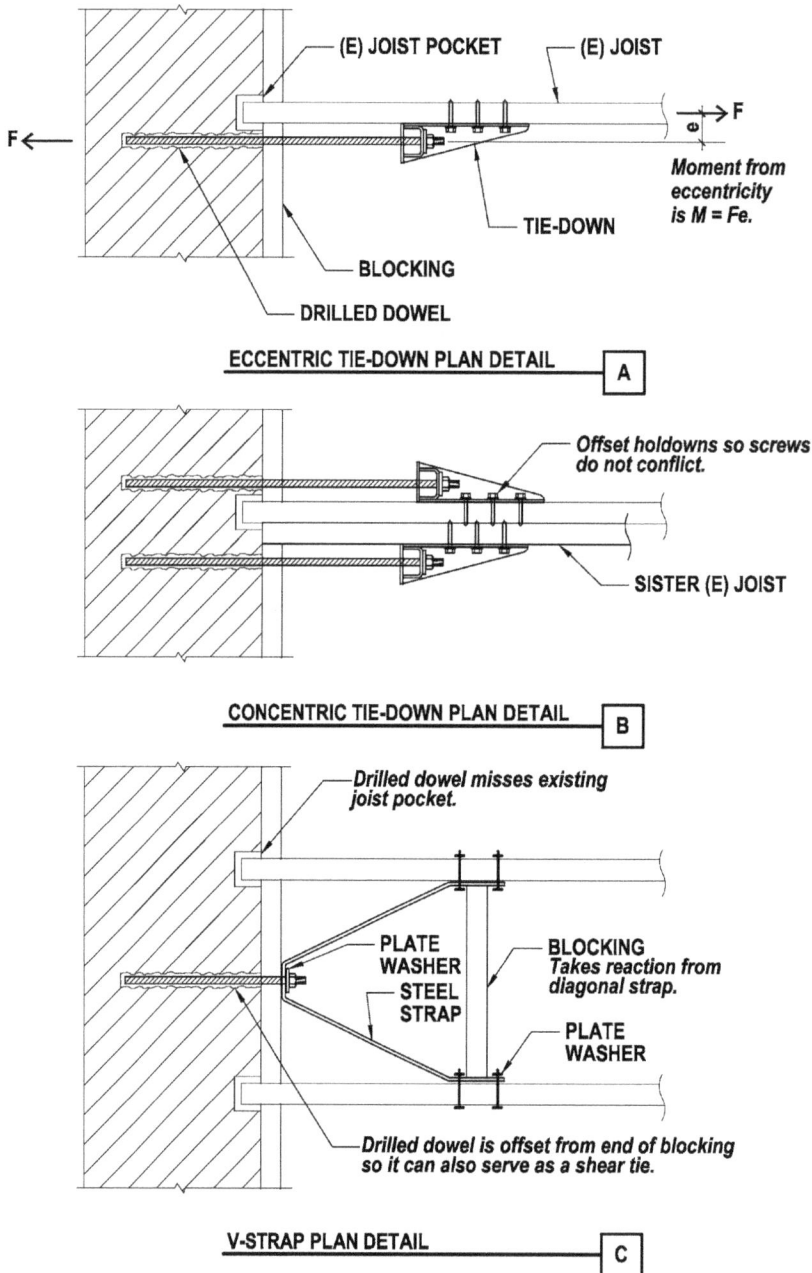

Figure 21.4.2-8: Tension Tie Connection Issues

more concentric. This detail, however, puts a large number of screws into the existing joist, so a sistered joist is shown. Adding the sister also permits the nailing into the diaphragm to be into each joist, reducing the nailing demand on the joists. Bolted tie-downs, instead of tie-downs with screws, can be used with through bolts placed in double shear. Traditional bolted tie-downs have greater slip than the more recent tie-downs using screws. There are proprietary connectors using tubes as tie-downs on each side without oversize holes that bolt eliminate eccentricity and reduce bolt slip. Both Figure 21.4.2-8A and 21.4.2-8B have dowels adjacent to the joist. This means the dowel will enter the wall next to or in the weakened area of joist pocket and at the end of new blocking used for shear transfer, where there is insufficient end distance to use the dowel as a shear tie. Figure 21.4.2-8C shows a V-strap detail where the drilled dowel is placed between joists, away from the joist pocket and with plenty of end distance. When the strap is in tension, forces perpendicular to the joists are produced that are resisted by the added blocking and plate washers.

Truss anchorage: In some **URM** buildings, there will be large gravity elements that bear on the wall, such as girders or trusses. These also become concentrated points of stiffness in the diaphragm. Since the relative rigidity of the elements cannot be easily quantified, it is usually prudent to use an enveloping or "belt and suspenders" approach of assigning demand, so that typical anchors between trusses take the uniform load and the ties connecting the wall and trusses take additional load.

New ties vs. reuse of existing ties: In many older **URM** buildings, there are existing ties called government or "dog" anchors. These anchors typically only occur in the direction where the joists are perpendicular to the face of the wall, and they may not be at sufficient spacing. The 1997 UCBC and 2003 IEBC permit use of these anchors as wall-to-diaphragm tension anchors if tested in accordance with certain standards and capacities are sufficient.

Dowel spacing and edge distance: The 1997 UCBC and 2003 IEBC have maximum spacing requirements on shear and tension dowels. When walls become thick, the out-of-plane demands and the relatively low ICC Evaluation Service report capacity values can lead to fairly tight spacing of dowels. The UCBC and IEBC do not have minimum spacing requirements. From a practical point of view, dowels should not be placed closer than 12" o.c. Some ICC reports provide minimum spacing limits as well, like those commonly employed for drilled dowels in concrete. For one vendor, these spacing limits are 16" o.c. in the horizontal and vertical direction, and there is 16"minimum for edge distance as well.

Corrosion considerations: Drilled dowels are typically installed from the interior. The masonry cover and epoxy serve as corrosion protection, so mild steel anchors are typically considered sufficient. For increased corrosion protection, stainless steel dowels and screen tubes can be used. When through bolted connections are installed, there is a more direct path for moisture intrusion. The anchor plate can be painted with exterior grade paint, galvanized or be made from stainless steel, and the through bolt can be made from stainless steel as well.

Screen tubes: The purpose of the screen tube is to prevent loss of epoxy into cracks or unfilled collar joint voids within the wall. Screen tubes vary somewhat from vendor to vendor and should be considered part of the manufacturer's assembly. Nylon screen tubes have begun to be supplied by many vendors as they are more economical than stainless steel and more corrosion resistant than carbon steel. They do have a much larger coefficient of thermal expansion than both steel screen tubes and masonry.

Hollow masonry: Anchorage of hollow clay tile, ungrouted concrete masonry units and other hollow masonry systems to diaphragms is particularly challenging. When forces are large, grouting in the region of the anchor is usually required. When forces are small, use of screen tubes may be acceptable. The screen tube is filled with adhesive, inserted into the wall and as the dowel is pushed into it, the adhesive seeps through the screen tube forming a key behind the face shell of the masonry. Capacities are small and the connection is nonductile. This type of connection may be viable for out-of-plane wall strengthening (see Section 21.4.3) where the demands are lower, but it is not recommended for wall-to-diaphragm connections. Figure 21.4.2-9 shows a method of connecting a floor to an ungrouted CMU wall. Even in ungrouted CMU, a grouted bond beam is usually found beneath the floor, and it helps provide bearing support for the floor joists. Figure 21.4.2-9 involves locally grouting the courses at and just above the floor to install a new anchor. Figure 21.4.2-10 shows an alternative that avoids working from above and uses the existing bond beam. Sistering and a nailer help get the new anchor to the proper elevation. If a grouted bond beam is not present, it may be necessary to create one to make the proper anchorage, similar to the top courses in Figure 21.4.2-9.

Drilling: Holes need to be drilled with a rotary drill or a rotohammer drill with the percussion setting turned off to limit vibration into the wall. This can slow drilling significantly. In some cases, coring with a diamond tipped blade is more efficient. This may be the only way some hard masonry, like granite, can be drilled. Sometimes water is used to cool the bit, and the slurry produced by the water, mortar and masonry can stain the face of the wall.

Cost/Disruption

Considerations for cost depend on the number, type and depth of dowels; the difficulty of access; and the extent of finishes that are impacted. Through bolts are usually less expensive than adhesive anchors.

Drilling is loud and can be disruptive to occupants. Typically, either the floor or ceiling has to be removed to install the dowels. Thus, it is usually not practical to install dowels in occupied rooms, though the work can be phased by building area so disruption is minimized.

Proprietary Issues

Values for anchor capacity come from individual vendors, but there are no known concerns with use of a properly procured product.

REMOVE FACE SHELL
TO PLACE DRILLED
DOWEL, THEN GROUT

DRILLED DOWEL
Used to engage
(E) bond beam.
Rotate hook after
placement in hole.

B

(E) BOND BEAM

TIE-DOWN AND
THREADED ROD
TIE

SECTION   A

(E) UNGROUTED
CMU

CHIP INTO FACE
SHELL TO PLACE
TIE

PLAN   B

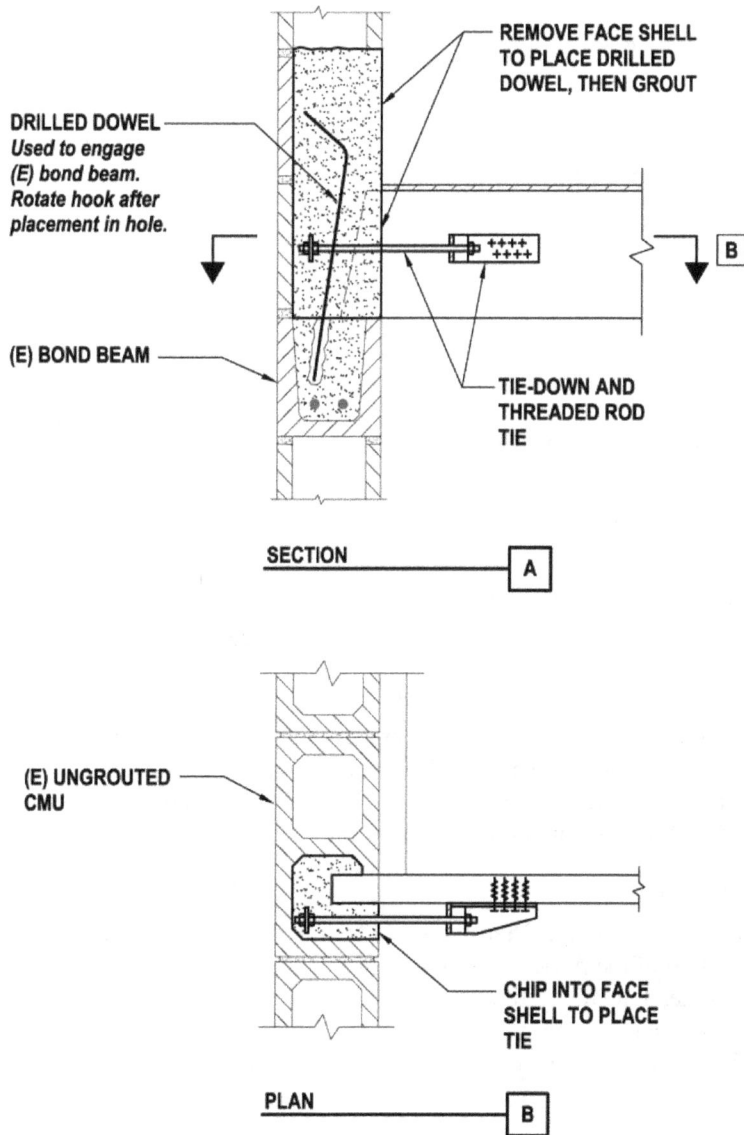

Figure 21.4.2-9: Wall-to-Floor Tension Tie in Hollow Masonry

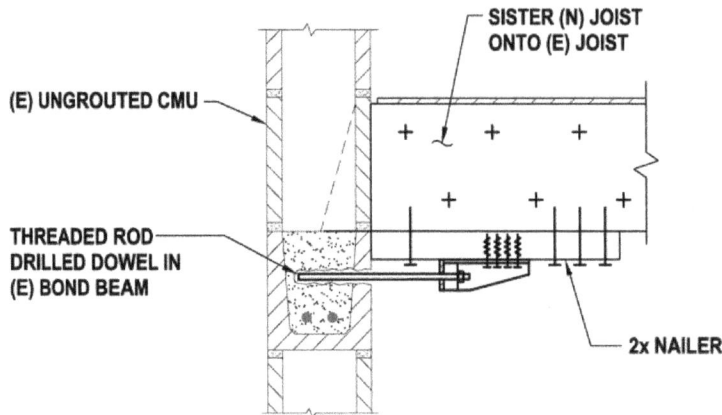

Figure 21.4.2-10: Wall-to-Floor Tension Tie in Hollow Masonry Alternate

21.4.3   Add Out-of-Plane Bracing for URM Walls

Deficiency Addressed by Rehabilitation Technique

Inadequate out-of-plane bending resistance of an unreinforced masonry wall.

Description of the Rehabilitation Technique

Two types of bracing can be used: diagonal braces that reduce the effective height of the masonry wall (Figure 21.4.3-1A) and vertical braces or strongbacks that span the full height of the inside face of the wall (Figure 21.4.3-1B).   Vertical braces can be surface mounted or, when aesthetic considerations are paramount, recessed into the wall; see Figure 21.4.3-2.

Design Considerations

Research basis: The most comprehensive set of testing done to date on out-of-plane response of URM walls was part of the ABK research program in the 1980s, and it is documented in ABK (1981c).   Full-scale, dynamic testing of 20 wall specimens was conducted. Specimens were 6' wide, 10' to 16' tall, and had height-to-thickness (*h/t*) ratios that varied from 14 to 25. Superimposed axial loads were varied; and materials included brick, grouted CMU, and ungrouted CMU.

H/t limits: It is tall, narrow walls that have been found to be susceptible to out-of-plane wall demands. The 1997 UCBC and 2003 IEBC provide maximum *h/t* requirements. Walls with larger *h/t* ratios must be braced.

Spacing: For strongbacks, such as shown in Figures 21.4.3-1B and 21.4.3-2A, the maximum spacing requirements are set by the 1997 UCBC or 2003 IEBC at the minimum of 10 feet or half the unsupported height of the wall. For diagonal braces, the maximum spacing is set at 6 feet.

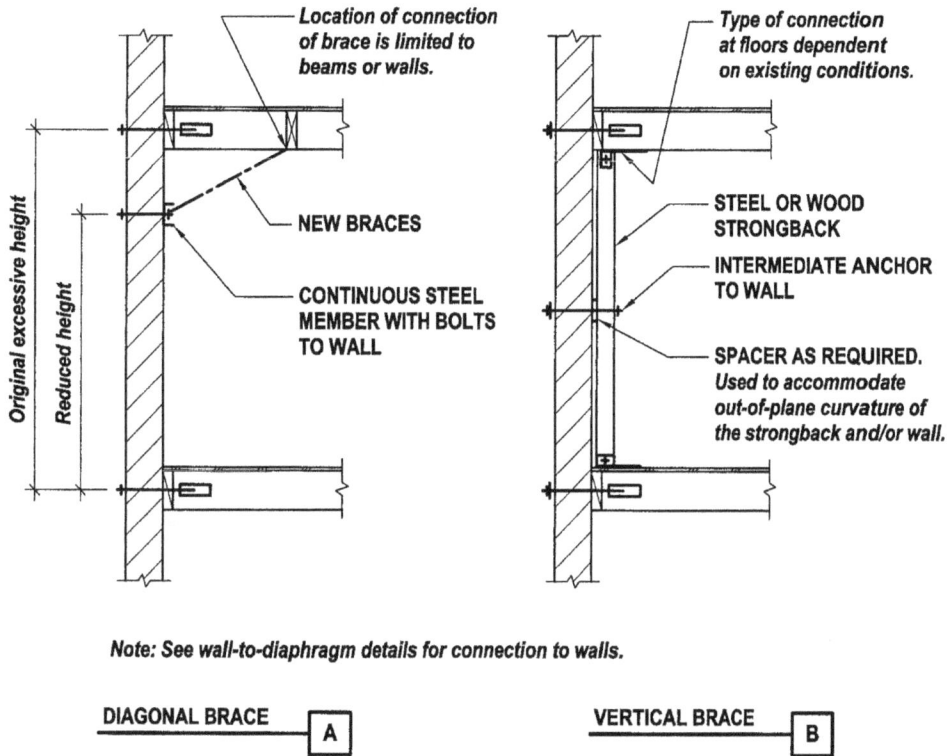

Figure 21.4.3-1: Exposed Out-of-Plane Wall Bracing

Stiffness: For strongbacks, such as shown in Figures 21.4.3-1B and 21.4.3-2A, the 1997 UCBC limits deflection of the wall at ASD demands to one tenth of the wall thickness. This is not a particularly stringent requirement. Say that the first story of a multistory building in Seismic Zone 4 is 13" thick and 18' tall and its resulting *h/t* ratio of 16.7 exceeds the *h/t* limit of 16 in the UCBC. Bracing would be need to be stiff enough to keep deflections down to 10% of 13" or 1.3". This is L/166, which is comparatively low to most masonry design requirements, which are typically L/360 or higher, up to even L/600. Kariotis (1982) notes that the goal of a flexible vertical brace is to keep the brace elastic and provide a predictable restoring force during cracked excursions of the masonry wall. For diagonal braces, the UCBC encourages detailing to minimize vertical deflections.

Diagonal braces loading vs. bracing the wall: If the roof deflects downward on a diagonal brace, a horizontal reaction is imparted to the wall. One concern with diagonal braces is that vertical vibration of the roof in an earthquake can contribute to the out-of-plane inertial forces on the wall. This concern, combined with the difficulty of making the roof stiff enough for against vertical deflections, makes vertical bracing a preferred engineering choice over diagonal bracing.

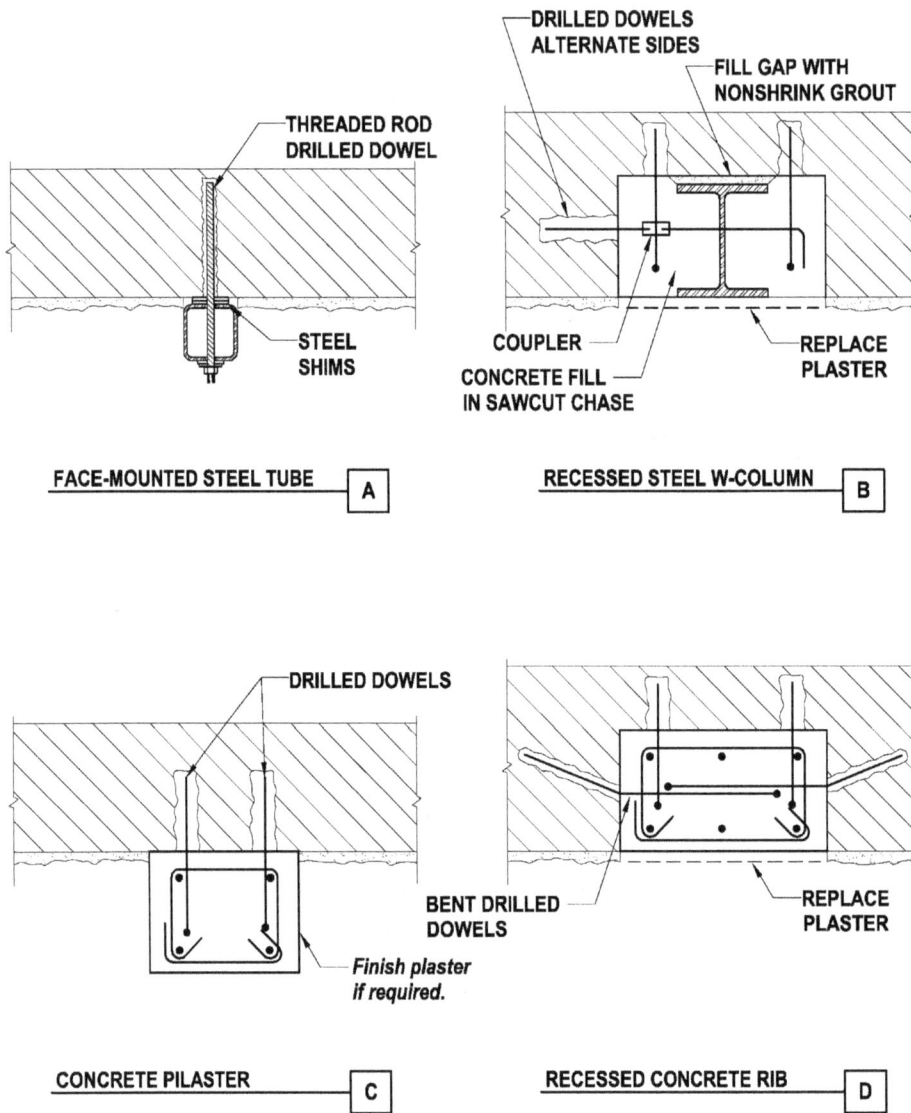

Figure 21.4.3-2: Vertical Bracing Alternatives

Recessed steel and concrete and surface-mounted concrete: Provisions in the 1997 UCBC and 2003 IEBC do not explicitly consider the approaches shown in Figures 21.4.3-2B, 21.4.3-2C and 21.4.3-2D. These approaches are unusual, but they can be used when a more sensitive aesthetic approach or higher loads are needed.

Detailing and Construction Considerations

Materials: Braces are typically done with steel as shown in Figures 21.4.3-1 and 21.4.3-2A, but strongbacks can also be done with wood posts or with concrete pilasters (Figure 21.4.3-2C).

Aesthetics: Figure 21.4.3-1 shows exposed braces. This is the least expensive approach and is appropriate for certain occupancies. When there is architectural desire to hide the steel, the bracing can be furred at added cost and impact on the usable space. To minimize the impact on the space, the vertical brace can recessed into a cavity cut in the wall with either a steel or a concrete member. See Figure 21.4.3-2. Recessing the steel or concrete requires significantly more work and raises the potential for cracking to propagate from the inside of the recess to the masonry face.

Strongback anchor spacing: Figure 21.4.3-1B shows only a central anchor at midheight of the wall. Often demand/capacity ratios for anchorage to the wall with through bolts or drilled dowels (see Section 21.4.2) will dictate a tighter spacing of anchors.

Floor/roof framing capacity: Figure 21.4.3-1 shows anchorage to joists oriented perpendicular to the wall. When joists are parallel to the wall, the horizontal anchorage force must be developed out into the diaphragm. In Figure 21.4.3-1A, the existing roof beams may need to be strengthened to provide adequate strength to resist downward loading.

Hollow masonry: Figures 21.4.3-1 and 21.4.3-2 apply to solid masonry. When the existing masonry is hollow, alternative connection methods are needed. Figure 21.4.3-3 shows use of vertical concrete ribs. A chase is created by removing the face shell on one side of the wall. Reinforcing steel is added and then grout or concrete fill. There is typically insufficient space for ties. This approach is messy and noisy. Figure 21.4.3-4 shows an alternative where steel strongbacks are bolted to the wall with either drilled dowels or through bolts. The screen tube anchor of Figure 21.4.3-4A relies on mechanical keying action from the spreading adhesive to engage the face shell. The capacity is limited to the face shell of the masonry and can be quite low, in the low hundreds of pounds at allowable stress design levels. It is also nonductile as the failure mechanism is spalling of the face shell. The through bolt in Figure 21.4.3-4B provides increased capacity and locally grouting in the anchor provides additional capacity.

Cost/Disruption

Diagonal bracing is usually less expensive, but is considered less reliable than vertical bracing. Furring can be used to cover the braces at added cost. Exposed braces are typically less expensive than more architecturally sensitive alternatives like recessed vertical braces or reinforced cores (See Section 21.4.4). Installation of bracing is fairly disruptive since it must occur around the entire perimeter; and it involves drilled dowels, and accessing and connecting to horizontal diaphragms.

Proprietary Issues

There are no known proprietary concerns with bracing of URM walls.

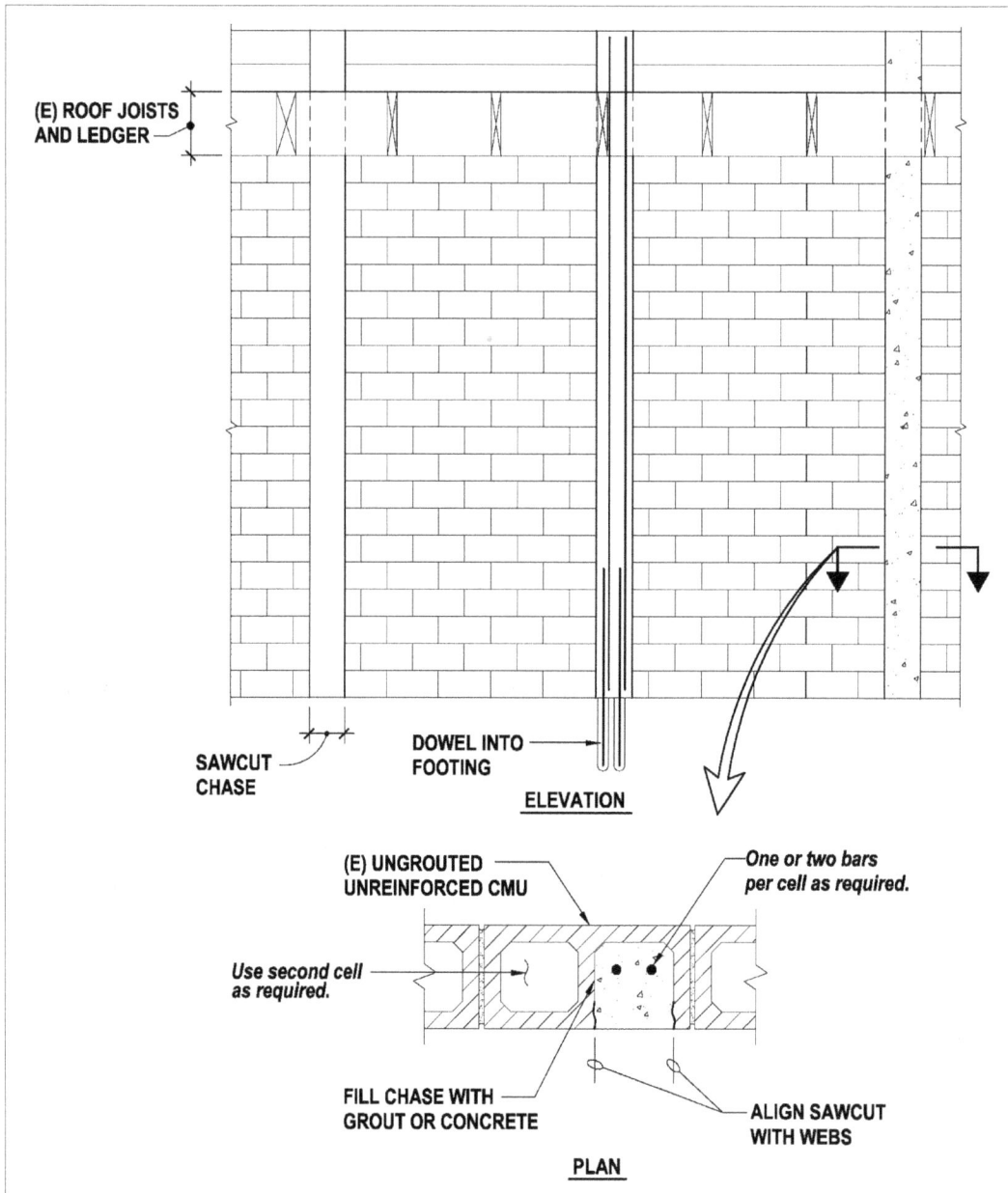

Figure 21.4.3-3: Concrete Ribs in Hollow Masonry

(E) UNGROUTED CMU,
HOLLOW CLAY BRICK
OR HCT

STEEL
STRONGBACK

ADHESIVE GROUT
*Provides mechanical
keying action behind
face shell. Capacity
is relatively low,
compared to* B .

THREADED ROD

SCREEN TUBE

DRILLED DOWEL TIE — A

ANCHOR PLATE
*As large as possible
to spread load.*

*When higher capacity is
needed, grout void locally
to encase anchor.*

*Locate dowel next
to masonry web
where possible*

THROUGH BOLT — B

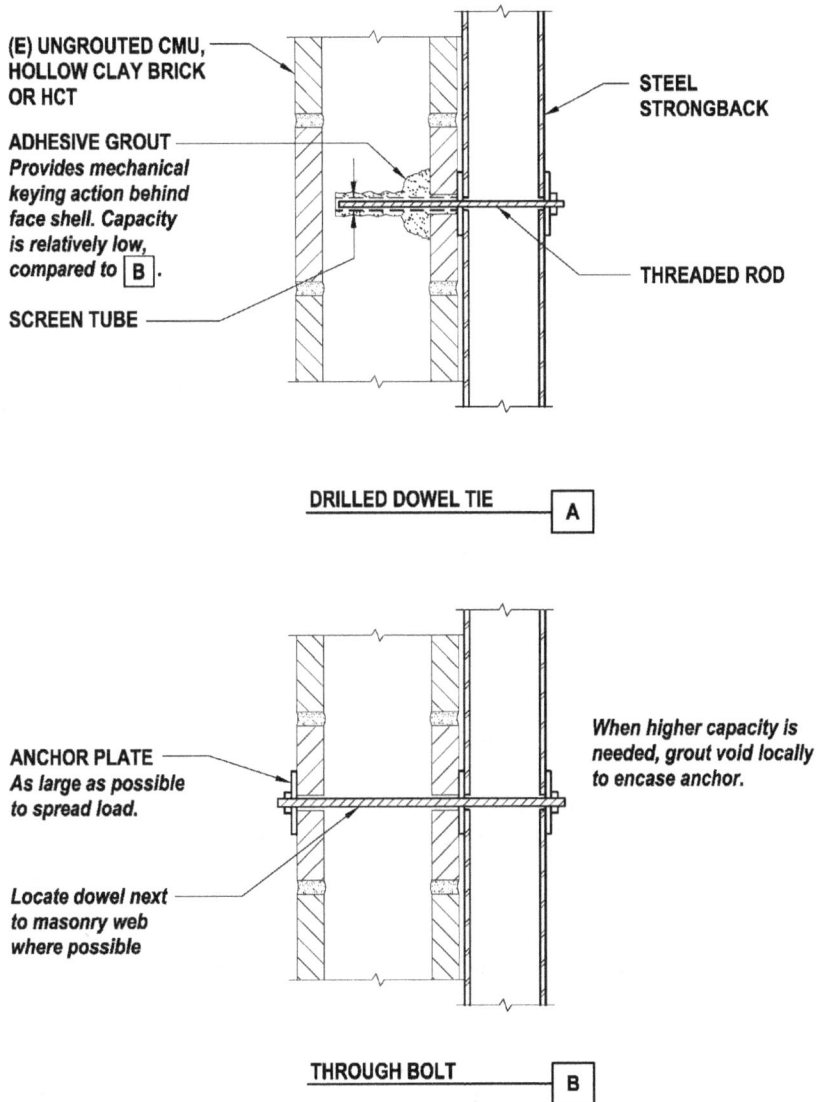

Figure 21.4.3-4: Connection of Strongback to Hollow Masonry

21.4.4   Add Reinforced Cores to URM Walls

Deficiencies Addressed by Rehabilitation Technique

Inadequate out-of-plane capacity and in-plane capacity of unreinforced masonry wall.

Description of the Rehabilitation Technique

Installing the reinforced core involves drilling a core from the roof down the inside of an unreinforced masonry wall. A steel reinforcing bar and grout are placed inside the hole to increase the wall strength. See Figures 21.4.4-1A and 21.4.4-1B. This process is used to avoid the aesthetic impact of exposed bracing described in Section 21.4.3.

Design Considerations

Research basis: The original research at CSU Long Beach and North Carolina State University for reinforced cores is summarized in (Plecnik, Cousins, and O'Conner, 1986) and Plecnik (1988). It covered both out-of-plane and in-plane loading. Subsequent vendor tests for in-plane loading were done at UC Irvine but have not been published. More recent in-plane testing is summarized in Abrams and Lynch (2001).

Out-of-plane capacity: When reinforced cores are used for enhancing out-of-plane bending capacity, the wall is analyzed as a reinforced masonry element. Some engineers have used a traditional allowable stress design code format for masonry. Another common approach is to use factored design methods. Plecnik, Cousins and O'Conner (1986) provided an ultimate strength design formulation. As with concrete design or typical reinforced masonry design, the compressive strength of the masonry, f'_m, is needed. Default values are available in FEMA 356, but it is important not to over-reinforce the masonry section and cause a brittle failure of the masonry, so it is often prudent to obtain the masonry strength.

In-plane capacity: Plecnik, Cousins and O'Conner (1986) tests showed significant increase in in-plane loading from the reinforced cores, but a design methodology was not provided. Breiholtz (1987) suggested using the test results as well, but did not provide a complete design methodology. One design approach is simply to extrapolate the test results on a per lineal foot basis. Another is to consider the vertical bars as the tie element in a strut-and-tie methodology, with diagonal struts in the masonry connecting the cores.

Post-tensioned masonry: Post-tensioned masonry has been used in a few instances. The goal of post-tensioning the bars can be to add compressive stress to the masonry wall to increase the effective shear stress, since unreinforced masonry shear capacity formulations, such as those in FEMA 356 or the ICBC, provided increased shear strength with higher compressive strength. Recent research by Rosenboom and Kowalsky (2004) provides some results from cyclic testing, but the specimens have central cavities filled with grout rather than solid multi-wythe brick, and design equations are not provided.

Detailing and Construction Considerations

Detailing and construction considerations for reinforced cores include the following.

CORE CENTERED IN
WALL, FULL HEIGHT OF
WALL WITH
REINFORCING BAR.
EXTEND INTO
FOUNDATION

*Patch top of wall after
coring and grouting.*

*RESET LOOSE MASONRY
AND REPOINT MORTAR
CRACKS WHERE GROUT
CAN LEAK*

FILL CORE AROUND
REINFORCING WITH
POLYESTER GROUT.
VERIFY CORE IS
COMPLETELY DRY
PRIOR TO GROUTING.
*If nonshrink grout
is used, hole is to be
pre-wetted.*

*Provide verification
ports to check grout
flow, one each story.*

B

*For wet coring, provide
drainage relief port.
Extend core through
footing or use vacuum
equipment to remove
excess water.*

SECTION   A1

*Minimum edge
distance of half
the wall thickness
is recommended.*

CORE AT ENDS
OF PIERS AND
AT 6'-0" MIN.
THROUGHOUT
WALL

ELEVATION   A2

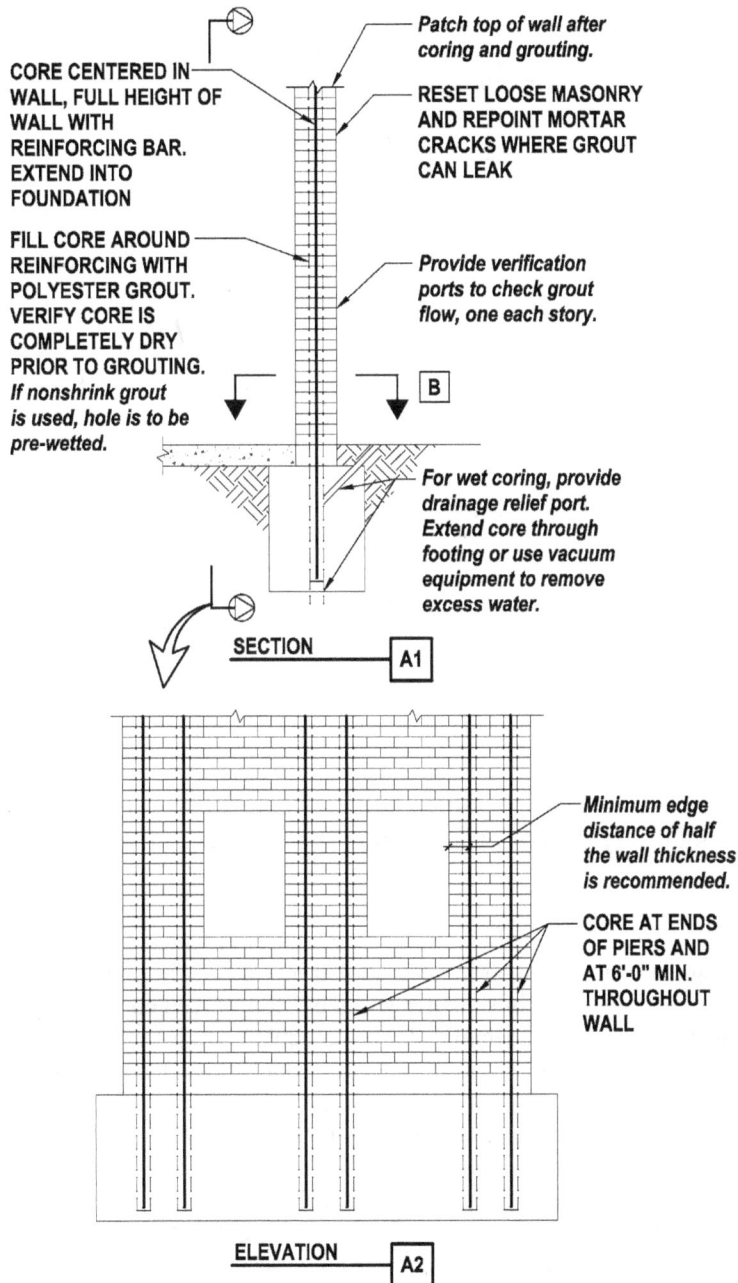

Figure 21.4.4-1A: Reinforced Cores

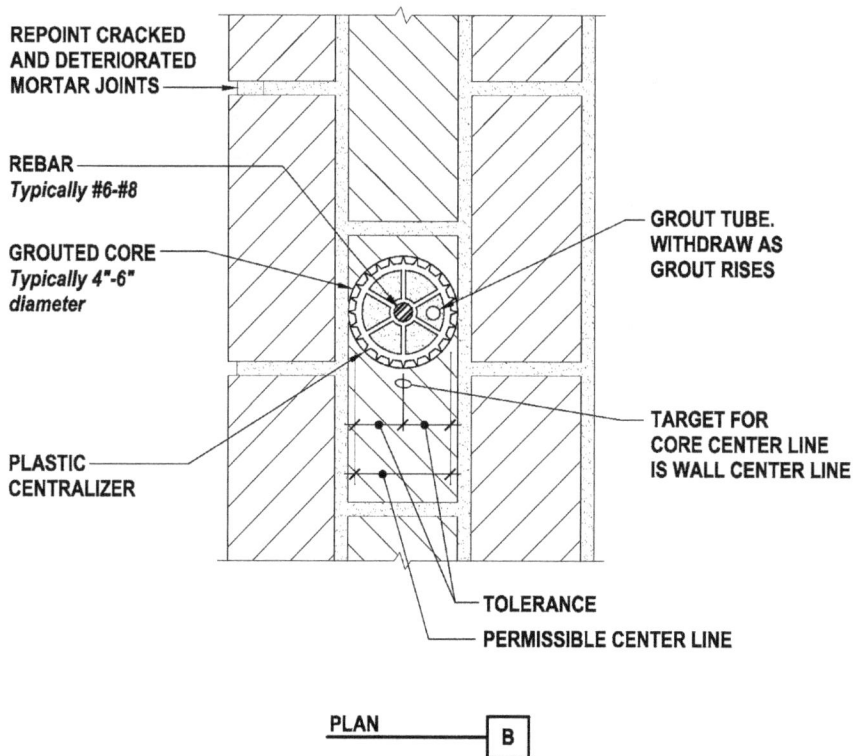

Figure 21.4.4-1B: Plan Detail of Reinforced Core in Masonry Wall

Wet vs. dry drilling: Traditionally, holes cored in masonry were done similarly to those in concrete, using diamond tipped coring bits cooled by water. The slurry created by the water and brick dust can lead to staining of sensitive surfaces. Reinforced cores gained popularity when drilling companies developed specialized drilling equipment that did not need water to cool the bit. In many cases, this involves coring drills that rotate quite slowly compared to traditional coring equipment. The material within the core comes out in cylindrical chunks or in small pieces of debris or dust that are vacuumed into debris containers.

Reducing leaks: To minimize leaks during wet drilling and during grouting, loose masonry should be repaired and cracks repointed. To limit the extent of repointing, consideration can be given to monitoring the location of dust clouds that escape cracks during dry drilling and repointing those locations.

Obstructions: Drilling progresses most rapidly if the masonry is neither too hard nor too soft and is relatively homogeneous. Encountering wood debris inside walls or metal veneer ties can slow or stop the drilling.

Angled drilling: Drilling is typically done from the top and straight down. Occasionally, special situations arise where angled drilling might be necessary. This can be done, but requires much greater skill from the driller.

Hole diameter: Hole diameters are typically about four inches, but can range from three to six inches.

Drilling tolerances: The wider and shorter the wall, the easier it is to drill because it provides better tolerance against drilling inaccuracy. Tolerances of about ±2" in reasonably tall walls are usually achievable. A simple way to check this is as follows. Say the hole is four inches in diameter. Attach a small penlight flashlight to the end of a string, creating a lighted plumb bob. Drop the plumb bob down from the center of the hole at the top. If it does not hit the side of the cored hole at any point on the way down, a two-inch tolerance has been met.

Bar material type: The reinforcing bars used in reinforced cores are typically regular ASTM A615 mild steel, as they are protected by the grout. For increased corrosion protection, stainless steel or epoxy coated rebar can be used. If post-tensioned center coring is done, high strength ASTM A722 threaded bars can be used.

Bar size: Bar size depends on the demand/capacity ratios, but typically ranges from #5 to #8.

Centralizers: In order to keep the rebar centered in the hole, a plastic centralizing wheel is used by some engineers. Others consider this an obstruction limiting the flow of grout.

Grout type: The original research by Plecnik, Cousins and O'Conner (1986) evaluated several grouts and concluded that a formulation using polyester grout provided the best dispersion into the masonry. Polyester has some offgassing concerns from styrene vapors and requires the hole to be dry before installing the grout. It also does not have the long-term track record of other more widely used grout materials. Some engineers, as a result, use a high quality nonshrink cementitious grout. With cementitious grouts, the hole needs to be prewetted prior to grout placement.

Grouting process: Center coring can be done in multistory buildings, so the depth of holes can get quite large. A tremie grouting technique can be used to assure placement of grout. A grout tube can be tied loosely to the bar and/or centralizer(s) as the bar is lowered into the hole. As grout is pumped into the hole, the tube is slowly withdrawn as the level of grout rises.

Verification ports: Grout will leak into voids in the masonry. To confirm that the grout is rising in the core, horizontal holes can be drilled into the wall. When the grout reaches the port, the port is plugged, and the grout is allowed to continue to rise to the top of the core.

Bottom of holes: Coring usually goes into the foundation. In a concrete foundation, there will not be any place for water used to cool the bit in wet coring to escape. The hole can be vacuumed out, or the core can be continued all the way down to the bottom of the foundation.

Post-tensioning: When reinforced cores are post-tensioned, several additional issues come into play. First, grouting is done in two stages. The first stage is in the foundation where the post-tensioning is anchored. After the grout cures and the bar is stressed, the second stage of grouting occurs up to the top of the hole. Post-tensioning vendors provide proprietary anchorage hardware for the bar at the top of the wall. A concrete cap or bond beam may be necessary or desirable to distribute the load on the top of the wall to reduce the stress on the masonry.

Access to top of wall: The top of the wall must be accessible to drilling equipment. Temporary scaffolding or work platforms will usually need to be erected adjacent to the hole. Bracing of drilling equipment back to the wall and a point in the roof is necessary to keep the drill plumb.

Spacing: Reinforced core spacing will depend on demand/capacity ratios, but a minimum spacing of six to ten feet is desirable.

Cost/Disruption

Adding reinforced cores can be considerably more expensive than exposed bracing, so it is usually only performed in historically and architecturally sensitive buildings. Since the core is placed inside the wall, the disruption to interior and exterior faces is limited to sealing cracks, access to the roof to place the drilling equipment, and drilling noise and vibration.

Proprietary Issues

Research for reinforced cores is in the public domain. Some drilling contractors reportedly have patents on certain types of proprietary drills. Some of the terms used with the process have been trademarked by some of the first engineers to implement the technique. As a result, the generic term "reinforced core" is used in this document.

21.4.5   Add Concrete Overlay to Masonry Wall

Deficiency Addressed by Rehabilitation Technique

Improving inadequate in-plane wall capacity is the primary purpose of a new concrete overlay, but the concrete can also improve inadequate out-of-plane bending capacity.

Description of the Rehabilitation Technique

New concrete is applied against an existing unreinforced masonry wall to increase the shear capacity of the wall. The new concrete is attached to the old wall with adhesive anchors and can either be cast-in-place concrete or sprayed-in-place. In rehabilitation work, sprayed concrete, known as shotcrete, is more commonly employed than cast-in-place construction, since the existing wall provides the back-side form. The thickness of the new concrete varies with strength requirements, but it is usually from four to 12 inches. See Figure 21.4.5-1.

Design Considerations

Research basis: A fair amount of research in the use of shotcrete overlays on masonry has been done. A summary is given in El Gawady, Lestuzzi and Badoux (2004). Early diagonal tension testing was done by Kahn (1984), and more recently static cyclic tests were done by Abrams and Lynch (2001). Kahn (1984) showed significant increases in strength from the shotcrete and that adding drilled dowels between the shotcrete and masonry or an epoxy bonding agent did not lead

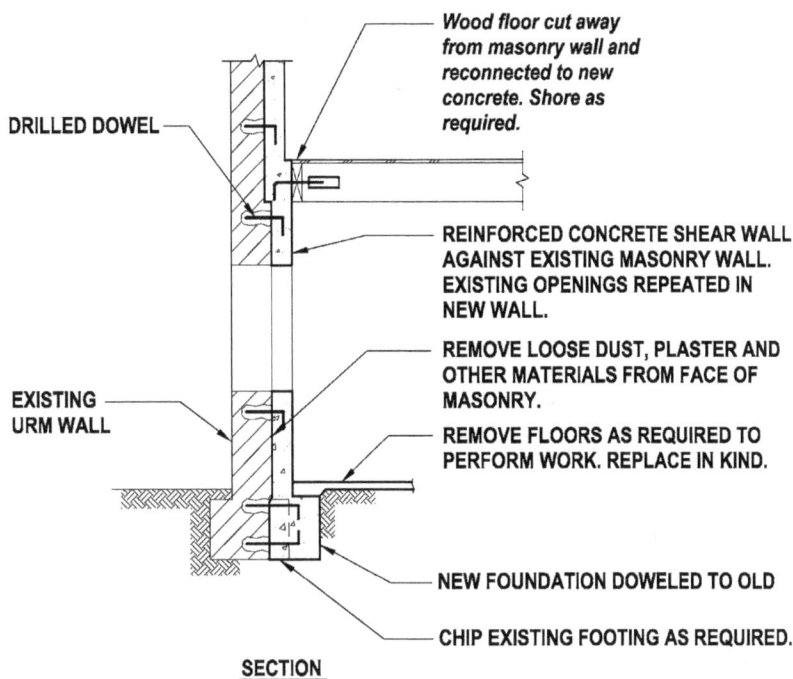

DRILLED DOWEL

Wood floor cut away
from masonry wall and
reconnected to new
concrete. Shore as
required.

REINFORCED CONCRETE SHEAR WALL
AGAINST EXISTING MASONRY WALL.
EXISTING OPENINGS REPEATED IN
NEW WALL.

REMOVE LOOSE DUST, PLASTER AND
OTHER MATERIALS FROM FACE OF
MASONRY.

EXISTING
URM WALL

REMOVE FLOORS AS REQUIRED TO
PERFORM WORK. REPLACE IN KIND.

NEW FOUNDATION DOWELED TO OLD

CHIP EXISTING FOOTING AS REQUIRED.

SECTION

Figure 21.4.5-1: Concrete or Shotcrete Wall Overlay

to significant improvements. A saturated masonry surface was recommended. Testing by Abrams and Lynch (2001) aimed at increasing the shear capacity to lead to flexural yielding of the tension bars in the shotcrete. Strength increased by about a factor of 3, but displacement capacity did not increase.

Design criteria: When a concrete overlay is used, there are several common force-based design approaches for the wall, due to the relatively high strength of the concrete compared to the masonry. One is to take 100% of the demand tributary to the strengthened wall line in the concrete overlay itself and ignore the masonry. While this may sound conservative, it can mean that the masonry will be significantly damaged before the concrete ever sees the majority of its design load. Another approach is to share the load, by relative rigidity, between the masonry and the concrete. When this is done, both the masonry and the concrete must be checked to confirm they are not overstressed. The most conservative approach is to use the overlay to resist 100% of the tributary load, but to also check that the masonry can resist the loads it will actually attract. Displacement-based design approaches inherently consider the relative rigidity of the concrete and masonry, but they are less commonly employed.

Discretely applied overlays: The URM walls to which the overlay is applied are typically punctured with window and door openings. It can be tempting to apply the concrete to wide piers. The comparatively high strength of the concrete means it can take high loads, but it is unlikely to have sufficient stiffness to actually attract the load it was intended to take. Eventually, though, when the masonry cracks, the load will find the concrete, but this can lead to significant cracking at the ends of the concrete in the masonry spandrels. See Figure 21.4.5-2A. This can be addressed by spreading out the influence of the overlay by using top and/or bottom spandrels or grade beams, such as shown in Figure 21.4.5-2B. Alternatively, drilled piers can be placed at the ends of the new walls to add stiffness, as in Figure 21.4.5-2C. Or most simply, a continuous overlay can be used, enabling a reduced thickness and consistent finish surface, as shown in Figure 21.4.5-2C.

Collector load transfer pathways: With discretely applied overlays in the field of the masonry wall, the question arises of how the load in the floors and the masonry wall will reach the overlay. Some engineers ignore this issue and assume the masonry wall will serve as the collector. Others provide an explicit steel collector in the edge of the diaphragm or a bond beam on top of the wall.

Out-of-plane load resistance: When the overlay is added to the wall, its additional inertial load must be considered in the seismic weight of the structure. Out-of-plane anchorage requirements at the overlays are thus larger, and wall-to-diaphragm tie spacing often decreases there.

Detailing and Construction Considerations

Detailing and construction considerations for concrete overlays include the following.

Drilled dowel spacing: It is common practice to connect the overlay to the masonry wall with drilled dowels. The drilled dowels transfer shear between the two materials, and they also serve as out-of-plane ties for the masonry. Spacing of two feet to three feet on center is typical. Good detailing involves drawing an elevation and showing the location of dowels around openings since the nominal spacing will typically change there.

New foundation at the base of wall: The base of the shotcrete can be set on the ledge of the existing footing if conditions permit or a new footing can be provided. The added load from the shotcrete and distribution of stresses on the existing footing must be considered. See Section 23.6.2 for issues involved in adding a new footing next to an existing footing.

Interface between the wall and diaphragms: Where the new wall meets the existing floors is usually the location where special consideration must be given to detailing. Conditions where the floor joists are parallel to the wall are the easiest to address. The first and/or second joists in from the masonry are removed to install the overlay, a ledger is placed back, and the floor sheathing run up to the ledger. If shotcrete is used, sufficient clearance must be provided to avoid shadowing during spraying. This should be checked as part of the preconstruction test panel.

NEW CONCRETE OVERLAY

EXISTING MASONRY WALL

WALL

SLAB FOOTING

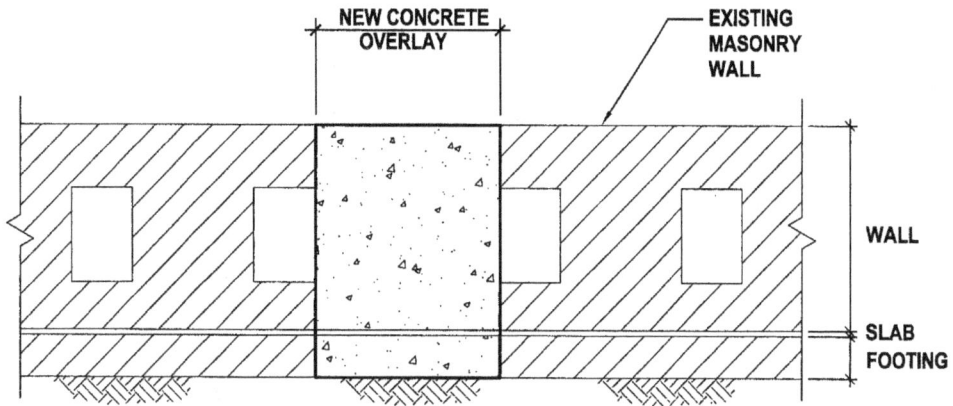

LOCALIZED CONCRETE OVERLAY ELEVATION A1

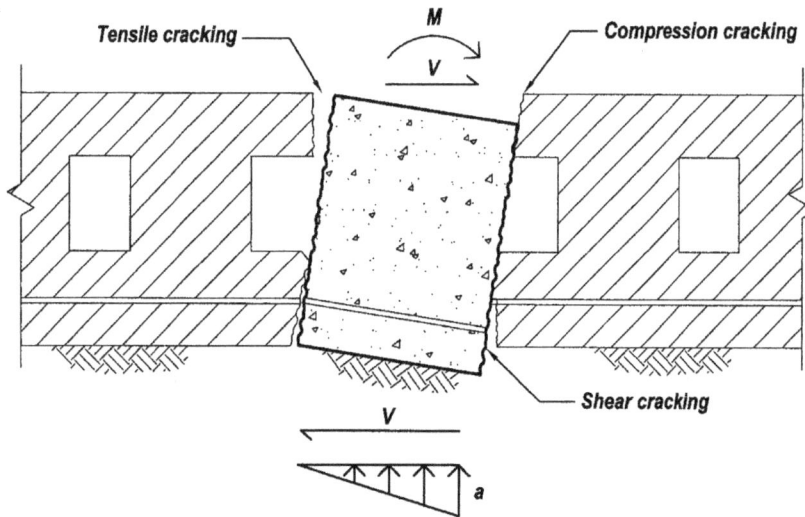

Tensile cracking

M

V

Compression cracking

Shear cracking

V

a

DAMAGE DUE TO STIFFNESS INCOMPATIBILITY A2

Figure 21.4.5-2A: Potential Damage at Ends of Narrow Concrete Overlay

TOP SPANDREL
*Extend as far
as required.*

CONCRETE OVERLAY

TOP SPANDREL   B1

BOTTOM SPANDREL
OVERLAY

*Use only grade beam
when its depth and
capacity are sufficient.*

GRADE BEAM

BOTTOM SPANDREL/GRADE BEAM   B2

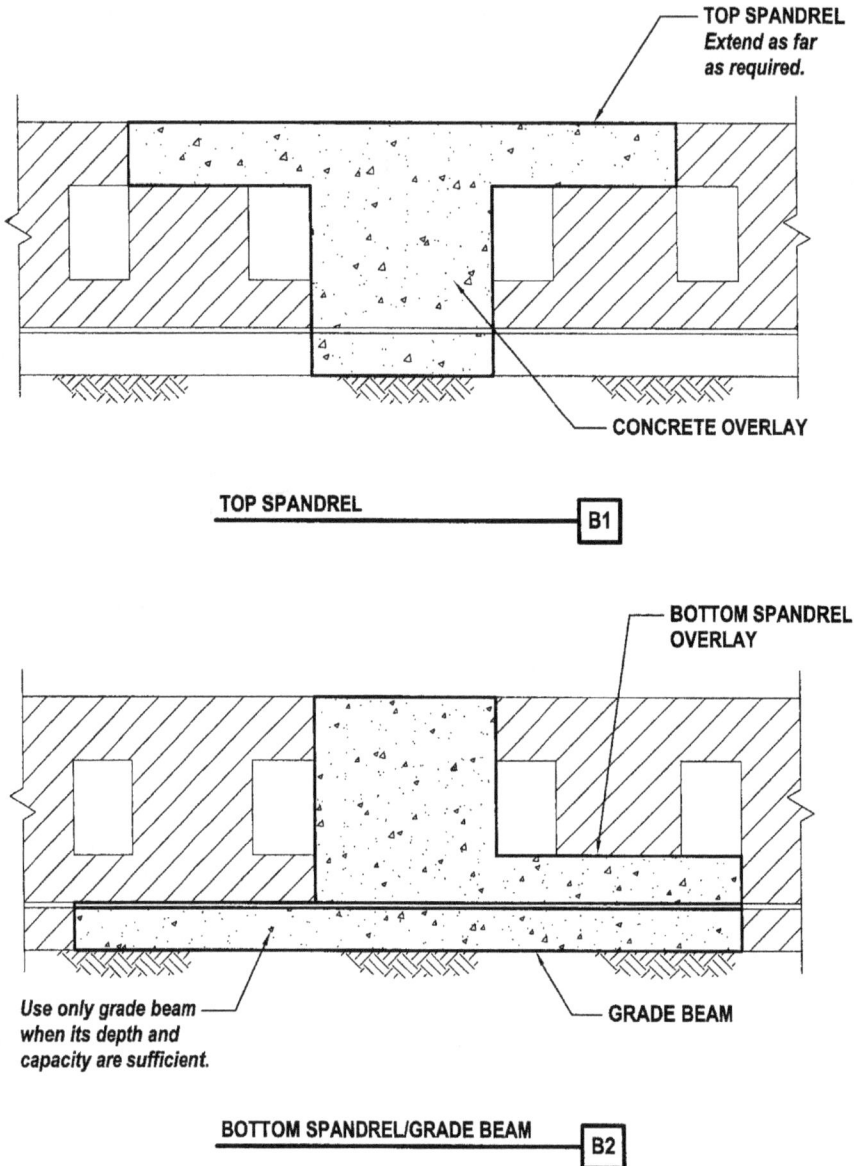

**Figure 21.4.5-2B: Alternatives to Distribute Overturning Loads
in Concrete Overlay**

With full overlay, thickness and
reinforcing can be reduced
compared to partial overlays.

FULL OVERLAY   C1

CONCRETE OVERLAY

GRADE BEAM

DRILLED PIER

DRILLED PIERS   C2

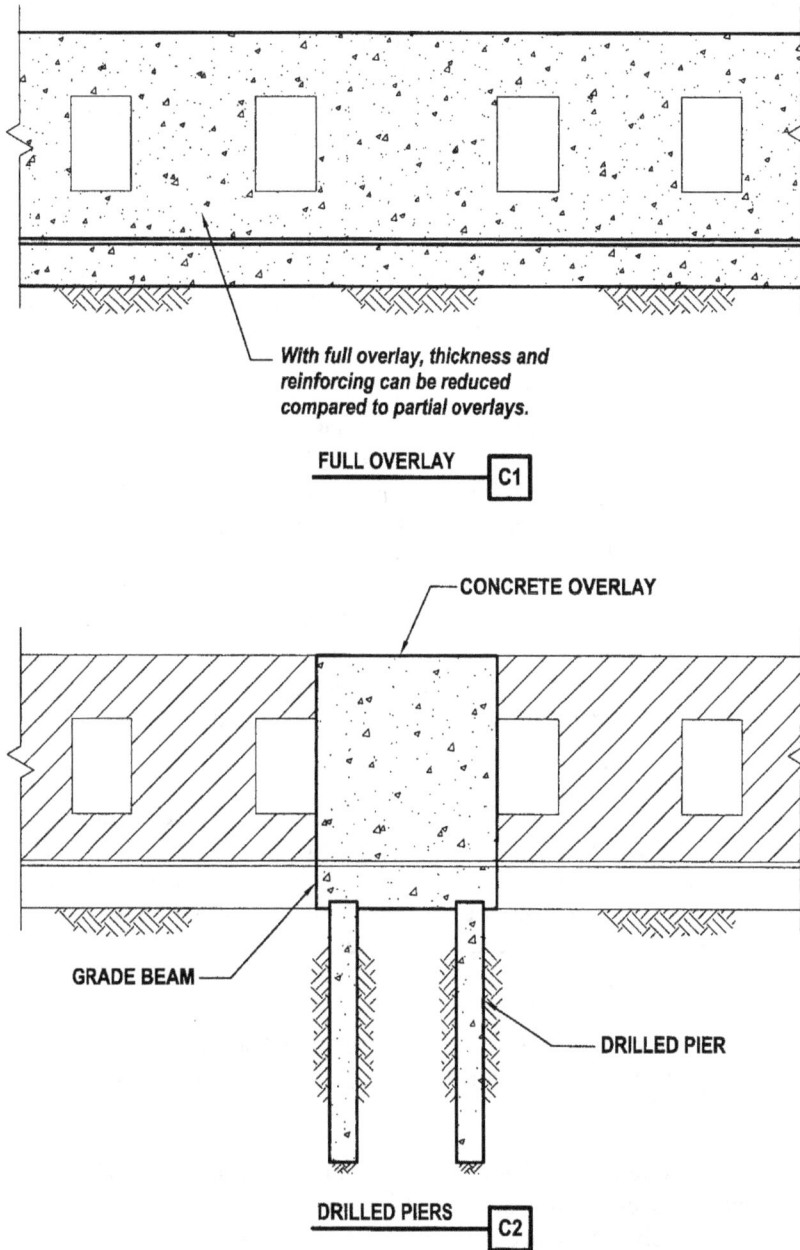

**Figure 21.4.5-2C: Alternatives to Distribute Overturning Loads
in Concrete Overlay**

When the joists are perpendicular to the wall, they can be cast into the wall, but when this is done, the joists should be treated with preservative, an air gap provided on the top and sides of the joists and building paper placed on the bottom to minimize moisture entering the lumber. Special rebar detailing will be needed to transfer shear from above the floor to below the floor through the weakened area of the joists. Given these issues, the perpendicular joists are often headed off and supported off a ledger on the face of the new wall. This requires shoring, but simplifies the remaining construction and provides for a better wall. This approach is shown in Figure 21.4.5-1.

Pre-wetting the masonry wall: Cast-in-place walls and, to a lesser extent, shotcrete walls have moisture in them during placement that will be absorbed by the masonry. Some engineers require wetting the masonry wall, just prior to placement.

Curing considerations in existing building: Curing concrete emits moisture. If the building has finishes that are sensitive to moisture emission, precautions will need to be taken to protect the finishes. This can be particularly critical if shotcrete is used. Curing of the face of the concrete is either done with curing compound, continuous spraying or a moisture-retaining cover. Finding a curing compound that will later be acceptable for certain adhered finishes can be difficult. Continuous spraying, however, adds substantial moisture to the interior space.

Efflorescence concerns from additives and moisture: Like all concrete, overlays can be susceptible to alkali salts leaching to the surface, usually leading to white streaks or spots. These stains can come from additives within the concrete or from salts within the masonry wall. Use of low-alkali concrete is recommended, and additives should be limited to those known not to lead to efflorescence.

Protection of existing masonry substrate: The main advantage of shotcrete is that the existing masonry serves as the backside form and forming the front is unnecessary. In typical stone masonry and brick masonry situations, the wall will be adequate to serve as the backside form. When the wall is particularly thin or in poor condition, the contractor will have to take care to brace the masonry to resist the force of the applied shotcrete.

Cost/Disruption

Adding new concrete, particularly with shotcrete, can be quite disruptive. Where access is sufficient, shotcrete is typically chosen as it is less expensive than cast-in-place work which requires front-side formwork.

Shotcrete: Placing shotcrete requires access for the hose and concrete truck and sufficient room (several feet) to spray the concrete. It is desirable to shoot downward, so scaffolding is needed at upper portions of walls to achieve the necessary angle. Spraying is noisy and very dusty. If an indoor wall is being shot, the room will usually be sealed off with plastic sheeting to control dust. During shooting, residue—known as rebound—forms at the base of the shoot and must be cleaned away so that it does not become part of the overlay. Protection against rebound on existing floor and wall surfaces is needed.

Cast-in-place concrete: Placing cast-in-place concrete also requires access for the hose and concrete truck. Less front-side access is needed than shotcrete, but a front-side form is required with the associated sawing and hammering noise of construction. Concrete placement is noisy, and in addition to workmen and concrete truck noise, there is the vibrator used to consolidate the concrete.

Proprietary Issues

There are no known proprietary concerns with shotcrete or cast-in-place overlays on existing masonry walls.

21.4.6   Add Fiber-Reinforced Polymer Overlay to Masonry Wall

Deficiency Addressed by Rehabilitation Technique

Improving inadequate in-plane wall strength is the primary purpose of a new fiber-reinforced polymer (FRP) overlay, but the overlay can also improve out-of-plane bending capacity.

Description of the Rehabilitation Technique

An FRP overlay, typically made of glass or carbon fibers in an adhesive matrix, is applied against an existing unreinforced masonry wall to increase the shear strength of the wall. The existing wall surface must be prepared to receive the new material, and after application the fiber composite must be protected against ultraviolet rays.

Design Considerations

Research basis: Research in fiber composites is extensive, but has primarily been focused on enhancement of concrete elements or reinforced concrete masonry. There is, nonetheless, a growing body of research for unreinforced masonry strengthening. A partial listing of some papers is given here.

For out-of-plane strengthening of unreinforced masonry, tests include Reinhorn and Madan (1995a) on clay brick; Portland State University (1998) on hollow clay tile; Vandergrift, Gergely, and Young (2002) on hollow concrete masonry; Tumialan, Galati, Namboorimadathil, and Nanni (2002) on surface applied fiber reinforced bars to hollow concrete masonry; Tumialan, et al. (2002) on glass and aramid fiber reinforced polymer composites on both concrete and clay brick; Tumialan, Galati, and Nanni (2002) on situ field tests in an infill frame building being demolished of brick and clay tile walls strengthened using glass fiber strips; Ehsani, Saadatmanesh, and Velazquez-Dimas (1999); and Velazquez-Dimas, Ehsani, and Saadtmanesh (2000) on half-scale clay masonry.

In-plane testing for clay brick masonry includes Reinhorn and Madan (1995b); Ehsani and Saadatmanesh (1996); Ehsani, Saadatmanesh, and Al-Saidy (1997); Haroun and Mosallam (2002); Senescu and Mosalam (2004). Elgwady, Lestuzzi and Badoux (2003) performed dynamic tests on slender and squat hollow clay masonry piers with and without aramid, glass, and carbon fiber composite overlays. Vandergrift, Gergely, and Young (2002) performed tests on half-scale hollow concrete masonry. Schwegler and Ketterborn (1996) discusses in-plane masonry strengthening with carbon fiber, including straps at various angles of orientation. Reinhorn and Madan (1995b) found about a 120% increase in strength from the composite and some additional displacement capacity in a one-cycle reversed cyclic test. After the fiber ruptured, however, masonry cracks immediately widened in a brittle manner. Haroun and Mosallam (2002) found a minor increase (about 20%) in strength capacity and about a 30% increase in displacement capacity. Cracking in the masonry was the ultimate limit which occurred after the composite debonded. Senescu and Mosalam (2004) used monotonic diagonal tension tests with mixed results, some showing improvements in strength and displacement capacity, others actually showing reductions in strength and displacement capacity. Paquette, Bruneau, and Brzev (2004) and Paquette and Bruneau (2004) investigated strengthening or repairing a one-story building with rocking-critical piers using fiber composite chord strips at ends of piers.

Design basis: There are no code guidelines or FEMA 356 provisions explicitly addressing FRP overlays on unreinforced masonry. Information can be obtained from manufacturer literature and can be used in conjunction with criteria in the ICC-ES interim standard (ICC-ES, 2003). They focus only on the design of the fiber itself, not the fiber and masonry combined performance. Velazquez-Dimas and Ehsani (2000) provide modeling and design recommendations for out-of-plane strengthening.

Behavioral mode: It is important to understand the underlying governing behavioral mode of both the unstrengthened unreinforced masonry wall and the strengthened wall. Fiber composites have little ductility. Adding an FRP overlay to a rocking critical wall pier may be able to reduce cyclic degradation of the pier, but will not change the strength or behavior mode if the fiber is only applied to the pier. If the fiber crosses the top and bottom of the pier into the spandrel, in can inhibit or prevent formation of a rocking mode. Strength will be increased, but ductility will be reduced from that of a rocking-critical mode to that of a shear critical mode.

Detailing and Construction Considerations

Detailing and construction considerations for FRP overlays include the following.

Surface preparation: The surface of the masonry needs to be cleaned of loose material and finishes that prevent proper adhesion. Sandblasting of the masonry is not usually necessary; a wire brush is used instead. See Figure 24.4.6-1.

Complete overlay vs. strips: Both in research and in practice, both complete overlays over the full surface of the wall and use of strips are found. Vertical strips are used when only improvement to out-of-plane resistance is needed. Diagonal strips have been used to resist diagonal tension stresses from in-plane shear.

One side or both sides of wall: Applying fiber to both sides of the wall improves performance, particularly for out-of-plane resistance, but testing has been performed with fiber on only one side.

Continuity of fiber at top and bottom: Providing load transfer with the fiber can challenging, particularly at floor-to-wall interfaces. If the fiber is being used to resist out-of-plane loads and is transferring these loads back into the floor diaphragm, special details may be needed to turn the vertical fiber overlay into the horizontal diaphragm. Fiber cannot be bent at 90 degrees;

Apply fabric over top of wall
or extend up face of parapet

PREPARE SURFACE BY WIRE BRUSHING
AND REMOVING ALL LOOSE MATERIAL.
RESET LOOSE MASONRY, AND REPOINT
DETERIORATED OR CRACKED JOINTS.

Apply fabric between joist
or attach to bottom of slab.

Paint over fabric at exterior
with paint which protects
against ultraviolet light.

FASTEN EDGE OF FABRIC WITH
ANGLE AND BOLTS. EPOXY
BOLTS INTO SLAB OR FOOTING.

EXCAVATE GRADE AS REQUIRED
TO APPLY FABRIC AND COVER STEEL
WITH LEAN CONCRETE FOR
CORROSION PROTECTION

CHIP SLAB TO SET ANGLE. FILL
WITH GROUT TO LEVEL FLOOR

APPLY FABRIC TO TOP OF
FOOTING OR SLAB

SECTION   A

Step 1
Prepare wall
surface

Step 3
Apply fabric
over epoxy.
Apply second
coat epoxy.

Step 2
Apply two part
epoxy to prepared
wall surface

STEEL ANGLE
AND BOLTS

ELEVATION   B

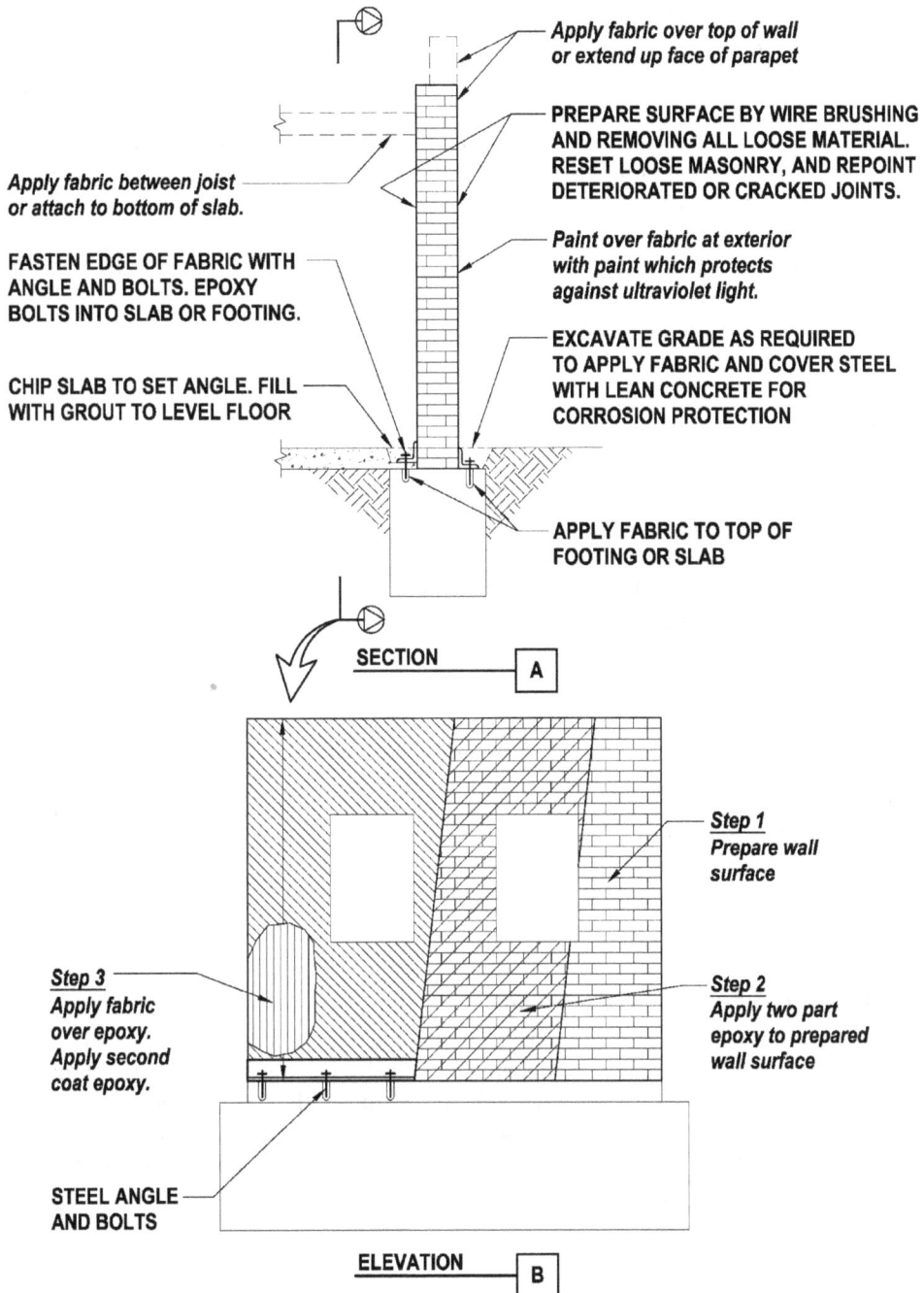

Figure 21.4.6-1: Fiber Composite Wall Overlay on URM Wall

rather a radius is needed. Sometimes, steel reinforcing plates are used to stiffen the turns. If the fiber is being used to transfer in-plane loads from one story to the next, continuity past the floor is needed and will require special details as does shear transfer out of the diaphragm into the wall

Moisture barrier: Fiber composites are impermeable. If continuous overlays are used, moisture transmission through the masonry wall will be stopped at the fiber. Eventually, the concern would be the moisture would build up and begin to delaminate the fiber bond and lead to general building concerns with excessive moisture.

Additional information: See Section 13.4.1 for more detailed discussion on FRP issues including:

> Composite makeup and application
> Mechanical properties
> Fiber and mechanical anchors
> Durability
> Constructability

Cost/Disruption

Fiber composites are relatively expensive as an application for masonry wall strengthening and have not seen significant use. Disruption comes from the sandblasting and surface preparation of the masonry wall, the fumes from the adhesives used in application of the fiber composite and removal and access requirements where the fiber transitions from story to story at the floor levels.

Proprietary Issues

Fiber composite materials are supplied by vendors. Capacities and design methods vary depending on the vendors.

21.4.7   Infill Opening in a URM Wall

Deficiency Addressed by Rehabilitation Technique

Inadequate unreinforced masonry in-plane wall strength.

Description of the Rehabilitation Technique

Window and door openings are filled to increase the shear capacity and reduce the shear stresses on the unreinforced masonry wall. The opening is typically filled with concrete, reinforced concrete masonry units, or reinforced clay brick, rather than with unreinforced masonry due to code concerns with adding unreinforced masonry. To provide adequate shear transfer between the existing wall and the new infill, the interface can be toothed, but more typically, drilled dowels are used. See Figure 21.4.7-1.

EXISTING
URM WALL

ROUGHEN EDGE OF
(E) OPENING

8" minimum thickness
concrete or use
CMU infill.

REINFORCING

Exterior window
glazing can be left
in place or modified
if necessary to retain
some aesthetic
similarity to existing
condition.

Plane of infill
can vary within
thickness of
masonry wall.

CLASS "B"
LAP SPLICE

DRILLED IN
AND EPOXIED
DOWELS AT 16"
ON CENTER

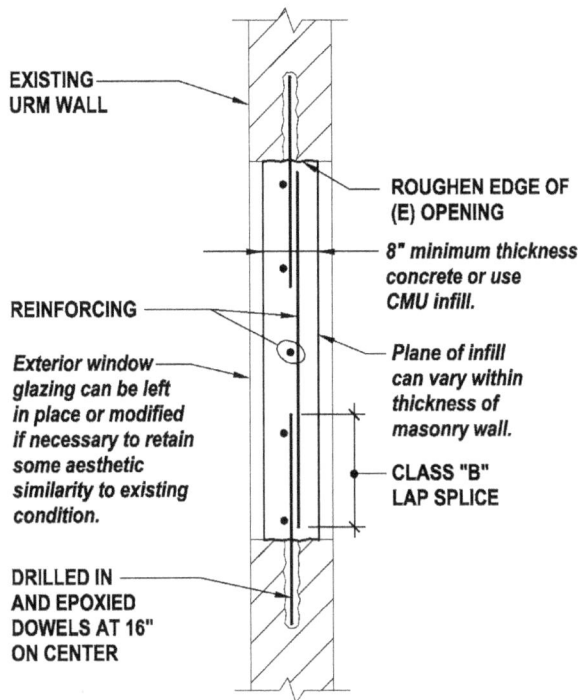

SECTION THROUGH INFILL

Figure 21.4.7-1: Infilling an Opening in a URM Wall

Design Considerations

Research basis: Research that directly addresses testing of infilled openings has not been identified.

Capacity of infill: The typical infill materials are stronger and potentially stiffer than the surrounding masonry. It is typical, however, to consider the infilled wall as solid unreinforced masonry in determining shear capacity of the composite wall. Since the increase in shear capacity from the infill is often not substantial, it is done when only moderate increases in capacity are needed.

Behavioral mode: It is important to understand the underlying governing behavioral mode of both the unstrengthened unreinforced masonry wall and the infilled wall. Infilling openings could change a rocking-critical wall line to a shear-critical wall line, which may be less desirable.

Detailing and Construction Considerations

Aethestics: Infilling openings can obviously have a significant visual impact. Sometimes new concrete is used and is set back from the exterior face so that a window can be placed. Lighting can be added between the glazing and infill to mitigate the opacity of the infill.

Cost/Disruption

Infilling an opening is relatively inexpensive if no architectural treatment is done to the face. Disruption is also more localized compared to other in-plane wall strengthening methods like concrete and fiber reinforced polymer overlays.

Noise will occur during drilling holes for drilled dowels and placing the infill.

Proprietary Issues

There are no proprietary concerns with infilling masonry wall openings.

21.4.8   Add Concrete or Masonry Shear Wall (Connected to a Wood Diaphragm)

Deficiency Addressed by Rehabilitation Technique

A new concrete or masonry wall provides additional global strength and stiffness, reduces demands on existing masonry walls and can reduce demands on diaphragms by cutting tributary spans.

Description of the Rehabilitation Technique

The new wall should be properly designed to meet current code detailing provisions. This section focuses on detailing at the interface between the new wall and the existing wood diaphragm. Figure 21.4.8-1 shows sample concepts both for joists parallel to the wall and joists perpendicular to the wall at an interior floor location; Figure 21.4.8-2 shows similar concepts at a roof.

Design Considerations

Research basis: Research specific to ledger connections between wood diaphragms and concrete or masonry walls has not been identified.

Capacity: The connection can be designed either for the code level demands or to develop the diaphragm, depending on where inelastic action is intended to occur.

Detailing and Construction Considerations

Detailing and construction considerations for connecting a new wall to an existing wood diaphragm include the following.

Masonry vs. concrete: Masonry is usually considered quicker to install and less expensive; concrete (or shotcrete) is stronger and stiffer and usually considered to have better earthquake performance. Making connections from concrete to wood diaphragms can be easier than with masonry.

REMOVE (E) FLOOR
SHEATHING AT (N) WALL

NEW TIE-DOWN ANCHOR
FOR TENSION TIE
Alternate: block each side.

*Where loads are large
and clearance for angle
is available above the
floor, install collector
angle with studs into
wall and lag screws to
joist below.*

*If loads permit,
joist can be used
as collector.*

THREADED ROD AS
SHEAR/TENSION TIE.
DRILL THROUGH
JOISTS AND CAST
CONCRETE AROUND
IT OR CAST WALL
AND INSTALL ROD
AS DRILLED DOWEL.

BLOCKING/STRAP
*To distribute tension
force back into
subdiaphragm*

ADD BLOCKING
TO FIT

JOISTS PARALLEL TO WALL   A

*Condition where existing
sheathing remains.*

*Condition where existing
sheathing is removed and
replaced with plywood.*

REMOVE (E) FLOOR
AT (N) WALL

BOUNDARY NAIL

SHORE, HEAD OFF
JOISTS, PLACE LEDGER
JOIST HANGERS, AND
CAST WALL

SHEAR TIE
SEE A

JOISTS PERPENDICULAR TO WALL   B

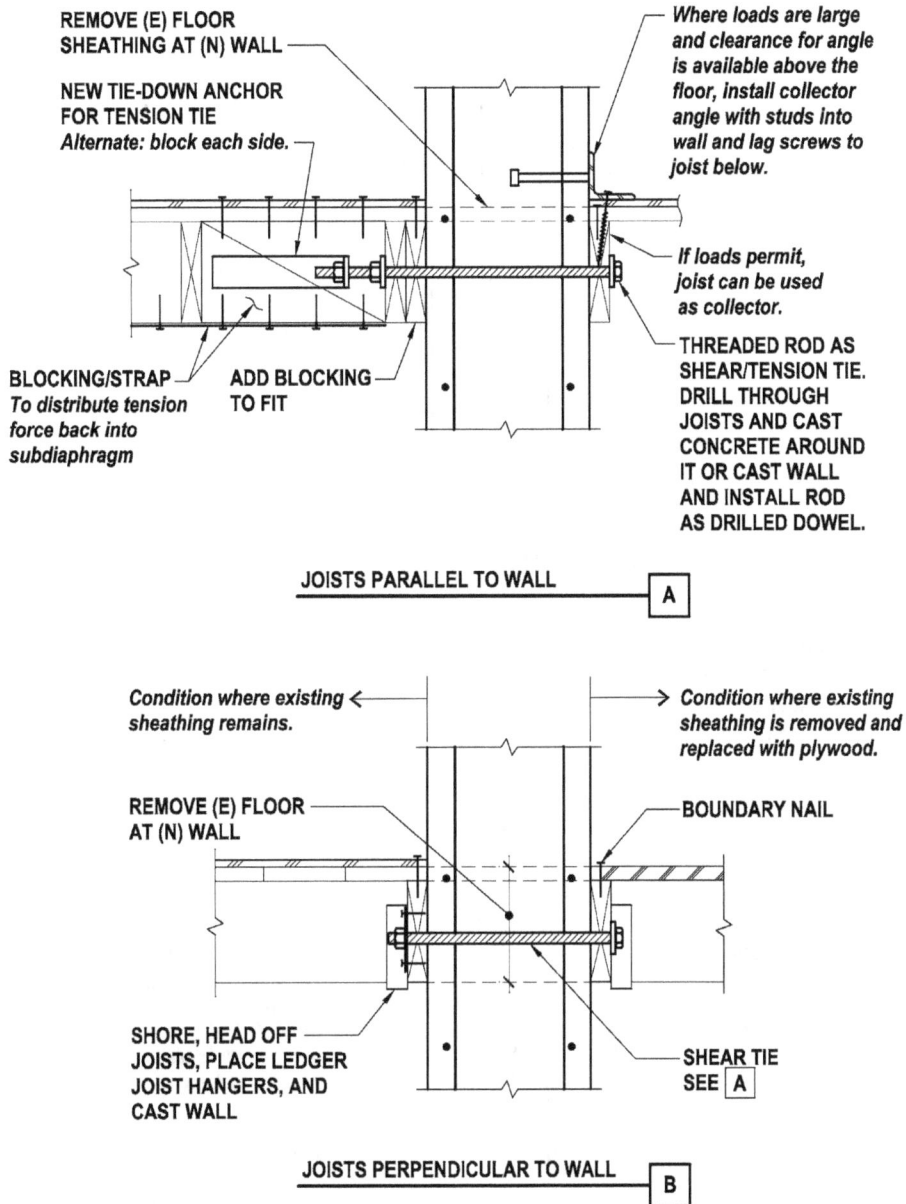

Figure 21.4.8-1: Connecting a New Concrete Wall to an Existing Wood Floor

STRAP NAILED TO BLOCKING
For use in out-of-plane tie

PLATE WASHER

NONSHRINK GROUT

B.N.

BLOCKING RIP TO FIT

2x BLOCKING
EACH SIDE
UNDER STRAP

*Note: This detail requires
placement of concrete from
above. An alternative is to
use an upturned block like* B
*and lowered nailer so wall
can be placed with shotcrete
under joists.*

JOISTS PARALLEL TO WALL   A

BLOCKING

FRAMING CLIP
For in-plane loads

PLYWOOD

B.N.

*Provide framing clips or toe nails at
end of block to take out rotation
from vertical eccentricity.*

FRAMING CLIP
*For out-of-plane loads,
shear transfer uses
perpendicular-to-grain
capacity of sill. If higher
capacity is needed, alternate
detail is required.*

SILL

NON-SHRINK GROUT.
*For shotcrete alternate:
Nail the sill to
bottom of joists
and use sill as part
of top side form.*

THREADED ROD WITH
PLATE WASHER AND
*DOUBLE NUTS.
Avoid use of J-bolts.*

0"

JOISTS PERPENDICULAR TO WALL   B

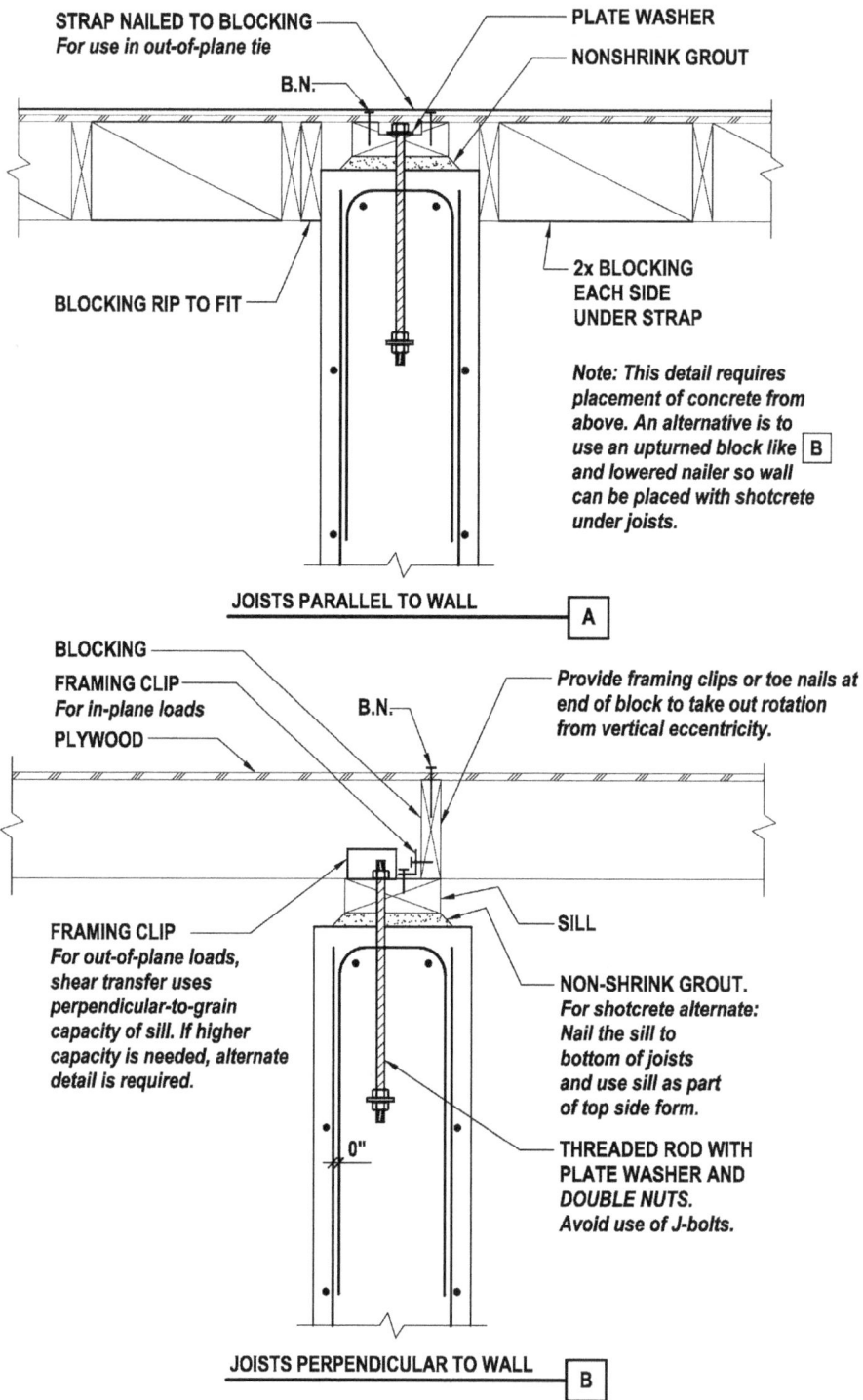

Figure 21.4.8-2: Connecting a Concrete Wall to an Existing Wood Roof at Interior

In-plane shear transfer: In Figure 21.4.8-1, shear transfer from the diaphragm to the wall goes from the diaphragm boundary nailing to the ledger and through the threaded rod into the wall. A tight fit on the rod and ledger is needed. The ledger should be dry dimensional lumber or glulam material to minimize vertical shrinkage of the ledger. When the wall is not as long as the diaphragm (a very common occurrence), a collector attachment into the wall will be needed. Figure 21.4.8-1A shows a steel angle with headed studs cast into the wall and diaphragm-to-collector connections using lag screws. The steel could go above or below the floor. When loads are relatively low, wood members such as the ledger can be used as the collector.

Out-of-plane tension transfer: In Figure 21.4.8-1, tension transfer of wall loads goes into the tie-down anchor, into the blocking, through straps in the blocking to additional blocks as required and eventually back into the diaphragm. Alternatively, blocking for a bay or two can be placed on both sides and out-of-plane resistance accomplished by compression bearing on the diaphragm joists.

Joist direction: When the wall can be fit in between existing joists, the amount of labor is reduced. When joists are perpendicular to the wall, the joists are typically headed off on each side of the wall to allow the wall to pass through. This requires temporary shoring of the floor around the wall. At the top of the wall, the wall can stop just under the joists and be blocked up to the diaphragm for shear transfer.

Shotcrete vs. cast-in-place concrete: See Section 21.4.5 for discussion of shotcrete vs. cast-in-place concrete issues.

Cost/Disruption

See Section 21.4.5 for discussion of cost and disruption issues.

Proprietary Issues

There are no proprietary concerns with connecting a concrete or masonry wall to a wood diaphragm.

21.4.9   Add Steel Moment Frame (Connected to a Wood Diaphragm)

Deficiency Addressed by Rehabilitation Technique

A new moment frame provides additional global strength, reduces demands on existing masonry walls and can reduce demands on diaphragms by cutting tributary spans.

Description of the Rehabilitation Technique

When a moment frame is added into a **URM** building, it typically goes either just behind a highly punctured street front façade or at an interior location within the diaphragm. Figure 21.4.9-1 shows the perimeter condition; Figure 21.4.9-2 shows interior conditions. A moment frame retrofit at a **W1A** building with a soft story is discussed in Chapter 6.

Design Considerations

Research basis: New steel moment frame issues are covered by FEMA 350 (FEMA, 2000). The CUREE woodframe project report on tuckunder building testing (Mosalam, et al., 2002)

documents quasistatic component testing of moment frame to wood diaphragm connections and
full-scale testing of a three-story tuckunder apartment building rehabilitated with a ground story
moment frame on the open front side.

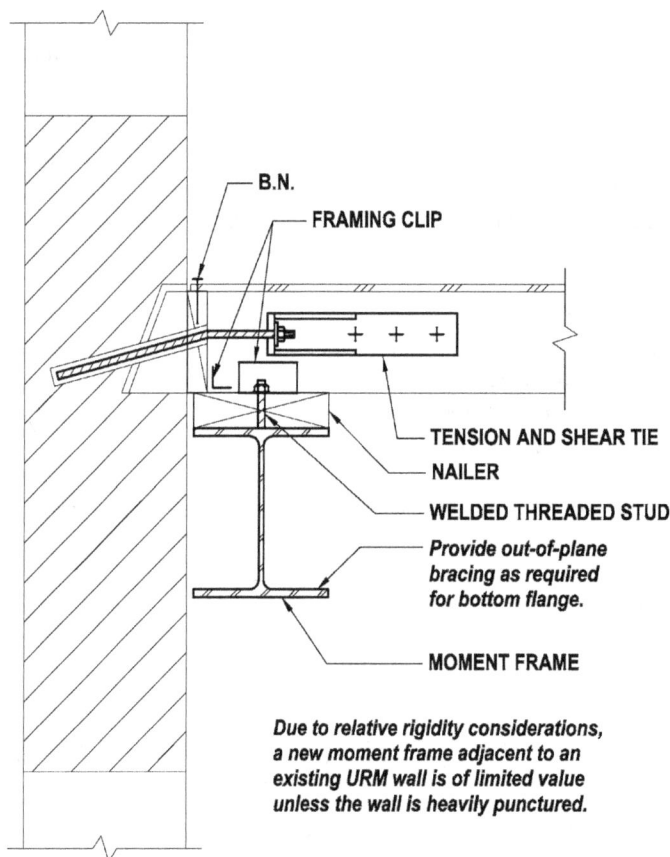

B.N.

FRAMING CLIP

+ + +

TENSION AND SHEAR TIE

NAILER

WELDED THREADED STUD

*Provide out-of-plane
bracing as required
for bottom flange.*

MOMENT FRAME

*Due to relative rigidity considerations,
a new moment frame adjacent to an
existing URM wall is of limited value
unless the wall is heavily punctured.*

SECTION

Figure 21.4.9-1: New Perimeter Steel Moment Frame to an Existing Wood Floor

Stiffness considerations: At either the perimeter or interior condition, reasonable stiffness of the
frame is desirable. At the perimeter, minimizing the amount of drift and resulting masonry
façade cracking is desirable. At the interior, if the moment frame does not have sufficient
stiffness, the diaphragm will span between the end walls with the moment frame taking out
relatively small loads due to its flexibility.

BLOCKING

FRAMING CLIP
*For in-plane
load transfer*

PLYWOOD

B.N.

*Provide framing clip
or toe nails at end of
blocking to take out
rotation from vertical
eccentricity.*

(E) JOIST

FRAMING CLIP
For top flange framing

*Provide out-of-plane
bracing as required
for bottom flange.*

NAILER.
*Add wood structural panel
shims for vertical tolerance
if required.*

THREADED WELDED
STUDS. *Coordinate
to miss existing joists
or countersink.*

JOISTS PERPENDICULAR TO FRAME   B

FRAMING CLIP
EACH END

FRAMING CLIP

BLOCKING

B.N.

STRAP

ANGLE BRACING
*This is an option for
bottom flange bracing
when bracing is required.*

WELDED THREADED
STUD

MOMENT FRAME BEAM

JOISTS PARALLEL TO FRAME   A

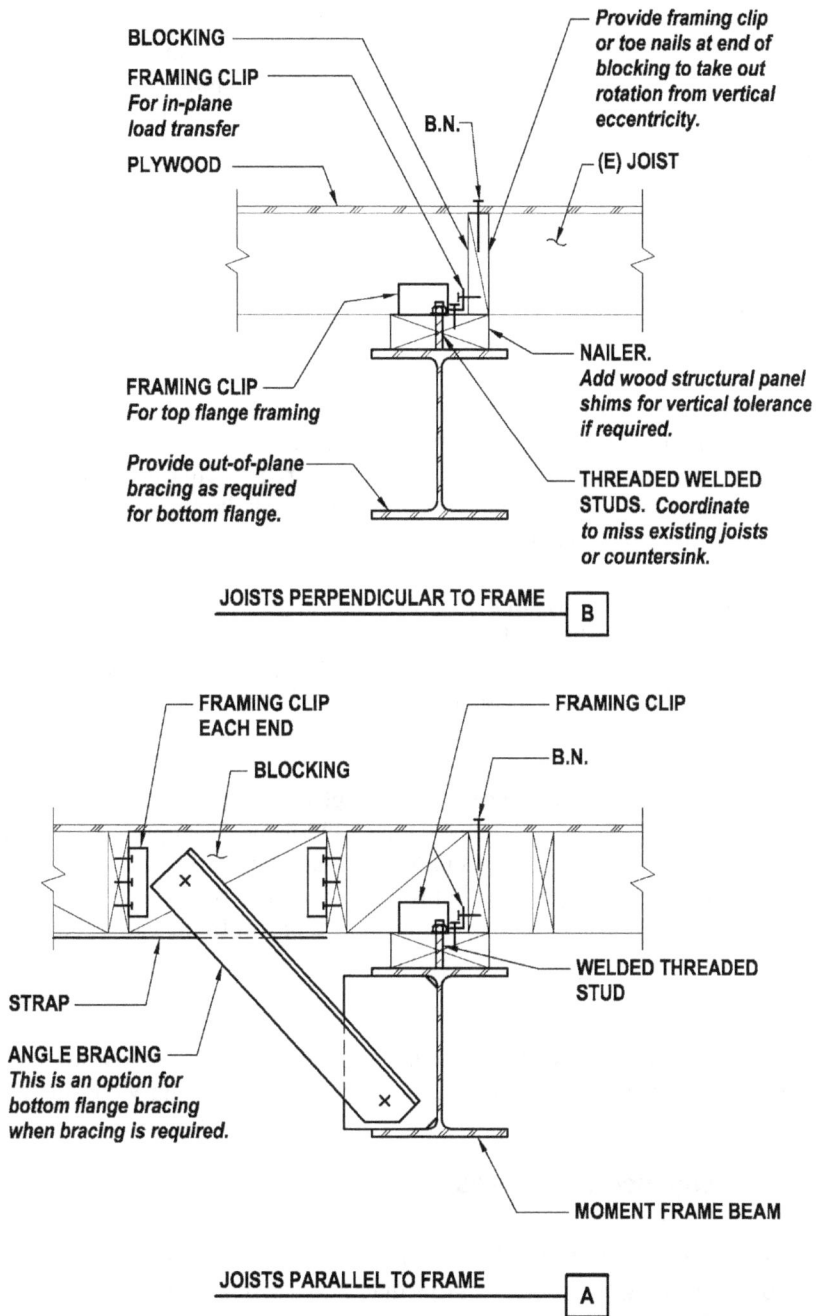

Figure 21.4.9-2: New Interior Steel Moment Frame to an Existing Wood Floor

Design forces: The new moment frame design can be governed by either stiffness or strength. Strength demands can either be minimum design loads or in some cases the moment frame can be designed to be stronger than the diaphragm so inelastic action happens in the diaphragm. For connection design of the frame to the diaphragm, it is particularly desirable to make sure the connections are stronger than the weaker of the diaphragm or the moment frame.

Pinned base: To minimize foundation demand requirements, new moment frames in retrofits are often designed with pinned bases.

Detailing and Construction Considerations

Detailing and construction considerations for connecting a new moment frame to an existing wood diaphragm include the following.

Welding vs. bolting: Welding adjacent to wood framing poses a very real fire hazard. Specifications and common sense usually dictate various fire watch provisions in these situations. Cases of hot welding slag lost from view and later reigniting wood material after the welding for the day was finished have been observed and are particularly troublesome. Where possible, detailing with shop welded connections, and then field bolting, is desirable. See Chapter 8 for additional comments on welding.

Connecting directly to the masonry: In Figure 21.4.9-1, the moment frame is connected to both the masonry façade and the diaphragm to take out load from the punctured wall into the frame and from the diaphragm into the frame. In alternative details, the load can be taken from the wall into the diaphragm and then through the diaphragm to the frame.

Cost/Disruption

Installation of a new moment frame can be fairly disruptive, though it is usually less disruptive than a new wall. The frame is chosen when existing window or door openings need to be preserved, but head height and visual issues must be considered. Adding new structural steel members can be comparatively expensive, but if the choice is to provide a wood structural panel overlay on a floor or add a new moment frame, the new moment frame can often be less expensive.

Proprietary Issues

There are no proprietary concerns with connecting a steel moment frame to a wood diaphragm. Certain moment frame beam-to-column connections may have proprietary considerations. See Chapter 8.

21.4.10   Add or Enhance Crosswalls

Deficiency Addressed by Rehabilitation Technique

Inadequate diaphragm strength and/or excessive diaphragm displacement.

Description of the Rehabilitation Technique

The ABK research program (ABK, 1984) showed that partition walls, called crosswalls, serve as energy-absorbing, displacement-limiting damping elements during seismic loading. The 2003

IEBC and 1997 UCBC permit certain qualifying buildings to use the "Special Procedure" with crosswalls as an integral part of the procedure. There are three basic types of crosswalls: existing partitions with various sheathing materials, new partitions, and new steel moment frames. Existing partitions may be adequate without any rehabilitation or they may need strengthening if they are not connected to the diaphragms or have insufficient capacity. New wood structural panel partitions must also be connected to the diaphragm and meet certain minimum capacities. New moment frames must meet minimum strength and stiffness criteria. See Figures 21.4.10-1 to 21.4.10-3.

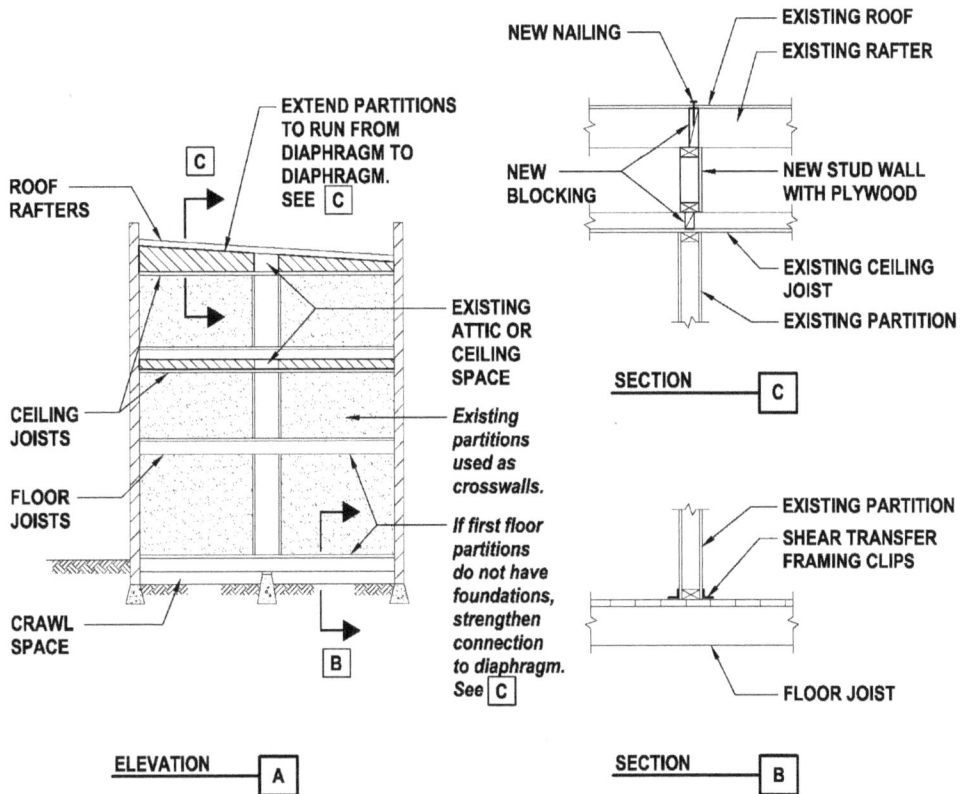

Figure 21.4.10-1: Strengthen Existing Crosswalls

Design Considerations

Research basis: ABK (1984) provides background for the basis of the crosswall concept.

Qualifying buildings: In the 2003 ICBC, the Special Procedure can be used only with buildings have flexible diaphragms at all levels and meet certain requirements regarding open fronts and number of wall lines in each direction. For the 1997 UCBC, these requirements apply, and buildings must be a maximum of six stories, and they cannot be essential or hazardous facilities. The crosswall concept and the Special Procedure were not adopted in FEMA 356.

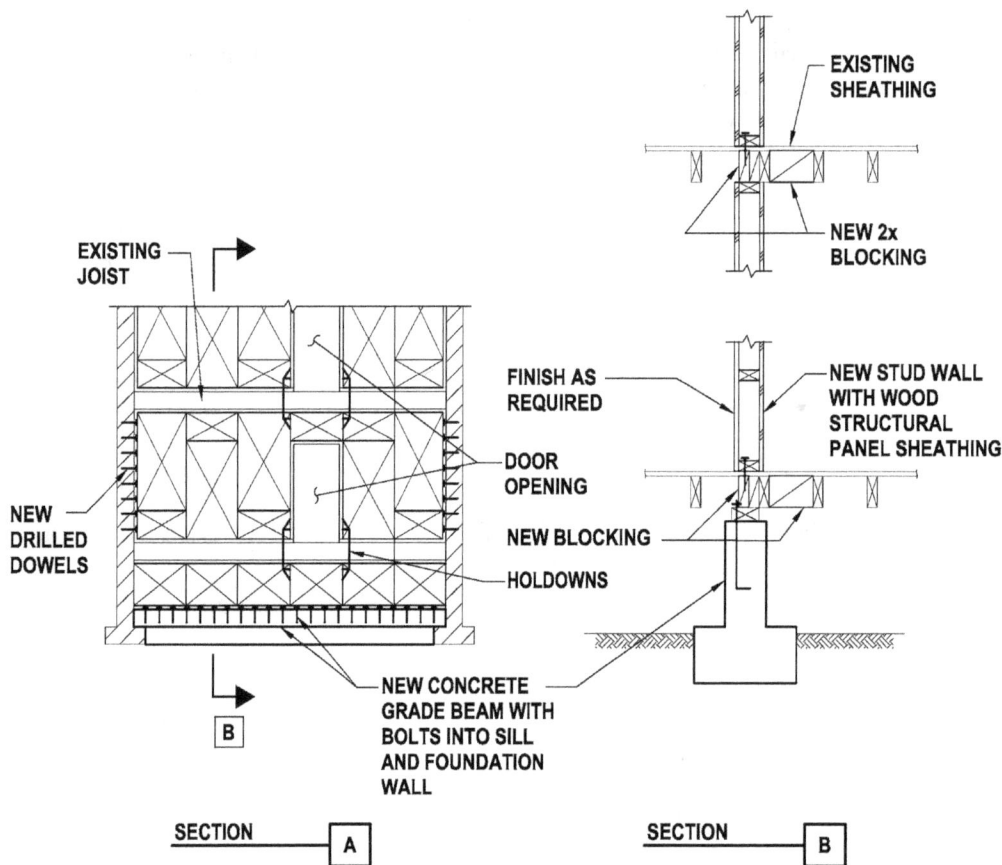

Figure 21.4.10-2: Add New Crosswalls

Crosswall requirements: The 1997 UCBC and the 2003 IEBC have a number of requirements on crosswall locations, aspect ratios, connection strength, spacing limits that need to be satisfied.

New crosswalls: New crosswalls will typically be done with structural wood panels and are similar to adding new wood structural panel shear walls. See Chapters 5 and 6 for additional information.

Steel moment frames: Adding moment frames for use as crosswalls is very similar to adding moment frames for use as new lateral force-resisting elements. See Section 21.4.9 for additional information.

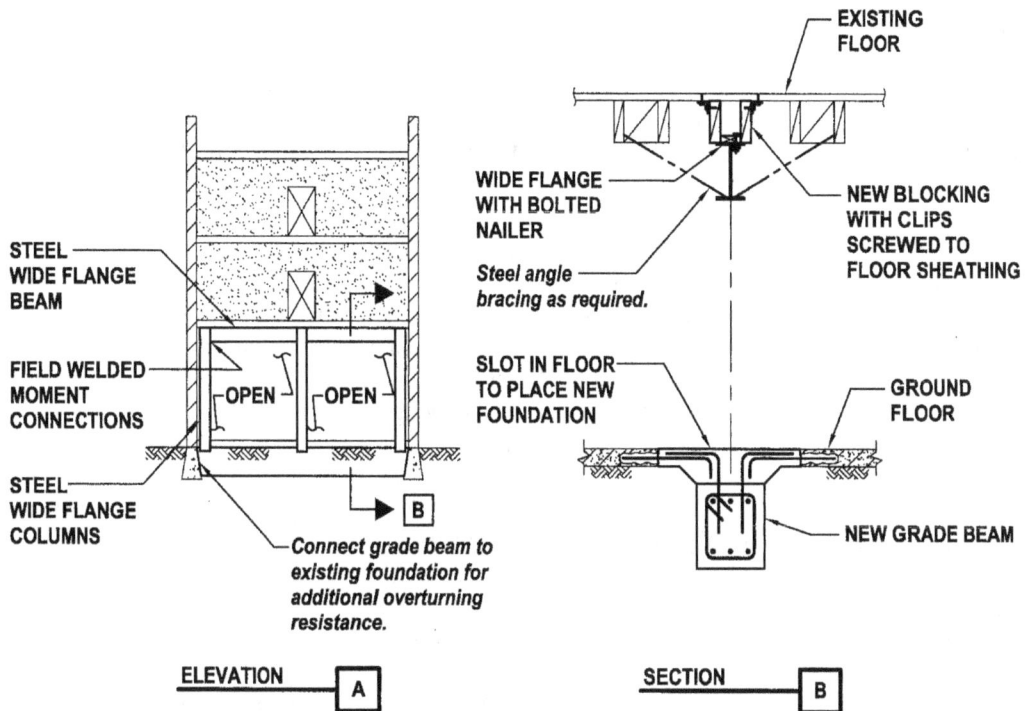

Figure 21.4.10-3: Add New Moment Frame as Crosswall

Detailing and Construction Considerations, and Cost/Disruption

Adding or enhancing woodframe crosswalls is similar to adding or enhancing woodframe shear walls; see Chapters 5 and 6. Adding a moment frame as a crosswall is similar to adding a moment frame as a lateral force-resisting element; see Section 21.4.9.

Proprietary Issues

There are no proprietary concerns with adding crosswalls. Certain steel moment frame beam-to-column connections are proprietary.

21.4.11   Add Supplemental Vertical Support for Truss or Girder

Deficiency Addressed by Rehabilitation Technique

Supplemental vertical supports provide a secondary load path for concentrated gravity loads on unreinforced masonry walls.

Description of the Rehabilitation Technique

A steel or wood post is added under existing trusses and girders.

Design Considerations

Research basis: There are no known tests of supplemental supports.

Purpose of the support: To some engineers, the goal of adding a supplemental support is to provide a back-up gravity load path if there is local deterioration of the masonry underneath a concentrated load like a truss bearing point. To others, it provides support if more wholesale failure of the wall occurs.

Independence of the support: The 1997 UCBC and 2003 IEBC both use the term "independent secondary columns" when referring to supplemental vertical supports. To some engineers, "independent" means separated from the wall. A post an inch or two away from the wall satisfies this requirement. To other engineers, "independent" simply means an alternative support, so that a ledger or pilaster on the wall is sufficient. In this scenario, the wall just beyond the damaged area is assumed to remain intact enough that gravity load resistance is not compromised. In Figure 21.4.11-1, a gap is shown.

Use as out-of-plane brace also: Some engineers like to take advantage of supplemental vertical support posts to serve as out-of-plane braces for the walls as well. Combined bending and axial demands must be considered.

Continuity of support: The 1997 UCBC and 2003 IEBC both do not specify whether the posts need to continue down to the next story. It is common to transfer loads at the base of upper story supplemental supports back to existing framing. This framing must be adequate to take the loads that would occur if they supplemental support began to take load.

Foundation support for posts: The 1997 UCBC and 2003 IEBC both do not explicitly specify whether the posts need a compliant new foundation or whether the posts can simply bear on an existing slab-on-grade. SEAOC (1992), however, states that a foundation is not required for the posts. Nonetheless, some engineers believe it is prudent to check the slab for bearing support and provide additional support if needed.

Triggering elements: The 1997 UCBC and 2003 IEBC only indicate that triggering elements are "trusses and beams, others than rafters and joists". The implication is that elements that support other structural members are the primary focus.

Detailing and Construction Considerations

Detailing and construction considerations for supplemental vertical supports include the following.

Steel vs wood: Supplemental support posts can either be of steel or wood. Steel members are smaller; wood members are less expensive.

Finish the new elements are leave bare: For certain architectural approaches, leaving the supplemental posts bare is compatible with the existing aesthetic. If it is not, the posts can be furred at added cost.

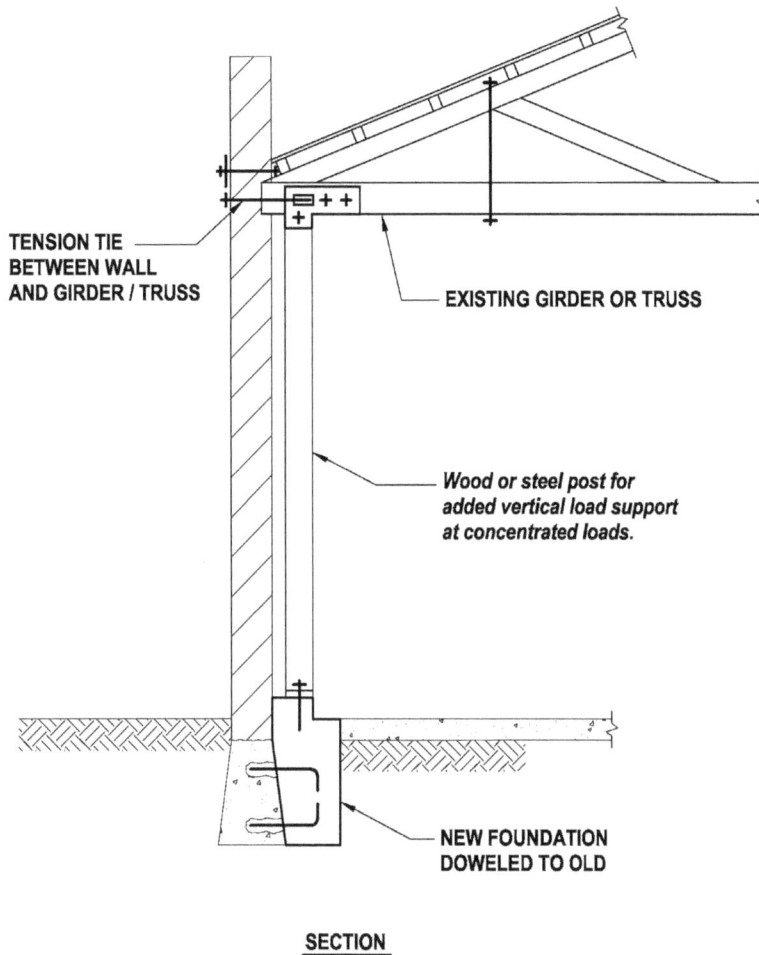

TENSION TIE BETWEEN WALL AND GIRDER / TRUSS

EXISTING GIRDER OR TRUSS

Wood or steel post for added vertical load support at concentrated loads.

NEW FOUNDATION DOWELED TO OLD

SECTION

Figure 21.4.11-1: Supplemental Vertical Support

Cost/Disruption

The relative cost of adding supplemental supports depends on the number used, whether they continue down to the ground and whether a new foundation is installed. Interior occupants will be disrupted locally as the posts are installed, and the usable space in the vicinity of the posts will be reduced.

Proprietary Considerations

There are no known proprietary concerns with employing supplemental vertical supports.

21.4.12   Add Veneer Ties in a URM Wall

Deficiency Addressed by Rehabilitation Technique

Missing or inadequate ties between a masonry veneer wythe and the backing wythes can lead to delamination of the veneer and a falling hazard.

Description of the Rehabilitation Technique

If the front or facing wythe of brick is not integrally tied into the interior wythes with header courses, it is considered a veneer. If sufficient metal veneer ties are not present to anchor the veneer to the backing wythe, new ties can be provided. Figure 21.4.12-1 shows anchorage using drilled dowels to connect the wythes, either from the exterior or interior, in a typical brick wall. Figure 21.4.12-2 shows anchorage between stone facing and brick backing.

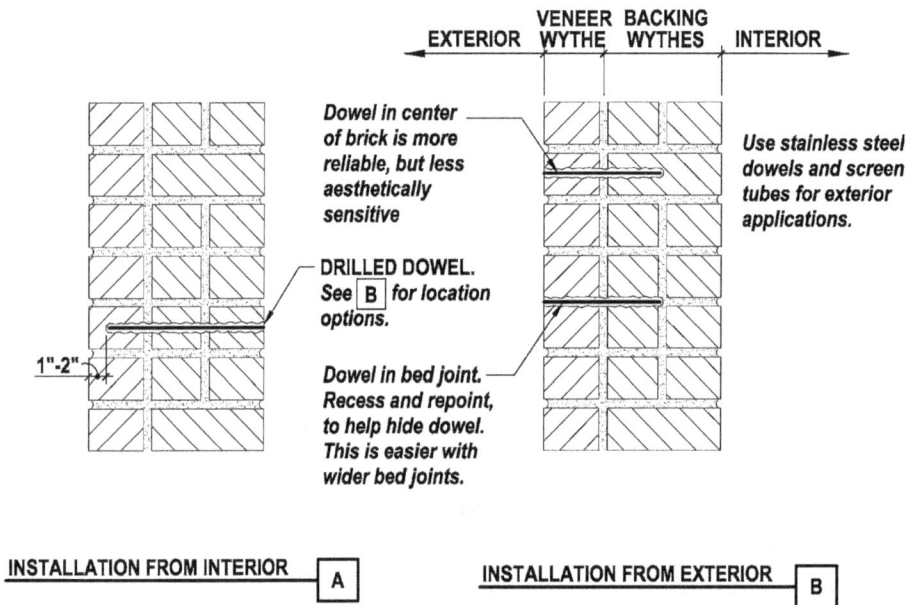

Figure 21.4.12-1: Veneer Ties in Brick Masonry

Design Considerations

Research basis: No academic research on veneer ties has been identified. Individual vendors have performed internal testing of their own products.

Veneer definition: Codes such as the 2003 ICBC and 1997 UCBC give minimum lay-up requirements for multi-wythe solid brick. The facing and backing wythes are to be bonded so than not less than 10 percent of the exposed face area is composed of solid headers extending less than 4 inches into the backing. The clear distance between adjacent full-length headers shall not exceed 24 inches vertically or horizontally. Facing wythes that do not meet these lay-up requirements are to be considered as veneer. Veneer wythes are not used in resisting shear forces in the wall and do not count in the thickness used for determining out-of-plane bending resistance.

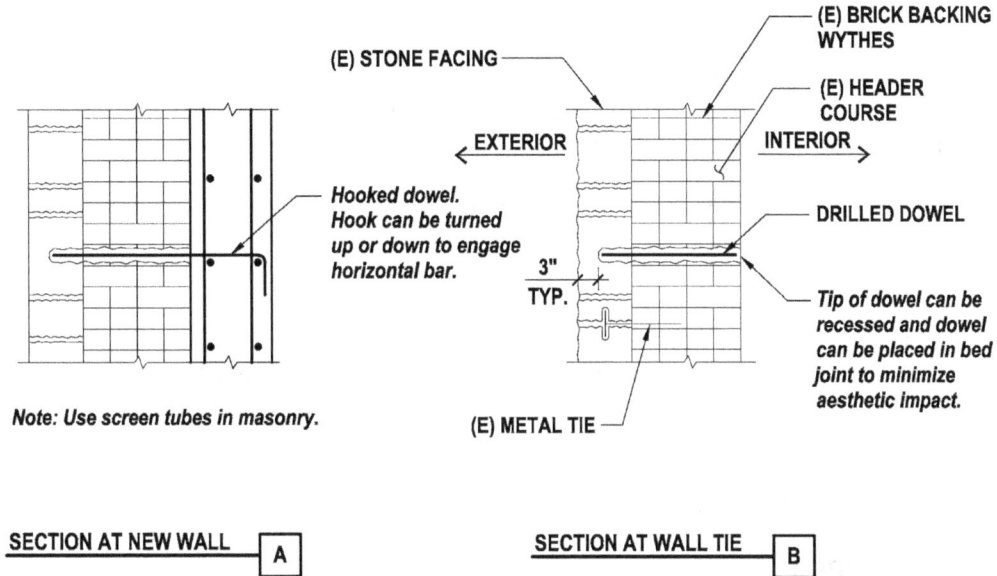

Figure 21.4.12-2: Veneer Ties for Stone Masonry Facing

Veneer tie requirements: Codes such as the 2003 ICBC and 1997 UCBC provide acceptance criteria for existing ties. They are to be corrugated galvanized iron strips shown to be in good conditions, with dimensions no less than 1" wide, 8" long and 1/16" thick and with a maximum spacing of 24" on center and a maximum supported area of four square feet. Veneer ties not meeting these requirements are to be strengthened.

Detailing and Construction Considerations

Detailing and construction considerations for veneer ties include the following.

Types of ties: The ties used in new construction are typically inappropriate for rehabilitation since they are installed as the wall is built up. There are, however, a fair number of proprietary products made by masonry accessory manufacturers that can be used in retrofit applications. Some involve expansion anchors in the backing wythes. Others involve helical anchors or spiral ties that "screw" into the backing wythes. Figures 21.4.12-1 and 21.4.12-2 show traditional drilled dowels. When drilled dowels are used, the tie diameter does not need to be large since loads are low and minimizing the size of the hole is important.

Drilled dowels installation face: Drilled dowels can be installed from the interior or exterior depending on which face is more sensitive.

Location of drilled dowel: The most reliable location of the drilled dowel is in the center of the brick, but this is will cause the largest aesthetic impact. The dowel can be placed in the bed joist or bed and head joint intersection to minimize the impact. Recessing the tip of the dowel and covering the end with repointing mortar is recommended.

Brick veneer vs. stone veneer: Anchoring thick stone is usually easier to do from the interior because the thickness of the stone permits greater cover on the face, reducing the likelihood of spalling. With random ashlar layout and interior installation, however, trying to locate the dowel away from the edges of the stone is not practical.

Corrosion considerations: Any dowel installed from the exterior should be done in stainless steel to minimize corrosion.

Cost and Disruption Considerations

Veneer anchorage can be relatively expensive depending on the number of new ties added. Ties installed from the inside are much more disruptive. See Section 21.4.2 for additional information on drilled dowels.

Proprietary Considerations

Many of the veneer tie anchorage systems are proprietary.

21.5   References

ABK, 1981a, *Methodology for Mitigation of Seismic Hazards in Existing Unreinforced Masonry Buildings: Categorization of Buildings*, a joint venture of Agbabian Associates, S.B. Barnes and Associates, and Kariotis and Associates (ABK), Topical Report 01, c/o Agbabian Associates, El Segundo, CA.

ABK, 1981b, *Methodology for Mitigation of Seismic Hazards in Existing Unreinforced Masonry Buildings: Diaphragm Testing*, a joint venture of Agbabian Associates, S.B. Barnes and Associates, and Kariotis and Associates (ABK), Topical Report 03, c/o Agbabian Associates, El Segundo, CA.

ABK, 1981c, *Methodology for Mitigation of Seismic Hazards in Existing Unreinforced Masonry Buildings: Wall Testing (Out-of-Plane)*, a joint venture of Agbabian Associates, S.B. Barnes and Associates, and Kariotis and Associates (ABK), Topical Report 04, c/o Agbabian Associates, El Segundo, CA.

ABK, 1984, *Methodology for Mitigation of Seismic Hazards in Existing Unreinforced Masonry Buildings: The Methodology*, a joint venture of Agbabian Associates, S.B. Barnes and Associates, and Kariotis and Associates (ABK), Topical Report 08, c/o Agbabian Associates, El Segundo, CA.

Abrams, D.P. and J.M Lynch, 2001, "Flexural Behavior of Retrofitted Masonry Piers," *EERC-MAE Joint Seminar on Risk Mitigation for Regions of Moderate Seismicity*, IL.

Breiholtz, D., 1987, "Centercore Strengthening System for Seismic Hazard Reduction of Unreinforced Masonry Bearing Wall Buildings," *Proceedings of the 56th Annual Convention of the Structural Engineers Association of California*, San Diego, CA.

Breiholz, D.C., 1992, "Center Core Seismic Hazard Reduction System for URM Buildings," *Proceedings of the Tenth World Conference on Earthquake Engineering*, Vol. 9, pp. 5395-5399.

Ehsani, M.R. and H. Saadatmanesh, 1996, "Seismic Retrofit of URM Walls with Fiber Composites," *The Masonry Society Journal*, Spring.

Ehsani, M.R. H. Saadatmanesh, and A. Al-Saidy, 1997, "Shear Behavior of URM Retrofitted with FRP Overlays," *Journal of Composites for Construction*, ASCE 1(1), 17-25.

Ehsani, M.R. H. Saadatmanesh, and J.I. Velazquez-Dimas, 1999, "Behavior of Retrofitted URM Walls Under Simulated Earthquake Loading," *Journal of Composites for Construction*, ASCE 4(3), 134-142.

ElGawady, M., P. Lestuzzi, and M. Badoux, 2003, "Rehabilitation of Unreinforced Brick Masonry Walls Using Composites," *Architectural and Structural Design of Masonry With Focus on Retrofitting of Masonry Structures and Earthquake Resistant Design*, International Short Course, Dresden University of Technology, December 7-18, Dresden, Germany.

ElGawady, M, P. Lestuzzi, and M. Badoux, 2004, "A Review of Conventional Seismic Retrofitting Techniques for URM, 13th International Brick and Block Masonry Conference, Amsterdam, July 4-7.

FEMA, 1997a, *NEHRP Guidelines of the Seismic Rehabilitation of Buildings,* FEMA 273, Federal Emergency Management Agency, Washington, D.C., October.

FEMA, 1997b, *NEHRP Commentary on the Guidelines of the Seismic Rehabilitation of Buildings,* FEMA 274, Federal Emergency Management Agency, Washington, D.C., October.

FEMA, 1999a, *Evaluation of Earthquake Damaged Concrete and Masonry Wall Buildings: Basic Procedures Manual*, FEMA 306, May.

FEMA, 1999b, *Evaluation of Earthquake Damaged Concrete and Masonry Wall Buildings: Technical Resources,* FEMA 307, May.

FEMA, 2000, *Prestandard and Commentary for the Seismic Rehabilitation of* Buildings, FEMA 356, Federal Emergency Management Agency, Washington, D.C., November.

Ganz, H.R., 1993, "Strengthening of Masonry Structures with Post-Tensioning," *Proceedings of the Sixth North American Masonry Conference*, Philadelphia, Pennsylvania, June 6-9, pp. 645-655.

Ghanem, G., M.A. Zied and A.E. Salama, 1994, "Retrofit and Strengthening of Masonry Assemblages Using Fiberglass," *Proceedings of the Tenth International Brick and Block Masonry Conference,* Vol. 2, pp. 499-508.

Haroun, M.A and A.S Mosallam, 2002, *Cyclic Shear Test of Multi-Wythe Existing Brick Wall: Retrofitted by a Single Layer of TYFO SHE-51A Applied on One Side*, University of California, Irvine and California State University Fresno, February.

Hutchison, D.L., P.M.F. Yong and G.H.F. McKenzie, 1984, "Laboratory Testing of a Variety of Strengthening Solutions for Brick Masonry Wall Panels," *Proceedings of the Eighth World Conference on Earthquake Engineering*, Vol. 1, pp. 575-582.

ICBO, 1997, *Uniform Code for Building Conservation, 1997 Edition*, International Conference of Building Officials, Whittier, CA.

ICC, 2003, *International Existing Building Code*, *2003 Edition*, International Code Council, Country Club Hills, IL.

ICC-ES, 2003, "Interim Criteria for Concrete and Reinforced and Unreinforced Masonry Strengthening Using Fiber-Reinforced Polymer (FRP) Composite Systems," Acceptance Criteria AC-125, June, International Code Council Evaluation Service, Whittier, CA.

ICC-ES, 2005, *Acceptance Criteria for Anchors in Unreinforced Masonry Elements* (AC60), Approved April 2005, Whittier, CA.

Jurukovski, D., L. Krstevska, R. Alessi, P.P. Diotallevi, M. Merli and F. Zarri, 1992, "Shaking Table Tests of Three Four-Storey Brick Masonry Models: Original and Strengthened by RC Core and RC Jackets, *Proceedings of the Tenth World Conference on Earthquake Engineering,* July, Madrid, A A Balkema, Rotterdam, Vol. 5, pp. 2795-2800.

Kahn, L.F., 1984, "Shotcrete Retrofit of Unreinforced Brick Masonry," *Proceedings of the Eighth World Conference on Earthquake Engineering*, San Francisco, CA, July, pp. 583-590.

Kariotis, J., 1982, "Bracing of Unreinforced Masonry Walls, Out-of-Plane," *Earthquake Hazard Mitigation of Unreinforced Masonry Buildings Built Prior to 1934*, Seminar Proceedings, Structural Engineers Association of Southern California, April, Whittier, CA.

Mosalam, K., et al., 2002, *Seismic Evaluation of an Asymmetric Three-Story Woodframe Building*, CUREE Publication No. W-19, Consortium of Universities for Research in Earthquake Engineering, Richmond, CA.

Paquette, J., M. Bruneau, and S. Brzev, 2003, "Pseudo-Dynamic Testing of Unreinforced Masonry Building with Flexible Diaphragm," *Journal of Structural Engineering*, Vol. 129, No. 6, June, pp. 708-716.

Paquette, J., M. Bruneau, and S. Brzev, 2004, "Seismic Testing of Repaired Unreinforced Masonry Building Having Flexible Diaphragm," *Journal of Structural Engineering*, Vol. 130, No. 10, October, pp. 1487-1496.

Paquette, J. and M. Bruneau, 2004, "Pseudo-Dynamic Testing of Unreinforced Masonry Building with Flexible Diaphragm, *Proceedings of the 13th World Conference on Earthquake Engineering*, Vancouver, B.C., Paper 2609.

Plecnik, J., T. Cousins and E. O'Conner, 1986, "Strengthening of Unreinforced Masonry Buildings," *Journal of Structural Engineering*, American Society of Civil Engineers, Vol. 112, No. 5, May, pp. 1070-1087.

Plecnik, J.M., 1988, "Preliminary Report on "Summary of the Second Meeting of the Advisory Committee for NSF Research Project on the Centercore Rehabilitation Technique," *Structures Laboratory Report 88-8-18*, California State University, Long Beach, August 17.

Portland State University, 1998, *Full Scale Tests of Retrofitted Hollow Clay Tile Walls*, Final Report Prepared for Contech, Department of Civil Engineering, June.

Reinhorn, A.M. and A. Madan, 1995a, *Evaluation of Tyfo-W Fiber Wrap System for Out of Plane Strengthening of Masonry Walls*, Report No. AMR 95-2001, Department of Civil Engineering, State University of New York at Buffalo, Buffalo, NY, March.

Reinhorn, A.M. and A. Madan, 1995b, *Evaluation of Tyfo-W Fiber Wrap System for In Plane Strengthening of Masonry Walls*, Report No. AMR 95-2002, Department of Civil Engineering, State University of New York at Buffalo, Buffalo, NY, August.

Rosenboom, O.A. and M.J. Kowalsky, 2004, "Reversed In-Plane Cyclic Behavior of Posttensioned Clay Brick Masonry Walls," *Journal of Structural Engineering*, Vol. 130, No. 5, May, pp. 787-798.

Rutherford & Chekene, 1990, *Seismic Retrofitting Alternatives for San Francisco's Unreinforced Masonry Buildings: Estimates of Construction Cost & Seismic Damage*, for the Department of City Planning of the City and County of San Francisco, Oakland, CA, May.

Rutherford & Chekene, 1997, *Development of Procedures to Enhance the Performance of Rehabilitated Buildings*, prepared by Rutherford & Chekene Consulting Engineers, published by the National Institute of Standards and Technology as Reports NIST GCR 97-724-1 and NIST 97-724-2.

Schwegler, G. and P. Kelterborn, 1996, "Earthquake Resistance of Masonry Structures Strengthened with Fiber Composites," *Proceedings of Eleventh World Conference on Earthquake Engineering*, Acapulco, June, Elsevier Science Ltd., Paper No. 1460.

Senescu, R. and K.M. Mosalam, 2004, *Retrofitting of Unreinforced Masonry Walls Using Fiber Reinforced Polymer Laminates*, Report UCB/SEMM 2004/3, University of California, Berkeley, CA, May.

SEAOC (Structural Engineers Association of California), 1992, *Commentary on Appendix Chapter 1 of the Uniform Code for Building Conservation, Seismic Strengthening Provisions for Unreinforced Masonry Bearing Wall Buildings*, June 20.

SEAOSC (Structural Engineers Association of Southern California),1982, *Earthquake Hazard Mitigation of Unreinforced Masonry Buildings Built Prior to 1934*, Seminar Proceedings, April, Whittier, CA.

SEAOSC (Structural Engineers Association of Southern California), 1986, *Earthquake Hazard Mitigation of Unreinforced Masonry Buildings Pre-1933 Buildings*, Seminar Proceedings, October, Whittier, CA.

Tumialan, J.G., N. Galati, and A. Nanni, 2002, "Field Assessment of URM Walls Strengthened with FRP Laminates," *Journal of Structural Engineering*, American Society of Civil Engineers, Vol. 129, No. 8, August.

Tumialan, J.G. N. Galati, S. Namboorimadathil, and A. Nanni, 2002, "Strengthening of Masonry with FRP Bars," *Proceedings of the Third International Conference on Composites in Infrastructure ICCI'02*, June 10-12, San Francisco.

Tumialan, J.G., A. Morbin, F. Micelli, and A. Nanni, 2002, "Flexural Strengthening of URM Walls with FRP Laminates, *Proceedings of the Third International Conference on Composites in Infrastructure ICCI'02*, June 10-12, San Francisco, CA.

Vandergrift, Gergely, and Young, 2002, "CFRP Retrofit of Masonry Walls," *Proceedings of the Third International Conference on Composites in Infrastructure ICCI'02*, June 10-12, San Francisco, CA.

Velazquez-Dimas, J.I. and M.R. Ehsani, 2000, "Modeling Out-of-Plane Behavior of URM Walls Retrofitted with Fiber Composites," *Journal of Composites for Construction*, ASCE 4(4), pp. 172-181.

Velazquez-Dimas, J.I., M.R. Ehsani, and H. Saadatmanesh, 2000, "Out-of-Plane Behavior of Brick Masonry Walls Strengthened with Fiber Composites," *ACI Structural Journal*, 97(3), American Concrete Institute, Farmington Hills, MI, pp. 377-387.

Chapter 22 - Diaphragm Rehabilitation Techniques

22.1   Overview

Diaphragm failures are less commonly observed in earthquakes, and the disruption caused by strengthening the diaphragm can be quite significant, so diaphragm rehabilitation is less commonly employed than adding global strength and stiffness, or improving connection paths. Some diaphragms are inherently less likely to be an issue, such as cast-in-place concrete flat slabs or waffle slabs; others like straight sheathed wood or poorly connected precast floors are of greater concern.   This chapter provides examples of various diaphragm systems and their strengthening techniques. They are organized here in a single chapter for convenience and because many of the diaphragms can be found in different building types. For discussion of diaphragm-to-wall connection issues, see individual building type chapters.

22.2   Detailed Description of Diaphragm Rehabilitation Techniques

22.2.1   Wood Diaphragm Strengthening

Deficiency Addressed by Rehabilitation Technique

Inadequate diaphragm strength and/or stiffness

Description of the Rehabilitation Techniques

The addition of new wood structural panel sheathing is a traditional and common approach to diaphragm strengthening. Adding fastening and blocking to existing wood structural panel sheathing can also be done. Specifically, this section covers:

> Replacing existing sheathing with new wood structural panel sheathing
> Wood structural panel sheathing overlays with new blocking
> Wood structural panel sheathing overlays without new blocking
> Improving strength and stiffness of an existing wood structural panel sheathed diaphragm

Each of these techniques aims to improve the shear strength and lateral stiffness of the existing diaphragm. Figure 22.2.1-1 shows the replacement of existing sheathing with new sheathing directly onto the existing joists. Figure 22.2.1-2 shows a wood structural panel overlay on existing straight sheathing floors or roofs when new blocking is added below the existing sheathing. Figure 22.2.1-3 shows an overlay when blocking is not added, and Figure 22.2.1-4 shows a similar overlay to use when the bottom of the existing sheathing is to remain exposed to view and penetrations through it would not be acceptable. Figure 22.2.1-5 shows how shear transfer can be made to get past an existing partition sill that is to remain in place.

E.N.

2x BLKG

16d AT 6" O.C.
STAGGERED

(E) 2x

E.N.

PANEL
JOINT

2x4 FLAT BLKG

USE $\boxed{A}$ INSTEAD OF $\boxed{B}$
AT CONTRACTOR'S OPTION

SECTION $\boxed{A}$

SECTION $\boxed{B}$

PANEL JOINT

E.N.

2x4 FLAT BLKG

WOOD STRUCTURAL
PANEL

E.N.

SHAPE TOP OF
BLKG. TO MATCH
SLOPE

(E) TRUSS. VERT.
MEMBER WHERE
OCCURS

(E) TRUSS TOP
CHORD OR RAFTER

SECTION $\boxed{C}$

SECTION $\boxed{D}$

WOOD STRUCTURAL PANEL

24" MIN.
PANEL
WIDTH,
TYP.

FACE
GRAIN

STAGGER NAILS AT
ADJACENT PANEL
EDGES

B.N. ALL BOUNDARIES

RIDGE LINE

(E) TRUSS VERT.
MEMBER, TYP.

E.N. ALL PANEL ENDS

FIELD NAIL AT ALL
INTERMEDIATE
BEARINGS

(E) 2x TRUSS TOP
CHORD OR RAFTER

STAGGER PANEL
END TYP.

INSTALL SHTS
2 SPAN MIN.

Notes:

1. *Remove existing sheathing.*
2. *Use 3x and 4x blocking for higher capacity if necessary.*
3. *Use of $\boxed{A}$ requires wood structural panels to be field cut to fit existing joist spacing. Detail $\boxed{B}$ permits nominal wood structural panel size.*

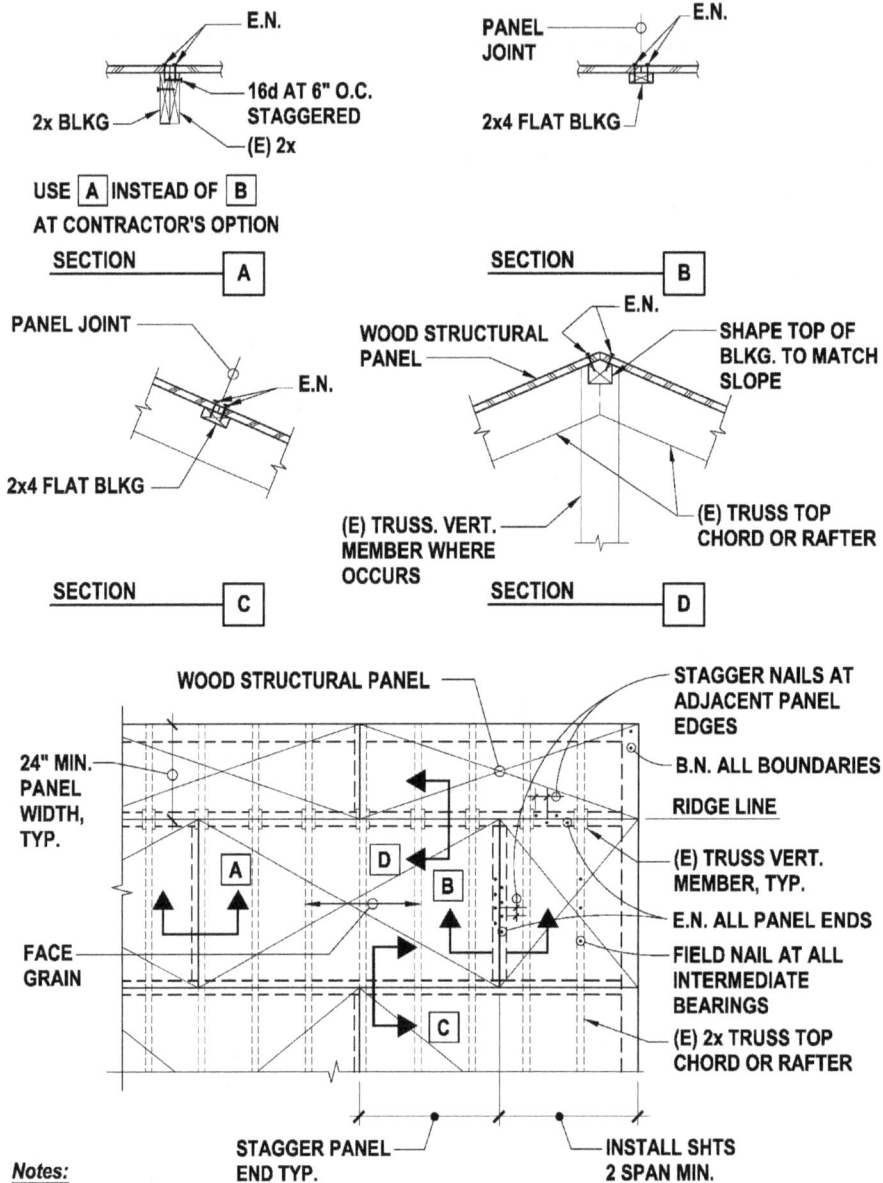

**Figure 22.2.1-1: Remove and Replace Existing Wood Sheathing
with Wood Structural Panel at a Roof**

ALTERNATE BLOCKING LOCATION [A]

PANEL JOINT

(E) SHEATHING

(E) JOIST

PREFERRED BLOCKING LOCATION [B]

[C]

BUTT PANEL EDGES
ON (E) RAFTERS, TYP

[C]

24" MIN. PANEL
WIDTH, TYP.

FACE GRAIN

WOOD STRUCTURAL
PANEL, TYP.

STAGGER NAILS AT
ADJACENT PANEL EDGES

B.N. ALL BOUNDARIES

E.N. ALL CONTINUOUS
PANEL EDGES

FIELD NAIL EACH DIRECTION
AT SPACING NOTED

E.N. ALL STAGGERED
PANEL ENDS

(E) SHEATHING

PROVIDE 2-16d AT ENDS & AT
INTERMEDIATE SUPPORTS OF
ALL (E) SHEATHING BOARDS
UNLESS NAILS ARE ALREADY
PROVIDED.

[A] OR [B]

(E) RAFTERS

STAGGER
PANEL END
TYP.

INSTALL
SHEETS
2 SPAN MIN.

PLAN

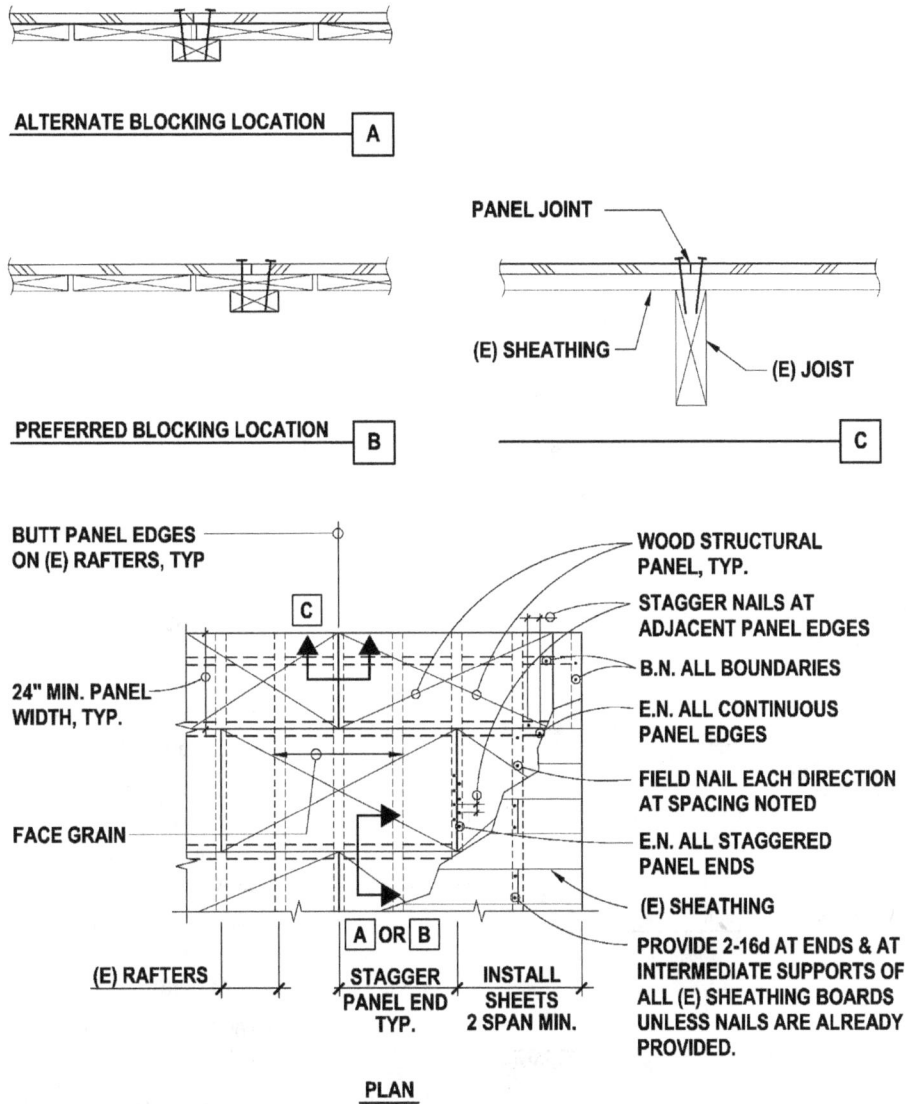

Figure 22.2.1-2: Wood Panel Overlay with Blocking Over Existing Sheathing

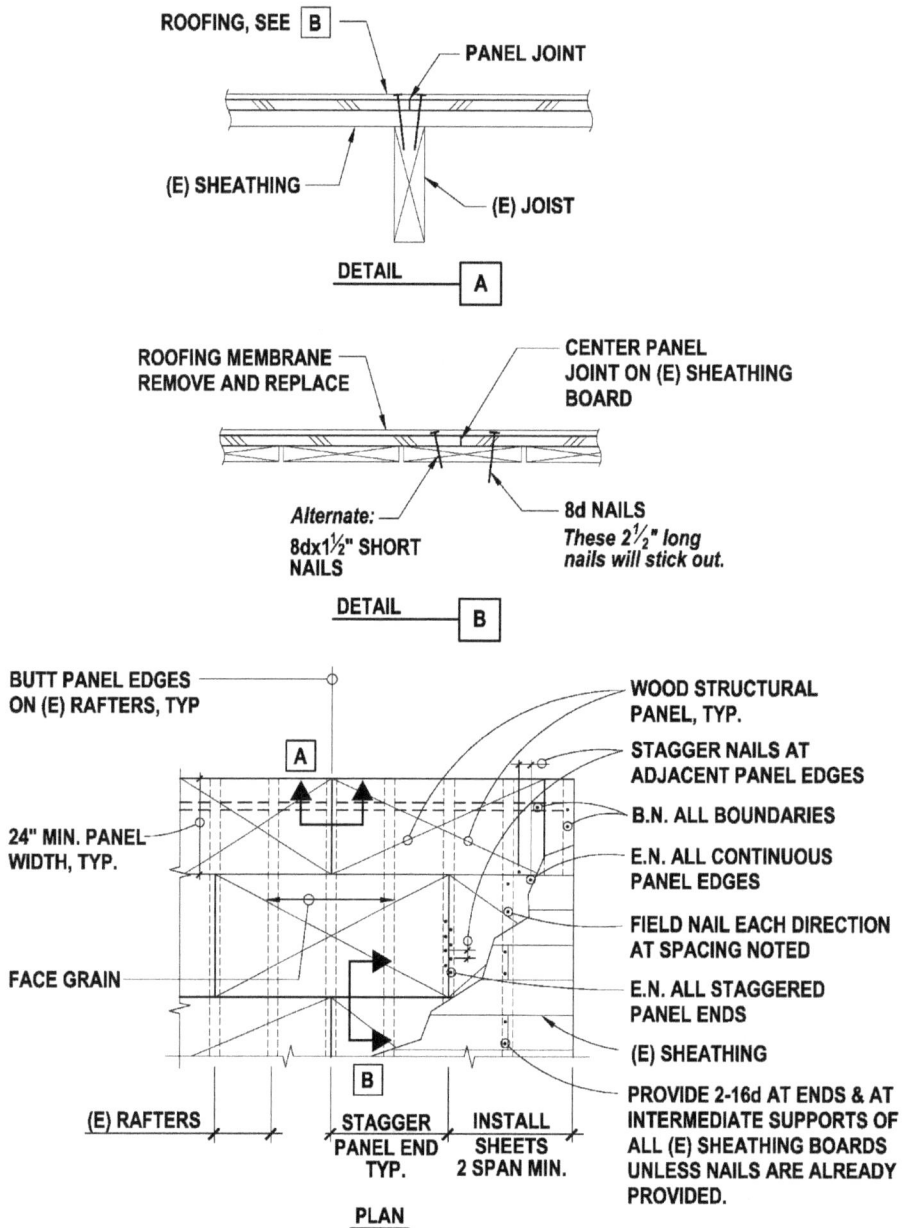

Figure 22.2.1-3: Wood Panel Overlay without Blocking Over Existing Sheathing

NAILS OR STAPLES

DETAIL A

CENTER PANEL
JOINT ON (E) SHEATHING
BOARD STAPLE ——————   ——STAPLE

Exposed face
of T&G sheathing
to be preserved.

DETAIL B

BUTT PANEL EDGES
ON (E) RAFTERS, TYP

A

24" MIN. PANEL
WIDTH, TYP.

FACE GRAIN

B

(E) RAFTERS

STAGGER
PANEL END
TYP.

INSTALL
SHEETS
2 SPAN MIN.

WOOD STRUCTURAL
PANEL, TYP.

STAGGER NAILS AT
ADJACENT PANEL EDGES

B.N. ALL BOUNDARIES

E.N. ALL CONTINUOUS
PANEL EDGES

FIELD NAIL EACH DIRECTION
AT SPACING NOTED

E.N. ALL STAGGERED
PANEL ENDS

(E) SHEATHING

PROVIDE 2-16d AT ENDS & AT
INTERMEDIATE SUPPORTS OF
ALL (E) SHEATHING BOARDS
UNLESS NAILS ARE ALREADY
PROVIDED.

PLAN

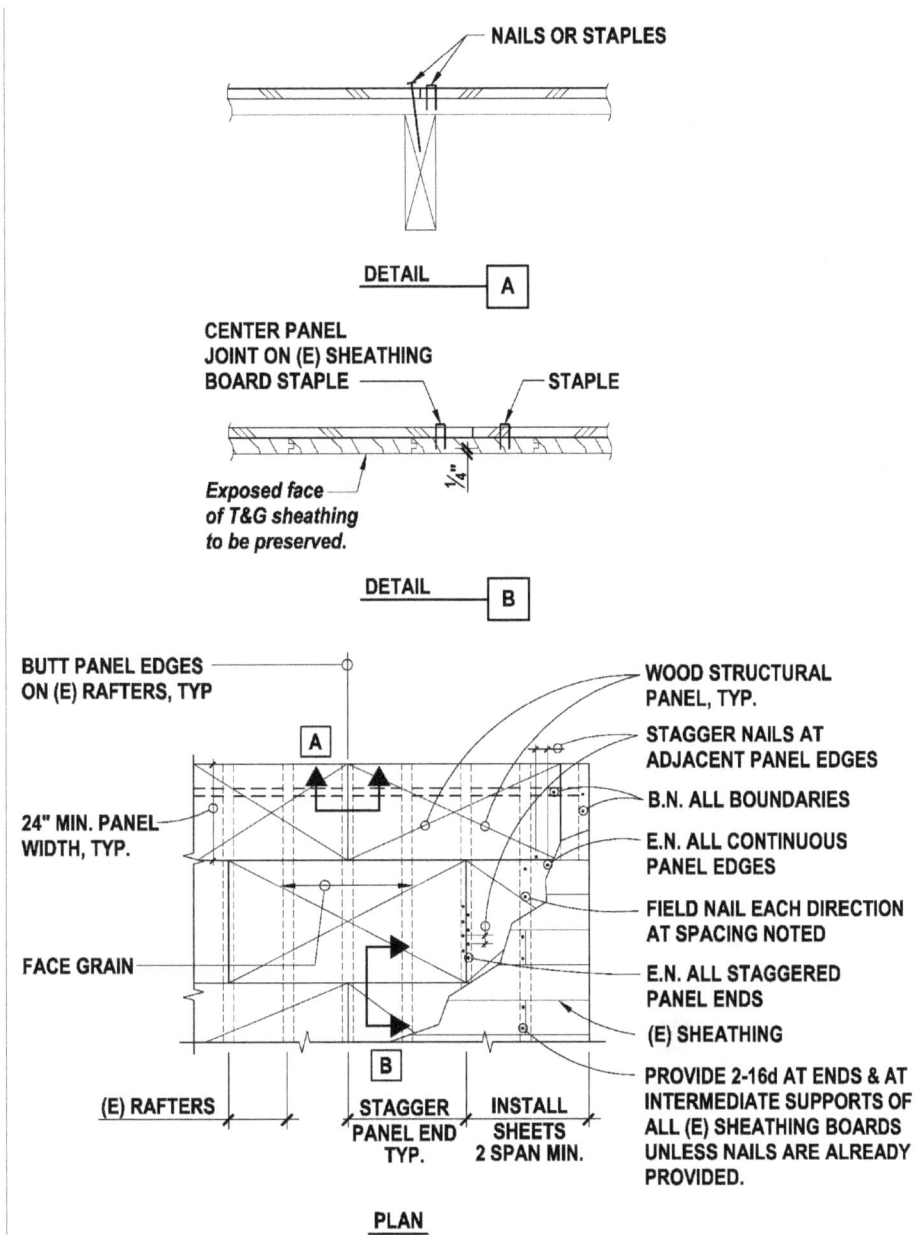

**Figure 22.2.1-4: Wood Panel Overlay without Blocking Over Existing Sheathing
When the Bottom of the Existing Sheathing is Visible**

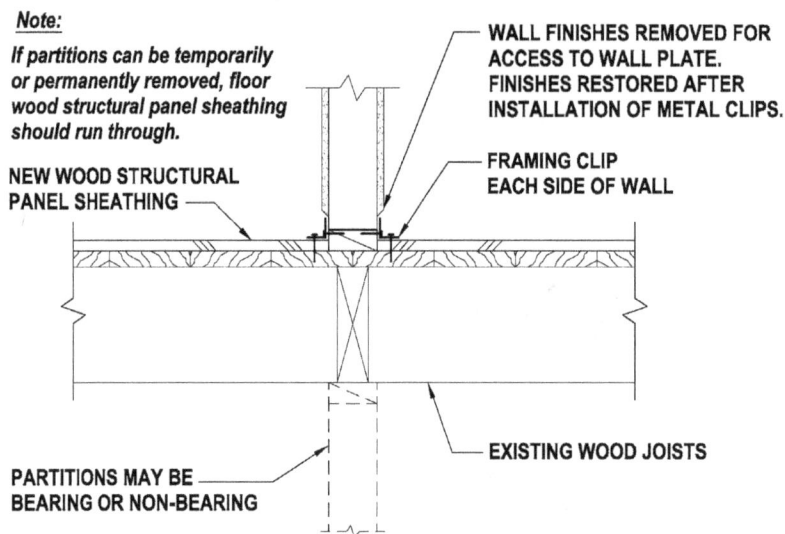

Note:

If partitions can be temporarily or permanently removed, floor wood structural panel sheathing should run through.

NEW WOOD STRUCTURAL
PANEL SHEATHING

WALL FINISHES REMOVED FOR
ACCESS TO WALL PLATE.
FINISHES RESTORED AFTER
INSTALLATION OF METAL CLIPS.

FRAMING CLIP
EACH SIDE OF WALL

EXISTING WOOD JOISTS

PARTITIONS MAY BE
BEARING OR NON-BEARING

Note: This detail is for transfer of relatively low shear forces. When boundary nailing and higher capacity are needed, alternate details are required.

SECTION

Figure 22.2.1-5: Shear Transfer in New Overlay at Existing Partitions

Design Considerations

Research basis: When new wood structural panel sheathing replaces existing sheathing, then the basic research for panel sheathing used to develop diaphragm capacities is applicable, and values would be taken from the relevant building code. When structural panel sheathing is used as an overlay, there is less research available. Values that have made it into model codes such as the UCBC (ICBO, 1997) and IEBC (ICC, 2003b) are based in part on the ABK research program for URM bearing wall strengthening, including ABK (1981). In this program, a series of 14 full-scale, 20'x60' horizontal diaphragm specimens were subjected to quasi-static, cyclic, in-plane displacements and dynamic, in-plane earthquake shaking. Specimens include filled and unfilled steel deck, blocked and unblocked plywood, and straight and diagonal sheathing with and without plywood overlays and with roofing material. More recent tests include Peralta, Bracci, and Hueste (2004) where a series of twelve 12'x24' horizontal diaphragm specimens were subjected to quasi-static, reversed cyclic in-plane displacements. Specimens included tongue groove sheathing retrofit with strapping and with an underlying steel truss, straight sheathing with and without openings retrofit with a steel truss and with blocked and unblocked plywood overlays. Results were compared with both FEMA 273 (1997a) and FEMA 356 (2000).

Types of diaphragms: Approaches to diaphragm rehabilitation can be categorized as follows:

Structural wood panel sheathing where the existing sheathing is replaced: This is the approach typically used when high capacities are needed.

- o "High load" diaphragms where 3x and 4x blocking is added and multiple lines of nailing are used: This may be done in accordance with provisions in the IBC; additional detailing information in ICC-ES Legacy Report 1952 (ICC-ES, 2004) is highly recommended. See APA (2000) for testing results.
- o Traditional diaphragms with 3x and 2x blocking and various panel layouts: The relevant building code capacities are used. An issue that often arises is whether existing joists, which are typically thicker than the code assumed 1-1/2", can count as 3x blocking. Some engineers ratio values between 2x and 3x code capacities.
- o Unblocked diaphragms: It is relatively unusual to remove existing sheathing only to replace it with unblocked wood structural panels as the capacities are not substantially different.

Wood structural panel sheathing overlays over existing 1x nominal sheathing: In the 1997 UCBC, there are values given for the following three approaches. The 2003 IEBC only lists the first type. Inherent in these approaches is the assumption that existing lumber sheathing is one-inch nominal (commonly 5/8-inch to 7/8-inch actual) thickness.

- o Wood structural panel overlays nailed directly over existing straight sheathing with ends of the panels bearing on joists or rafters and edges of the panels located on center of individual sheathing boards: The lack of blocking makes this a relatively weak diaphragm.
- o Wood structural panel overlays nailed directly over existing diagonal sheathing with ends of wood structural panel sheets bearing on joists or rafters: Diagonal sheathing provides increased strength compared to the overlay of straight sheathing.
- o Wood structural panel overlays nailed directly over existing straight or diagonal sheathing with ends of panels bearing on joists or rafters with edges of panels located over new blocking and nailed to provide a minimum nail penetration into framing and blocking of 1-5/8": The 1997 UCBC limits this to 75% of code values for wood structural panel overlays without the existing sheathing, due in part to the potential for bending of the nail in the existing sheathing before it reaches the main member blocking and the risk of the nailing being near the edges of the existing sheathing.

Wood structural panel sheathing overlays over existing lumber planking (2-inch nominal or thicker) or laminated decking; the IBC (ICC, 2003a) and the AF&PA (2005) permit wood structural panel diaphragm sheathing to be fastened over solid lumber planking or laminated decking using full tabulated values for new construction. Inherent is the assumption that the sheathing nail will have a penetration of not less than 10 diameters (1-3/8 inches for 8d common and 1-1/2" for 10d common) into the planking or decking. Special attention is needed at all diaphragm boundaries to ensure shear transfer from the sheathing, through the planking or decking to the boundary members below.

Wood structural panel sheathing overlays over existing spaced (or skip) sheathing: A common roof framing system is to span 1x nominal boards across rafters. Building paper is placed on top of the boards and under the final roofing layer such as shakes or shingles. Wide spaces of several inches are left between the 1x boards both to save sheathing material and to permit air flow to help dry the roofing sandwich. This construction is the most flexible and the weakest type of existing wood diaphragm and has no code values. Wood structural panel overlays can be placed across the skip sheathing. Care should be taken to align the panel edges atop the spaced sheathing. Due to the 1x thickness of the spaced sheathing, full development of the nail will not be achieved. With the gaps between sheathing boards, two edges of the wood structural panels will not be blocked. 1x sheathing or wood structural panel nailing strips with matching thicknesses can be placed atop the rafters in the gap to serve as "blocking" at these edges. Direct code values for these overlays are not available, though some engineers use code values reduced down by the amount of actual vs. full nail development length. Alternatively, staples can be used to help address the shallow sheathing depth.

Wood structural panel sheathing overlays over existing wood structural panel sheathing: Two layers of wood structural panel sheathing have been tested and documented in APA (2000). The tested configuration used overlays at panel ends in high-load regions.

Existing wood structural panel diaphragm enhancement without overlays: A wide variety of rehabilitation measures are available for existing wood structural panel diaphragms that do not involve new overlays. These include:
   o   Addition of 2x wood blocking to an unblocked diaphragm (Dolan et al., 2003)
   o   Addition of sheet steel blocking to an unblocked diaphragm (APA, 2000)
   o   Addition of nailing to existing blocked diaphragm (allows limited improvement because framing member requirements change at closer nail spacing)
   o   Adding staples to existing wood structural panel diaphragm. Staples are designed to carry entire seismic unit shear
   o   Stapling of tongue and groove sheathing joints (APA, 2000).
   o   Addition of a wood structural panel soffit in local areas of high diaphragm shear (see Section 22.2.2)

Existing diaphragms without overlays: In the 1997 UCBC and 2003 IEBC, there are values for the following existing materials:
   o   Roofs with straight sheathing and roofing applied directly to the sheathing
   o   Roofs with diagonal sheathing and roofing applied directly to the sheathing
   o   Floors with straight tongue-and-groove sheathing
   o   Floors with straight sheathing and finished wood flooring with board edges offset or perpendicular: Values are relatively high for this combination
   o   Floors with diagonal sheathing and finish wood flooring: Values are also relatively high for this combination

FEMA 356 has its own extensive listing of diaphragm types, and there are examples and even tests in the literature exploring the influence of glue, double layers of panel sheathing, herringbone panel overlays. IBC, APA (2000), and ICC-ES (2004) provide techniques for calculating code level values, including stapled diaphragms.

In order to select and properly detail diaphragm rehabilitation measures, it is important to determine the layout and thickness of existing sheathing and framing. Significant attention is needed to transfer of shear at all diaphragm boundaries. This includes diaphragm chords (Section 22.2.2), subdiaphragms and cross-ties for flexible diaphragm/rigid wall buildings (Section 22.2.3), and collectors (Sections 6.4.5 and 7.4.2).

Condition assessment of the existing roof structure is important. It is common to find decay damage to existing framing and sheathing in the vicinity of roof drains.

Detailing and Construction Considerations

Detailing and construction considerations for wood diaphragm strengthening include the following.

Aligning panel edges: When the existing sheathing is removed, the joists or rafters typically remain in place. Their spacing will vary. To align the edges of new 4'x8' sheets of structural wood panels on top of the supporting framing requires field measuring and cutting the sheets. Alternatively, new blocking can be added between existing framing to reduce the need to cut the structural wood panels. See Figure 22.2.1-1 for examples of each approach.

Missing sheathing edges: To reduce the risk of splitting during installation or later during the earthquake, nailing through the center of existing joists is desirable. This can take considerable field effort, however, due to the need to field measure and cut the structural wood panels. See Figure 22.2.1-2 and 22.2.1-3 for examples.

Staples, short nails, regular length nails: When the existing sheathing is removed and the structural wood panel is placed directly on the framing, regular length nails are commonly used. When the structural wood panel is applied to the existing lumber sheathing without blocking, 8d and 10d nails will go well through the underside of the sheathing. "Short" or "diaphragm" nails can be used to reduce the amount of nail protrusion. See Figure 22.2.1-3. When the overlay is on a diaphragm that is architecturally exposed from below, nail penetrations are not desirable. Staples can also be used, such as shown in Figure 22.2.1-4; per IBC, 16 gage staples require one-inch penetration into framing for tabulated values.

The nail penetration into diaphragm framing members required to achieve code and standard tabulated allowable shear values has changed recently. In the past, a nail penetration of 1-5/8 inches was required to obtain full diaphragm capacity. As a result, allowable shear reductions were applied when only 1-1/2 inch penetration was provided, as commonly occurs with 2x flat blocking in diaphragms or engineered joist top chords. The 2003 IBC only requires 1-3/8-inch penetration for 8d common nails and 1-1/2-inch penetration for 10d common nails. APA T98-22 (APA, 1998) provides one explanation, based on calculation using yield-mode equations. The nail penetration requirements are stated specifically in the diaphragm tables, and methods to adjust for reduced penetration are not suggested. Reduction in penetration below the IBC minimums is not recommended; because considerable slip can occur between sheathing and framing as a diaphragm takes up load, reduced embedment may lead to premature withdrawal failure. These APA and IBC penetration requirements are applicable to sheathing-to-framing fastening.

Nail penetration requirements have also been changing in the NDS (AF&PA, 2005), where a nail penetration of 10 diameters is now adequate to develop tabulated nail capacities. This number has been 12 and 11 diameters in previous provisions. Nails with a penetration of less than six diameters are not permitted to be used. These NDS penetration requirements are applicable to framing-to-framing fastening.

Gluing of diaphragms: Adding glue between a wood structural panel and supporting framing in a diaphragm or shear wall assembly where inelastic behavior is anticipated is strongly recommended against, as glued sheathing has limited ductility or energy dissipation capacity. This applies whether or not nailing is provided in addition to the glue. Dolan et al. (2003) evaluated the effect of diaphragm gluing on strength and stiffness.

Partitions: A diaphragm that is continuous between walls provides the stiffest and most direct load path. In an existing building, however, there are almost always existing partitions on the floor. If they are to remain during the rehabilitation, Figure 22.2.1-5 shows a detail for shear transfer from one side to the other of the partition sill in an overlay. This approach is adequate when the value of the load transfer is relatively low; when higher capacities are needed such as for boundary nailing or double rows of nails, alternative details will need to be developed and typically include blocking down and around the partition.

Weight: Adding structural wood panel sheathing over existing sheathing adds weight to diaphragm. This rarely poses a problem, but the engineer should consider the issue.

Location of diaphragm: Figures 22.2.1-1 through 22.2.1-5 all show the structural wood panel added to the top of the floor. In many situations, due to finishes on the top of the floor or usage of a particular story, enhancing the underside of the diaphragm is a less disruptive approach.

Cost/Disruption

Adding structural wood panel overlays can be a significant disruption to occupants, just from the need for access to either the top or underside of the floor, as well as from the noise of sawing and hammering. If the building is to remain occupied during rehabilitation, work is sometimes phased by floor or wing to minimize the number of impacted occupants at any one time. Many existing buildings have had roof strengthening done from above with the occupants in place. Sometimes the work is limited to certain hours that are considered less disruptive. When improvements or overlays are installed on top of the roof, it may be necessary to develop detailing to allow work around existing roof top equipment platforms and curbs, skylights, etc.

Proprietary Issues

There are typically no proprietary concerns with wood diaphragm strengthening.

22.2.2   Add or Enhance Chord in Existing Wood Diaphragm

Deficiency Addressed by Rehabilitation Technique

This rehabilitation technique addresses inadequate, incomplete or missing chords in buildings with reinforced concrete or masonry shear walls; also addressed is inadequate shear transfer into chord members. Provision of chord members is specifically not required for diaphragms in unreinforced masonry buildings, where wall bed joint shear is thought to provide some chord member capacity. See Chapter 21 for additional discussion of URM buildings.

Rehabilitation approaches discussed may also be applicable to collectors and detailing at re-entrant corners. While systematic evaluation may identify the need for chord enhancement, it is also often provided in conjunction with diaphragm enhancement, as discussed in Section 22.2.1.

Description of the Rehabilitation Technique

The purpose of a diaphragm chord is to act as a tension or compression member resisting diaphragm flexural forces; this requires both an adequate member and adequate transfer of shear from the diaphragm to the chord member along the full member length. In buildings with wood diaphragms and reinforced concrete or masonry walls, the most common chord members are reinforcing steel placed in the wall at or near the roof diaphragm elevation and a structural steel angle bolted to the wall.

Where the existing chord member is adequate, rehabilitation may be limited to enhancing shear transfer. Figures 22.2.2-1A and 1B show added fastening at the roof diaphragm boundary and added adhesive anchors to the concrete or masonry wall, where the existing reinforcing steel is adequate. The reader is cautioned to check the adequacy of the reinforcing as-built conditions at tilt-up concrete walls and reinforced masonry walls with movement joints. See Chapters 16 and 19 for further discussion.

Where additional chord capacity is needed, it is most practical to add a new steel angle on the surface of the existing concrete wall, as shown in Figures 22.2.2-2A and 22.2.2-2B.

Diaphragm chords may be incomplete when vertical offsets occur in the roof diaphragm. When this occurs, it may be possible to use a tilt-up panel to resolve the vertical offset, as shown in Figure 22.2.2-3. Where chords are not occurring at the roof diaphragm level, care should be taken in assessing the unsupported length for compression design.

Design Considerations

Research basis: No research applicable to this rehabilitation measure has been identified.

Enhancement to an existing chord member must be compatible with existing chord behavior. It is unlikely that any chord enhancement applied to the wall face can be compatible with an existing reinforcing steel chord, because of fastener slip required to develop forces in the new chord member. Where an existing reinforcing steel chord is being enhanced, it is suggested that the capacity of the existing reinforcing be neglected.

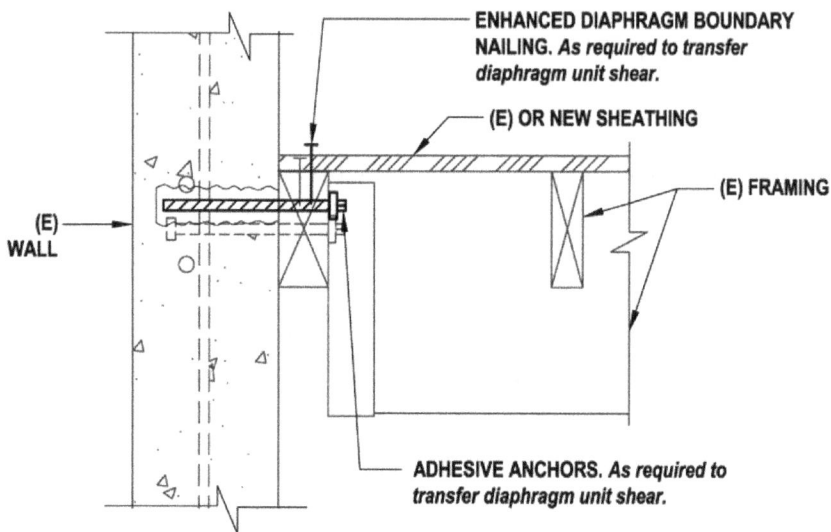

ENHANCED DIAPHRAGM BOUNDARY
NAILING. *As required to transfer
diaphragm unit shear.*

(E) OR NEW SHEATHING

(E) FRAMING

(E)
WALL

ADHESIVE ANCHORS. *As required to
transfer diaphragm unit shear.*

Note: *Load path from diaphragm sheathing to chord member must be adequate to transfer
diaphragm unit shear. Special attention is needed if diaphragm nailing varies along the chord
length. Anchorage for wall out-of-plane loads is not shown.*

SECTION
AT (E) WOOD LEDGER   [A]

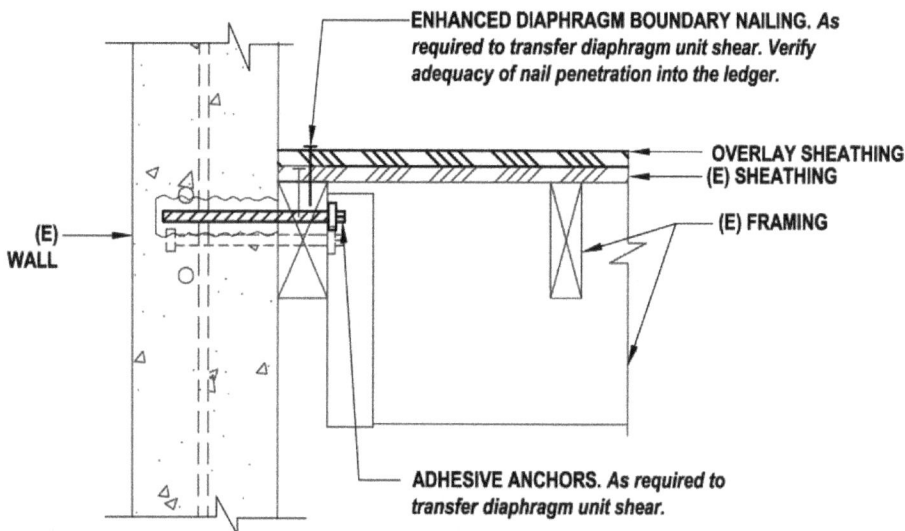

ENHANCED DIAPHRAGM BOUNDARY NAILING. *As
required to transfer diaphragm unit shear. Verify
adequacy of nail penetration into the ledger.*

OVERLAY SHEATHING
(E) SHEATHING

(E) FRAMING

(E)
WALL

ADHESIVE ANCHORS. *As required to
transfer diaphragm unit shear.*

Note: *Load path from diaphragm sheathing to chord member must be adequate to transfer
diaphragm unit shear. Special attention is needed if diaphragm nailing varies along the chord
length. Anchorage for wall out-of-plane loads is not shown.*

SECTION
AT SHEATHING OVERLAY   [B]

Figure 22.2.2-1: Enhanced Chord Member Fastening at Wood Diaphragm

Alternate location for chord
if roofing is to be removed.

(E) SHEATHING

(E) FRAMING

(E)
WALL

STEEL ANGLE CHORD
WITH ADHESIVE
ANCHORS

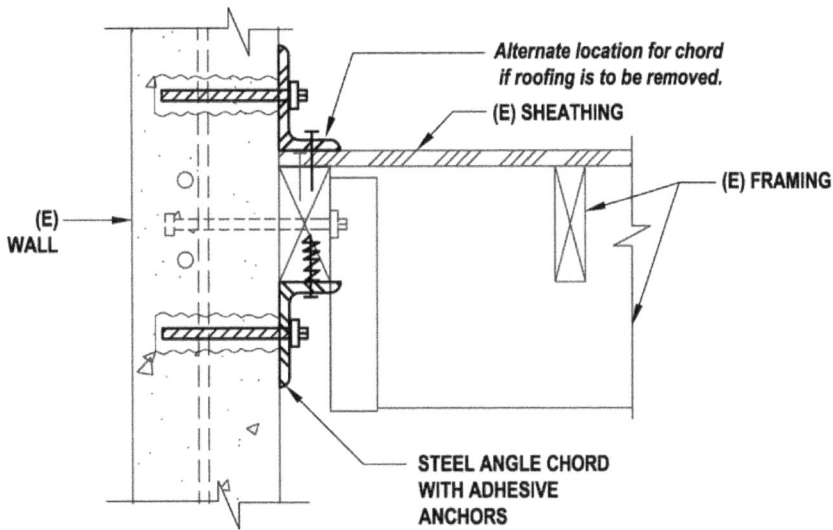

Note: Load path from diaphragm sheathing to chord member must be adequate to transfer
diaphragm unit shear. Special attention is needed if diaphragm nailing varies along the chord
length. Verify grout location at partially grouted wall. Anchorage for wall out-of-plane loads
not shown.

SECTION
AT (E) WOOD LEDGER   A

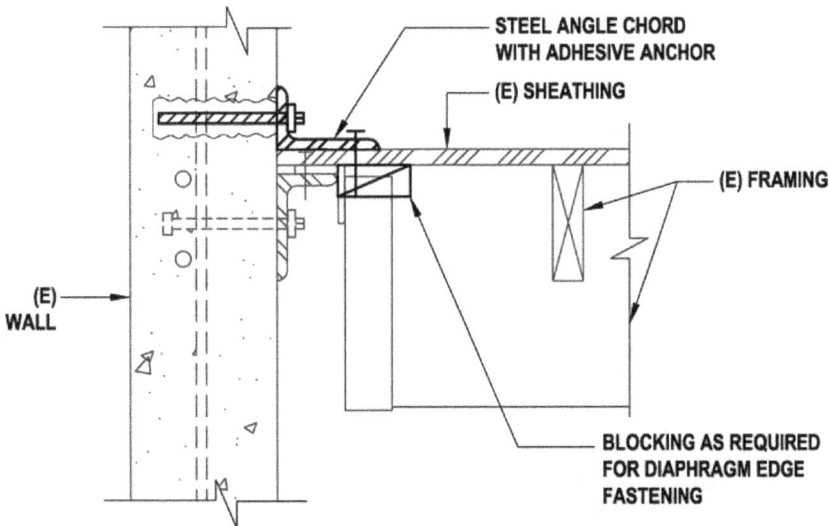

STEEL ANGLE CHORD
WITH ADHESIVE ANCHOR

(E) SHEATHING

(E) FRAMING

(E)
WALL

BLOCKING AS REQUIRED
FOR DIAPHRAGM EDGE
FASTENING

Load path from diaphragm sheathing to chord member must be adequate to transfer diaphragm
unit shear. Special attention is needed if diaphragm nailing varies along the chord length. Verify
grout location at partially grouted wall. Anchorage for wall out-of-plane loads not shown.

SECTION
AT (E) STEEL LEDGER   B

Figure 22.2.2-2: Enhanced Chord Member and Fastening at Wood Diaphragm

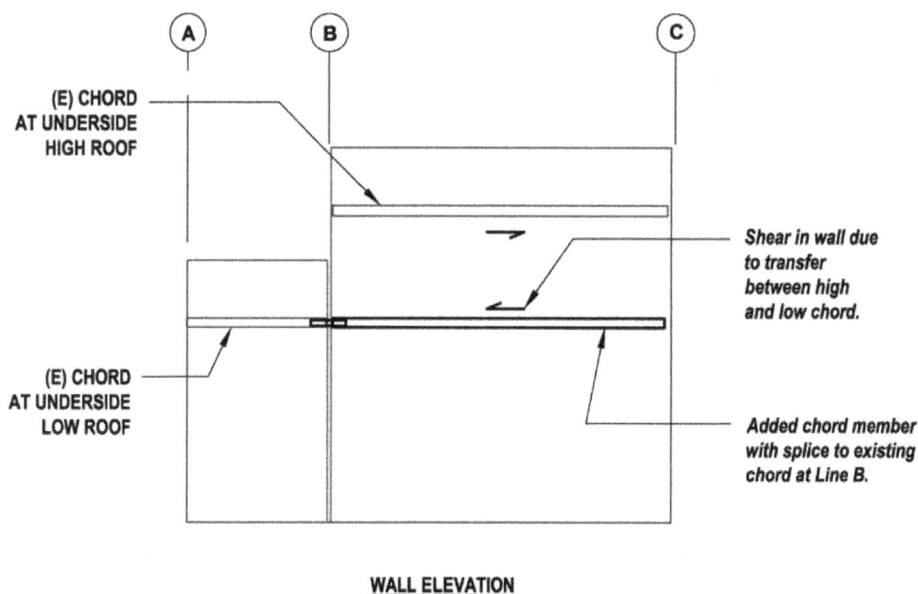

WALL ELEVATION

**Figure 22.2.2-3: Elevation of Wall Panels with Incomplete Chord
Due to Vertical Offset in Roof Diaphragm**

The new or enhanced chord member must be anchored into the diaphragm for unit shear transfer. Anchorage for shear transfer is also discussed in Section 22.2.3. As a wood sheathed diaphragm is loaded, slip will occur between the perimeter framing member and the sheathing. Fastening of the chord or chord enhancement should not inhibit this slip. If the slip is not permitted, premature failure at the opposite side of the sheathing panel could occur. This is not a concern with a welded steel deck diaphragm, which has limited slip.

Chord stresses due to shrinkage and temperature change have been identified as a concern for connections between tilt-up panels (SEAOSC, 1979), as discussed in Chapter 16, and these stresses should be considered in chord design.

Detailing Considerations

It is desirable to keep the chord elevation as close as possible to the elevation of the diaphragm in order to minimize secondary stresses and additional deformation. At the edge of the diaphragm this is most easily accomplished by putting a new chord member on the top of the diaphragm (shown as an alternate location in Figures 22.2.2-2A and 22.2.2-2B). This is only possible when re-roofing will occur at the time of rehabilitation work. Otherwise added chord members must be located below existing perimeter members and connections. Splicing of the new or enhanced chord member needs to be specifically detailed.

Diaphragm boundary fastening: Based on observed shear wall test behavior (Gatto and Uang, 2002), providing extra nailing at the diaphragm boundary will likely not provide extra diaphragm capacity, and it may result in premature failure at the first interior joint due to shifting of the center of the fastener group. As a result, sheathing fasteners should be placed symmetrically around the panel edge where possible, and care should be taken to not arbitrarily put extra rows of fasteners at the boundary chord and collector members.

It is preferable to use the same type and size of sheathing fastener at the diaphragm boundary as at the diaphragm interior; however, this may be difficult where new steel chord members are being added on top of the diaphragm, as shown in the alternate location in Figures 22.2.2-2A and 22.2.2-2B. Although graphically shown as a nailed connection from the steel angle chord member to the diaphragm, it may become necessary to use wood screws or lag screws for higher-load diaphragms. Testing of this mix of fasteners has not been identified, so behavior is not known. Behavior of cut-thread wood screws in sheathing to framing fastening has been observed to be problematic, as discussed in Section 6.4.2.

Partially grouted masonry walls: Where shear transfer is being provided into partially grouted masonry walls, it is necessary to verify that the existing wall is grouted at the anchorage location. It is generally anticipated that the existing masonry will be grouted and reinforced at the existing roof ledger location. If, however, anchorage to the wall needs to occur above or below this location, presence of grout will need to be verified. Although methods of anchoring only to the face shell are available, these have very low capacities and should never be mixed with anchors to grouted masonry. So, it is recommended that anchorage to grouted cells be provided. It may be possible to grout cells at desired anchor locations, particularly if just above the roof line and accessible from at the parapet. Care should be taken so that the anchor force in a grouted cell does not exceed the force that can be transferred by the unit bed joint.

Collector connections: Where possible, it is desirable for the collector member to be located at the face of the shear wall and extend the full length of the shear wall, matching the chord detailing shown in Figures 22.2.2-1A and 22.2.2-1B and Figures 22.2.2-2A and 22.2.2-2B. This detailing approach is often but not always possible. Great care should be taken when a significant collector load needs to be transferred into the very end of a concrete or masonry wall. The load needs to be transferred far enough into the wall that wall reinforcing can develop adequate capacity. Edge and center to center spacing requirements need to be met for anchorage to the wall.

Cost, Disruption and Construction Considerations

When rehabilitation work is undertaken on the roof diaphragm, it is important that the cost and the preferred location for work take into account the combination of work, rather than considering one portion at a time. If several diaphragm measures will be undertaken, it will quickly become cost-effective to remove the roof and allow work from the top.

Proprietary Concerns

There are no proprietary concerns with this rehabilitation technique other than the use of proprietary connectors and adhesives as part of the assemblage.

22.2.3   Add or Enhance Diaphragm Cross-ties for Out-of-Plane Wall-to-Diaphragm Loads in Flexible Wood and Steel Diaphragms

Deficiency Addressed by Rehabilitation Technique

This rehabilitation technique addresses inadequate or missing diaphragm cross-tie systems, as part of wall anchorage requirements for flexible diaphragm / rigid wall buildings. This rehabilitation technique is used when diaphragm cross-tie systems have not been provided, or do not provide adequate strength. Both wood and steel flexible diaphragms are addressed. The diaphragm cross-tie system is an extension of wall to diaphragm anchorage for out-of-plane loads, as addressed in Chapter 16 for **PC1** buildings, Chapter 18 for **RM1t** buildings, and Chapter 21 for **URM** buildings.

The addition or enhancement of the diaphragm cross-tie system is recommended as a high priority for rehabilitation for wood diaphragm **PC1, RM1t,** and for **URM** buildings. Due to limited earthquake experience to date, the vulnerability of and need to rehabilitate cross-tie systems in flexible steel diaphragms is not known; however, vulnerabilities similar to wood diaphragms buildings might occur. This section illustrates the basic rehabilitation concepts. *SEAONC Guidelines* (SEAONC, 2001) provides exhaustive treatment of detailing for **PC1** buildings.

Description of the Rehabilitation Technique

A system of continuous ties between exterior walls of flexible diaphragm / rigid wall buildings is now a requirement for new construction in areas of high seismic hazard. The concept is to tie all the way across the diaphragm to opposing walls. The wall anchorage will generally occur at four, six or eight feet on center.

Cross-ties at each wall anchor location can be fairly easily accommodated in new steel deck diaphragm buildings. The steel deck is permitted to be used as the cross-tie in the direction of its span, provided it can be shown to be adequate for tension and compression forces. See Chapter 16 for further discussion. Perpendicular to the decking span, with relatively long-span steel joist members it is practical to provide diaphragm cross-ties at each joist. The number of cross-tie splices required is not excessive, and wall anchorage forces do not greatly change the open web joist design. This is also the preferred approach for rehabilitation of cross-ties in steel deck construction, where the forces can be accommodated by decking and joists.

Cross-ties at each wall anchor location are not as easily accommodated in wood diaphragm systems, particularly in panelized wood diaphragm systems with eight foot subpurlins spans, due to the number of breaks in framing members across which connectors would have to be provided. A cross-tie system using subdiaphragms has been developed for wood diaphragm buildings. This same approach can be used in steel diaphragm buildings. Rather than representing anticipated building behavior, subdiaphragms need to be viewed as a computational tool. Unit shears from subdiaphragm design are not intended to be added to main diaphragm shears. Design in each area of the diaphragm needs to be for the more critical of subdiaphragm or main diaphragm seismic forces.

Figure 22.2.3-1A illustrates a roof plan for a wood diaphragm that uses subdiaphragms as part of the cross-tie system. For loading in the east-west direction, subdiaphragms are provided between Lines A and B and Lines G and H. Similarly, for loading in the north-south direction, subdiaphragms are provided between Lines 1 and 2 and Lines 3 and 4. The depth of the subdiaphragm is selected based on the unit shear at the subdiaphragm reaction, as well as having a member available to act as a subdiaphragm chord. The wall anchor force is transferred into the subdiaphragm over the full subdiaphragm depth. For east-west loads subdiaphragms span between Lines 1 and 2, 2 and 3, and 3 and 4. Subdiaphragm reactions are resisted at the exterior walls at Lines 1 and 4, and interior cross-ties are provided on Lines 2 and 3. Boundary nailing must be provided for each subdiaphragm on Lines 1, 2, 3, 4, A and B. The cross-tie provides a continuous tie between exterior walls with a capacity not less than the subdiaphragm reaction. This pattern is repeated for subdiaphragms between Lines G and H, 1 and 2, and 3 and 4.

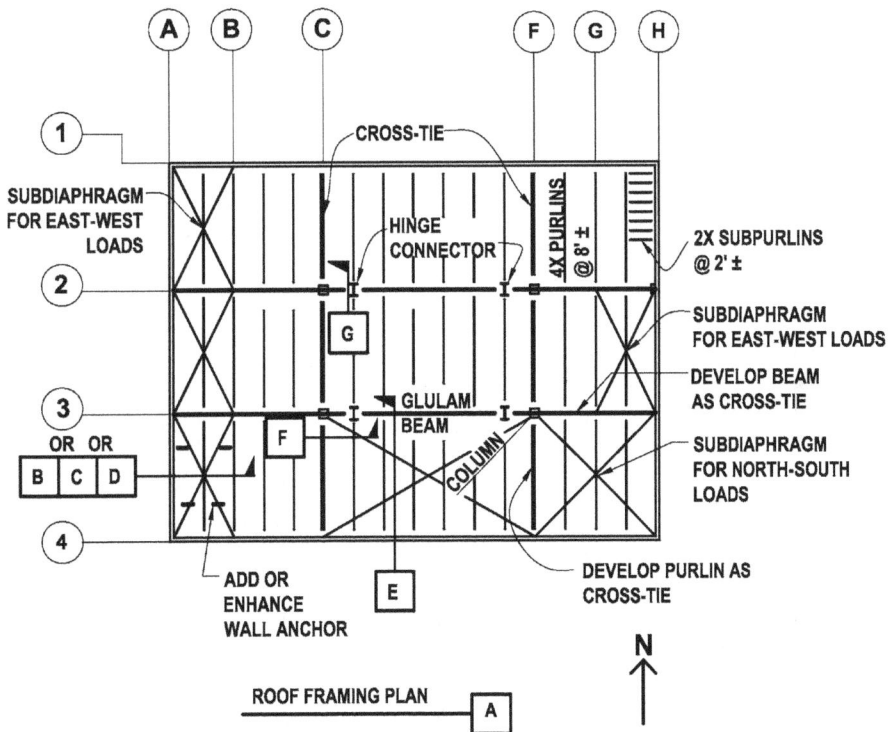

**Figure 22.2.3-1A: Roof Plan with Diaphragm Cross-Tie System
Using Subdiaphragms, Shown for Wood Diaphragm**

Figures 22.2.3-1B, 22.2.3-C, and 22.2.3-D depict sections through the subdiaphragm extending between Lines A and B. Figure 22.2.3-1B shows the assumed subdiaphragm where existing roof sheathing is not being modified. The subdiaphragm depth will be controlled by the capacity of the existing sheathing. The wall anchor engages each wall purlin across the subdiaphragm depth.

Existing subpurlin-to-sheathing nailing must be adequate to transfer the wall anchor force to the subdiaphragm. In Figure 22.2.3-1B the added wall anchor is located between existing subpurlins in order to engage more existing sheathing nailing. Sheathing fastening to subpurlins must be assumed to be field nailing unless edge nailing has been confirmed.

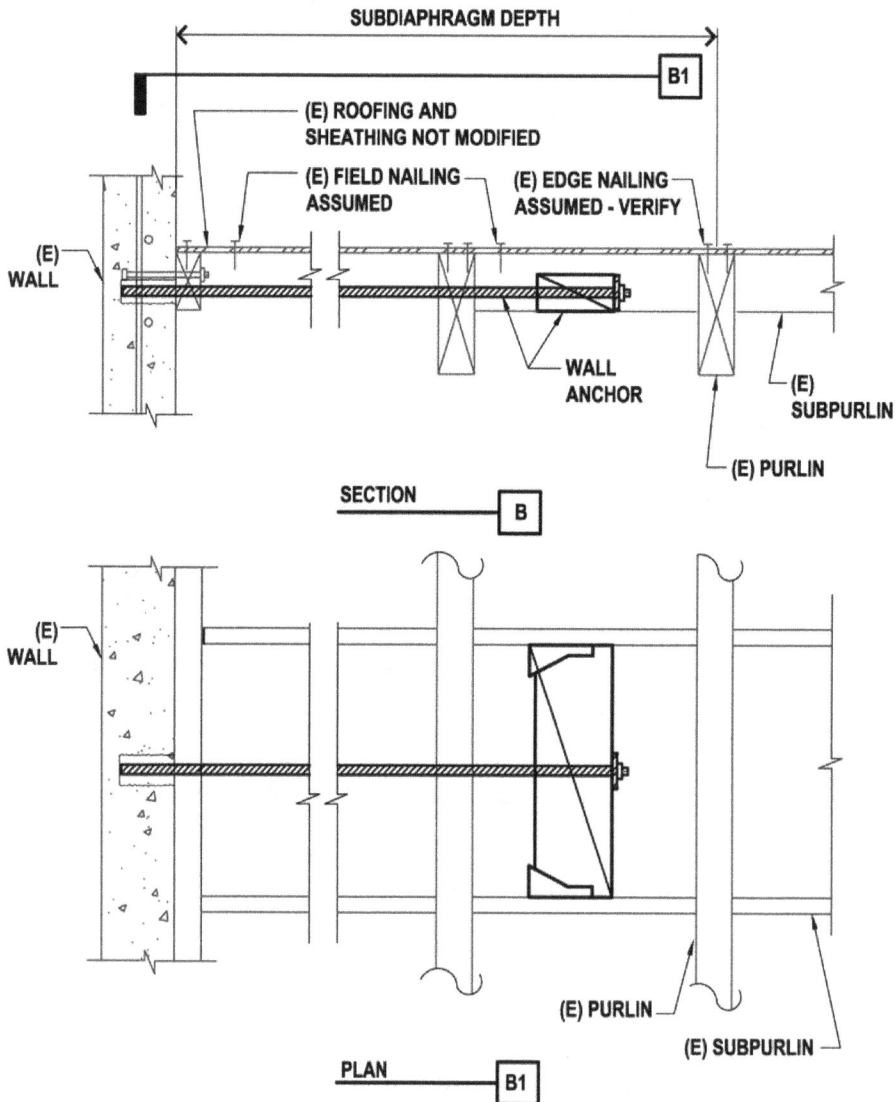

**Figure 22.2.3-1B: Subdiaphragm for Flexible Wood Diaphragm –
Roofing Not Removed**

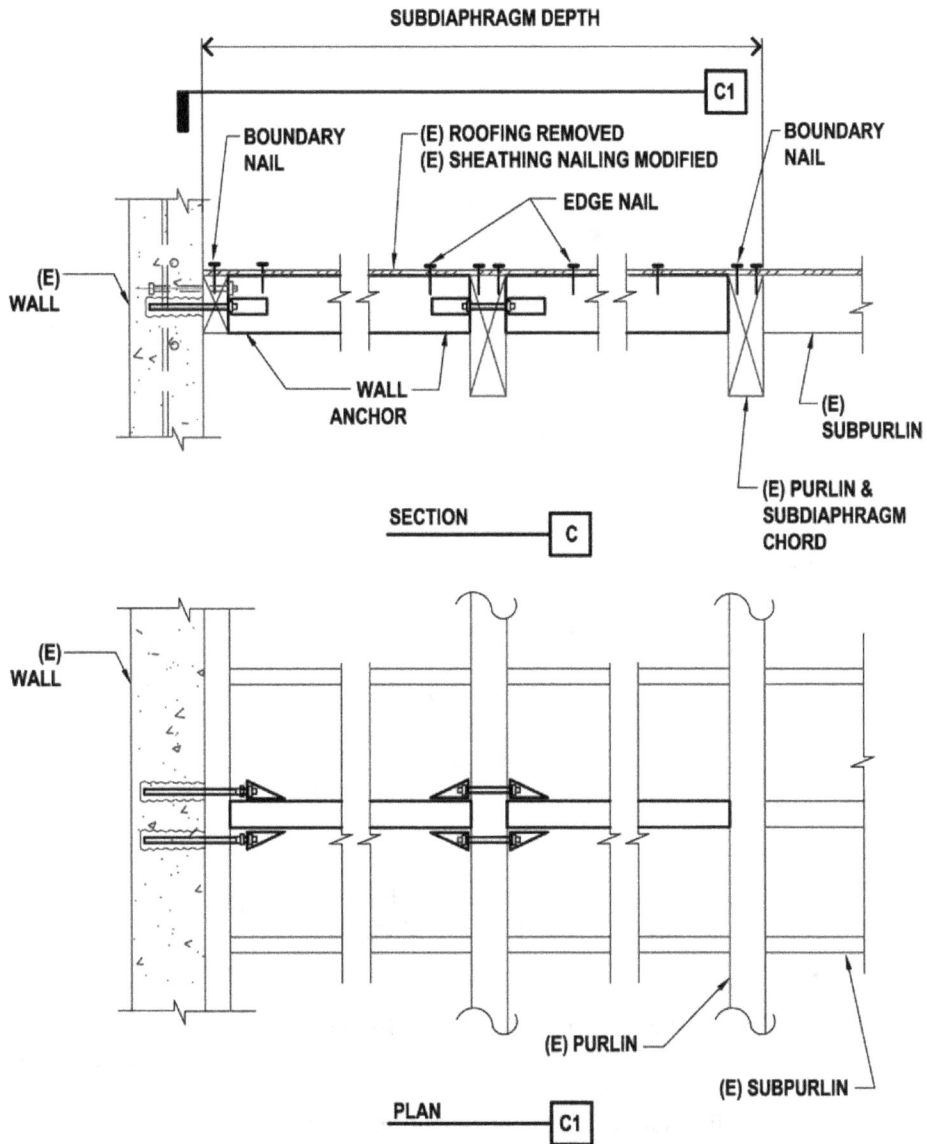

**Figure 22.2.3-1C: Subdiaphragm for Flexible Wood Diaphragm –
Roofing Removed**

SUBDIAPHRAGM DEPTH

(E)
WALL

(E) ROOFING AND SHEATHING NOT
MODIFIED

(E) FIELD NAILING
ASSUMED - VERIFY

(E) EDGE NAILING
ASSUMED - VERIFY

WOOD STRUCTURAL
PANEL SHEATHING
AT SOFFIT

(E) SUBPURLIN

(E) PURLIN

BOUNDARY AND FIELD
NAILING

CORE 2 HOLES EACH PURLIN
BAY TO ALLOW VENTILATION

SECTION   D

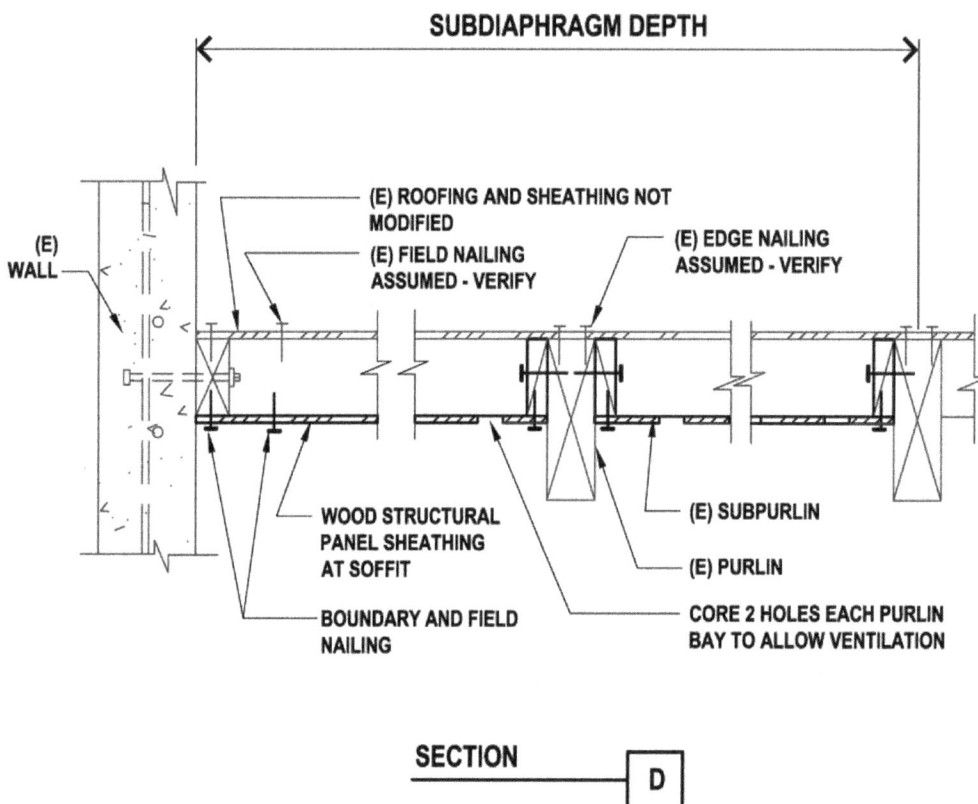

**Figure 22.2.3-1D: Enhanced Wood Subdiaphragm with
Added Wood Structural Panel Soffit**

Figure 22.2.3-1C depicts a subdiaphragm where access from the top is assumed, and the
subdiaphragm can be renailed to meet required demands. A new member is provided at the wall
anchor. Tie-downs are added to carry the wall anchorage force across the entire subdiaphragm
width. Figure 22.2.3-1D illustrates a third subdiaphragm alternative where new subdiaphragm
sheathing is provided as a soffit at the underside of the roof framing. Wall out-of-plane
anchorage is not shown, but would be similar to Figure 22.2.3-1C. Attention is needed to
providing shear transfer into the main diaphragm at all subdiaphragm boundaries. See other
chapters for additional discussion of wall anchorage.

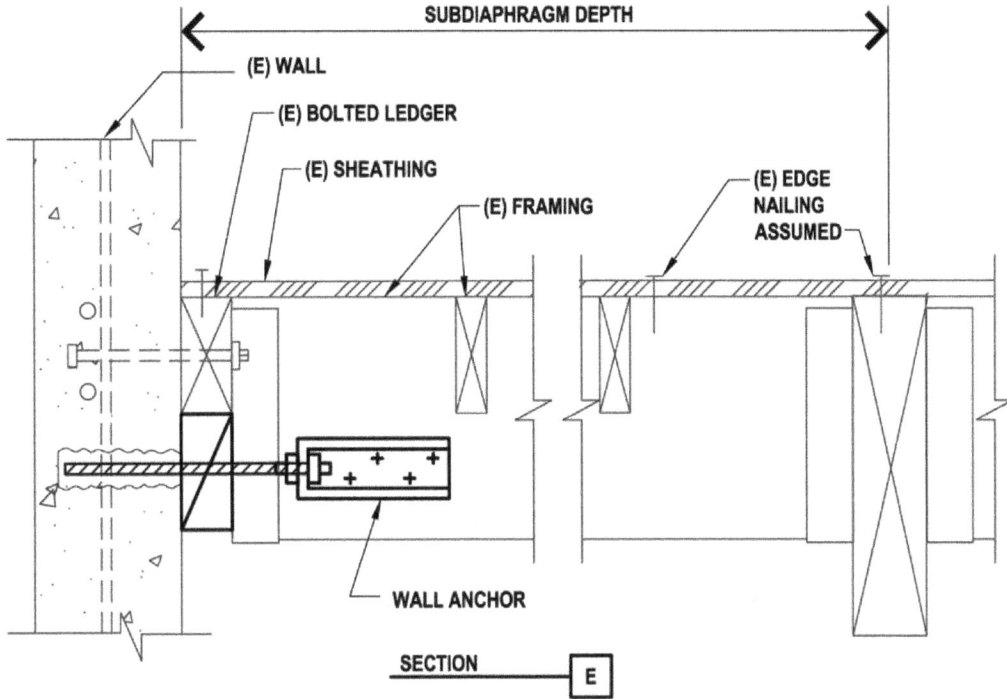

Figure 22.2.3-1E: Subdiaphragm for Flexible Wood Diaphragm at Purlins

Work can be conducted either from the underside or the top of the diaphragm. Location of access needs to be decided early on in the design process and will drive both calculations and detailing of the rehabilitation work. Where the roofing is not going to be removed, it is possible to strengthen the diaphragm in local areas by sheathing the underside of the roof subpurlins, as shown in Figure 22.2.3-1D. This is expensive and tedious work that should not occur over large areas, but may be advantageous for reinforcing of subdiaphragms in combination with wall anchorage.

Figure 22.2.3-1E illustrates anchorage of the north and south walls into subdiaphragms extending between Lines 1-2 and 3-4.

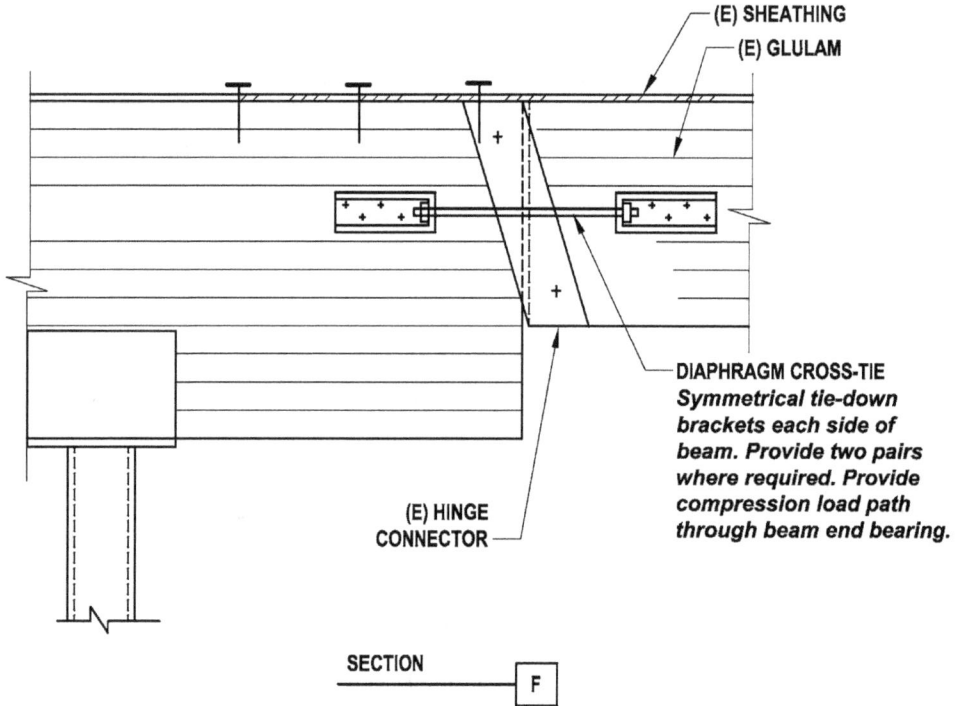

Figure 22.2.3-1F: Cross-Tie for Flexible Wood Diaphragm at Glulam Beams

Figure 22.2.3-2A illustrates a similar roof plan with a steel diaphragm. Figures 22.2.3-2B through 22.2.3-2D provide details. Instead of using subdiaphragms, direct ties are provided. Alternative connections locations for field welded connections between joists (Figures 22.2.3-2C and 22.2.3-2D) include the joist top chord, vertical and horizontal legs. The alignment of joists at support locations will greatly affect the connection detail used, so field determination of detail and alignment should be made. See Section16.4.1 for additional discussion.

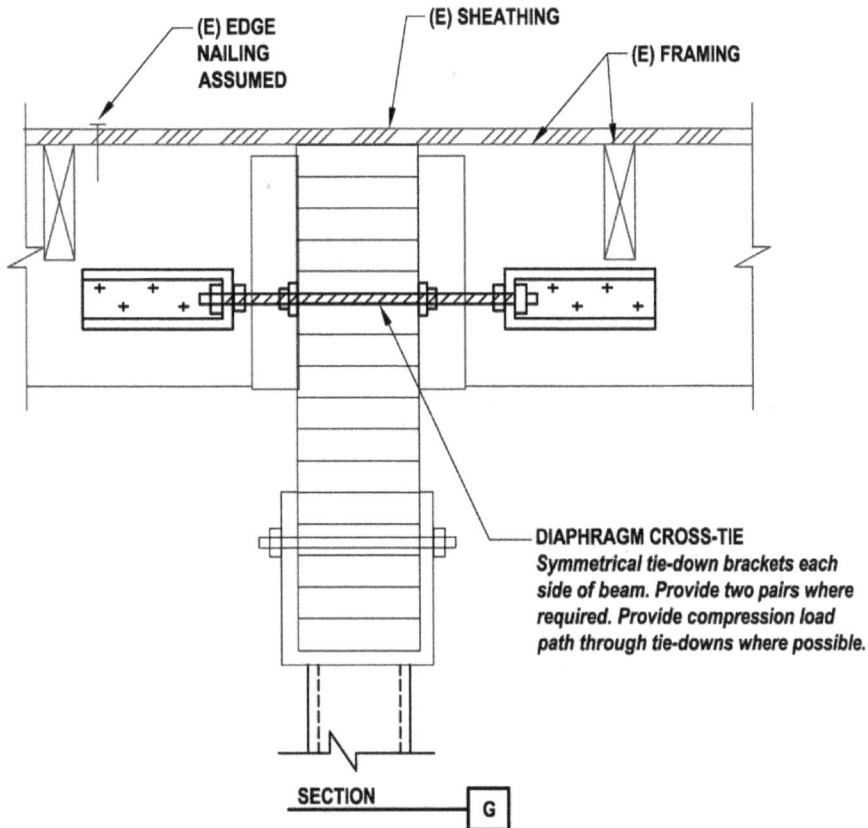

Figure 22.2.3-1G: Cross-Tie for Flexible Wood Diaphragm at Purlins

Design and Detailing Considerations

Research basis: No research relating to the performance or adequacy of enhanced anchorage methods has been identified; however, the demands created in flexible diaphragms have been studied by Fonseca, Wood and Hawkins (1996); Hamburger and McCormick (1994); and Ghosh and Dowty (2000).

The reader is referred to the extensive discussion in the *SEAONC Guidelines* for design and detailing considerations for the wood diaphragm.

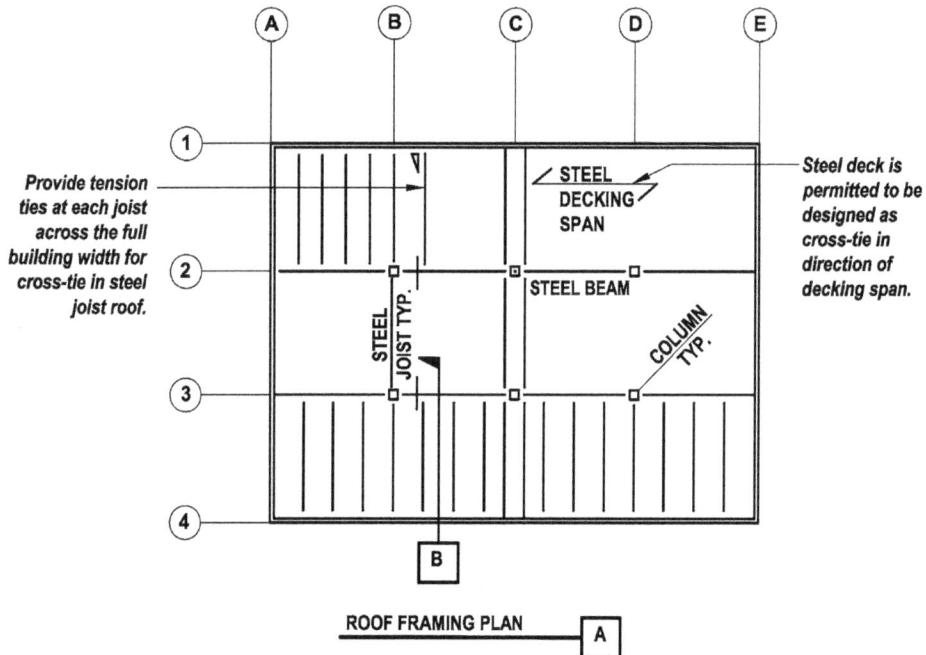

Figure 22.2.3-2A: Roof Plan with Diaphragm Cross-Tie System Using Direct Ties, Shown for Steel Diaphragm

Cost, Disruption and Construction Considerations

When rehabilitation work is undertaken on the roof diaphragm, it is important that the cost and the preferred location for work take into account the combination of work, rather than considering one piece at a time. If several diaphragm measures will be undertaken, it will quickly become cost-effective to remove the roof and allow work from the top. This is particularly true if a steel deck requires several rehabilitation measures.

Proprietary Concerns

There are no proprietary concerns with this rehabilitation technique other than the use of proprietary connectors and adhesives as part of the assemblage.

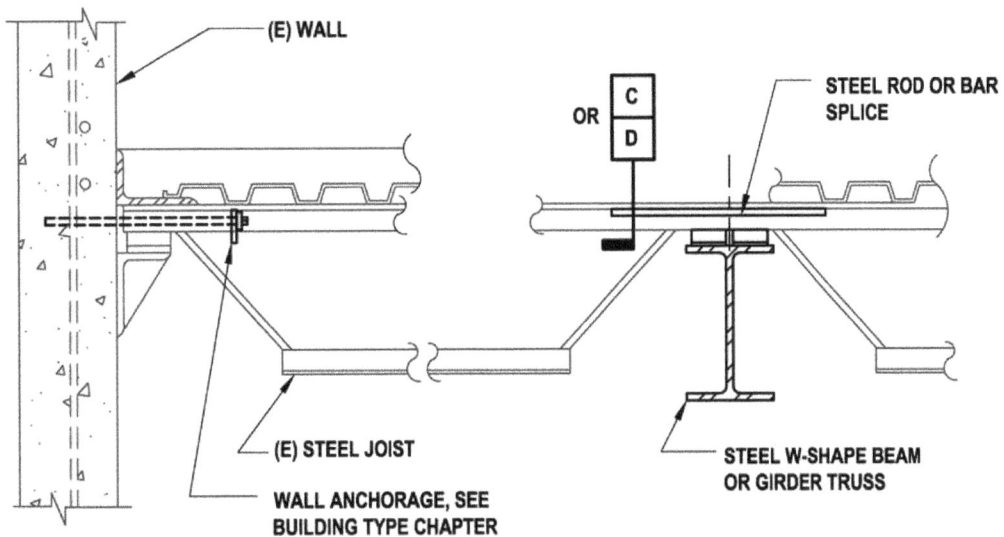

(E) WALL

OR

STEEL ROD OR BAR SPLICE

(E) STEEL JOIST

STEEL W-SHAPE BEAM OR GIRDER TRUSS

WALL ANCHORAGE, SEE BUILDING TYPE CHAPTER

Note: Verify that steel joist top chord is adequate for combined gravity and wall anchorage loads.

SECTION B

Figure 22.2.3-2B: Cross-Tie for Flexible Steel Diaphragm

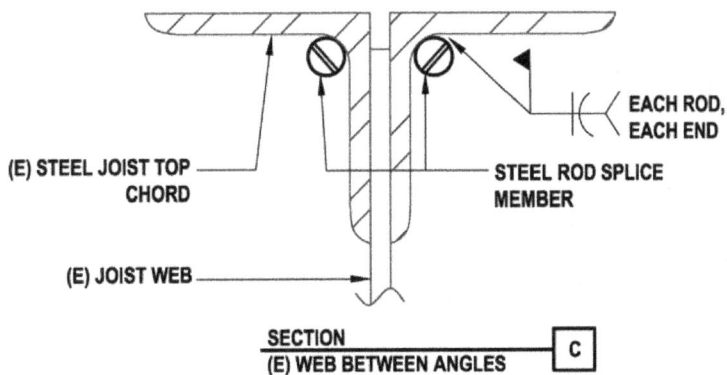

EACH ROD, EACH END

(E) STEEL JOIST TOP CHORD

STEEL ROD SPLICE MEMBER

(E) JOIST WEB

SECTION
(E) WEB BETWEEN ANGLES   C

**Figure 22.2.3-2C: Steel Open Web Joist Connection
for Diaphragm Cross-Ties**

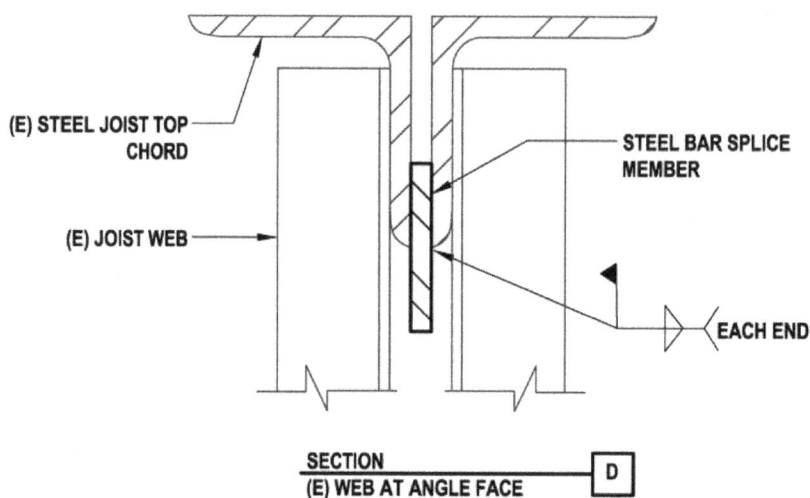

(E) STEEL JOIST TOP CHORD

STEEL BAR SPLICE MEMBER

(E) JOIST WEB

EACH END

SECTION
(E) WEB AT ANGLE FACE   D

**Figure 22.2.3-2D: Steel Open Web Joist Connection
for Diaphragm Cross-Ties**

22.2.4   Infill Opening in a Concrete Diaphragm

Deficiencies Addressed by the Rehabilitation Technique

Inadequate diaphragm shear or chord capacity at existing opening.

Description of the Rehabilitation Technique

Addition of a structural infill to close an existing opening is a relatively simple method of correcting this type of local diaphragm deficiency in a concrete diaphragm. The new infill will reduce concentrated shear and chord force demand in the surrounding diaphragm and eliminate the need for often nonexistent local chords around the edges of the opening. In almost all cases, the new infill will be made with cast-in-place reinforced concrete or shotcrete. While it is conceivable, and perhaps possible in some unusual cases, to close the opening with steel plate or a precast concrete "plug," the connections to the surrounding slab are very problematic, and their effectiveness as a mitigation measure is doubtful.

Design Considerations

Gravity load support: In addition to diaphragm shear demand, a new infill of an existing opening will create new floor or roof area which must be designed to support its self weight and the associated live load. In addition, the surrounding floor or roof system must be capable of supporting the gravity loads delivered from the newly infilled area. For larger infills, new beams may be required, both in the infill area and at the affected surrounding slabs, to provide this capacity.

Detailing Considerations

Connection to existing concrete floor and roof diaphragms: Typical details of a reinforced concrete (cast-in-place or shotcrete) infill are indicated in Figure 22.2.4-1. Sufficient dowels must be placed into the existing diaphragm slab on all sides of the opening to transfer the required shear demand to and from the infill section. Forms may be supported from the floor below or suspended from the surrounding floor or roof. This latter option is much more common for smaller openings or for openings surrounded by waffle ribs, pan joists or beams. Since the concrete infill will shrink relative to the surrounding slab, some care should be given to use shrinkage compensated mix.

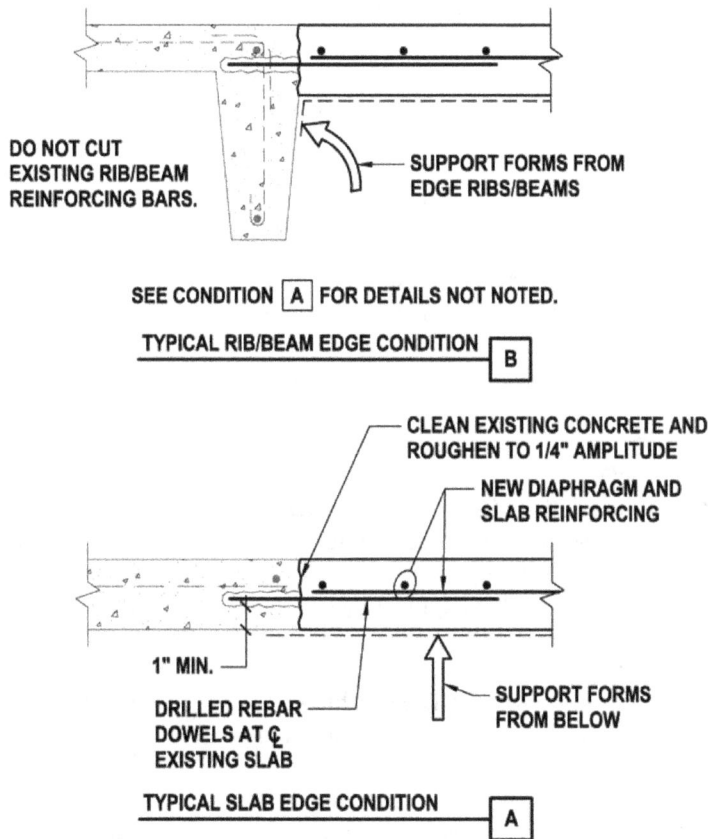

Figure 22.2.4-1: Typical Infill Opening in a Concrete Diaphragm

Cost/Disruption Considerations

The cost of this type of infill is very modest and will generally be a very small component in the overall retrofit project. Except for the noise and vibration associated with the dowel drilling, disruptions associated with this type of infill will be very localized, affecting only the immediate surrounding floor area and the area on the floor below.

Construction Considerations

The existing concrete surfaces around the entire perimeter of the existing opening to be in contact with the new concrete infill should be thoroughly cleaned of all finishes, paint, dirt, or other substances and then be roughened to provide ¼" minimum amplitude aggregate interlock at joints and bonded surfaces. Alternatively, a lower μ-factor and more dowels can be used with less roughening.

For shotcrete applications, separate test "panels" should be made to represent the slab infill work in addition to the normal test panels for shear walls. Nozzle operators should have several years experience with similar structural seismic improvement applications.

22.2.5   Add Fiber-Reinforced Polymer Composite Overlay to a Concrete Diaphragm

Deficiencies Addressed by the Rehabilitation Technique
Inadequate shear capacity in a slab

Description of the Rehabilitation Technique
The use of an FRP overlay with slabs for in-plane shear strength (diaphragm shear) enhancement is a very new technique that has had limited implementation. For shear enhancement of monolithic slab construction, the fibers are oriented parallel to the applied shear direction. The technique is also used for precast floor systems, where the shear plane is the joint between panels. Joint strengthening usually employs bi-directional fibers orientated at 45 degrees to the shear plane.

Design Considerations
Research basis: Although there has been a significant amount of research conducted on flexural strengthening of concrete slabs or strengthening of bridge decks using FRP overlays, published research focused specifically on strengthening of concrete diaphragms using FRP overlays has not been identified. Designers have typically considered results of tests performed on FRP composite strengthened shear walls relevant for diaphragm strengthening applications.

Chord and collector considerations: The diaphragm usually resists seismic loads in both directions, which requires bi-directional fiber orientation.

While shear transfer between two concrete elements has been tested and proved to be reliable, there are diaphragm internal forces termed *chord* and *collector* forces. This rehabilitation technique, which may have been intended solely as a shear enhancement may, in fact, have chord and collector force demands.

Chord actions, which develop from in-plane flexing of the full diaphragm depth, are developed in boundary elements gradually over the span length. These forces can be very high. The limited bond capacity and difficulty of anchoring the FRP composite may prohibit development of such large forces. Further, the strain limitations of the FRP composite prevent significant yielding; hence, the diaphragm chord forces should be based on the diaphragm forces required to yield the vertically-oriented elements of the lateral force-resisting system. This force level would be similar to a code level force multiplied by omega, an over-strength factor, which is the same force level used to design diaphragm collectors.

The use of FRP composite overlay to provide collector type load transfer is more difficult than that for chords. The collector force is usually being transferred from the diaphragm to a concentrated location, such as a brace frame or shear wall element. Strain compatibility and anchorage issues discussed with the bond-critical application (see Section 13.4.1, "Enhance Shear Wall with Fiber-Reinforced Polymer Composite Overlay, Fiber-Reinforced Polymer Overview, Requirements at the FRP-to-Substrate Interface") prevent reliable transfer of the collector force to the frame of wall element.

If, however, this technique must be used, then the bond, load transfer, strain compatibility, uncertainty in diaphragm demand forces, etc. must be carefully considered and reflected in the design and details.

See Section 13.4.1, "Enhance Shear Wall with Fiber-Reinforced Polymer Composite Overlay, Fiber-Reinforced Polymer Composite Overview," for background information.

Detailing Considerations

Given the high dependence on the bond strength of the FRP overlay to the substrate, in situ bond testing is recommended as part of the contract documents. A testing program will verify the design assumptions and assist in providing quality assurance. The vertical offset between the two slabs should be minimized. This can be achieved by removing surface projections and applying leveling compound to ensure that the FRP composite overlay does not exceed the 1-2% out-of-plane angle. Offsets exceeding this limit or lack of bond between the leveling compound or substrate and the polymer may cause premature delamination.

In many situations, improvement in shear transfer capacity at the edge of the diaphragm will be needed in addition to enhancement of the capacity of the diaphragm itself. Transfer details from the slab to the wall using FRP need careful consideration. See Figure 22.2.5-1. Typically, the fiber is lapped from the slab to the wall, and fibers are oriented at 45 degrees (in plan view) to the length of the wall. The 90 degree bend in the fiber at the turn to the wall creates several issues. First, preparation of the existing sharp corner with resin putty is needed to allow a reasonable radius for the fiber. Second, when shear forces develop, they create tensile forces in the fiber. Because of the bend in the fiber, a substantial out-of-plane component is developed which must be resisted. The bond stress of the fiber has limited capability to take this force, usually leading to the need to reinforce the bend with mechanical means. A cut pipe placed against the corner, matching the radius of the curve, can be anchored with drilled dowels through the fiber to the wall or slab. Finally, testing to date of slab-to-wall shear transfer details is limited, necessitating increased caution.

Construction Considerations

Should underside of slab strengthening be used, the utilities at this location may need to be removed and reinstalled. This could impact building function during the construction period, and will add to the construction cost. For above slab strengthening architectural finishes, thresholds, and slopes will need to be considered.

Proprietary Concerns

See Section 13.4.1 for brief discussion of proprietary concerns.

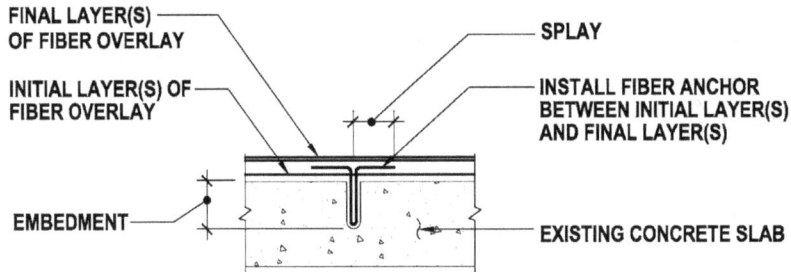

FINAL LAYER(S) OF FIBER OVERLAY

SPLAY

INITIAL LAYER(S) OF FIBER OVERLAY

INSTALL FIBER ANCHOR BETWEEN INITIAL LAYER(S) AND FINAL LAYER(S)

EMBEDMENT

EXISTING CONCRETE SLAB

FIBER ANCHOR DETAIL

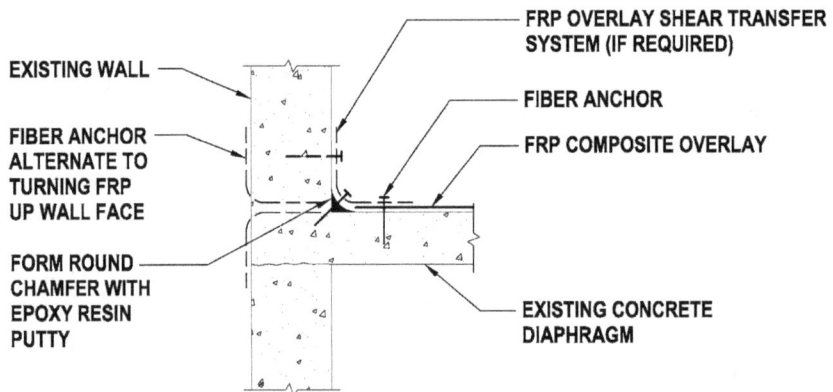

EXISTING WALL

FRP OVERLAY SHEAR TRANSFER SYSTEM (IF REQUIRED)

FIBER ANCHOR

FIBER ANCHOR ALTERNATE TO TURNING FRP UP WALL FACE

FRP COMPOSITE OVERLAY

FORM ROUND CHAMFER WITH EPOXY RESIN PUTTY

EXISTING CONCRETE DIAPHRAGM

DIAPHRAGM STRENGTHENING

Figure 22.2.5-1: Shear Strengthening of Concrete Diaphragm Using FRP Composite

22.2.6　Infill Opening in a Concrete Fill On Metal Deck Diaphragm

Deficiency Addressed by Rehabilitation Technique

Increase diaphragm shear and/or chord capacity by infilling opening.

Description of the Rehabilitation Technique

Adding infill to an existing opening is a simple method of reducing local stresses around the opening as well as the demand on the diaphragm. However, this technique can only be employed if an existing opening is no longer necessary for the function of the building. Thus, it would likely have to coincide with other building renovations that eliminate the function of the opening. The opening may have been used for stairs, an elevator shaft, a pipe and conduit shaft, or an atrium. The infill should be constructed in a similar manner as the existing diaphragm when possible, using similar types of metal deck and concrete as well as reinforcing steel layout. This ensures that the infill matches the strength and stiffness of the surrounding diaphragm. The new metal deck can be connected to the existing deck with welds or fasteners while the new reinforcing steel bars are doweled into the edges of the opening. The edges of the opening should be roughened to ensure adequate bond between the new and existing concrete. For smaller openings, it may be acceptable to span the opening with a flat piece of gauge steel instead of metal deck, provided that proper measures are taken to fill the openings between the deck flutes.

Design and Detailing Considerations

Gravity loads: The infill has to support its self-weight and additional dead and live loads. The surrounding floor system should also be evaluated for these new loads. At larger infills, new steel framing may be required either directly below or at the edge of the infill.

Metal deck attachment: The new metal deck should overlap the existing metal deck around the perimeter of the opening. The deck can be attached to one another with puddle or seam welds, or mechanical fasteners, which may include expansion anchors, screws, or shot pins.

Bar development: Details of the reinforcement are similar to that for infilling an opening in a concrete diaphragm shown in Figure 22.2.4-1. Development lengths for the same size reinforcing bar will vary depending on the grout or adhesive product used to dowel the bar into the existing concrete. Bars on opposite sides of the openings should be spliced inside the opening. At smaller openings, the splice lengths will be limited by the size of the opening. The bars can be hooked in these cases for development. Adding bars to thin slabs will be difficult, particularly in the direction perpendicular to the metal deck flutes. Existing bars that are parallel to the flutes may be damaged while drilling holes for the new dowels. As an alternative, it may be easier to use welded wire fabric (WWF) instead of reinforcing steel. The slab would have to be chipped back around the opening to allow for development of the WWF.

Cost/Disruption

The cost associated with this technique is minimal compared to other diaphragm strengthening techniques, such as adding concrete overlays or horizontal braced frames. Since the infilling of an opening is likely related to other changes to a building, the disruption caused by the other changes are often more significant.

Construction Considerations

See Section 8.4.1 for general discussions of welding issues, removal of existing nonstructural and structural elements, and construction loads.

Proprietary Concerns

Many grout and adhesive products are available.

22.2.7   Increase Shear Capacity of Unfilled Metal Deck Diaphragm

Deficiency Addressed by Rehabilitation Technique

Strengthen inadequate bare metal deck diaphragm.

Description of the Rehabilitation Technique

Metal deck diaphragms are governed by either the capacity of the deck or its connection to other components of the lateral force-resisting system. Connection capacity is limited by the strength of the welds or other mechanical fasteners. At locations where welds or fasteners cannot be directly added, such as concrete walls, the addition of a steel angle connected with expansion anchors or adhesive dowels to a wall and diaphragm is often feasible. The capacity of a longitudinal joint between deck units is limited by the strength of the crimps or seam welds. These connections should be upgraded to the strength of the metal deck to achieve ductile diaphragm behavior during an earthquake. If the connections can develop the metal deck capacity, but the deck is found to be inadequate, significant increases in capacity may be obtained by adding a reinforced concrete fill or horizontal braced frame (Section 22.2.9).

Design and Detailing Considerations

Connections: In order to enforce deformation compatibility, new connections should be constructed similarly to the existing connections. Thus, puddle welds should be used if the existing diaphragm is welded to the steel framing. Similarly, the same types of mechanical fasteners should be used to match the existing fasteners when screws, shot pins, or expansion anchors are found at the connections.

Deck stiffeners: Some deck manufacturers fabricate stiffeners specifically intended for use with unfilled metal decks. The stiffeners are constructed to match the profile of the decks, which provide additional stiffness at the supports and in turn, increase the strength of the diaphragm. The stiffeners are typically welded to the deck and the steel beams.

Concrete fill: When reinforced concrete is added over metal deck, a shear transfer mechanism from the concrete to the lateral force-resisting system is required, e.g. welded shear studs at steel beams and drilled dowels at concrete members. Since the addition of a concrete overlay will increase the dead weight of the structure, the existing forces, members, connections, and foundation must be checked to determine whether they are capable of resisting the added loads.

Cost/Disruption

Diaphragm connection upgrades can be performed efficiently to minimize disruption and are cost effective if upgrades to other parts of the lateral force-resisting system are not required. If concrete fill is added, cost and disruption could increase significantly if upgrades are required to other parts of the lateral force-resisting system. Also, nonstructural elements such as insulation fill, roofing, and partitions would all require temporary removal.

Construction Considerations

See Section 8.4.1 for general discussions of welding issues, removal of existing nonstructural and structural elements, and construction loads.

Proprietary Concerns

Metal deck stiffeners are only provided by some manufacturers for use with their decks.

22.2.8   Enhance Masonry Flat Arch Diaphragm

Deficiency Addressed by Rehabilitation Technique

A relatively common type of floor in a masonry building or steel frame infill building, particularly outside the West Coast, uses narrowly spaced steel beams to support shallow or "flat" arches of masonry. The masonry can be made of hollow clay tile or brick. It is usually bearing on the bottom flange of the steel beam and supports nonstructural and acoustic fill above it. The horizontal kick from the base of the arch is balanced in the diaphragm interior by the adjacent arch. At the exterior, this kick either goes into the wall, or a tension tie of steel is provided at the bottom of the beams. In some cases, the steel strapping or bars run the full width of the diaphragm. When a tension tie is missing at the base of the arch and the diaphragm vibrates and expands, localized gravity failure can result when loss of arching action occurs. At the exterior of the diaphragm, the unbalanced kick of the arch can add to out-of-plane demands on the wall, contributing to out-of-plane wall failure and loss of vertical support. See Figure 22.2.8-1 for examples of failure scenarios.

Description of the Rehabilitation Techniques

There are several rehabilitation techniques for masonry flat arches floors. They can be combined for economy of scale.

Wall-to-diaphragm tension ties: Figure 22.2.8-2 shows the addition of tension ties from the wall to the steel beams for conditions when the beams are perpendicular to the wall and when they are parallel. When beams are perpendicular, an angle and drilled dowel is sufficient. When beams are parallel, strapping back to joists inside the floor is necessary. Figure 22.2.8-3 shows an example of placing the strapping on top of the beams, in case this is the preferred location for work.

Wall-to-diaphragm shear ties: The drilled dowels in Figure 22.2.8-2 also serve as ties for transferring shear forces from the edge of the diaphragm into the wall.

Chord: If the angle in Figure 22.2.8-2 is continuous, it can serve as a diaphragm chord.

Interior tension: While providing a tension tie for the case when the beams are parallel to the wall next to the wall is the most critical priority, it is desirable as well to continue the strapping all the way across the floor so local interior failure does not occur. Figure 22.2.8-4 shows the straps, plus notes the tension and shear ties and the chord.

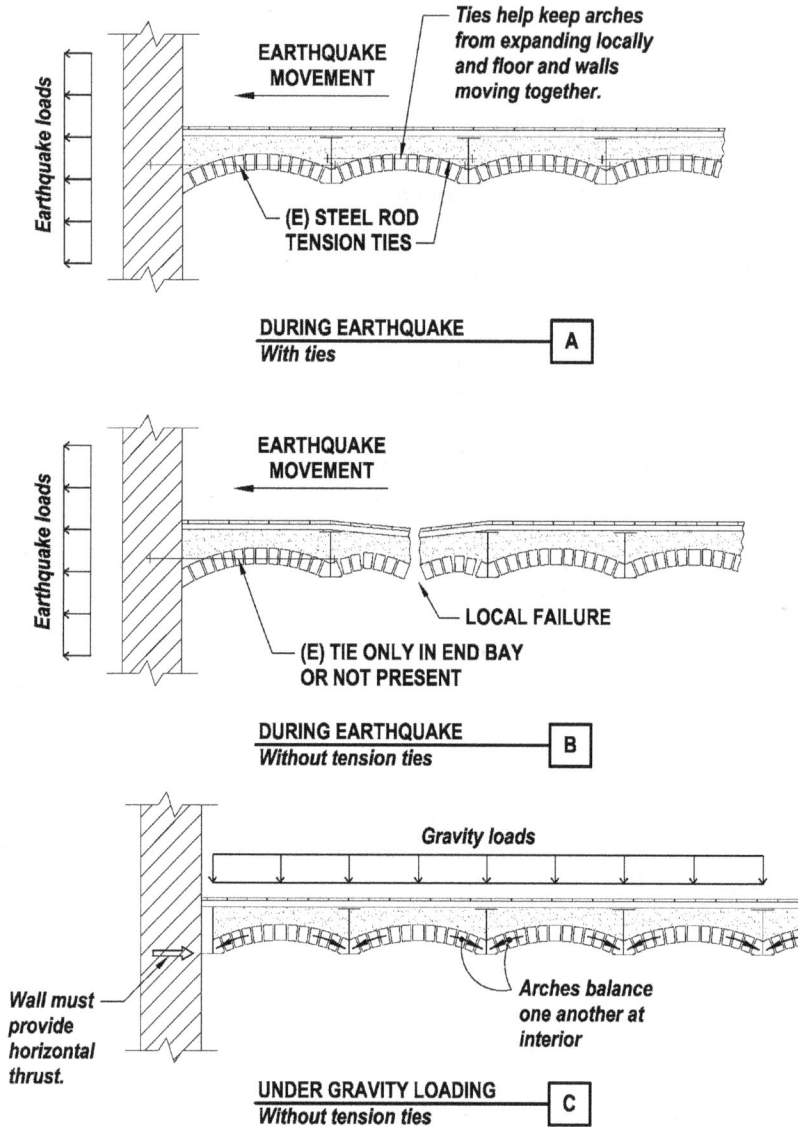

Figure 22.2.8-1: Failure Scenarios for Masonry Flat Arch Floors

WELD
WASHER

CONTINUOUS
PLATE OR STRAP.
REMOVE EXISTING
PLASTER AND BRICK AS
NEEDED TO INSTALL

DRILLED DOWEL

ANGLE. REMOVE EXISTING PLASTER
& BRICK AS NEEDED TO INSTALL

BEAMS PARALLEL TO WALL   A

DRY PACK (OR FILL WITH
NONSHRINK GROUT) BEAM
POCKET WHERE IT IS NOT
BRICKED IN SOLID

EXISTING BEAM

ANGLE. REMOVE EXISTING
PLASTER AND BRICK AS
NEEDED TO INSTALL

DRILLED DOWEL

WELD WASHER

VARIES

BEAMS PERPENDICULAR TO WALL   B

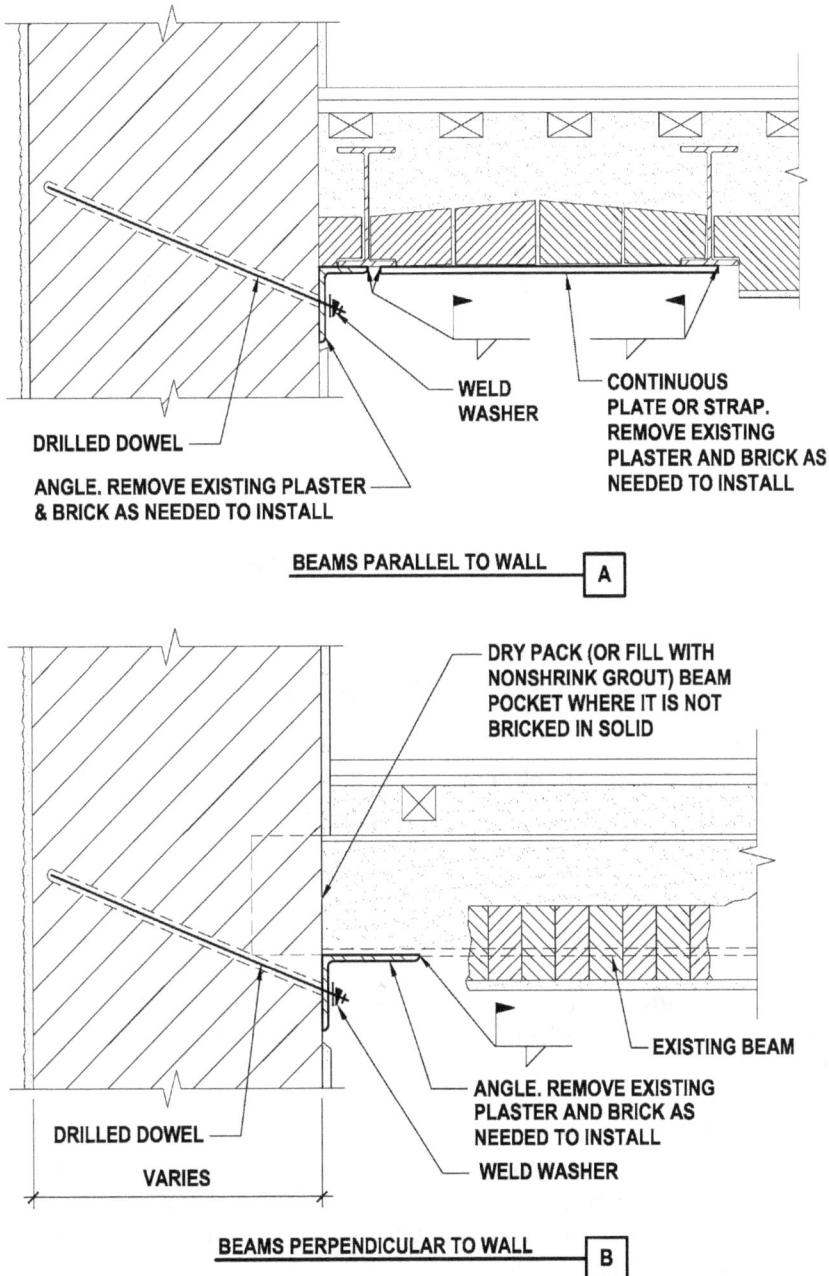

**Figure 22.2.8-2: Add Wall-to-Diaphragm Ties and Chord for Masonry Flat Arch Floor -
Access from Below the Floor**

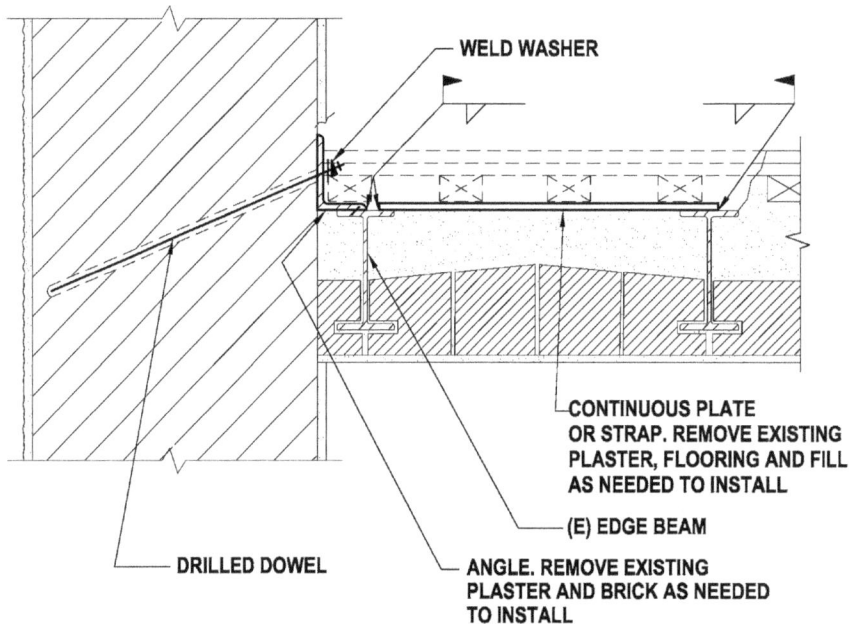

JOISTS PERPENDICULAR TO WALL

**Figure 22.2.8-3: Add Wall-to-Diaphragm Ties and Chord for Masonry Flat Arch Floor -
Access from Above the Floor**

Diaphragm strengthening: Figure 22.2.8-4 also shows how adding diagonal bracing can be combined with existing beams and straight to create a horizontal braced frame diaphragm.

Topping slab: Theoretically, part of the flooring substrate can be replaced with a reinforced concrete diaphragm, though the vertical capacity of the floor would need to be sufficient and the weight of the new concrete adds to the inertial weight of the building.

Design Considerations

Research basis: No research specific to seismic rehabilitation of flat arch floors has been identified. There is also very limited information about how the floors have performed in actual earthquakes. There was some damage in the 1906 San Francisco Earthquake reported for these floors (Himmelwright, 1906) though much of the damage was due to fire. There are photos of the flat arch roof failures and reports of significant damage in Iranian earthquakes when tension ties are not present (Alimoradi, 2005).

EXISTING
URM WALL

EXISTING
STEEL JOISTS

Diagonal
strapping
and existing
joists act
as horizontal
braced frame
diaphragm.

SPANNING
DIRECTION
OF MASONRY
ARCH

Steel angle
acts as chord.

Crosstie strap.
It also serves as
tension tie to
prevent local
arch failure.

Drilled dowels
for shear and
tension ties

PLAN

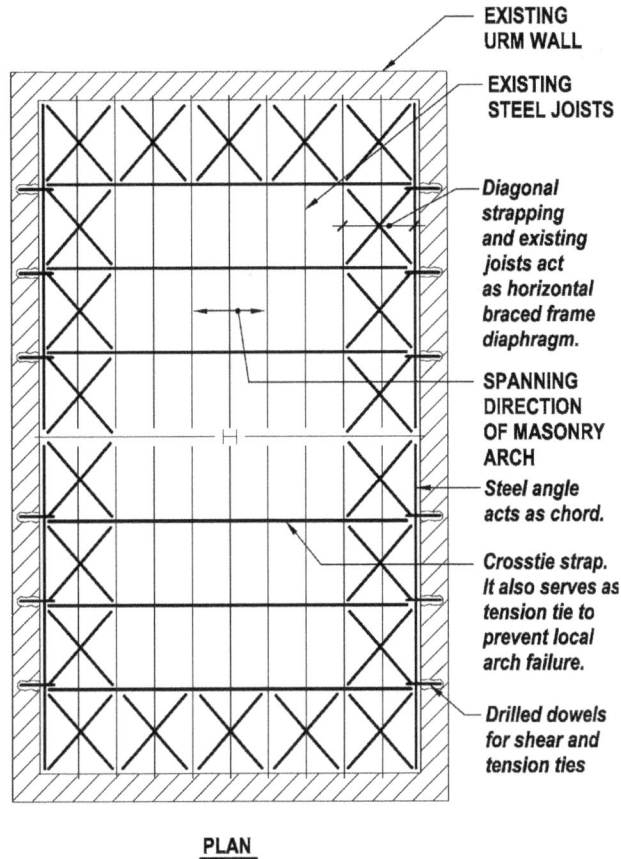

Figure 22.2.8-4: Masonry Flat Arch Floor Strengthening

Shear capacity: This type of floor has not been addressed by recent evaluation publications like FEMA 273 (FEMA, 1997a), FEMA 274 (FEMA, 1997b), FEMA 356 (FEMA, 2000), or ASCE 31-03 (ASCE, 2003), so capacity evaluations are from first principles. One strategy is to take all of the lateral force resistance in the new diaphragm strengthening due to the lack of interconnections in the diaphragm. Another approach is to develop strut-and-tie models in the diaphragm with the new and existing steel as ties and the masonry as a strut.

Stiffness: Although this floor lacks interconnections, it is likely to be quite stiff, as well as extremely heavy.

Detailing and Construction Considerations
Floor types: Lavicka (1980) is a reprint of an 1899 textbook on turn-of-the-century construction techniques and has an excellent summary of masonry flat arch variations. The system was intended to provide improved fireproofing and acoustic benefits. Flat tile arches were popular and had flat top and bottom surfaces to the tile, but beveled edges to create internal arching

action. Side method arches had the voids in the hollow clay tile parallel to the beams; end method arches oriented the voids perpendicular to the beams. There were combinations of the orientations as well. The tile at the steel beam was usually notched around the bottom flange to provide masonry cover of the bottom of the bottom flange. Tile depths range from 6" to 12" with beams spaced from 3'6" to 7'6". Segmental tile arches had shallow arches of several inches at the crown, the voids were parallel to the beams, and the end tile would bear on top of the bottom flange. Other systems have been observed to include clay bricks oriented with the long direction of the brick perpendicular and parallel to the beams. The masonry arches often supported a fill of cinders, sometimes mixed with mortar. This in turn would support wood sleepers spanning over the top of the steel beams and a wood floor. Tension ties were recommended; they were to be ¾" diameter rods placed near the bottom of the steel beam web and at about a spacing of 7'-8'.

Bottom cover: Figure 22.2.8-2 shows clay tile floors covering the bottom of the bottom flange. There is typically plaster adhering to the masonry. To install steel strapping, the plaster and masonry must be notched. Figure 22.2.8-3 shows an alternative to avoid damaging the underside by adding steel plate or straps, but working from the top. Of course, this is quite disruptive to occupants as well. In some arch types, though, the bottom flange is not covered and adding steel from below is much less disruptive.

Cost/Disruption

Rehabilitation of a masonry flat arch floor can be quite disruptive and expensive, particularly when ties are necessary at the building interior and if plaster ceilings and masonry or floors must be temporarily removed and patched.

Proprietary Issues

There are no proprietary concerns with diaphragm improvements in masonry flat arch floors.

22.2.9   Add Horizontal Braced Frame as a Diaphragm

Deficiency Addressed by Rehabilitation Technique

Strengthen inadequate diaphragm.

Description of the Rehabilitation Technique

Providing a horizontal braced frame as a diaphragm strengthening technique is useful if the existing floor cannot be disturbed for functional reasons or the cost of replacing the existing diaphragm is more expensive (e.g., a sloped roof). This is also an alternative when concrete overlays add too much mass or lead to other construction complications. The existing diaphragm could be constructed of concrete filled or unfilled metal deck, or wood. The new horizontal bracing is added under the existing diaphragm, in which the existing framing with new diagonal members forms the horizontal bracing system. The diaphragm shears are shared with the existing diaphragm in proportion to the relative rigidity of the two systems. The design philosophy is generally to have the diaphragm remain essentially elastic, with the goal of achieving ductile inelastic behavior in the vertical lateral force-resisting elements. See Chapter 9 for a general discussion of braced frames.

Design Considerations

Force distribution: The diaphragm strength could be evaluated by considering boundary solutions. First, its capacity including both the existing diaphragm and the horizontal braced frame is determined based on their relative rigidities. This alternative may not be always be fully effective if the existing diaphragm has much greater rigidity of that of the bracing system, such as metal deck with heavily reinforced concrete fill. Thus, an evaluation should also be performed assuming failure of the concrete fill. The diaphragm strength would only include that of the braced frame with minimal contribution from the metal deck without the concrete fill. If the latter solution yields a greater value, extensive cracking of the concrete fill and greater diaphragm displacements would be assumed to be acceptable.

Sloped roofs: The horizontal braced frames could be sloped to match the roof slopes, which would require proper consideration of the slopes and their effects on the diaphragm forces. Alternatively, the braced frames could have a flat layout, but this may affect the functional space as well as aesthetics.

Brace members: Similar to the selection of members in braced frames, compact and non-slender sections are preferred for their ductility. Installation of the braces should be factored into their selection due to the logistics associated with delivering and attaching the braces to their final locations. Note the self-weight of the braces adds a component to the flexural forces that may be reduced by adding hanger rods.

Chords and collectors: The new horizontal bracing system requires continuous chord and collector members to receive the brace forces and transfer these forces to the lateral force-resisting elements. The existing members that serve this purpose should be used when possible, as shown in Figure 22.2.9-1.

Detailing Considerations

Connections: For steel structures, the braces can be welded or bolted with or without gusset plates to the existing framing. An example of a welded connection is shown in Figure 22.2.9-2. Bolting eliminates welding issues that include space restrictions and venting weld fumes while welding may permit smaller and more compact connections. In concrete structures, connection of the new horizontal bracing system to the existing vertical system is accomplished by welding braces to plates that connect to the walls or frames with mechanical fasteners, such as threaded dowels and expansion anchors.

Cost/Disruption

These costs of adding horizontal bracing must be weighed against that of a concrete overlay. Temporary removal or relocation of nonstructural elements such as piping and partition walls are required and should be included in the cost evaluation for both options. The horizontal braced frame requires connection modifications, which are locally very disruptive.

Construction Considerations

The engineer's involvement during the construction phase is critical during a seismic rehabilitation. The design of the retrofit scheme must not neglect the construction phase and should consider these issues at a minimum:

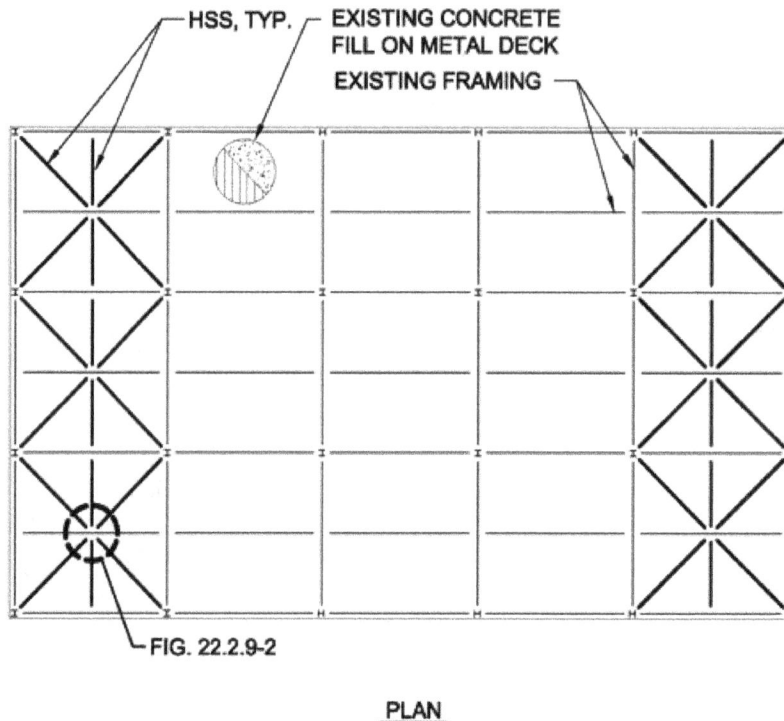

PLAN

Figure 22.2.9-1: Diaphragm Strengthening using Horizontal Braced Frame

Welding/bolting issues: See general discussion in Section 8.4.1. Primary issues associated with bolting consist of typical field bolting issues such as set up, fit-up, and alignment.

Removal of existing nonstructural elements: This technique requires access to the underside of the floor or roof framing and may require relocation of piping, ducts, or electrical conduits as well as difficult and awkward connections to the existing framing. See Section 8.4.1 for discussions of fireproofing, asbestos, and concrete encasement.

Removal of existing structural elements: Existing structural elements do not typically have to be removed to add horizontal steel bracing. However, if required, shoring and temporary bracing may be necessary.

Construction loads: See general discussion in Section 8.4.1.

Proprietary Concerns

There are no known proprietary concerns with this technique.

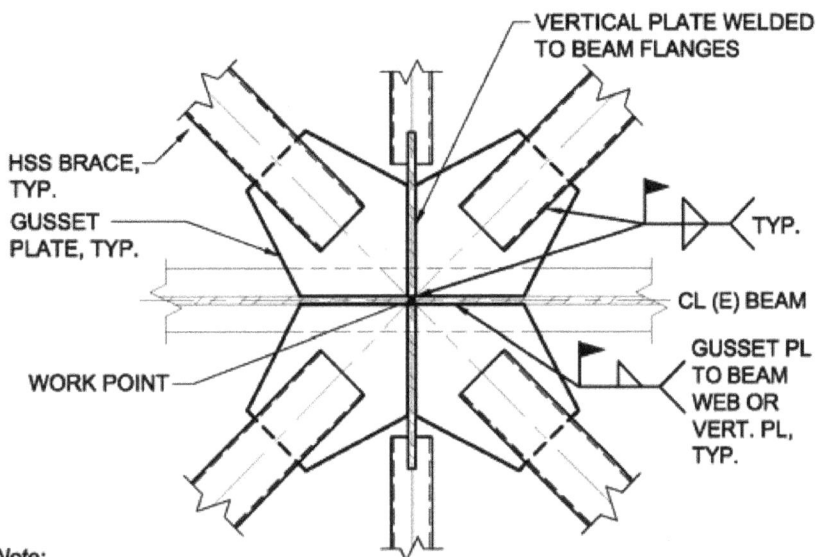

Note:
Elevation of horizontal braced frame with respect to diaphragm is a balance between minimizing eccentric forces and allowing for construction access from below.

Figure 22.2.9-2: Horizontal Braced Frame Connection

22.2.10   Improve Tension Rod Horizontal Steel Bracing

Deficiency Addressed by Rehabilitation Technique

Repair nonductile tension rod bracing and/or connections

Description of the Rehabilitation Technique

Tension rod bracing consist of rods that are spliced together by turnbuckles and connected to clevis pins at the ends. The clevis pins are bolted to typical gusset plates. Tension rods that are inadequate for the seismic demands should be replaced entirely since it would probably be more complicated to upgrade existing rods. Increasing the rod size also requires replacing the turnbuckles and clevis pins. Connections that are inadequate can be upgraded similarly as typical braced frame connections. An example of a typical rod connection to a concrete or CMU wall is shown in Figure 22.2.10-1. The connection to the wall should develop the strength of the rod.

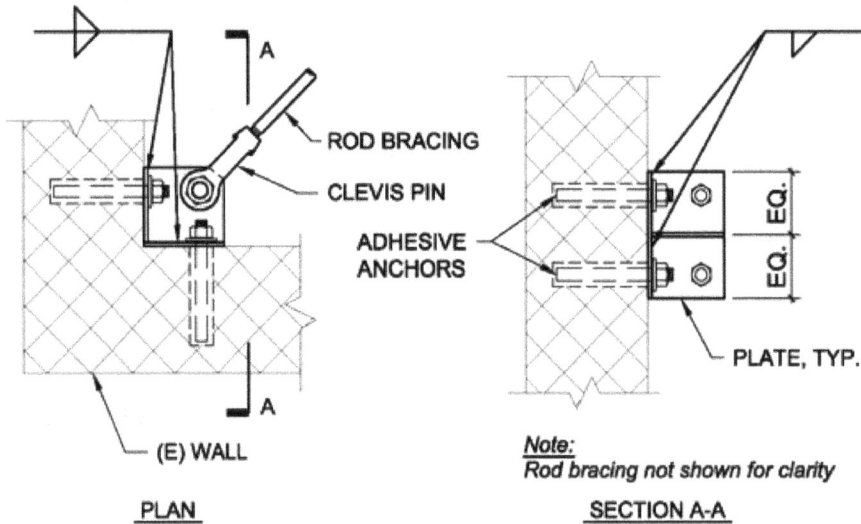

Figure 22.2.10-1: Tension Rod Connection at Wall

Design and Detailing Considerations

Tension rod bracing is used in applications where seismic forces are relatively low. It would be most appropriate for unfilled metal deck or wood diaphragms. The rod upgrades may increase the stiffness of the existing diaphragm and the total diaphragm force. Thus, all other elements of the lateral force-resisting system—connections, chords, collectors, frames or walls, and foundations—should be evaluated and upgraded accordingly.

Cost/Disruption

Replacing tension rods is fairly efficient on both a cost and time basis compared to other types of diaphragm upgrades. Connection modifications will only be locally disruptive and can be performed rapidly.

Construction Considerations

See Section 8.4.1 for general discussions of welding issues, removal of existing nonstructural and structural elements, and construction loads. Also see Section 22.2.9 for a discussion of construction issues related to modification of horizontal steel bracing.

Proprietary Concerns

There are no known proprietary concerns with this technique.

22.2.11   Improve Shear Transfer in Precast Concrete Diaphragm

Deficiency Addressed by Rehabilitation Technique

Inadequate diaphragm strength and/or stiffness

Precast diaphragm deficiencies and observed behavior have been discussed in some detail in Chapter 17; the reader is referred to these discussions. To date, construction of precast buildings in areas of high seismic hazard in the U.S. has been of limited quantity, resulting in limited opportunities to observe earthquake performance.   The poor performance of some long-span precast diaphragms in parking structures in the 1994 Northridge earthquake has raised questions about shear capacity in diaphragms with topping slabs, excessive diaphragm deformation due to performance of chords and collectors and the interaction of shear and flexure. The complete lack of connection between hollow core floor planks within diaphragms appears to have been a primary contributor to collapse of nine-story residential precast concrete frame buildings in the 1988 Armenia earthquake (EERI, 1989).

Description of the Rehabilitation Techniques

There are three types of precast concrete diaphragms commonly used: topped precast tee-beam, untopped precast tee-beam, and untopped precast hollow-core plank. Topped precast hollow-core may be used on occasion, but is not as common.

To date, very little rehabilitation of precast diaphragms has occurred in the U.S. As a result, the following discussion of rehabilitation measures draws from limited available research, suggested details for new precast construction, and application of rehabilitation techniques for concrete buildings to the specific configurations of precast elements.

Fiber-reinforced polymer (FRP) composite overlays provide one possible approach to shear connections between adjacent precast diaphragm members, and overlays could be used for any of the three common systems noted above. For parking structures, attention to both ultraviolet (UV) ray exposure and wearing under vehicle loads would be important to performance. The FRP overlay could be applied continuously over the area of high diaphragm shear and then used to transfer loads into supporting shear walls or frames, or applied locally at each member joint. Research by Pantelides, Volnyy, Gergeley, and Reaveley, (2003) on FRP composite connection between wall panels may be of interest; however, the reader is cautioned to consider the effects of simultaneous shear and tension at joints. See Section 17.4.2.

For untopped hollow-core diaphragms it may be possible to add construction roughly equivalent to that used for new construction. In new design, where diaphragm shear stresses exceed those allowed for grout key shear transfer, the cast concrete beams at the diaphragm perimeter or interior are used as flexural elements, resisting horizontal diaphragm forces. New beams could be added to serve this purpose. Connection between the precast sections and the new beams, either by bearing or mechanical connection is required. Attention to adequate strength and stiffness is also required.

Bolted steel plate connections providing shear connections from panel to panel are another possible approach. This involves use of a continuous plate or series of plates crossing the precast panel joint, with adhesive or expansion anchors on each side of the joint. Steel plate thickness must be selected in order to avoid plate bucking between connections. See Chapter 17 for discussion of anchors.

Design Considerations

Research basis: No research applicable to rehabilitation of precast diaphragm strength and stiffness has been identified; however, the following research for new construction may provide some guidance for rehabilitation:

> A significant integrated analytical and experimental research program is currently underway to develop a comprehensive design methodology for precast concrete diaphragm systems. The project intends to address the discrepancy between current design practice, based on inelastic behavior concentrating in vertical elements, and observed performance in which substantial inelastic behavior has occurred in diaphragms (Wan et al., 2004; and Naito and Cao, 2004). The project proposes to determine force and deformation demands required for design, connection details to support the performance, and address deformation relative to the gravity load-carrying system. This information will be invaluable for both new design and rehabilitation. Testing will include individual connections, joints, and half-size components. Analytical modeling of full buildings is being used to identify critical demands. Of particular interest is the simultaneous occurrence of shear and tension or compression on connections normally considered to carry only shear. Published information to date (Naito and Cao, 2004) provides a database of connector properties from existing literature and suggests a simplified analysis model based on initial finite element testing. Additional information should be available over the next several years.
>
> *Shear Diaphragm Capacity of Untopped Hollow-Core Floor Systems* (Concrete Technology Associates, 1981) describes testing of grouted hollow-core joints. Note that issues raised by the Northridge earthquake might imply modification of testing approach.
>
> Research by Pantelides, Volnyy, Gergeley, and Reaveley, (2003) on FRP composite connection between wall panels.
>
> Research by K.S. Elliott, University of Nottingham, on untopped hollow-core diaphragms.

Basic design approach: The *PCI Handbook* (PCI, 1999) and *Design and Typical Details of Connections for Precast and Prestressed Concrete* (PCI, 1988) are basic references for design and detailing of precast concrete structures. These documents discuss the use of grouted keys for diaphragm-to-diaphragm connections, and they also recognize use of friction connections, without positive anchorage for wall-to-diaphragm connections. Use of these mechanisms must be given very careful consideration for possible inelastic seismic demands.

Shear and flexure interaction: One of the issues identified from the performance of parking structures in the Northridge earthquake is the interaction of shear and flexural deformations. Diaphragm deformations will result in tension and compression forces between adjacent diaphragm members. As a result, shear connections between members will need to accommodate simultaneous tension or compression plus shear. It is recommended that the diaphragm chord and collector members also be evaluated, and rehabilitated if necessary to control tension forces. It is also recommended that the rehabilitation measure chosen be capable of withstanding anticipated simultaneous forces. Methods of estimating diaphragm demands are proposed in Nakaki (1998) and Naito and Cao (2004).

Transfer into and out of diaphragm reinforcing: Transfer of loads is key to the use of fiber composites or steel plate for connecting between precast diaphragm segments. Where panel-to-panel connections are made, it is necessary to transfer the full design load in and out at each connection. In some cases it may become more practical to provide reinforcing over the entire diaphragm or highly loaded sections of the diaphragm; the load portions of load carried in the existing diaphragm and the reinforcing would need to be determined by deflection compatibility.

Topping slabs: The addition of a topping slab is seldom a practical approach because of the added weight for vertical loads and mass for seismic loads. In rare cases where additional vertical load capacity has been provided, this may be possible. The additional capacity is needed not only in the diaphragm slab and beam system, but in all of the vertical support system through the foundation. The removal and replacement of a topping slab could permit the addition of reinforcing and connections without increasing gravity or seismic loads. This, however, is a costly process.

Proprietary Concerns

Fiber composite materials and adhesive and mechanical anchors are proprietary and must be used in accordance with manufacturer and ICC-ES requirements.

22.3 References

ABK, 1981, *Methodology for Mitigation of Seismic Hazards in Existing Unreinforced Masonry Buildings: Diaphragm Testing*, A Joint Venture of Agbabian Associates, S.B. Barnes and Associates, and Kariotis and Associates (ABK), Topical Report 04, c/o Agbabian Associates, El Segundo, CA.

ASCE, 2003, *Standard for the Seismic Evaluation of Buildings*, ASCE 31-03, Structural Engineering Institute of the American Society of Structural Engineers, Reston, VA.

AF&PA, 2005, *National Design Specification for Wood Construction, ASD/LRFD*, American Forest and Paper Association, Washington, D.C.

Alimoradi, A., 2005, "Steel Frame with Semi-Rigid 'Khorjini' Connections and Jack Arch Roof 'Taagh-e-Zarbi'," *EERI World Housing Encyclopedia*, www.world-housing.net; Report 25.

APA, 2000, *Research Report 138, Plywood Diaphragms*, APA The Engineered Wood Association, Tacoma, WA.

Concrete Technologies Associates, 1981, *Shear Diaphragm Capacity of Untopped Hollow-Core Floor Systems* (TCA 38), Concrete Technology Associates, Tacoma, WA.

Dolan, J.D., D. Carradine, J. Bott, and W. Easterling, 2003, *Design Methodology of Diaphragms*, CUREE Publication No. W-27, CUREE, Richmond, CA.

EERI, August 1989, *Armenia Earthquake Reconnaissance Report*, Earthquake Spectra Special Edition, Earthquake Engineering Research Institute, Oakland, CA.

FEMA, 1997a, *NEHRP Guidelines for the Seismic Rehabilitation of Buildings*, FEMA 273. Federal Emergency Management Agency, Washington, D.C.

FEMA, 1997b, *NEHRP Commentary on the Guidelines for Seismic Rehabilitation of Buildings*, FEMA 274, Federal Emergency Management Agency, Washington, D.C.

FEMA, 2000, *Prestandard and Commentary for the Seismic Rehabilitation of Buildings*, FEMA 356, Federal Emergency Management Agency, Washington, D.C.

Fonseca, F., S. Wood and N. Hawkins, 1996, "Measured Response of Roof Diaphragms and Wall Panels in Tilt-Up Systems Subject to Cyclic Loading," *Earthquake Spectra,* Volume 12, Number 4, Earthquake Engineering Research Institute, Oakland, CA.

Ghosh, S.K. and S. Dowty, 2000, *Anchorage of Concrete or Masonry Walls to Diaphragms Providing Lateral Support*, Draft 2000, Not Published.

Hamburger, R.O. and D. McCormick, 1994, "Implications of the January 17, 1994 Northridge Earthquake on Tilt-up Wall and Masonry Wall Buildings with Wood Roofs," *Proceedings of the 1994 Convention of the Structural Engineers Association of California*, Sacramento, CA.

Himmelwright, A., 1906, *The San Francisco Earthquake and Fire, A Brief History of the Disaster, A Presentation of Facts and Resulting Phenomena, with Special Reference to the Efficiency of Building Materials, Lessons of the Disaster*, The Roebling Construction Company, New York, NY.

ICBO, 1997, *Uniform Code for Building Conservation*, 1997 Edition, International Conference of Building Officials, Whittier, California.

ICC, 2003a, *International Building Code*, International Code Council, Country Club Hills, IL.

ICC, 2003b, *International Existing Building Code*, International Code Council, Country Club Hills, IL.

ICC-ES, 2004a, *303 Siding and High-Load Diaphragms*, ICC-ES Legacy Report ER-1952, ICC Evaluation Service, Inc, Whittier, CA.

Lavicka, W., 1980, *Masonry, Carpentry, Joinery*, reissued version of 1899 textbook of the same name, Chicago Review Press: Chicago, IL.

Naito, C. and L. Cao, 2004, "Precast Diaphragm Panel Joint Connector Performance," *Proceedings of the 13th World Conference on Earthquake Engineering*, Vancouver, B.C. Canada.

Nakaki, S., 1998, *Design Guidelines: Precast and Cast-in-Place Concrete Diaphragms*, Earthquake Engineeering Research Institute, Oakland, CA.

Pantelides, C. P., V. A. Volnyy, J. Gergeley, and L.D. Reavely, *Seismic Retrofit of Precast Concrete Panel Connections with Carbon Fiber Reinforced Polymer Composites*, PCI Journal, Vol. 48, No. 1, January/February 2003, Precast/Prestressed Concrete Institute, Chicago, IL.

PCI, 1988, *Design and Typical Details of Connections For Precast and Prestressed Concrete*, Prestressed Concrete Institute, Chicago, IL.

PCI, 1999, *PCI Design Handbook, Precast and Prestressed Concrete*, Fifth Edition, Prestressed Concrete Institute, Chicago, IL.

Peralta, David, Joseph Bracci and Mary Beth Hueste, 2004, "Seismic Behavior of Wood Diaphragms in Pre-1950s Unreinforced Masonry Buildings," *Journal of Structural Engineering*, ASCE: Reston, VA, Volume 130, Number 12, December, pp. 2040-2050.

SEAOSC (Structural Engineers Association of Southern California), 1979, *Recommended Tilt-up Wall Design* (Yellow Book), SEAOSC, Los Angeles, CA.

SEAONC (Structural Engineers Association of Northern California), 2001, *Guidelines for Seismic Evaluation and Rehabilitation of Tilt-Up Buildings and Other Rigid Wall/Flexible Diaphragm Structures*, SEAONC, San Francisco, CA.

Wan, G., R. Fleishchman, C. Naito, J. Restrepo, R. Sause, L. Cao, M. Schoettler, and S.K. Ghosh, 2002, "Integrated Analytical and Experimental Research Program to Develop a Seismic Methodology for Precast Diaphragms," *SEAOC 2004 Convention Proceedings*, Structural Engineers Association of California, Sacramento, CA.

Chapter 23 - Foundation Rehabilitation Techniques

23.1   Overview

While the need to add or supplement existing foundations for new superstructure elements such as shear walls and braced frames is relatively common in seismic rehabilitation, rehabilitation of existing foundation deficiencies is comparatively less common. There are two basic reasons for this: foundation work in existing buildings is quite expensive, and there has been relatively little note in earthquake reconnaissance reports of life loss and property damage resulting from foundation failures in buildings.

Foundation analysis can be one of the most challenging areas of seismic rehabilitation. Different assumptions regarding base conditions of restraint, soil properties, and locations and types of potential nonlinearity can lead to widely varying results. For many buildings, it can take significant analytical effort in modeling and evaluating interim results to understand how the foundation interacts with the superstructure and surrounding soil under earthquake loading. Often, the weakest link or governing mechanism may be a foundation element or soil yielding, but it is only after looking at the substructure and superstructure as a whole that the sequence and nature of element behavior can be determined.

In the past, force-based analytical techniques placed emphasis on strength capacity and whether the foundation and underlying soils were "overstressed". With the advent of displacement-based analytical techniques, the extent of soil movement is acknowledged as more critical. Due to the cost and disruption of foundation rehabilitation work, the consequences of foundation deflection should be carefully evaluated to determine if there are actually going to be unacceptable movements. Large soil movements from rigid body rotation of a shear wall, for example, may have minimal consequences if the entire structure rotates, but they may have significant consequences to attached adjacent elements which are not rotating in phase or at all.

When careful analysis reveals that new foundations must be added or that existing foundations must be enhanced, the structural engineer must have a good understanding of soil engineering issues; rehabilitation goals, performance criteria, and assumptions; and construction techniques and limitations. Obviously, it is usually much more difficult to perform work inside an existing structure than it is in a new building when the site is open. Because of the cost of foundation rehabilitation, other options should be fully explored, and the need for foundation modification should be thoroughly investigated.

There are relatively few, if any, proprietary issues associated with foundation rehabilitation, though some equipment used to install new elements in limited access areas may have been developed by a specialty contractor and thus not widely available.

This chapter provides a short discussion of general goals for foundation rehabilitation, brief mention of some key analytical considerations, and general construction issues; then provides discussion of structural rehabilitation techniques for foundations; reviews common ground improvement techniques; and ends with a short discussion of other ground hazards such as fault rupture, lateral spreading, and seismic-induced landsliding.

23.2   General Goals for Seismic Rehabilitation of Foundations

The goal of any seismic evaluation is to identify deficiencies, their relative likelihood of occurrence, and the hazards they pose. The foundation must not be ignored during the evaluation, and foundation behavior response must be placed in the context of the overall performance of the building. If the foundation is identified as the weak link, the type of foundation mechanism needs to be identified A shear wall might be overstressed in shear or bending if assumed to have a fixed base, but when its small foundation is considered, rocking or overturning might be the governing mechanism. A braced frame might have adequate strength and stiffness, but the pile caps its columns sit on may not have any reinforcing to take uplift forces that occur beyond code level forces. Existing drilled piers may lack adequate confining ties in the top of the pier or insufficient lateral resistance in general or their connections to the pier cap may be insufficient.

Consideration of the foundation is an integral part of the overall rehabilitation strategy for the structure. It may be possible to change the building behavior response by superstructure rehabilitation to preclude undesirable foundation modes. When foundation work is necessary, goals for rehabilitation design include providing sufficient strength, stiffness, and ductility for compression, tension, and lateral loading; identifying a defined and ductile mechanism of energy dissipation; and minimizing gravity stress redistribution within the existing foundation system. New foundations should not undermine existing foundations, either during construction or over the long-term. Moreover, the relative lower stiffness of unconsolidated soil under new foundations versus the higher stiffness under existing older foundations needs to be considered.

23.3   Construction Issues

Construction issues are quite critical during foundation work in existing buildings and will often drive the systems and techniques being considered. Issues include:

Access and height restrictions: Installing shallow foundations, such as spread footings or grade beams, is usually done with hand methods or small excavation equipment and will rarely be a problem, though it will take longer than it would in a new building. Installing deep foundations, however, can run into several construction limitations. Drill rigs for piers, for example, are much more efficient when they are larger. Getting a drill rig into a building may require enlarging existing openings. Once inside, story heights will usually significantly limit the size of the drill rig that can be used. Special drills have been developed for use in existing buildings, but they often require at least 9 feet to 12 feet of vertical clearance. Drilling next to adjacent walls may limit the size of the pier or lead to shifting it inboard of the wall creating a horizontal eccentricity to be addressed.

Noise and vibrations limits: Pile driving imparts significant noise and vibration. Even if there were clearance outside the building for a pile driving rig, the vibration is usually too significant. Drilled piers impart less vibration, though the noise requires consideration. Micropiles have even less vibration and noise, so they are a common rehabilitation technique.

Restrictions imposed by existing utilities: Most buildings will have utilities beneath the existing ground level suspended floor or slab-on-grade. The locations and depths may not be fully known. Excavating below grade requires careful effort, often with hand methods, so that utilities are not damaged.

Restrictions associated with ongoing operations: As with any rehabilitation work in the superstructure, if the building is occupied with people or equipment, foundation demolition, drilling, and excavation work will have to be coordinated.

Contaminated soil: There can be contaminated soil underneath the existing building, particularly if it has or had industrial uses. Removal of contaminated soil requires special techniques and must be taken to special landfills, increasing costs.

23.4   Analytical Issues

This document's focus is on detailing of rehabilitation techniques, not on analyzing the existing or rehabilitated structure, but it still worth pointing out a few analytical considerations in foundation modeling that often arise in seismic rehabilitation since codes and design guidelines provide limited guidance.

Modeling the base of the building: The most basic question to be established is: Where is the dynamic base of the building? If there is no basement, this is straightforward. When there is a basement, partial basement, or sloped site, this is not a simple issue. Say that the building is four stories above grade and has a one-story full basement. Figure 23.4-1 shows several possible modeling approaches. Model A is probably the most common approach—to stop the model of the superstructure at grade on a fixed base and take the results and impart them separately to the foundation walls and other elements. Model B is to ignore the ground entirely and put the base of the building at the bottom of the basement. When this is done, the inertial loads of the ground floor are usually not included. Model C is the same as Model B, except the ground floor loads are conservatively included. Model D changes the base conditions to account for vertical flexibility of the soil under the building. Significant modeling effort and variability have to be considered when springs are used. None of these models captures the "backstay" effect caused by the embedded foundation and the potential for shear reversals in the basement shear walls from soil pressures. To evaluate this effect, horizontal springs must be added to simulate the strength and stiffness of the surrounding soil, as shown in Model E. Note that this type of effect is similar but not the same as the backstay effect resulting from upper levels landing on larger, stiffer podium bases, which distribute local overturning loads out to other resisting elements using the diaphragms at the top and bottom of the podium. Nonlinearity can be added to the superstructure, substructure, and soil springs in these models as well. See below.

Modeling soil stiffness: With displacement-based analytical seismic rehabilitation methodologies, understanding and quantifying displacements has becoming increasingly necessary. In the past, when displacement was considered, it usually was in the form of construction and long-term differential settlements between columns or the modulus of subgrade reaction for gravity loading under a mat or grade beam on soft soil. During seismic loading, we need stiffness values relevant to the short-term nature of earthquake demands. ATC-40 and FEMA 356 provide detailed advice on these issues, but there

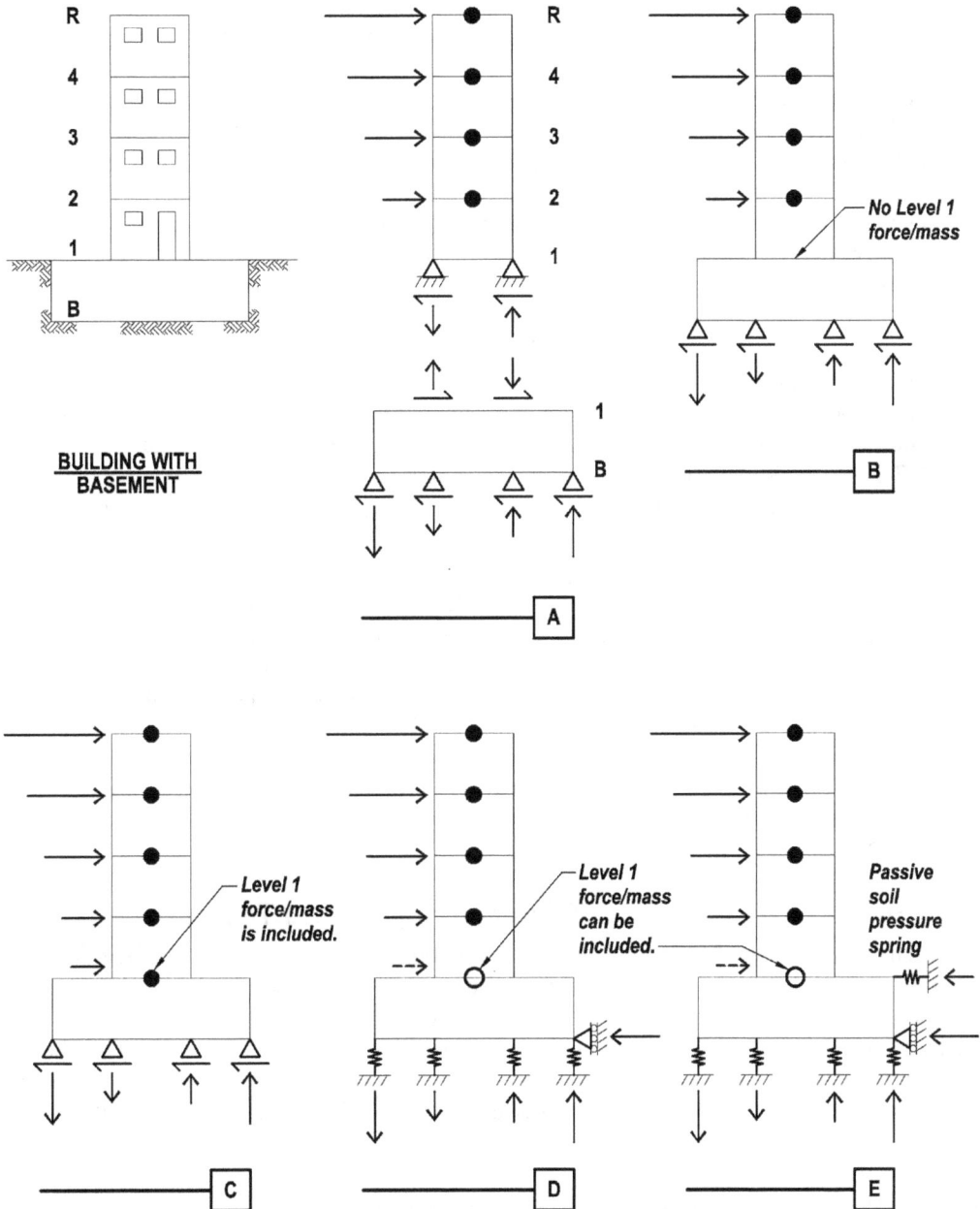

Figure 23.4-1: Modeling Approaches for Buildings with Basements

remains relatively limited data on short-term stiffnesses, particularly under high loads, and wide ranges of potential properties must be considered. These documents recommend taking half and twice the target stiffness estimates (i.e., a factor of four on the range). Key issues include whether to model soils springs with initial high stiffness relevant before yielding, a lower secant stiffness for some larger displacement, or to use nonlinear models that account for the expected nonlinear force-displacement curve of the soil. While this is the most accurate, it can take a significant analytical effort in any moderate to large building. Quantitative information on soil nonlinearity at high strains is limited. Some geotechnical engineers continue to use linear models, even in soils like clay. Significantly different results can occur if strains are sufficient to reach the point of nonlinearity. In fact, nonlinearity in the soil can lead to the accumulation of permanent deformation.

Damping, basement embedment and base slab averaging: Soil-structure interaction generally tends to reduce the input motion to the building as does an embedded basement and a slab or other foundation system that can distribute or average peak motions over the site. The input motion reduction is higher for buildings with a fundamental period below 0.7-1.0 seconds and not that significant for longer period buildings. These effects are now being considered in seismic evaluation and rehabilitation, and they are the subject of FEMA 440 (FEMA, 2005).

Second opinions: In some situations, the lower bound and higher bound of geotechnical strength and stiffness properties that are being provided can lead to significantly different results. Alternative opinions or geotechnical peer review can be advantageous in identifying alternative sources of information, narrowing the range of assumptions, or increasing the strength and displacement capability.

For detailed information on evaluation, analytical and design for foundation elements see ATC 40 (ATC, 1996), FEMA 274 (FEMA, 1997), FEMA 356 (2000), and ASCE 31-03 (ASCE, 2003).

23.5   Increasing Estimates of Capacity by In-Situ Testing

Existing shallow and deep foundations might have as-built capacities that exceed their design capacities. If these higher capacities can be confirmed, additional loads can be imposed on these foundations without any modifications to the existing foundations. Alternatively, the estimated capacity of new micropiles or drilled piers installed as part of a rehabilitation project can be verified or increased by performing in-situ load tests on them. The most common direct method for confirming these higher capacities is by performing in-situ load tests of the foundation elements.

The plate bearing test is probably the most common direct in-situ test for estimating the capacity of an existing shallow foundation. It involves the determination of the load-deformation characteristics of the soil directly below the shallow foundation. It is worth noting that indirect methods involving in-situ (instead of laboratory determination) of the settlement characteristics of foundation soils are sometimes employed. These indirect methods, which are not covered in this document, allow one to more accurately estimate settlement associated with additional loads to be imposed on the existing foundation. Examples of such methods include 1) the use of dilatometer tests to define the in-situ deformation characteristics of sandy and clayey soils and 2) the use of pore pressure dissipation techniques during cone penetration tests to estimate in-situ settlement characteristics of soft clays.

The most common direct method for estimating the load-deformation characteristics of deep foundation elements is static load tests in tension or compression.

23.5.1   Plate Bearing Tests

Preparation

Plate bearing tests are generally performed on existing shallow foundations to determine their capacities. Access to the bottom of the existing foundation, which is used as a reaction element, is required for the test to be performed. Access to the bottom of the foundation is facilitated via an access pit, which is at least 3 feet by 3 feet in plan view and extends at least 18 inches below the bottom of the foundation. The access pit is located in such a way that the exterior edge of the foundation is exposed in the pit. This access pit can be dug with a backhoe. Depending on the depth of the pit and the materials that are exposed in the pit, shoring may or may not be required.

From the bottom of the pit, a rectangular mini-tunnel that extends from the exposed to the opposite edge of the foundation is dug underneath the foundation using handmining techniques. The tunnel has to be at least 18 inches wide in cross section to facilitate the placement of bearing plates and hydraulic jacks for the test. Sometimes, it is necessary to chip off excess concrete from the bottom of the foundation to create a flat surface for the placement of the upper bearing plate. Also, the bottom of the mini-tunnel must be prepared to create a flat surface for the lower bearing plate. The access pit and the mini-tunnel are depicted in Figure 23.5.1-1. It must be noted that sometimes it is more economical to dig the access pit from the crawl space side of the existing foundation in Figure 23.5.1-1.

Equipment, Set-Up and Testing

The minimum required equipment includes the following:

> A hydraulic ram that is capable of imposing load exceeding the design capacity of the existing foundation. The pressure gage of the ram must be calibrated to allow the load imposed by the ram to be estimated. A load cell can be used in addition to the pressure gage for more accurate determination of imposed loads.
> One-inch thick steel 12-inch square bearing plate.
> Minimum of four dial gages for measuring soil deformation. Linear variable displacement transducers or transformers (LVDT)s could be used in lieu of dial gages for measurement of deformations.

The hydraulic ram and bearing plates are set up as depicted in Figure 23.5.1-1. The test is performed by imposing load incrementally on the soil below the lower bearing plate and measuring the corresponding deformations. The procedure has been standardized as ASTM D1194.

FLOOR

Excavate and shore test pit.
Use pit for access to find
bottom of footing details,
review soil conditions and
install plate bearing rig.

CRAWL
SPACE

Place bearing test rig
jacks between bottom
of footing and soil.

Monitor load
vs. displacement

SECTION

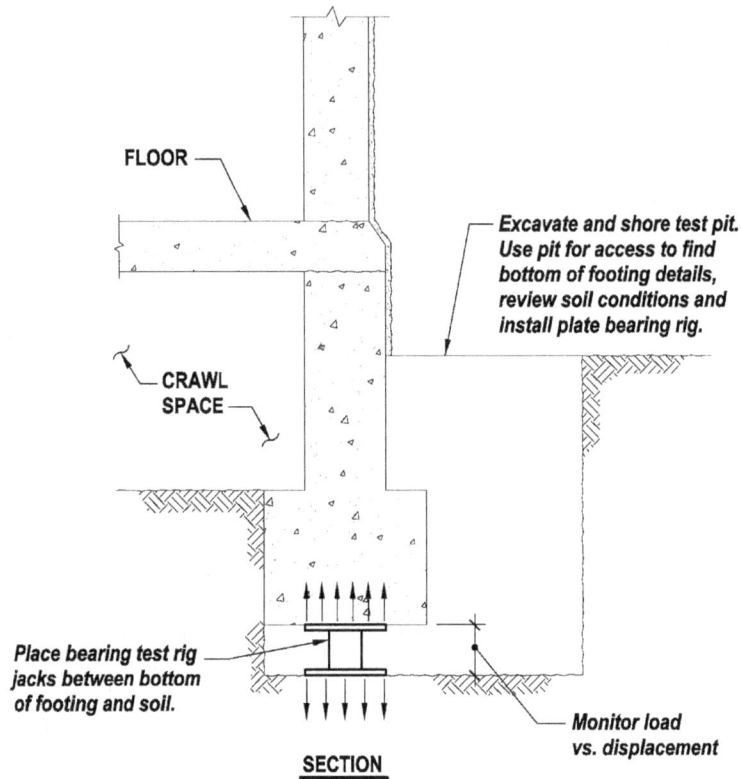

Figure 23.5.1-1: Plate Bearing Tests for In-Situ Bearing Capacity Determination

After the test is completed, the mini-tunnel and access pit are usually partially backfilled with lean concrete or controlled density fill to the top of the foundation, and the balance of the pit is backfilled with either the same material or compacted native soil.

Test Results

The load-deformation data that are recorded are applicable to the 12-inch square lower bearing plate. The data must be corrected for scale effects to apply them to the prototype foundation. Reasonable results are usually obtained when plate bearing tests are performed on very stiff clays or sandy or gravelly soil. Poor results are usually obtained when tests are performed on soft to stiff clays. Refer to Bowles (1996) for discussion on extrapolating test results.

Cost/Disruption/Challenges

Digging the access pit and mini-tunnel can be somewhat disruptive and costly, though much less expensive than the cost of foundation rehabilitation. If the pit and mini-tunnel are dug from the crawl space side, usually only handmining techniques can be employed, and the hauling of excavated spoils becomes time-consuming.

If groundwater is encountered, the conditions in the pit are mucky. Even without groundwater, the conditions in the pit are damp and cramped for the individual who has to set up the plates, hydraulic ram and dial gages in the mini-tunnel as well as for the person who has to crouch in the pit to read the dial gages while a test partner applies the load and records readings from a position near the edge of the access pit. It is often necessary to provide a plastic covering at the bottom of the pit.

Plate bearing tests on clayey soils can be time consuming because it takes a longer time for the deflection under each load increment to level off.

23.5.2   Static Tests on New Deep Foundations

Preparation

Static load tests are usually performed on new micropiles to determine their axial capacities because of the potential effects of installation procedures on the capacities. The tests can be performed in compression or tension. For a compressive type load test, reaction micropiles must be installed at a distance of at least three times the diameter of the test or reaction micropile, whichever is greater, to minimize the potential for group effects between the test pile and the reaction elements. For tension tests, timber mats could be used as reaction elements in lieu of reaction micropiles. The test and reaction micropiles should be allowed to cure for at least seven days after installation before the load test is performed.

Static load tests can also be performed on new drilled piers installed as part of a rehabilitation project as a means of increasing their estimated axial capacities. The preparatory work described above for micropiles also applies to drilled piers. Because of the size of drilled piers in comparison to micropiles, the spacing between the test pier and the reaction piers is much larger. This implies that a much larger reaction beam is required for tests on drilled piers.

Equipment, Set-Up and Testing

The minimum required equipment includes the following:

A reaction beam spanning between the reaction elements and capable of sustaining the maximum test load without excessive deflection.

A hydraulic ram that is capable of imposing load exceeding the design capacity of the existing foundation. The pressure gage of the ram must be calibrated to allow the load imposed by the ram to be estimated. Usually, a load cell is used in addition to the pressure gage for more accurate determination of imposed loads.

An independent reference beam with supports that are located away from the test or reaction micropiles.

Minimum two dial gages for measuring the deflection of the pile head. LVDTs could be used in lieu of dial gages for measurement of deformations. Whether LVDTs or dial gages are used, a secondary system of deflection measurement is required as a back-up.

The reaction beam, the reference beam, hydraulic ram and dial gages are set-up as depicted in
Figure 23.5.2-1 for compression tests and in Figure 23.5.2-2 for tension tests. The test is
performed by imposing load incrementally on the soil below the lower bearing plate and
measuring the corresponding deformations. The maximum test load is usually about 1-1/2 to 2
times the design load. The test can be performed in accordance with ASTM D1143 (for
compression tests) and ASTM D3689 (for tension tests).

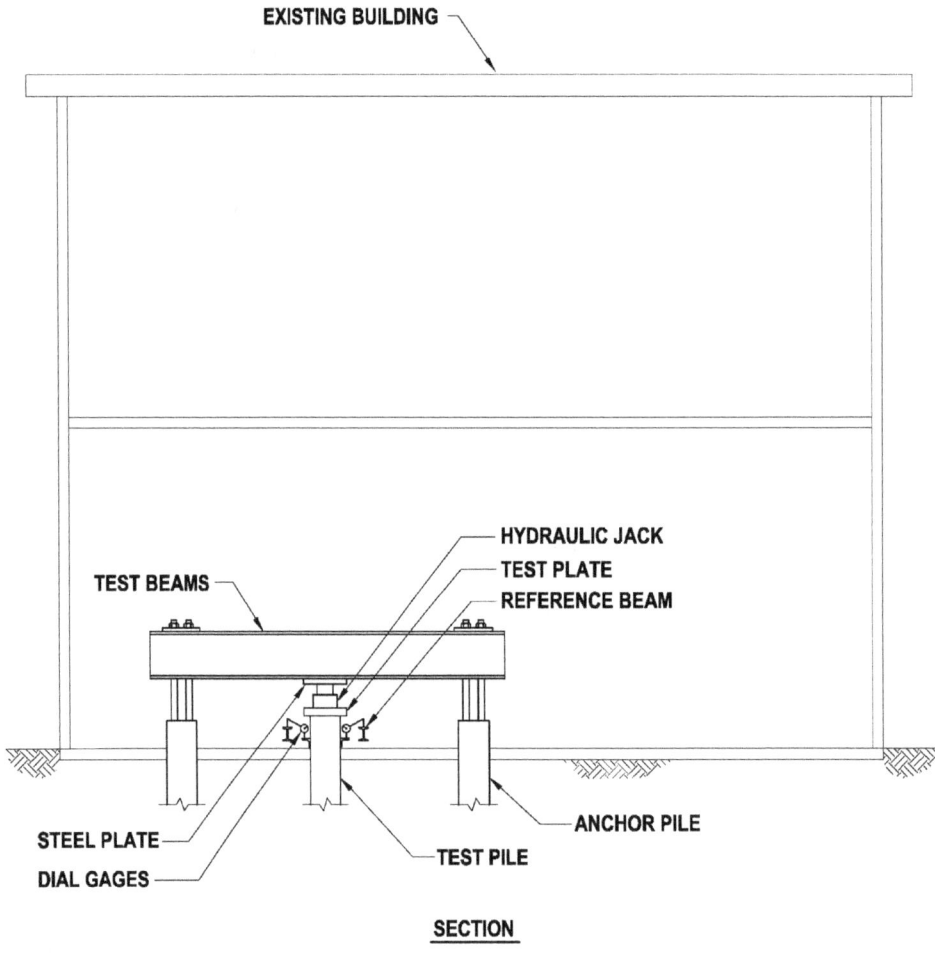

Figure 23.5.2-1: Static Pile Load Test in Compression

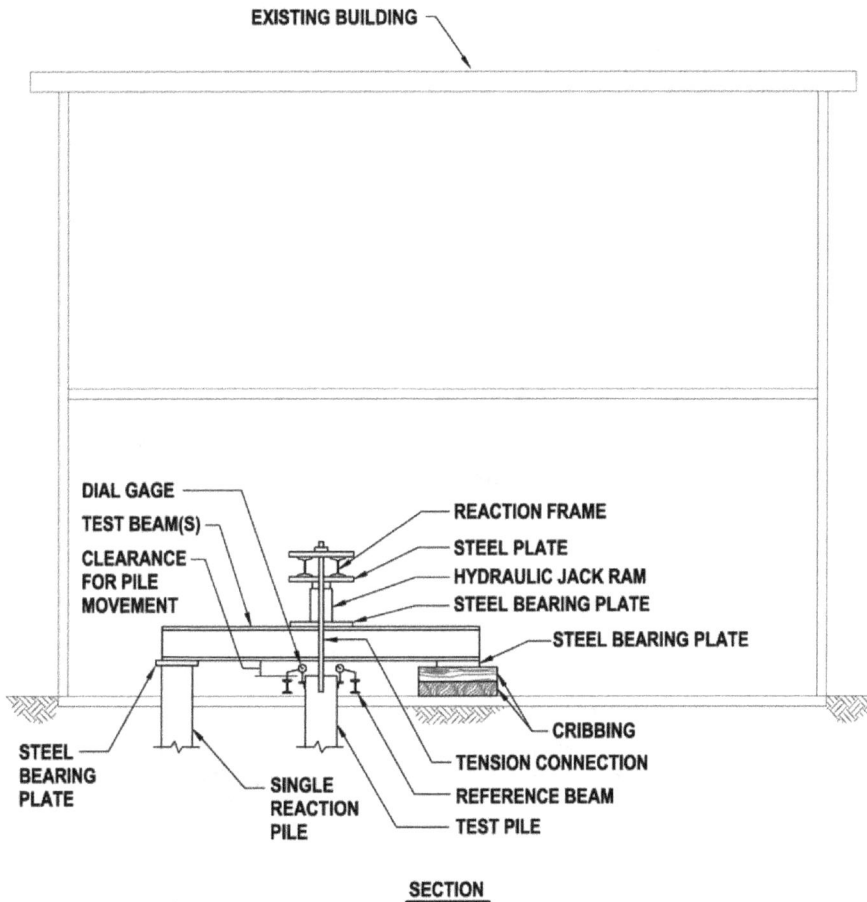

Figure 23.5.2-2: Static Pile Load Test in Tension

Test Results and Interpretation

The load-deformation data that are recorded must be interpreted in two phases. The first phase
involves the determination of the axial capacity in tension or compression of the test foundation
element. The next phase involves interpreting the axial capacity relative to the known
foundation conditions, such as the applicability of the axial capacity to a group of deep
foundation elements or the applicability of the observed settlement from the load tests, given its
short duration, to a production deep foundation element bearing in clayey soil that could
consolidate.

Cost/Disruption/Challenges

Setting up and performing a load test can be quite costly. Setting up can be time consuming because there is more manual labor involved in transporting test equipment from one test location to another in cramped situations. In the case of drilled piers, it might be impossible because of access related issues to set up an adequate reaction beam for a static load test. It may be possible, however, to find locations on site that are not within the building, such as parking lots or landscaped areas, with similar underlying soils and perform the test on elements that will not be used under the building.

23.5.3   Static Load Tests on Existing Deep Foundations

While theoretically possible, this approach is so disruptive and costly that it is generally not implemented in practice except where the existing foundation consists of timber piles or where the existing pier is located on the exterior of the building. This kind of testing would require temporary shoring of the column supported by the pier or pile to be tested and removal by cutting of the structural connection between the deep foundation element and the column. The top of the pier or pile must be accessible to allow for a load test set up. For a compressive type load test, reaction micropiles or piers must also be installed.

23.6   New Foundations

23.6.1   Types of New Foundations Commonly Used in Seismic Rehabilitation

Foundation elements can be broadly classified into two basic categories: shallow and deep foundations. Shallow foundations include continuous strip footings, isolated spread footings, grade beams, and mats. Deep foundations include drilled piers and micropiles. Driven piles are rarely used in existing construction due to access and vibration limitations. Figure 23.6.1-1 shows examples of these foundation types. Several excavation approaches are shown in the figures. In cohesive soils, the soil may be able to be cut without it sloughing into the hole. Metal stayforms (expanded metal lath forms) are sometimes used when there is some risk of the soil sloughing after the intial excavation. The stayforms are left in place when the concrete is poured. When the excavation gets to a certain depth, however, shoring can be required due to safety regulations or an open cut excavation can be used. In cohensionless soils, like sand, an open cut excavation will be necessary. A form can be placed, the concrete poured, the form removed, and then soil backfilled into the remaining open cut. Alternatively, the form can be left out and the concrete for the footing "overpoured" in the full open cut. The eccentricity of the overpour should be evaluated.

23.6.2   Add Shallow Foundation Next to Existing Shallow Foundation

Description of the Rehabilitation Technique

When a concrete overlay is placed against an existing wall, a new footing is typically needed. A common situation is the existing footing is a continuous strip footing and the new footing is either a strip footing or a grade beam. Figure 23.6.2-1 shows an example of a new concrete wall and footing against and existing unreinforced masonry wall and concrete strip footing.

MOMENT FRAME OR BRACED FRAME COLUMN

CONCRETE OR MASONRY SHEAR WALL

SPREAD FOOTING A

STRIP FOOTING, GRADE BEAM, OR MAT FOUNDATION B

Overexcavate and backfill or overpour concrete footing in cohesionless soil.

Vertical cut may be possible in cohesive soil.

DRILLED PIERS C

MICROPILES D

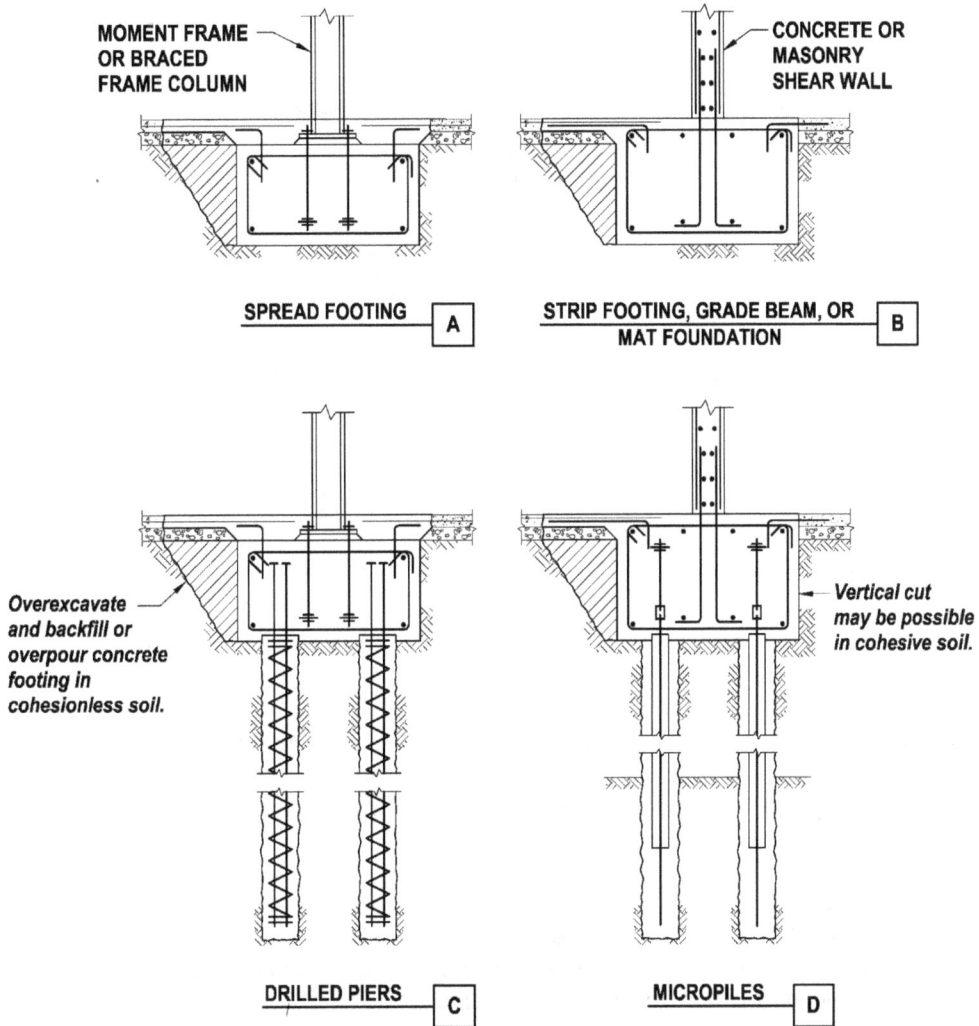

Figure 23.6.1-1: Types of New Foundations Commonly Used in Seismic Rehabilitation

Design Considerations

Effective footing width: Several approaches to footing design are used. One is to assume only the new footing resists the loads under the new overlay. Another is to share loads between the new and existing footing simply on the basis of area. The most sophisticated approach is to recognize the potentially different stiffness between the soil under the existing footing which has been consolidated already and the soil under the new footing which is likely to be more flexible since loading is likely to be lighter and only newly applied. Sometimes jacking is employed to transfer loads to new foundations.

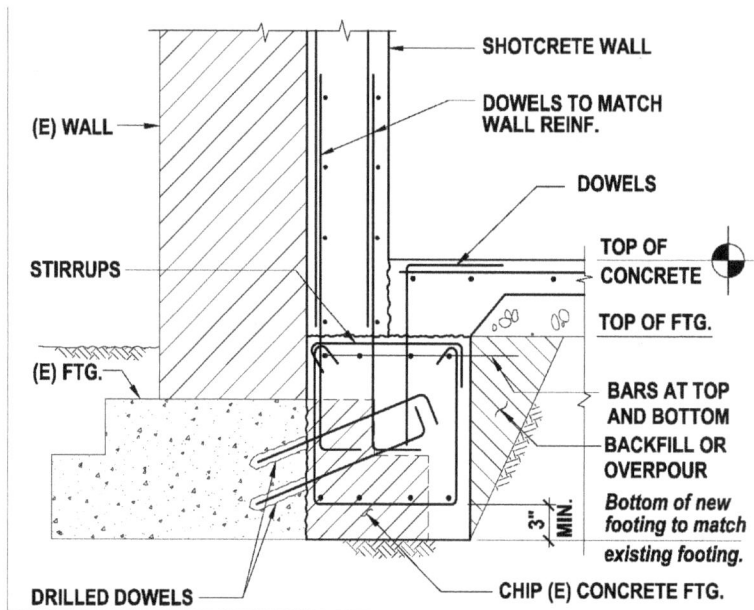

Figure 23.6.2-1: New Concrete Strip Footing Next to Existing Strip Footing

Shear transfer: It is standard practice to connect the new and existing footings with drilled dowels, though it useful to consider whether the dowels are actually necessary elements. Dowels in the footing and wall above should be designed to be sufficient to transfer the force intended to be resisted under the existing footing.

Unreinforced existing footings: The existing footing may be unreinforced masonry or poorly reinforced concrete. If the footing is wide enough so that so beam action will result under bearing pressure, the bottom drilled dowels can be extended deep into the existing footing near the base of the footing to serve as positive reinforcing.

Detailing and Construction Considerations

New footing is deeper than existing footing: A key goal when adding a new footing is not to surcharge or undermine an existing footing. The best approach, then, is to match the new and old footing depths. This is, of course, not always possible. Figure 23.6.2-2 shows the situation when the new footing needs to be deeper than an existing footing. If excavation proceeds without underpinning, particularly in soils with minimal cohesion, soil can slough away from under the existing footing into the new excavation leading to damaging footing movement. Underpinning is used to address this situation. Underpinning means digging a series of short length pits separated by a sufficient distance, digging under the existing footing adjacent to the pit, adding concrete to the base of the final excavation depth, and then going back and completing the underpinning in between the initial pits.   An alternative underpinning approach is to place long underpinning piers intermittently beneath the new footing to derive support at depth so that the typical new footing need not be deeper than the existing footing.

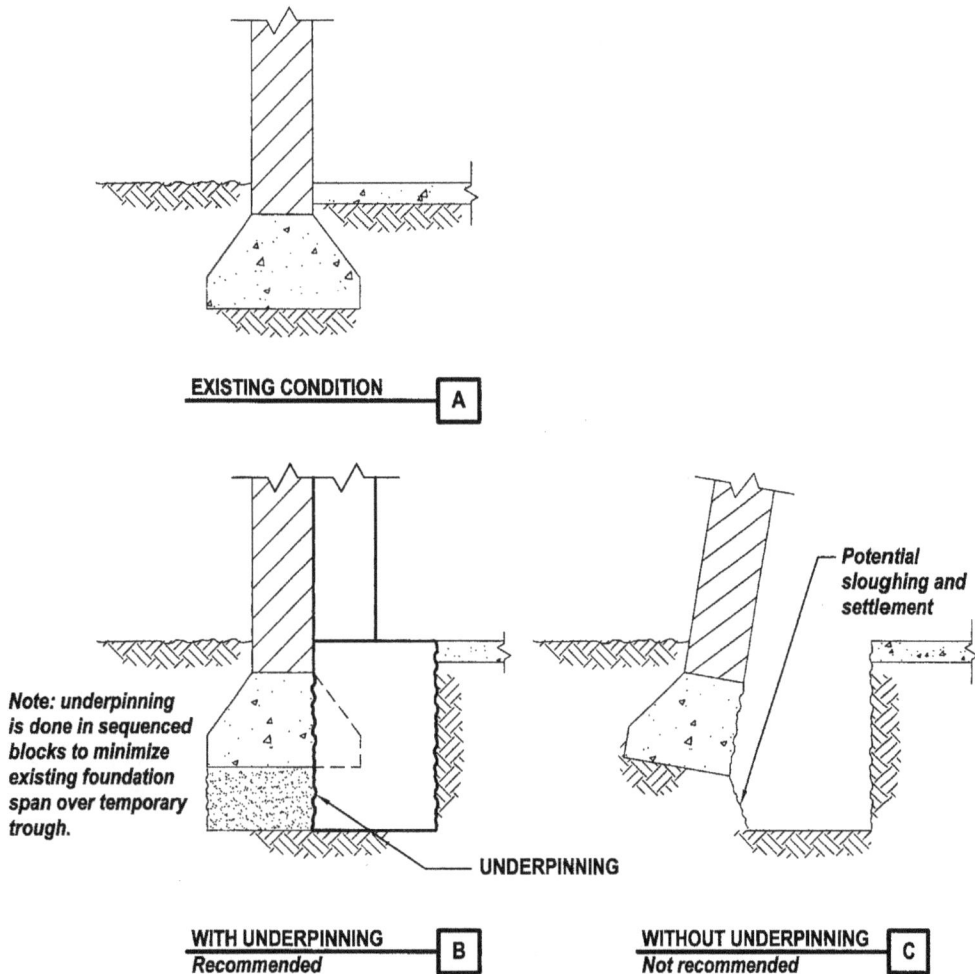

EXISTING CONDITION [A]

Note: underpinning
is done in sequenced
blocks to minimize
existing foundation
span over temporary
trough.

— UNDERPINNING

Potential
sloughing and
settlement

WITH UNDERPINNING [B]
Recommended

WITHOUT UNDERPINNING [C]
Not recommended

Figure 23.6.2-2: New Shallow Footing is Deeper than Existing Shallow Footing

New footing does not need to be a deep as existing footing: Figure 23.6.2-3 shows the situation when the new footing does not need to be as deep. If the excavation is kept shallow, the new footing when loaded can impart additional and eccentric loads into the existing footing that may not be desirable. As a result, it is common to extend the bottom of the new footing down to match the depth of the existing footing. The extension is often lightly reinforced.

Existing footing is in the way: As Figures 23.6.2-1 to 23.6.2-3 show, the existing footing will often extend inboard from the existing wall underneath the new wall. To place the new footing, the existing footing often must be chipped away to develop a properly reinforced footing. This can be done with jackhammering or sawcutting. The capacity of the existing footing during the temporary condition where it is smaller and eccentrically loaded should be verified as adequate.

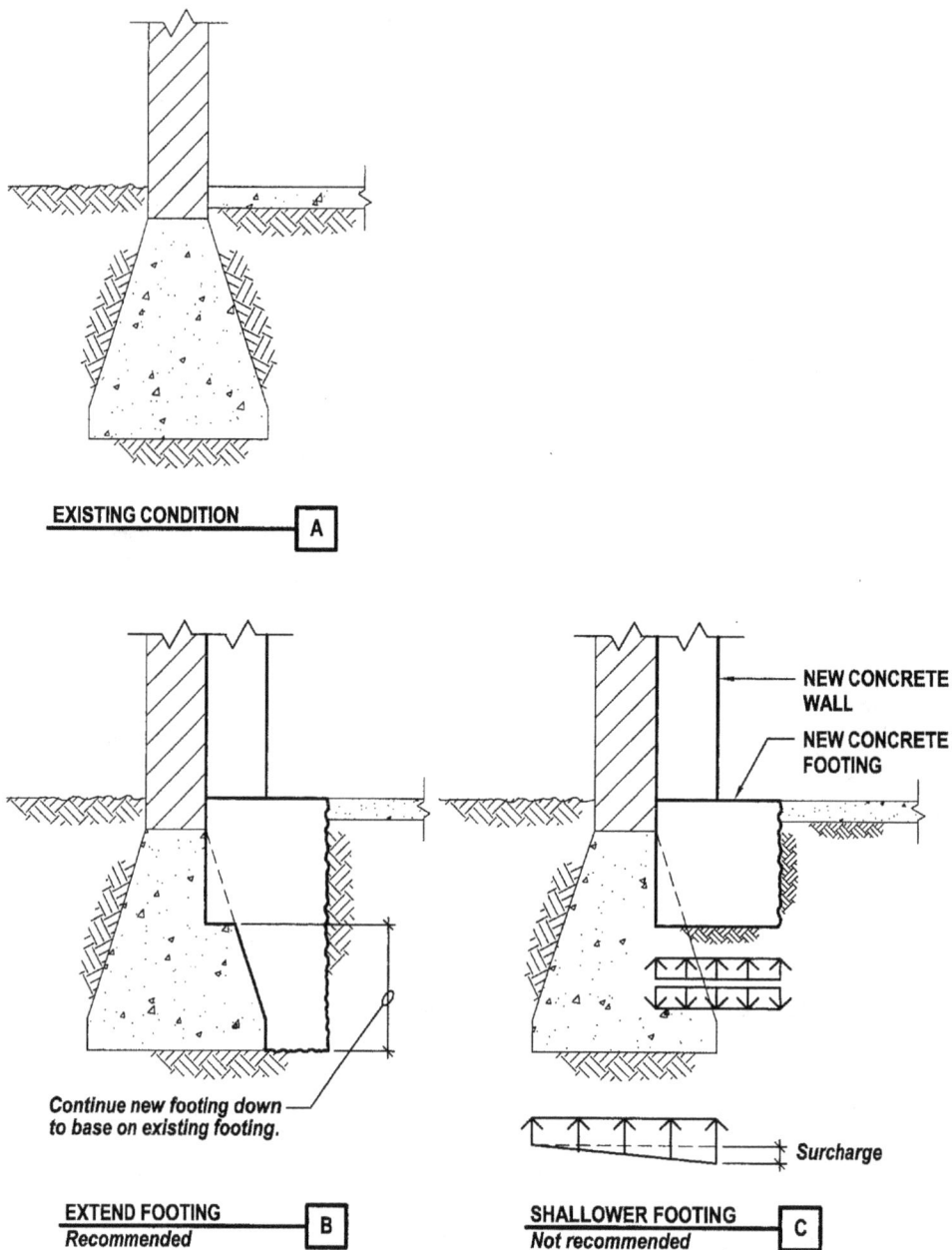

EXISTING CONDITION [A]

EXTEND FOOTING [B]
Recommended

Continue new footing down
to base on existing footing.

SHALLOWER FOOTING [C]
Not recommended

NEW CONCRETE
WALL

NEW CONCRETE
FOOTING

Surcharge

**Figure 23.6.2-3: New Shallow Footing Does Not Need to be as Deep as
Existing Shallow Footing**

Cost/Disruption

Adding a new footing is quite disruptive and costly. The existing slab-on-grade must be sawcut and removed, then the trench excavated, drilled dowels installed, rebar laid, debris in the footing removed, and concrete placed. This is all time-consuming, messy, and noisy.

23.6.3   Add Shallow Foundation Next to Existing Deep Foundation

Adding a new shallow foundation next to an existing deep foundation is relatively rare for two reasons. First, the existing foundation was deep because soil or structural loading conditions would not permit a shallow foundation. Without ground remediation, a new foundation would have the same issue. Second, as noted in Section 23.6.2, if the new foundation is higher than the existing foundation, the new foundation will impart gravity and earthquake loads to the existing foundation which is usually undesirable. With careful study of relative rigidity considerations, there can be situations where a new shallow foundation can be added adjacent to an existing deep foundation, such as a mat next to drilled piers. See Section 23.8.2.

There can be cases, though, when adding new deep foundations are very disruptive or not economical practicable due to existing access limitations. Sometimes a shallower foundation is added, such as a new mat next to an existing drilled pier foundation. The relative stiffness of each foundation then becomes the key consideration.

23.6.4   Add Deep Foundation Next to Existing Shallow Foundation

Adding a new deep foundation next to an existing foundation is occasionally done, such as drilled piers under a new wall next to an existing strip footing. Figure 23.6.4-1 shows an example of this technique. Drilling limitations can be significant, and they include access requirements for the drill rig, height restrictions for the drill rig, the offset needed to get the edge of the drill up against the existing wall, vibration during drilling, and utilities in the way of the drilling. Sometimes when the exterior face of the building is accessible, slanted drilling is done under the existing footing. Usually, the drilled piers are spaced at a sufficient distance that the existing footing and walls can span around or over the open hole. After the pier and new wall are installed and dowelled into the existing wall and footing, a composite system has been created. While many engineers simply take gravity in the existing spread footing, and overturning in the piers, live loads and earthquake loads are of course actually distributed throughout the system by relative rigidity.

23.6.5   Add Deep Foundation Next to Existing Deep Foundation

There will also be situations where new deep foundations are added next to existing deep foundations. New deep foundations include drilled piers and micropiles. See Section 23.8 for examples.

23.7   Structural Rehabilitation for Existing Shallow Foundations

23.7.1   Goals

Typical structural improvements to existing shallow foundations can be simplified into two basic categories: enhancing compression capacity and enhancing tension capacity.

Need suffient access through door opening for drill rig.

Property line limits access from exterior.

Check with driller on minimum distance from face of wall to pier centerline

Special drill rigs available for low overhead clerance

SHEAR WALL

Backfill or overpour

Shear key recommended

Tighter spiral

Typical sequence:

1. Drill hole.
2. Monitor vibration in walls.
3. Place cage and concrete for pier.
4. Excavate for new footing.
5. Pour new footing/slab and then wall above.

4" cover recommended

Wider spiral

CONCRETE BAR SUPPORTS WIRED TO CAGE

DRILLING HOLE FOR PIER　[A]

INSTALLED CONDITION　[B]

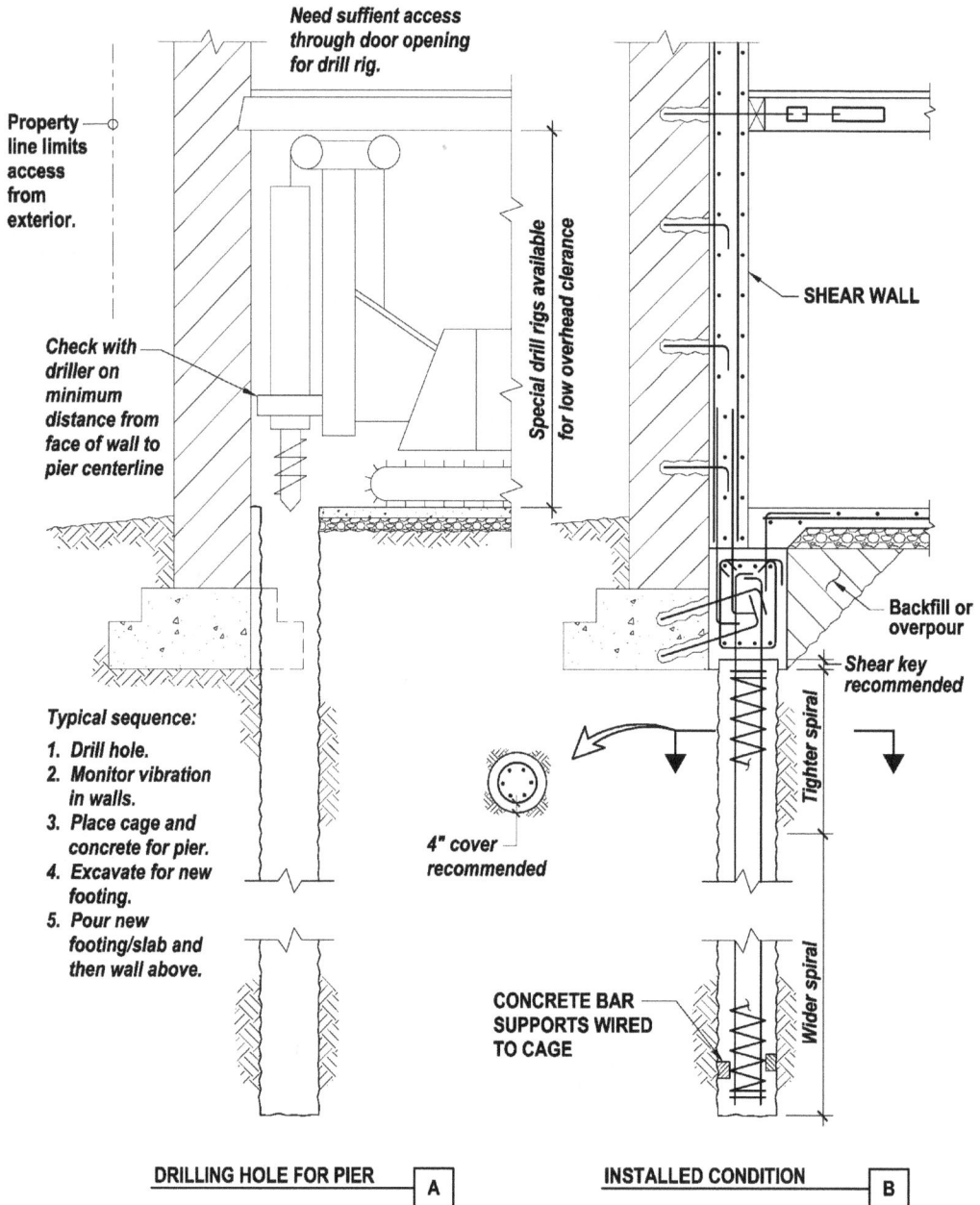

Figure 23.6.4-1: New Drilled Pier Next to Existing Strip Foundation

General techniques for improving inadequate compression capacity: Compression strength capacity of existing spread and strip footings can be addressed by widening the footing base; replacing the footing with an enlarged foundation; adding micropiles, screw anchors or drilled piers adjacent to the existing footing; adding micropiles through the existing footing; or adding grade beams to connect isolated spread footings together.

General techniques for improving inadequate tension capacity: Improving inadequate tension capacity of existing spread and strip footings uses similar techniques to those for improving compression capacity, including widening the footing base to increase the dead load; replacing the footing with an enlarged foundation; adding micropiles, screw anchors, or tie-downs adjacent to or through the existing footing; or adding grade beams to adjacent footings and columns to pick of dead load to resist uplift.

The following sections provide some examples of rehabilitation techniques for existing shallow footings.

23.7.2   Add Micropiles Adjacent to an Existing Strip Footing

Deficiencies Addressed by Rehabilitation Technique

  Inadequate compression capacity at the toe of strip footing beneath a wall
  Inadequate tension capacity at the heel of a strip footing beneath a wall

Description of the Rehabilitation Technique

To improve the compression and/or tension capacity of the existing footing, the footing is widened and micropiles, also known as pin piles, are added. Figure 23.7.2-1 provides an example.

Design Considerations

Research basis: FHWA (2000) provides guidelines for the design and construction of micropiles.

Compression strength and stiffness: When micropiles are added together with the strip footing, resistance is shared between the two different elements, depending on their relative rigidity. Micropile strength and stiffness are given in the geotechnical report. Governing strength depends on both the soil capacity and the structural capacity of the pile, including the pipe, grout, and reinforcing bar. Compression stiffness considers the pile elements and surrounding soil movement.

Tension strength and stiffness: Uplift resistance is taken by the micropiles. Structural tension strength is lower than compression strength in the micropiles and is usually based on just the reinforcing bar, unless special details are used to engage the top of the casing in tension. Tension stiffness is also usually lower; tension flexibility comes from the reinforcing bar elongation and surrounding soil movement.

Corrosion effects: Permanent casing associated with micropiles is typically uncoated. Depending on the corrosivity of the soil, corrosion of the permanent casing can occur over time. Techniques are available for estimating the extent of thickness of the steel pipe lost to corrosion;

with the estimates, a reduced thickness and reduced lateral and buckling capacities of the pile can be calculated.

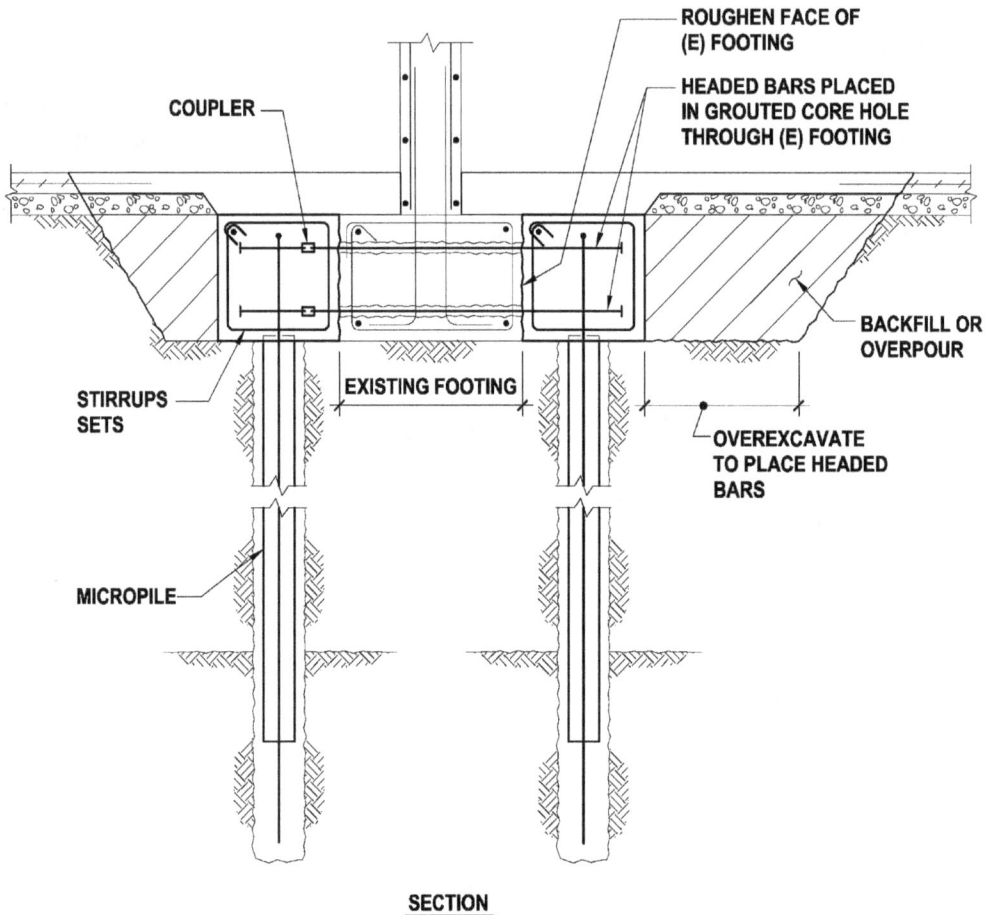

Figure 23.7.2.-1: Micropile Enhancement to Existing Strip Footing

Testing: Performance and proof load testing are performed at the start of and periodically during construction to verify that specified design capacities will be achieved. During performance testing, the test piles are usually loaded to 2.0 to 2.5 times the design load. Proof testing, on the other hand, involves testing the pile to 1.33 to 1.67 times the design load. Proof testing is usually limited to a percentage of the production piles. Creep tests are typically performed as part of the performance and proof tests, especially if the micropiles are to be bonded in clayey soils that are susceptible to creep. PTI (1996) provides guidelines on performing and evaluating performance, proof, and creep tests on foundation elements.

End bearing vs. friction: Because of its small size, micropiles generally derive most of their capacities from friction. The geotechnical compression capacity of the micropile is therefore generally equal to the axial tensile capacity.

Filling the annulus with grout: Where the cutting tool of micropile drilling equipment creates a hole slightly larger than the permanent casing, an annulus is created around the casing. Typically, this annulus is not grouted.

Detailing and Construction Considerations

Detailing and construction considerations for adding micropiles to an existing footing include the following.

Connecting to the new footing: Figure 23.7.2-1 shows bars drilled all the way through the existing footing. If this not done, the existing capacity of the footing for bending and the center where the moment is largest must be checked; it is unlikely to be acceptable. In the figure, the through dowels are installed from the right and coupled on the left. Headed bars are shown for ease in installation. Hooked bars could be used, but they would trigger a position coupler (one that eliminates the need to rotate the bar), at least for the bottom row of bars. To install the longer dowels, over excavation of the adjacent soil is needed. This needs to be understood during detailing, as their may be existing elements on top of that portion of the slab.

Access and height limitations: Adequate clearance must be available for the equipment used to install micropiles inside existing buildings.

Anchorage to the footing: Figure 23.7.2-2 shows a micropile and some of its details. In this figure, tension is taken by threaded rod and the plate at the top of the rod. Sufficient embedment of the plate above the base of the footing is needed to develop the strength of the rod. Similarly, the bottom plate is designed to take the compression and deliver it to the pipe. In some cases, the bottom plate may not be necessary as the grout diameter or top plate can be sufficient. If the top plate is used, it must be sufficiently deep below the top of the footing so it is not the weak link.

Bar types and size: Bar types include ASTM A722 high strength threadbar, with F_y = 150 ksi, with common sizes of 1", 1-1/4", 1-3/8" and 1-3/4" diameter. CALTRANS has typical details using #18 bars in ASTM A615 steel, where ends needing nuts or couplers are threaded.

Pipe types and sizes: API casing with F_y = 80 ksi is commonly used, with 7" diameter and 9-5/8" diameter pipes being common.

Depth of pipe and grouting: The pipe typically goes down into the bearing layer the requisite depth. The reinforcing bar usually continues deeper. Grouting fills up the hole at the base, the annulus around the pipe and the inside of the pipe. Post-grouting or secondary grouting can be used at the base to increase the bar capacity.

Strain limits at the top: To increase the length over which the bar is strained in tension, the top of bar below the anchorage plates are sometimes debonded with a greased PVC pipe.

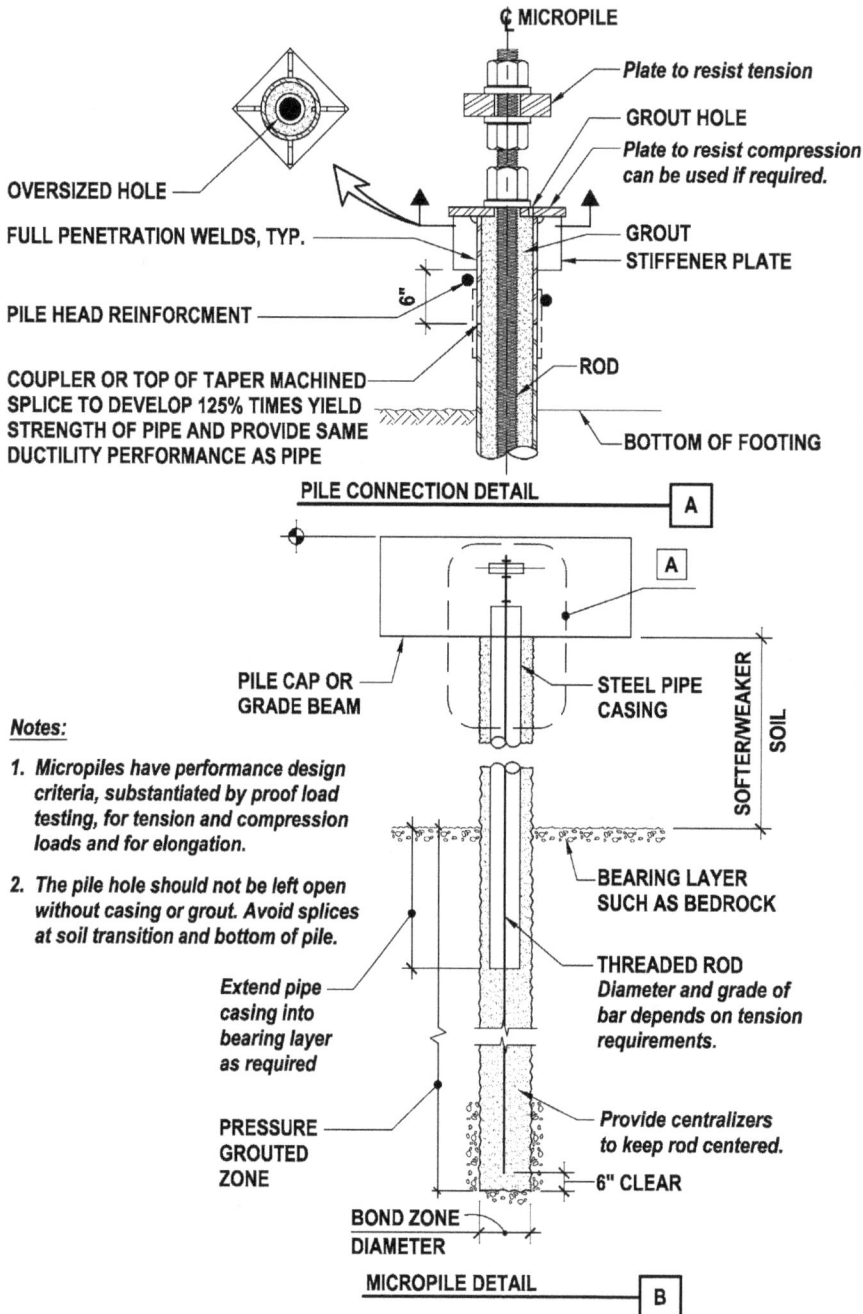

Ç MICROPILE

Plate to resist tension

GROUT HOLE

Plate to resist compression
can be used if required.

GROUT

STIFFENER PLATE

OVERSIZED HOLE

FULL PENETRATION WELDS, TYP.

6"

PILE HEAD REINFORCMENT

ROD

COUPLER OR TOP OF TAPER MACHINED
SPLICE TO DEVELOP 125% TIMES YIELD
STRENGTH OF PIPE AND PROVIDE SAME
DUCTILITY PERFORMANCE AS PIPE

BOTTOM OF FOOTING

PILE CONNECTION DETAIL   A

A

PILE CAP OR
GRADE BEAM

STEEL PIPE
CASING

SOFTER/WEAKER SOIL

Notes:

1. Micropiles have performance design
   criteria, substantiated by proof load
   testing, for tension and compression
   loads and for elongation.

2. The pile hole should not be left open
   without casing or grout. Avoid splices
   at soil transition and bottom of pile.

BEARING LAYER
SUCH AS BEDROCK

Extend pipe
casing into
bearing layer
as required

THREADED ROD
Diameter and grade of
bar depends on tension
requirements.

PRESSURE
GROUTED
ZONE

Provide centralizers
to keep rod centered.

6" CLEAR

BOND ZONE
DIAMETER

MICROPILE DETAIL   B

Figure 23.7.2.-2: Micropile Details

Cost/Disruption

Micropiles are typically less expensive than drilled piers, unless very large capacities are required. They require less headroom and smaller footings to receive the bars and pipe. Excavation noise and dust, and drilling and grouting noise must be considered as part of the rehabilitation strategy.

Proprietary Issues

Micropile specifications are often written like tiebacks, so that the contactor must design and build the micropile to meet performance requirements. Figure 23.7.2-2 shows a generic type of pile. There are other proprietary piles that use a pointed pipe casing as the drill.

23.7.3   Enlarge or Replace an Existing Spread Footing

Deficiency Addressed by Rehabilitation Technique

> Inadequate compression capacity at a spread footing
> Inadequate tension capacity at a spread footing

Description of the Rehabilitation Technique

An existing spread footing may be under a braced frame, moment frame or a concrete column below a discontinuous shear wall and be subjected to compression or tension forces that exceed the footing capacity. The existing footing can be enlarged or replaced to increase compression capacity or the dead load for resisting tension.

Design Considerations

Research basis: No research specific to enlarging or replacing existing footings has been identified.

Bending moment and shear checks in enlarged footing: Gaining large increases in compression capacity by enlarging an existing footing is often difficult given the limits of the existing footing. In Figure 23.7.3-1, reinforcing is drilled in from the sides, but does not go through to the other side. The shear capacity of the footing is not increased. The bending capacity has to be checked at critical locations "A" and "B". Location A will typically govern. If sufficient capacity cannot be achieved, the footing can be replaced as shown in Figure 23.7.3-2.

Tension capacity: Tension capacity can be quite limited if the existing spread footing only has bottom reinforcing bars which would be typical. Drilled dowels can be added to the top of the footing and top steel added in the slab-on-grade level. See Section 23.9 for a similar example in a pile cap.

Detailing and Construction Considerations

Detailing and construction considerations for enlarging or replacing an existing spread footing include the following.

Existing reinforcing: Existing reinforcing should be preserved in the footing. This will typically require placing new drilled dowels at a higher elevation, with a resulting lower moment capacity.

SLAB-ON-GRADE

SPREAD FOOTING

EXISTING SPREAD FOOTING   A

Notes:

1. Establish existing layers and organize with new bars so existing bars are not cut.

2. Hole for bar without coupler can be smaller.

3. Check new bar capacity at location Ⓑ. Check existing bar capacity at location Ⓐ.

CORE FROM BAR IN
CROSSING DIRECTION

CORE FOR DRILLED DOWEL

DRILLED DOWEL
TO (E) SLAB

BACKFILL OR
OVERPOUR

CLASS "B" LAP
W/ (E) BAR

Ⓐ   Ⓑ

Shorter over-excavation can be used when bar is coupled.

To improve shear transfer, new footing can extend under existing footing.

Longer overexcavation needed if drilled dowel bar is not coupled.

ENLARGED FOOTING   B

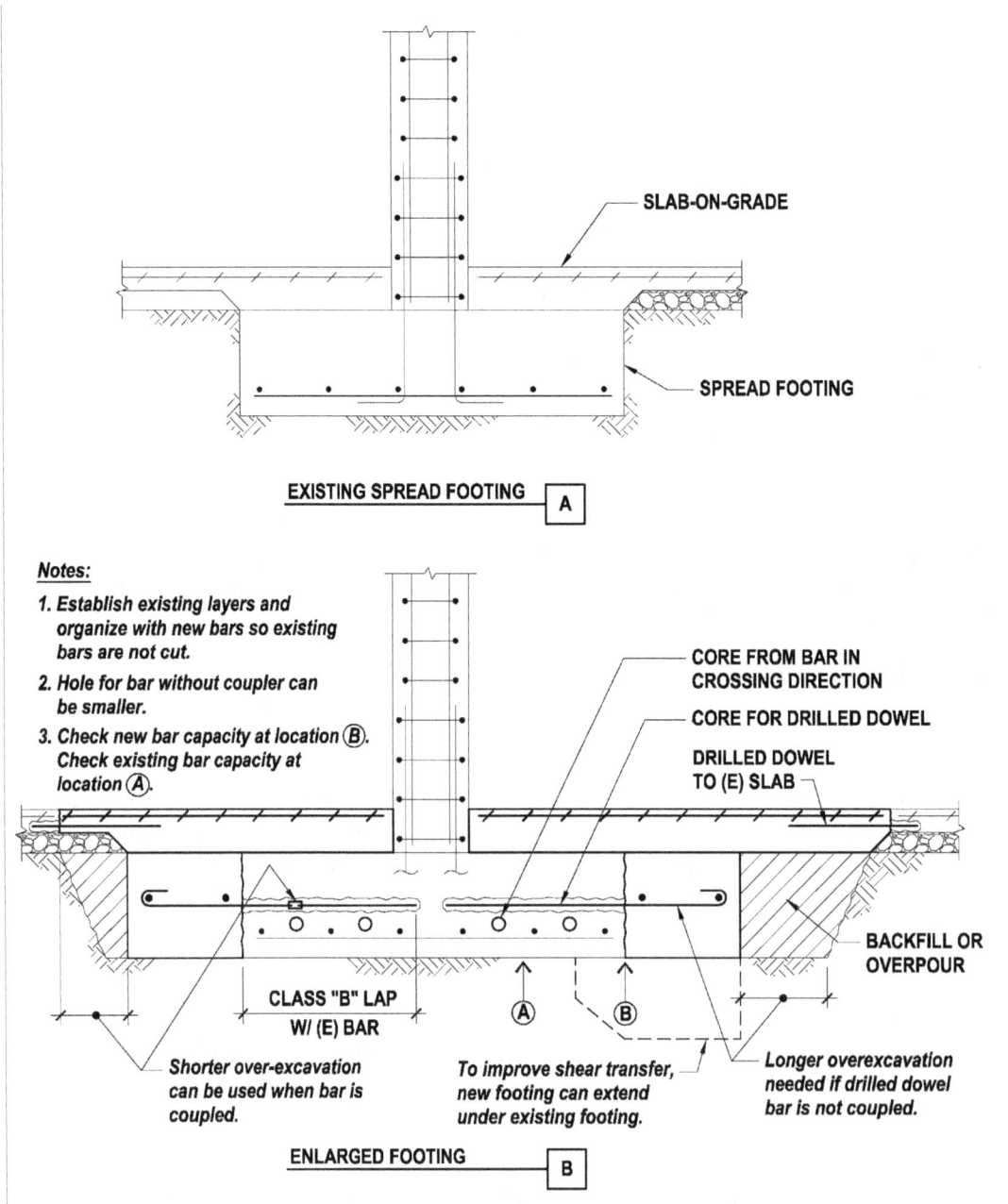

Figure 23.7.3-1: Enlarge Existing Spread Footing

COLUMN WITH
DISCONTINUOUS
SHEAR WALL
BEYOND

*Note: Similar approach can be
used in strip footing.*

WALL BEYOND

Provide shoring,
then remove
existing footing.

*See discussion on column
strengthening methods.*

COLUMN BENEATH
DISCONTINUOUS
SHEAR WALL

*Preserve existing column
bars. Supplement if required.*

PRESERVE
EXISTING
REINFORCING.

REMOVE (E)
FOOTING

REPLACEMENT FOOTING   A

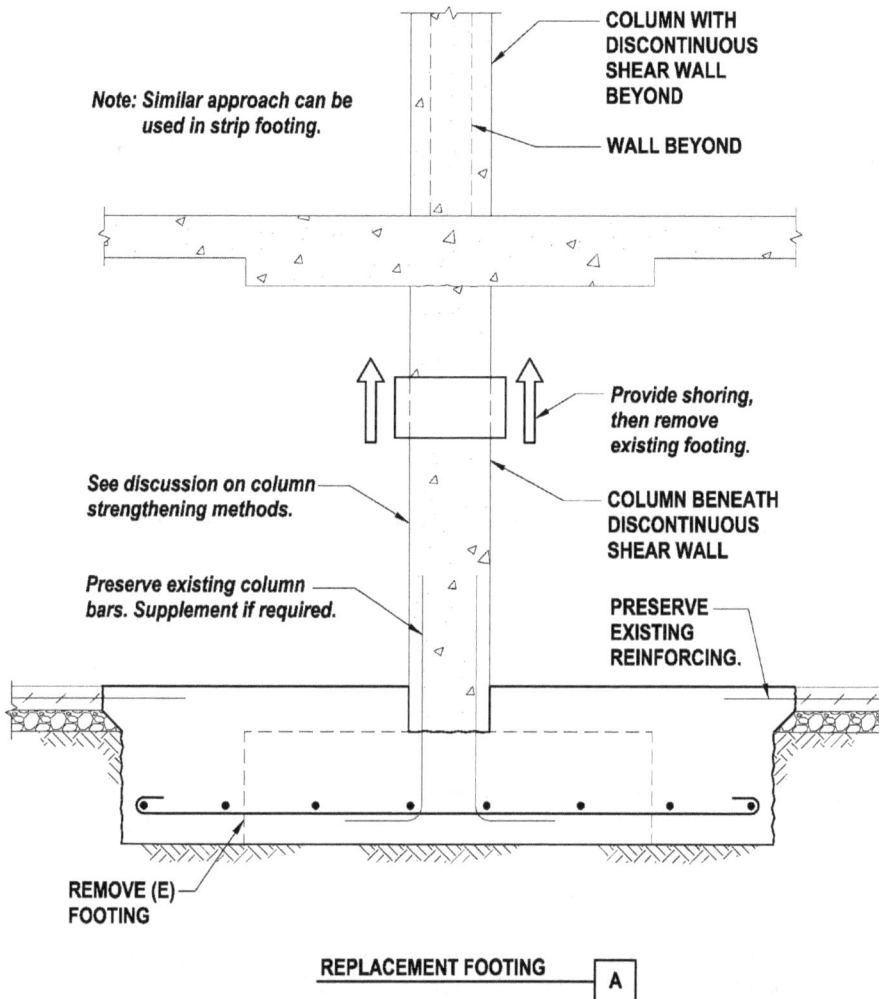

Figure 23.7.3-2: Replace Existing Spread Footing

Installing drilled dowels: Figure 23.7.3-1 shows two approaches for installing bars. On the left a coupler is used, permitting a smaller overexcavation past the footing, but triggering a hole large enough to accommodate the coupler. This is likely to force pressure grouting with nonshrink grout as the annulus will be too large for most adhesives like epoxy. On the right, a larger over-excavation is used and a single piece bar is installed.

Lapping with the existing slab-on-grade: The existing slab is likely to have wire mesh. To minimize vertical offsets the new and existing slabs should be dowelled together. Either the mesh in the existing slab-on-grade can be preserved when the slab is demolished or drilled dowels can be drilled into the edge of the existing slab.

Shear transfer between the new and existing footings: Transferring shear between the existing footings is necessary. This can be accomplished by roughening the existing footing face. Some engineers bevel the existing face as well with the top wider than the bottom, so outward pressure is exerted under compression loading. Some engineers dig the new footing slightly deeper than the existing footing and undercut the soil at the edge of the existing footing, so that the new footing acts as a corbel to resist downward pressure from the existing footing.

Shoring: If the existing footing is replaced, shoring will be needed. It is critical that the base of footing be properly compacted and the new concrete be tightly placed beneath the existing column to minimize or eliminate any settlement when the shores are removed.

Cost/Disruption

Enlarging or replacing an existing footing is a localized but disruptive process, involving excavation, dust, mud, drilling/jackhammering noise and concrete placement. Protection of existing finishes in the vicinity and in the working path is necessary.

23.8   Structural Rehabilitation for Existing Deep Foundations

23.8.1   Goals

Typical structural improvements to existing deep foundations can be simplified into several basic categories: enhancing the overall compression capacity, tension capacity, or lateral capacity of the foundation; and improving the ductility and detailing of specific elements or connections within the system.

General techniques for improving overall inadequate compression, tension and lateral capacity: Inadequate strength and deformation capacity of existing pile and pier foundations can be addressed by adding new shallow adjacent shallow foundations, and new piers or micropile foundations, either in vertical or battered orientations.

General techniques for improving inadequate improving ductility and detailing: Inadequate confinement can be improved with enlarged or replacement pier and pile caps; lack of top steel in pier and pile caps can be addressed with new concrete overlays on top of the cap.

The following sections provide some examples of rehabilitation techniques for existing deep footings.

23.8.2   Add a Mat Foundation, Extended Pile Cap or Grade Beam

Deficiency Addressed by the Rehabilitation Technique

The deficiency addressed by this technique is inadequate compression capacity of an existing deep foundation element. The technique involves taking advantage of the contributions of shallow foundation elements that are part of the overall foundation system.

Description of the Rehabilitation Technique

When existing piers or piles have inadequate capacity, the usual approach to increasing their capacity is to install new micropiles or piers connected by grade beam to the existing adjacent piles or piers. Where competent bearing soil is within five feet of building ground floor, an alternative approach to installing new piles or piers is to widen and deepen the cap or grade beams atop the pier or pile or connecting adjacent piles or piers. Figure 23.8.2-1 shows an example of existing piers whose capacities are augmented by installing a mat between the piers.

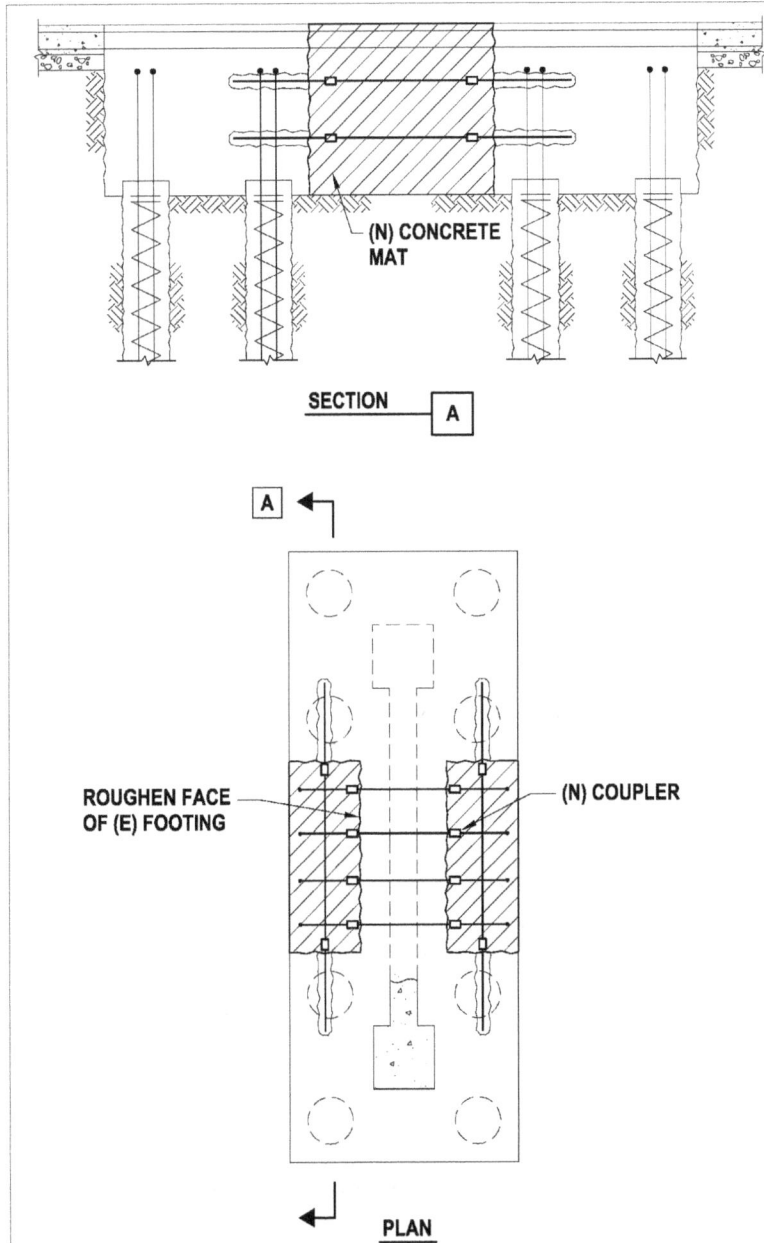

Figure 23.8.2-1: New Mat Foundation Between Existing Drilled Piers

Design Considerations

Analysis: Several approaches to design are used. One is to assume that only the new mat foundation resists the new loads imposed by the retrofit scheme. This assumption is inaccurate. The most sophisticated approach is to model the soil under the new mat as a spring with stiffness that is different from that of the spring representing the existing piers. The analysis would show that new loads are supported by both the new mat foundation and the existing piers based on the relative stiffnesses of the two sets of foundation elements.

Shear transfer: It is standard practice to connect the new mat or cap or grade beam and existing piers with drilled dowels. Dowels in the piers and grade beams above should be designed to be sufficient to transfer the force intended to be resisted under the new arrangement.

Detailing and Construction Considerations

The best approach is to match the new mat and old pier cap and grade beam depths. This is, of course, not always beneficial especially if the soils that the caps or grade beams are bearing on have low bearing characteristics. Sometimes the new mat needs to be deeper than the existing grade beams or pier.

Cost/Disruption

Adding a new mat or cap or grade beam is quite disruptive and costly. The existing slab-on-grade must be sawcut and removed, then the foundation excavation completed, drilled dowels installed, rebar laid, debris in the excavation removed, and concrete placed. This is all time-consuming, messy and noisy.

23.8.3   Add Drilled Piers to an Existing Drilled Pier Foundation

Deficiency Addressed by Rehabilitation Technique

    Inadequate compression capacity of a drilled pier foundation
    Inadequate tension capacity of a drilled pier foundation
    Inadequate lateral capacity or ductility of a drilled pier foundation

Description of the Rehabilitation Technique

An existing drilled pier footing beneath a shear wall may lack sufficient compression capacity at the toe, tension capacity at the heel, or the existing pier reinforcing may be inadequate for lateral demands. Adding new, well detailed drilled piers provides supplemental capacity to reduce the demands on existing elements or increase the overall capacity and ductility. Figure 23.8.3-1 shows an example where the "web" or center of the footing is widened or replaced and new drilled piers are added.

Design Considerations

Research basis: No research specific to supplementing existing drilled pier footings has been identified.

Relative rigidity: In Figure 23.8.3-1, all of the piers—new and existing—will participate in resisting axial and lateral demands and should be considered in modeling efforts. Demands in existing piers should be confirmed as adequate.

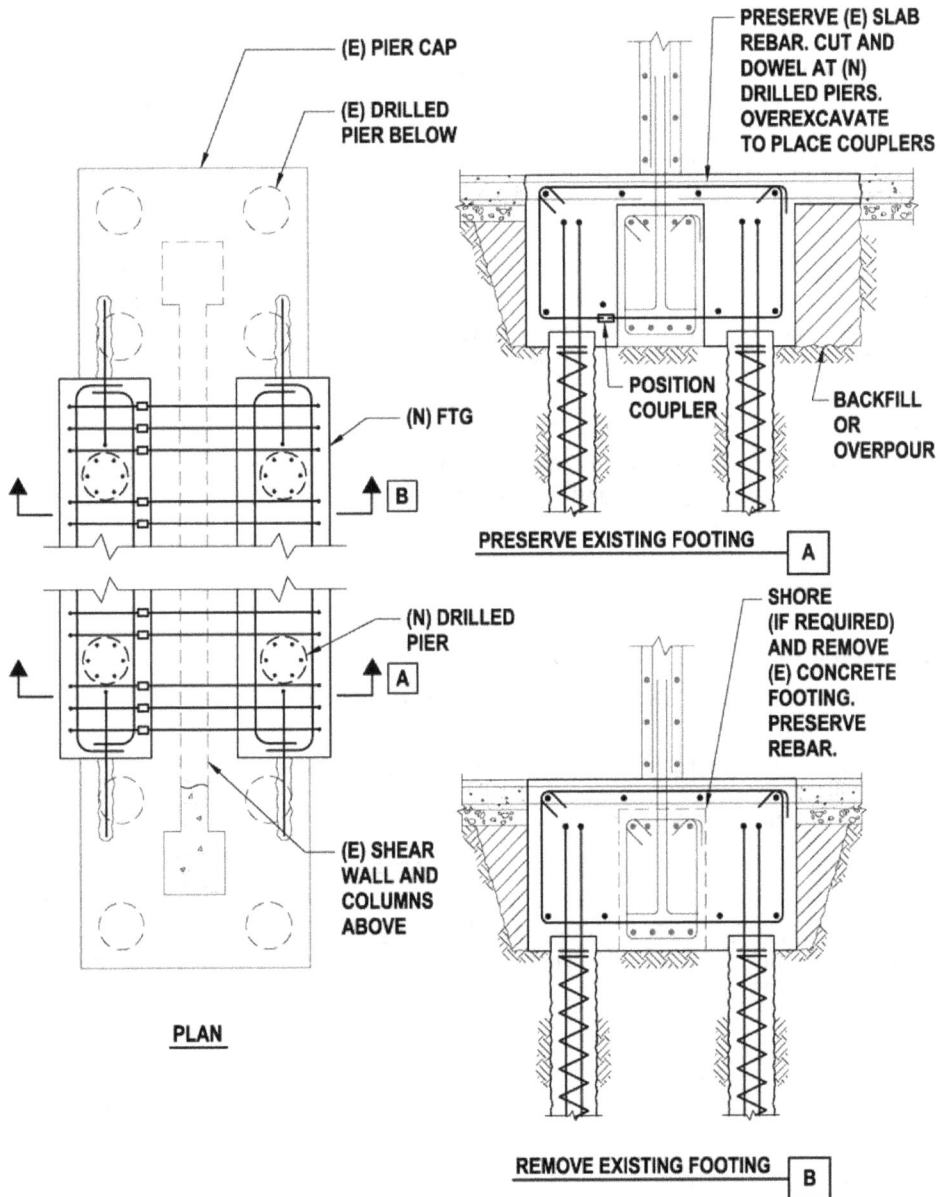

Figure 23.8.3.-1: Adding Drilled Piers to an Existing Drilled Pier Foundation

Detailing and Construction Considerations

Detailing and construction considerations for adding new drilled piers to an existing drilled pier footing include the following.

Access and height limitations: All of the drilling limitations noted in Section 23.8 for drilled piers apply here as well.

Spacing: Drilled piers typically have spacing limits of three times the pier diameter to avoid group effect reductions. This can limit the number of piers that can be installed.

Pier cap/thickened footing: The concrete above the piers will likely require widening as shown in Figure 23.8.3-1. Either drilled dowels can be installed in the existing footing, or the footing can be demolished and replaced. With a shear wall above, the wall may be able to bridge across to the belled ends of the footing without any shoring.

Cost/Disruption

Adding new drilled piers in an existing building is very disruptive, involving excavation, dust, mud, drilling/jackhammering noise and concrete placement. Protection of existing finishes in the vicinity and in the working path is necessary.

23.8.4   Add Micropiles to an Existing Drilled Pier Foundation

Deficiencies Addressed by Rehabilitation Technique

    Inadequate compression capacity of a drilled pier foundation
    Inadequate tension capacity of a drilled pier foundation
    Inadequate lateral capacity or ductility of a drilled pier foundation

Description of the Rehabilitation Technique

An existing isolated drilled pier footing supporting a braced frame column, moment frame column or concrete column under a discontinuous shear wall may lack sufficient compression capacity, tension capacity, or the existing pier reinforcing may be inadequate for lateral demands. Adding new micropiles provides supplemental capacity to reduce the demands on the existing drilled pier. Figure 23.8.4-1 shows an example.

Design Considerations

Research basis: No research specific to supplementing existing drilled pier footings with adjacent micropiles has been identified.

Relative rigidity: In Figure 23.8.4-1, both the new micropiles and existing drilled pier will participate in resisting axial and lateral demands and should be considered in modeling efforts. Demands in the existing piers should be confirmed as adequate. While the axial strength of the new micropiles may be comparable to the drilled pier, they will much lower lateral stiffness.

Detailing and Construction Considerations

Detailing and construction considerations for adding new micropiles to an existing drilled pier footing include the following.

MICROPILE (TYP.)

CORNER BARS IN
JACKET EXTEND
INTO FOOTING

CHIP (E) PIER TO
FACE OF (E) TIES
3" MAX.

₵ SYM.

Separation
between new
pile and existing
pier per project
soil report.

₵ COL., FTG. & SYM.

PLAN   A

USE CLASS "B" LAP SPLICE
AT CONTRACTOR'S OPTION

(E) COLUMN

CONCRETE COLUMN
See section on wrapping
a column.

CONC. SLAB

HEADED BARS
EACH SIDE TOP
AND BOTTOM

STIRRUPS
SEE A FOR LOCATION

A

Note:
Relative rigidity between
existing pier and new
micropile must be
considered.

BEARING
LAYER

SECTION   B

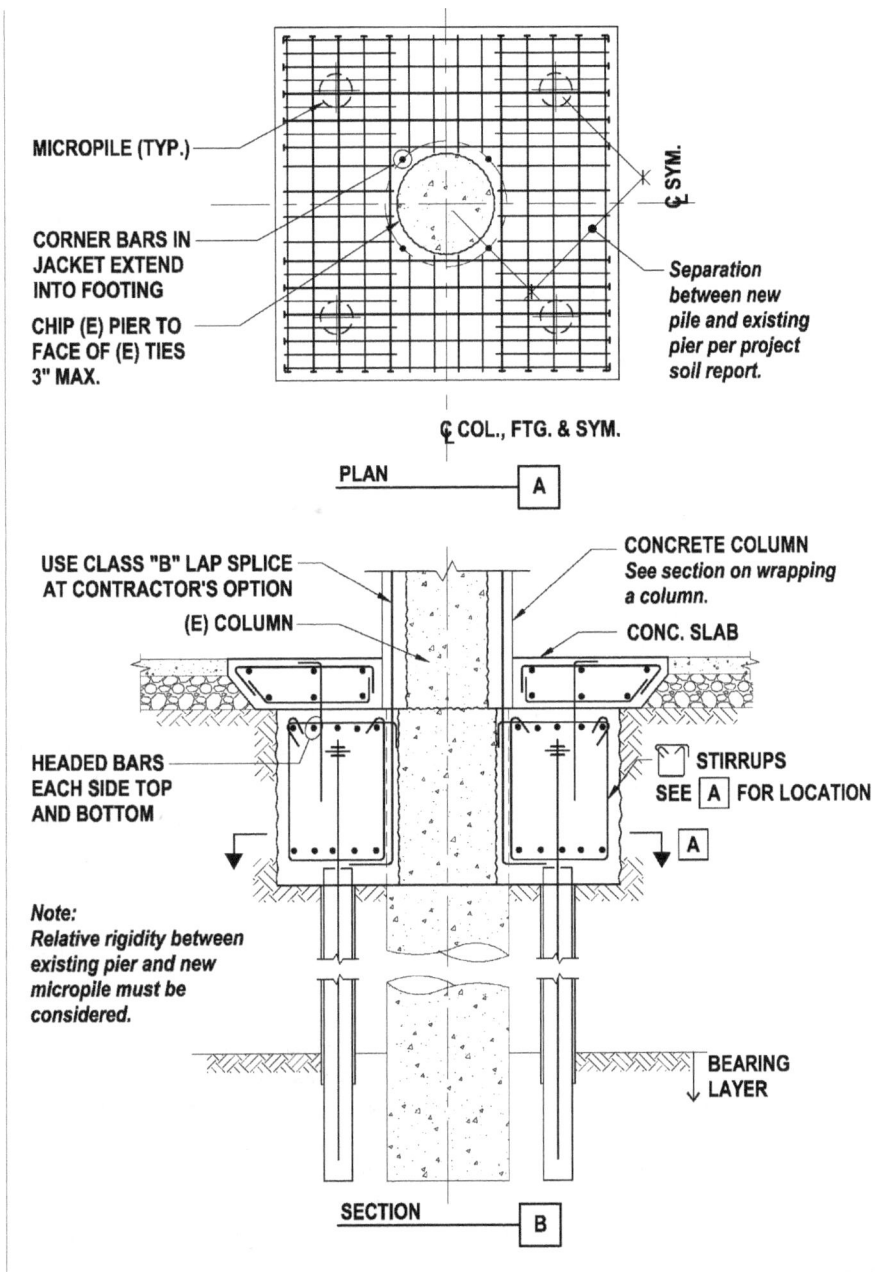

Figure 23.8.4.-1: Micropile Enhancement of an Existing Drilled Pier Footing

Access and height limitations: All of the drilling limitations noted in Section 23.7 for micropiles apply here as well.

Spacing: Spacing limits between the drilled pier and micropile to avoid group effects should be addressed.

Collar around drilled pier: In Figure 23.8.4-1, a concrete collar wraps the top of the drilled pier and provides the termination point for the pile anchors. It can also provide the starter bar location for the concrete jacket used to wrap the concrete column above.

Cost/Disruption

Adding new micropiles in an existing building is less disruptive than new drilled piers, but it still involves excavation, dust, mud, drilling/jackhammering noise and concrete placement. Protection of existing finishes in the vicinity and in the working path is necessary.

23.8.5   Add Top Bars to an Existing Pile Cap

Deficiency Addressed by Rehabilitation Technique

Inadequate bending capacity of the top of the pile cap to resist uplift forces

Description of the Rehabilitation Technique

An existing pile cap or pier cap may lack top reinforcing bars because the original design showed no net uplift. When a capacity design approach to evaluation or a pushover is conducted, it is likely that uplift will occur at some point and trigger the need for top bars in the pile cap to resist bending. Sometimes there is sufficient capacity in the reinforcing of the slab and its nominal connection to the top of the pile cap that it can serve the function of top steel. If not, top bars can be added as shown in Figure 23.8.5-1.

Design Considerations

Research basis: No research specific to adding top bars to existing pier or pile caps has been identified.

Anchorage of the pile to the pile cap: In Figure 23.8.5-1, it is assumed that the anchorage of the pile to the pile cap is adequate for uplift. If the foundation was not originally designed for uplift, only nominal anchorage between the pile and pile cap is likely to be found. This could be a single large bar or a bundle of two bars placed in a grouted hole in the top of the pile. It could be the pile reinforcing extended up into the pile cap. This would be more likely in end bearing piles where refusal is hit early and the top of the pile must be chipped down to the right elevation. It is important to realize that even a pile designed for a pin top with a central bar will resist moment unless special design considerations such as neoprene pads are added on top of the pile. This is highly unlikely in an older building. The tension in the pile anchorage under lateral loading has to be added to the uplift from the superstructure.

Anchorage of the column to the pile cap: In Figure 23.8.5-1, it is assumed that the anchorage of the column to the pile cap is adequate for uplift. If the foundation was not originally designed for uplift, anchor bolt embedments and diameters may not be sufficient.

Older braced frame
where lack of code
level tension led to
pile cap design only
for compression forces.

DRILL THROUGH
GUSSET AS REQUIRED

REMOVE (E) SLAB

(E) BOTTOM
BARS

INSTALL DRILLED
DOWEL "U" BARS
AS TOP BARS

Anchor bolts at
column and pile-to-cap
connection must be
adequate for uplift in
this approch.

SECTION

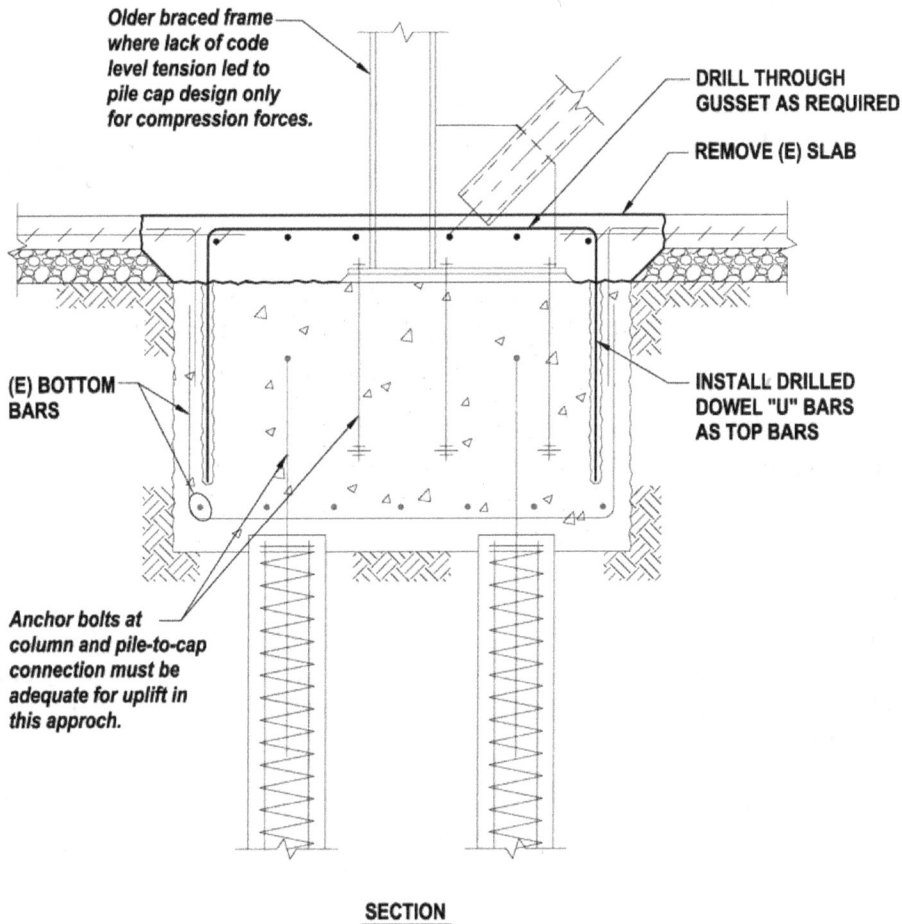

Figure 23.8.5-1: Adding Top Bars to an Existing Pile Cap

Pile cap modeling: The pile cap is typically fairly deep and may behave more as a deep beam. It can also be analyzed using strut-and-tie models.

Detailing and Construction Considerations

Detailing and construction considerations for adding top bars to an existing pile cap include the following.

Existing reinforcing: Existing reinforcing should be preserved in the footing. This will typically require placing new drilled dowels inboard of the edges of the existing footing. Added edge distance is also desirable if there or no or minimal side bars.

U bars or lapped L bars: Figure 23.8.5-1 shows new U bars, so that each leg of the U must be lowered simultaneously into the pile cap. L bars lapping over the top of the cap may make installation easier.

Confinement: The existing slab is likely to have wire mesh. To minimize vertical offsets the new and existing slabs should be dowelled together. Either the mesh in the existing slab-on-grade can be preserved when the slab is demolished or drilled dowels can be drilled into the edge of the existing slab.

Cost/Disruption

Adding top bars to the pile cap is a localized and less disruptive process than many foundation retrofits. It does involving dust, drilling/jackhammering noise and concrete placement. Protection of existing finishes in the vicinity is necessary.

23.9   Ground Improvement for Existing Shallow and Deep Foundations

23.9.1   Goals

Typical goals for ground improvement under existing shallow and deep foundations can be classified into two categories: mitigating the potential impacts of an identified geologic hazard, and enhancing the capacity of the foundation by changing the load-deformation characteristics of the foundation soil. The general techniques used to achieve these two goals separately or in combination include compaction grouting and permeation grouting. Warner (2004) is a good resource for both types of grouting.

Typical geological hazards that are mitigated using ground improvement include liquefaction and compaction settlement. These hazards have to be established by a Geotechnical Engineer who will define the recommended zone of geologic hazard mitigation.

23.9.2   Compaction Grouting

Description of the Rehabilitation Technique

Compaction grouting involves the injection of a very stiff grout at a high pressure into a layer of soil to force the individual soil particles into a tighter packing. The resulting increase in the density of the soil substantially increases its resistance to liquefaction as well as its bearing capacity. Compaction grouting can be performed in a wider range of soil types than other grouting methods. It can be performed in various types of sands, and clayey materials, but has limited effectiveness in clean coarse sands and gravels and in high plasticity soils.

The grout is required to have low flowability. This low flowability is necessary because the most important characteristic for effective densification is for the grout to form a controlled mass, which is columnar or tear-shaped, when injected. If it behaves instead like a fluid in the ground, it can create fractures in the soil, through which the grout can flow. Since the effectiveness of the grout is based on its ability to stay as a mass pushing soil particles together, that effectiveness is lost when the grout flows.

The grout—which consists of mostly of sand, cement, and water—is injected through grout holes that are drilled in a grid pattern of between 4 and 12 feet. Casing that typically has an internal diameter of 2 to 4 inches is usually installed in the grout holes. The injection pressure is directly proportional to the pumping rate, the optimal pumping rate being between 1 – 2 cubic feet per minute. Grout is usually injected in a strict primary-secondary pattern. Alternate primary holes are drilled and grouted first, followed by the secondary holes.

Grout is usually injected in stages. Staging involves the injection of only a few feet of grout hole at a time. Staging can proceed from top-down or bottom-up, the latter approach being the most commonly used.

The bottom-up grouting approach involves the following:

1. A hole is drilled to the bottom of the zone to be grouted.

2. Casing is installed to within a few feet of the bottom of the hole.

3. Grout is injected until refusal is reached. Refusal is assumed to have been reached if

    A slight movement of the ground surface or overlying improvement occurs.

    A predetermined amount of grout is injected.

    A given maximum pressure is reached at a given pumping rate.

4. The casing is raised one to two feet.

5. Grout injection is resumed until refusal is reached.

6. Steps 4 and 5 are repeated until the top of the grout zone is reached.

The level of densification achieved is verified by performing a cone penetration test or standard penetration test.

Specific Issues Relating to Grouting Under Shallow Foundations

Compaction grouting can be performed under shallow foundations in the manner described above except that the grout holes tend to be vertical rather than inclined. This is because inclined grout holes result in large horizontal areas that increase the likelihood of surface heave. The grout zone usually extends from the bottom of the existing shallow foundation to a dense or very stiff layer below the foundation.

Unlike compaction grouting performed in an open undeveloped area, the level of densification achieved in the soil below a shallow foundation cannot be verified using cone penetration or standard penetration tests. The level of densification is verified instead through monitoring the volume of grout injected in the holes.

To establish the relationship between the volume of grout and the level of densification, a pilot test program is performed in an open area adjacent to the existing building which will have compaction-grouted footings. The pilot test site is divided into segments where injection points at different spacings are laid out in a grid format as shown in Figure 23.9.2-1. In each hole,

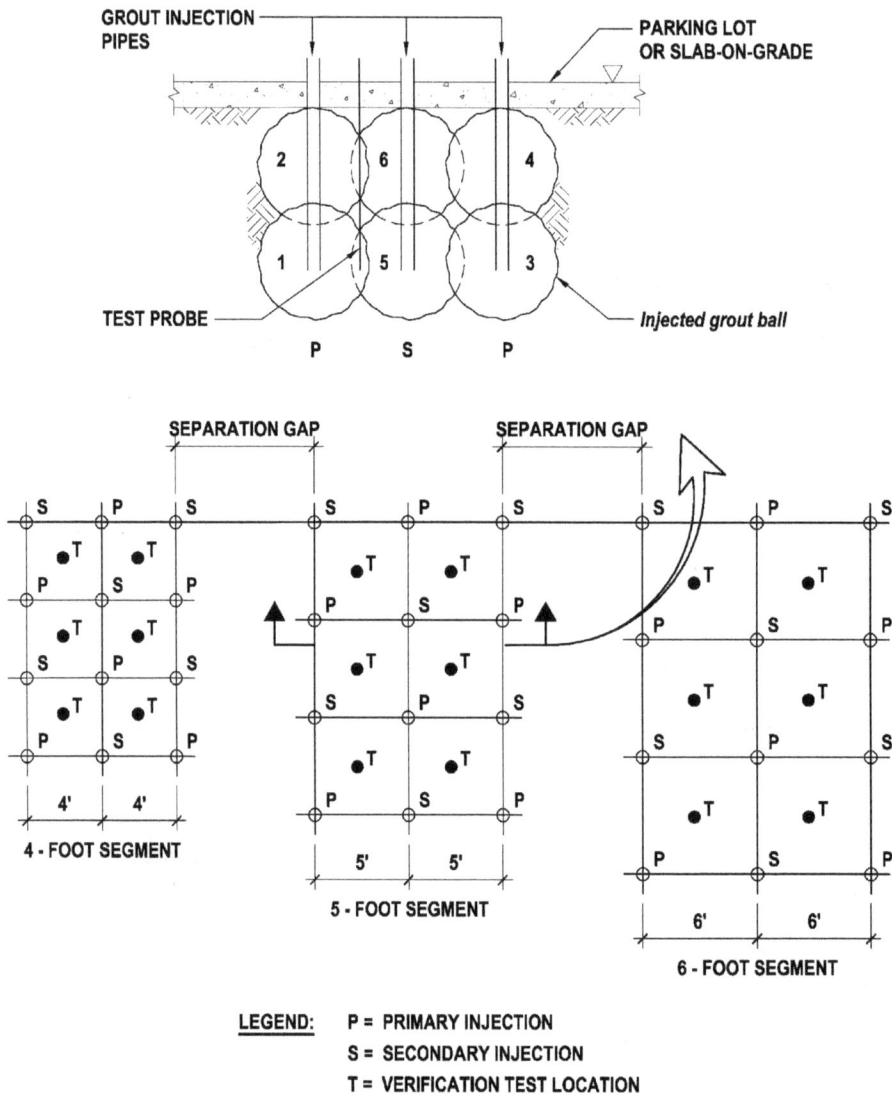

**Figure 23.9.2-1: Compaction Grouting Under Existing Shallow Foundations –
Pilot Test Program**

grout is injected within the upper and lower limits of the zone to be grouted. The volume of grout injected in each hole is recorded. After grout injection is completed, the level of densification achieved in each segment of the pilot test area is verified by performing cone penetration or standard penetration tests at the test locations as depicted in Figure 23.9.2-1. The spacing and the corresponding volume of grout injected in each hole that produced the acceptable level of compaction is selected for production grouting underneath the shallow foundations. The injection point grid pattern will be similar to the pattern in a segment in Figure 23.9.2-1 with the test location coinciding with the center of a square footing or the centerline of a continuous footing. Note that a separation gap equal to three times the minimum spacing of four feet is placed between the segments to minimize the impact of one segment on the other.

Specific Issues Relating to Grouting Under Deep Foundations

The goal of compaction grouting around deep foundations is to enhance the skin friction contribution from the soils surrounding the deep foundation element. As in the case of shallow foundations, compaction grouting around deep foundation elements is performed using vertical rather than inclined grout holes. The grouting zone extends from the top to the tip of the deep foundation element. The injection points are usually set up at least six feet away from the center of the deep foundation element. A pilot test program, similar to the one described above for shallow foundations is performed in an open area adjacent to the existing building, which will have compaction grouted deep foundation elements. See Figure 23.9.2-2. Verification tests are performed at the location marked "T/DF". The spacing and corresponding volume of grout injected in each hole that produced the acceptable level of compaction is selected for production grouting around the deep foundation elements. The injection point grid pattern for production grouting is set up similar to the pilot test program in Figure 23.9.2-2, except that the deep foundation element location will correspond to a location marked "T/DF."

Cost/Disruption

Compaction grouting is quite disruptive and costly especially if the creation of injection holes includes drilling through existing pile or pier caps, grade beams, or concrete footings and slab. This grouting process could also be time-consuming and messy. Disruption to the current operations of the building is usually minimized by performing the compaction grouting at night and cleaning up the work area before the start of work the next morning. Compaction grouting is generally less costly than permeation grouting for a given scope of work.

23.9.3   Permeation Grouting Under Existing Shallow and Deep Foundations

Description of the Rehabilitation Technique

Permeation grouting involves the injection of chemical or cement grout into the pore spaces of soils and aggregates without displacing the materials. This helps solidify the usually sandy soils that are amenable to this technique. The resulting increase in shear strength of the soil substantially increases its resistance to liquefaction as well as its bearing capacity. Permeation grouting can be performed in sands and sandy soils that contain minor amounts of fine particles. The structure and the size of voids in the soil structure dictate the type of grout that can be effectively used. In general, either micro-fine cement grout or a chemical grout. The use of chemical grouts has been diminishing for environmental reasons.

GROUT INJECTION PIPES

PARKING LOT
OR SLAB-ON-GRADE

TEST PROBE

Injected grout ball

*Foundation element
if this were not pilot
test program.*

SEPARATION GAP

SEPARATION GAP

4 - FOOT SEGMENT

5 - FOOT SEGMENT

6 - FOOT SEGMENT

LEGEND: P = PRIMARY INJECTION
           S = SECONDARY INJECTION
           T/DF = VERIFICATION TEST LOCATION

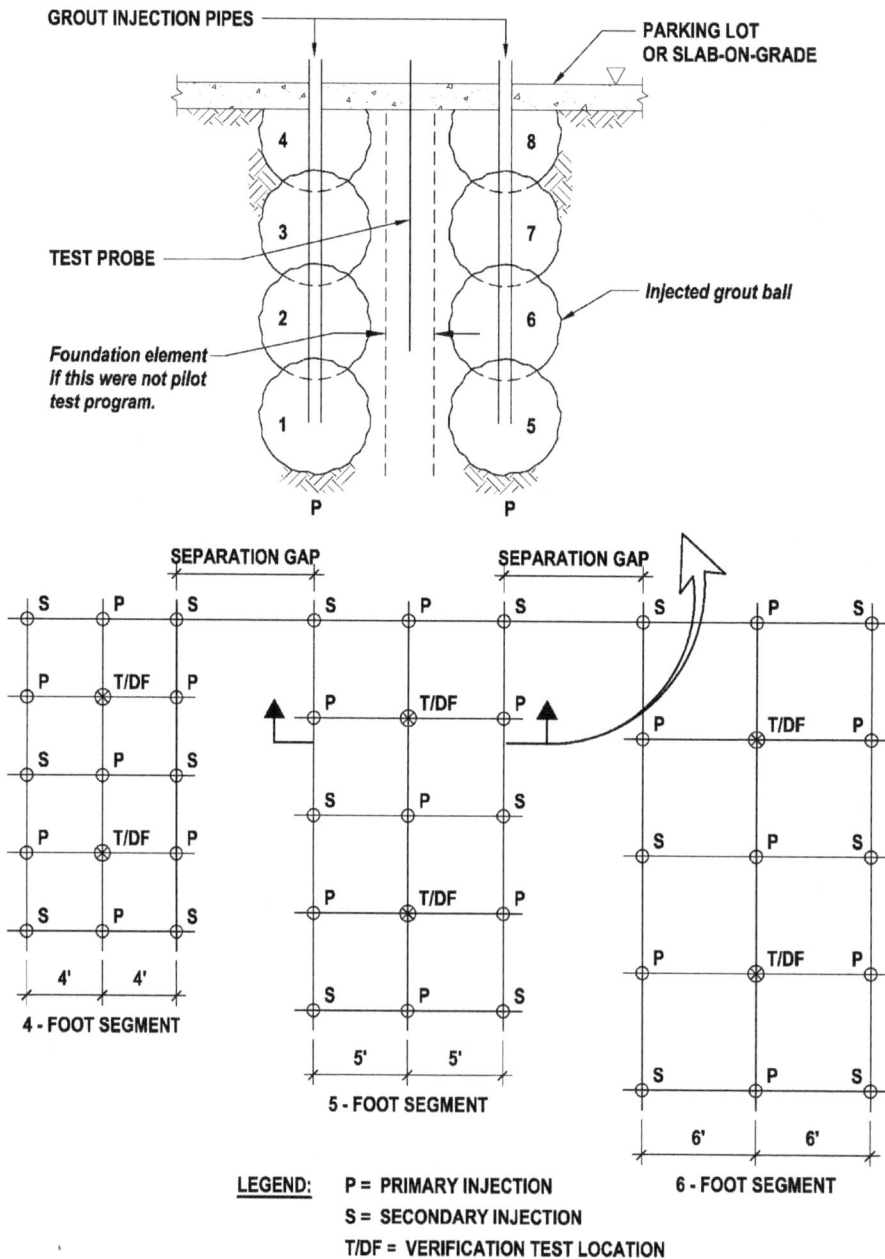

**Figure 23.9.2-2: Compaction Grouting Under Existing Deep Foundations –
Pilot Test Program**

The grout is injected through grout holes that are drilled in a grid pattern of between 2 and 6 feet. Casing that typically has an internal diameter of 2 to 4 inches is usually installed in the grout holes. Grout is usually injected in a strict primary-secondary pattern. Alternate primary holes are drilled and grouted first followed by the secondary holes. The level of solidification achieved is verified by exhuming grouted soil bulbs, taking samples of the grouted soil and performing unconfined compression tests on the samples.

Shallow foundations: The goal for the shallow foundation elements is to create a solidified mass of sandy soil below the footprint of the footing as a minimum. The solidified mass should extend from the bottom of the footing to the top of the dense sand layer as shown in Figures 23.9.3-1 and 23.9.3-2.

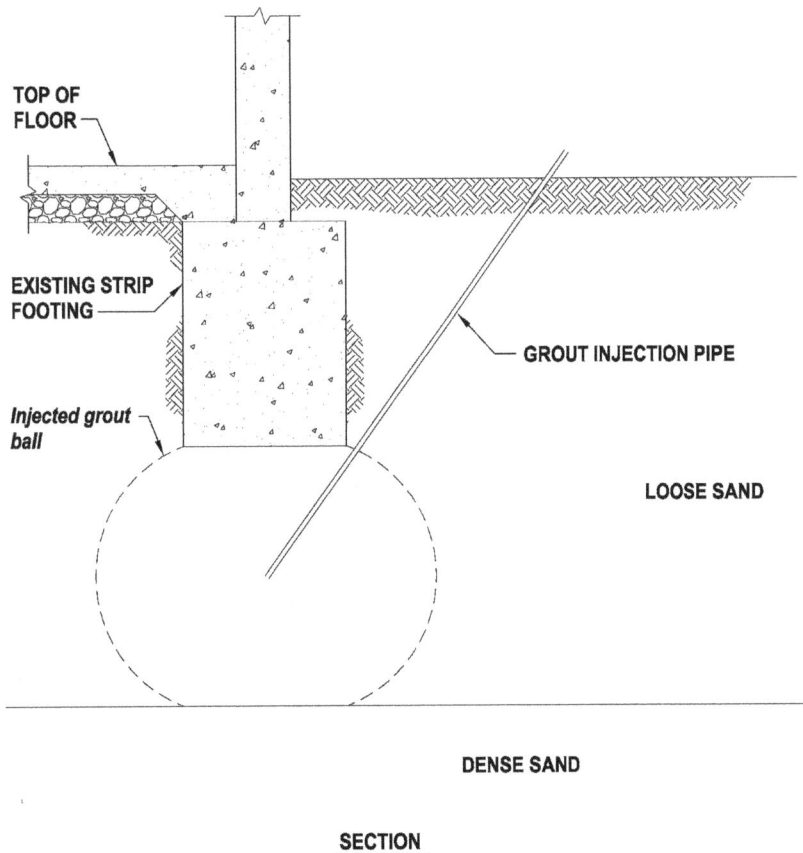

Figure 23.9.3-1: Permeation Grouting of Loose Sand Under Existing Shallow Foundation

Deep foundations: The goal for the deep foundation element is to create a zone of solidified sand around it. The injection points can be as close as three feet to the foundation elements. The zone of grouting should extend from the bottom of the grade beam or cap atop the deep foundation element to the top of the dense sand layer shown in Figure 23.9.3-3.

TOP OF
FLOOR

EXISTING STRIP
FOOTING

*Injected grout
ball*

LOOSE SAND
(DRY)

GROUT INJECTION PIPE

SATURATED LOOSE SAND

DENSE SAND

SECTION

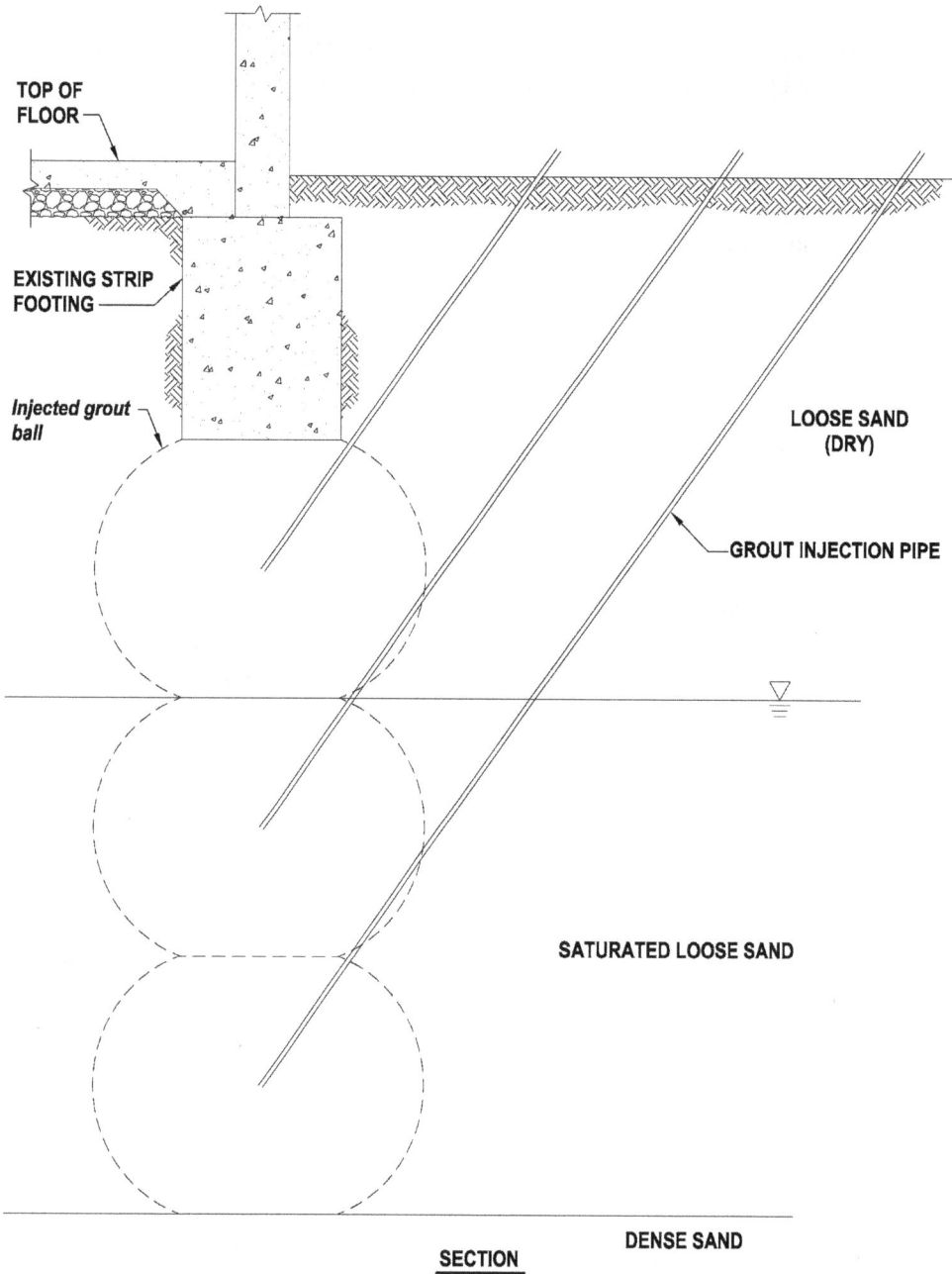

**Figure 23.9.3-2: Permeation Grouting of Liquefiable Layer
Under Existing Shallow Foundation**

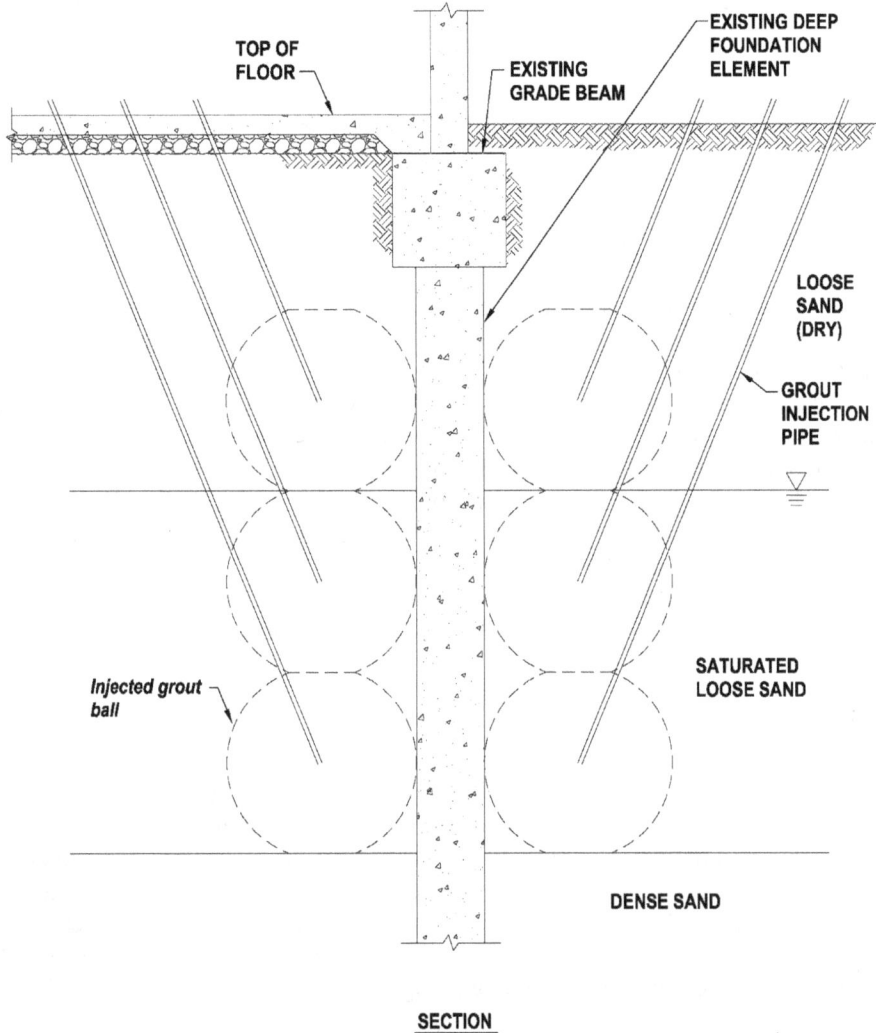

**Figure 23.9.3-3: Permeation Grouting of Liquefiable Layer
Around Existing Deep Foundation**

Cost/Disruption

Permeation grouting can be quite disruptive and costly especially if injection holes have to be drilled through existing concrete footings and slabs. If current operations in the building are to continue, the usual approach is to do the permeation grouting at night and clean up the work area before the start of work the next morning.

In the case of grouting under shallow foundations, there is a tendency for grout to migrate down, resulting in a weakly cemented lens of sand immediately below the shallow foundation elements. The tendency can be minimized by ensuring that grouting is performed in a strictly primary-secondary sequence.

23.10 Mitigating the Impacts of Other Ground Hazards on Existing Foundations

23.10.1   Issues to be Addressed

The mitigation measures described in the preceding sections of this chapter deal primarily with individual foundation elements in a building. Even if these mitigation measures are implemented, their usefulness can be negated by other ground hazards that tend to have global stability effects on the behavior of the entire foundation system and could lead to the collapse of the building to be rehabilitated. These other hazards must therefore be mitigated if they exist, for the intent of the mitigation methods described in the previous sections to be realized. This class of ground hazards includes fault rupture, lateral spreading, and seismic-induced landslide. The potential for these hazards, the level of severity, and the necessary mitigation measures to be implemented must be established by a geotechnical engineer and/or an engineering geologist. See FEMA 274 for additional information. Mitigating the potential impacts of these hazards can be very costly. Disruption to the existing building, however, should be minimal since the work is external to the building.

23.10.2   Fault Rupture

The potential for fault rupture exists when an engineering geologist establishes through fault trenching that an active fault trace traverses the footprint of the existing building that is to be rehabilitated as depicted in Figure 23.10.2-1. The rupture results in displacement along the fault trace, which depending on the type of fault, could be lateral or vertical movement. The magnitude of earthquake-induced displacement in the ground along the fault trace can range from a few inches to several feet. Such displacements can have the effect of tearing a building apart when they occur under or adjacent to a building.

It is difficult to upgrade a building straddling an active fault to accommodate such displacements without collapse. Options for mitigating the hazard include:

> Change the occupancy level from the current to a much lower level in an effort to minimize the potential for loss of life.
> Move the affected structure to a location at least 50 feet from the mapped fault trace, if feasible.

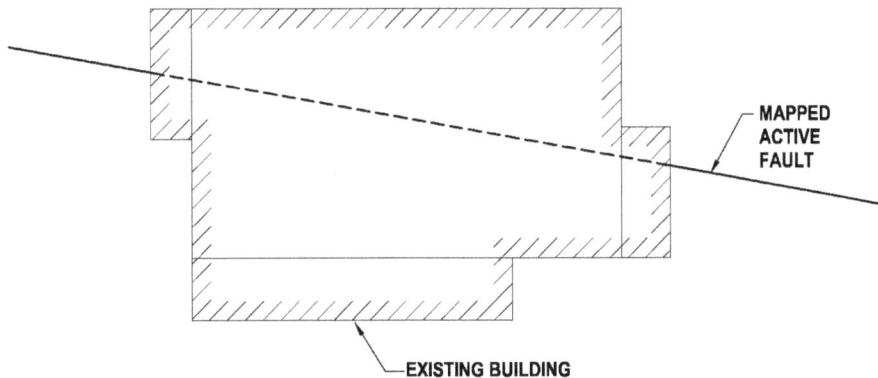

Figure 23.10.2-1: Active Fault Traversing the Footprint of Existing Building

23.10.3 Lateral Spreading

Definition of Hazard and Potential Impacts

Lateral spreading is one of the phenomena associated with liquefaction. It occurs when the blocks of non-liquefiable surface material above a layer of liquefiable soil move laterally towards an open face as depicted in Figure 23.10.3-1. The magnitude of lateral movement can range from a few inches to several feet.

Lateral spreading has the effect of globally moving the building laterally, or tearing it apart, if the building is supported on shallow foundations or on deep foundations that do not extend into stable material below the liquefiable layer. If the building is supported on deep foundations that extend into the stable material below the liquefiable layer, lateral spreading could result in loss of lateral capacity of the deep foundation elements.

Mitigation

The potential for lateral spreading can be mitigated by creating a stable mass of material near the open face. This can be accomplished by either densifying the layer of potentially liquefiable soil or solidifying the soil to prevent liquefaction. Techniques used to achieve this goal include compaction grouting (densifying) and permeation grouting (solidifying), which were described in Section 23.9. Alternately, vibrocompaction methods involving the installation of stone columns can be used to densify the potentially liquefiable layer as shown in Figure 23.10.3-2.

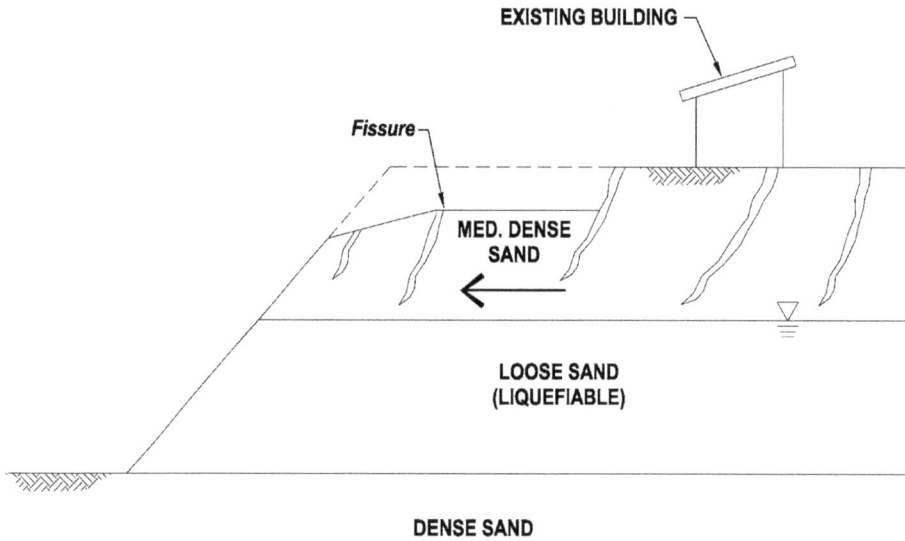

Figure 23.10.3-1: Lateral Spreading of Soil Layer Underlying Existing Building

23.10.4   Seismic-Induced Landslide

Definition of Hazard and Potential Impacts

Seismic-induced landslides result from the failure of an existing slope under earthquake loading. The landslide can occur under various sscenarios as shown in Figure 23.10.4-1. In Figure 23.10.10-1A, the landslide could undermine the building, causing it to collapse. In Figure 23.10.4-1B, the landslide displacement along the failure surface would be much greater than the displacement under the building at the ground surface. Extensional fissures, however, can occur under the building due to the displacement at depth. The geotechnical engineer and engineering geologist can estimate the magnitude of these fissures. In Figure 23.10.4-1C, the landslide could result in debris flow impact on the building.

Mitigation

Various mitigation schemes developed by the geotechnical engineer and engineering geologist can be implemented in each of the three scenarios. In the first scenario, potential mitigation measures include the construction of a stabilizing berm at the toe of the slope as shown in Figure 23.10.4-2A or the construction of a soil nail wall as shown in Figure 23.10.4-2B. In the second scenario, the existing foundation system can be enhanced to span or accommodate the estimated extensional fissures. In the third scenario, a debris wall, as shown in Figure 23.10.4-2C, can be built to protect the building or techniques in Figure 23.10.4-2B can be used to stabilize the slope.

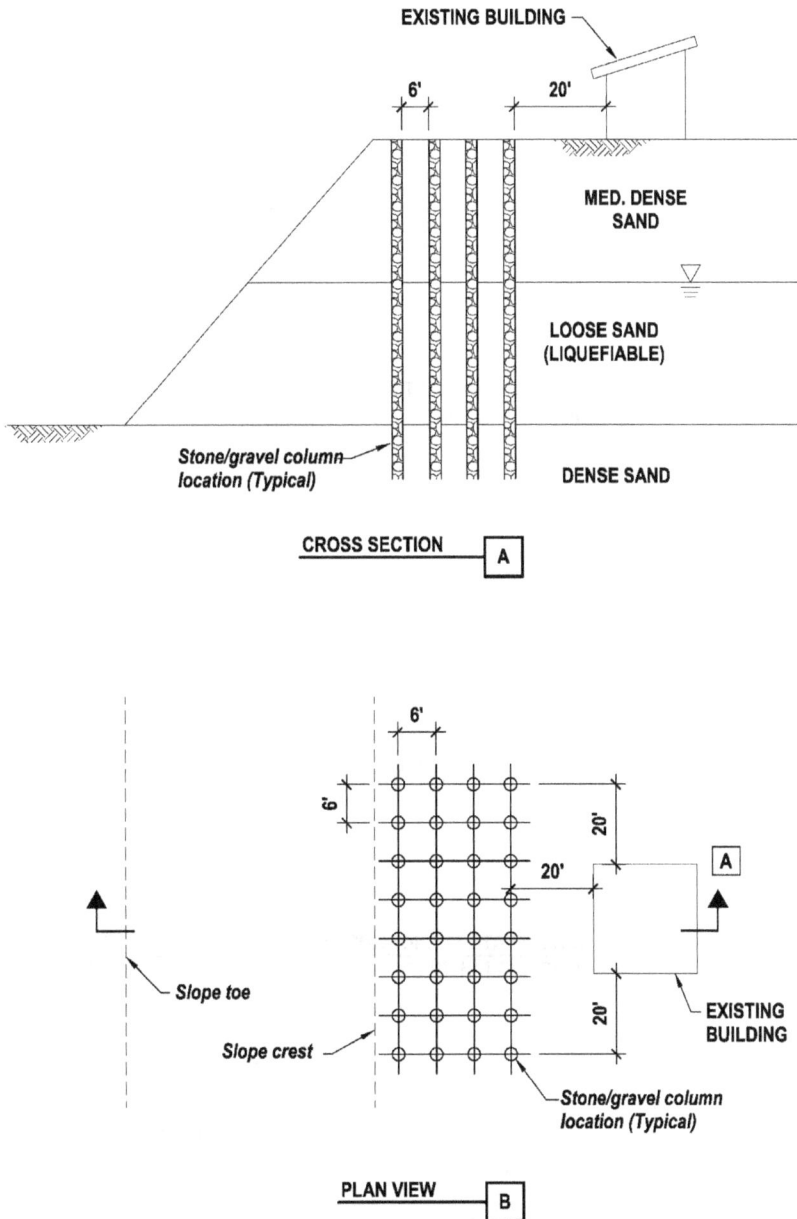

Figure 23.10.3-2: Mitigation Measure for Lateral Spreading – Stone/Gravel Column

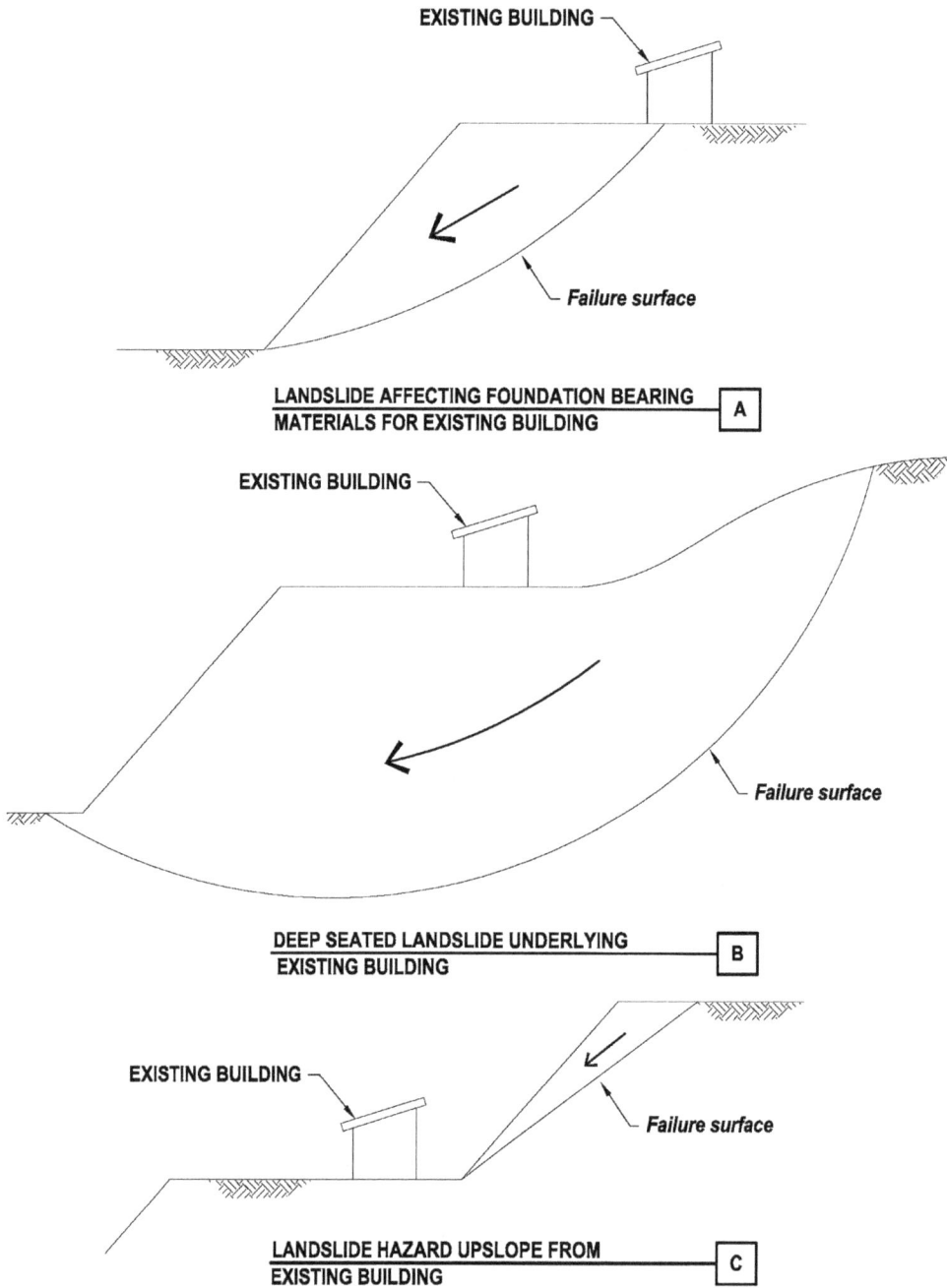

Figure 23.10.4-1: Landslide Hazards

EXISTING BUILDING

Berm

MITIGATION OF LANDSLIDE HAZARD WITH
TOE BEAM — A

EXISTING BUILDING

Backfill

Soil nail
wall

Original slope

MITIGATION OF LANDSLIDE HAZARD WITH
SOIL NAIL WALL — B

Debris wall

EXISTING BUILDING

MITIGATION OF LANDSLIDE HAZARD WITH
DEBRIS WALL — C

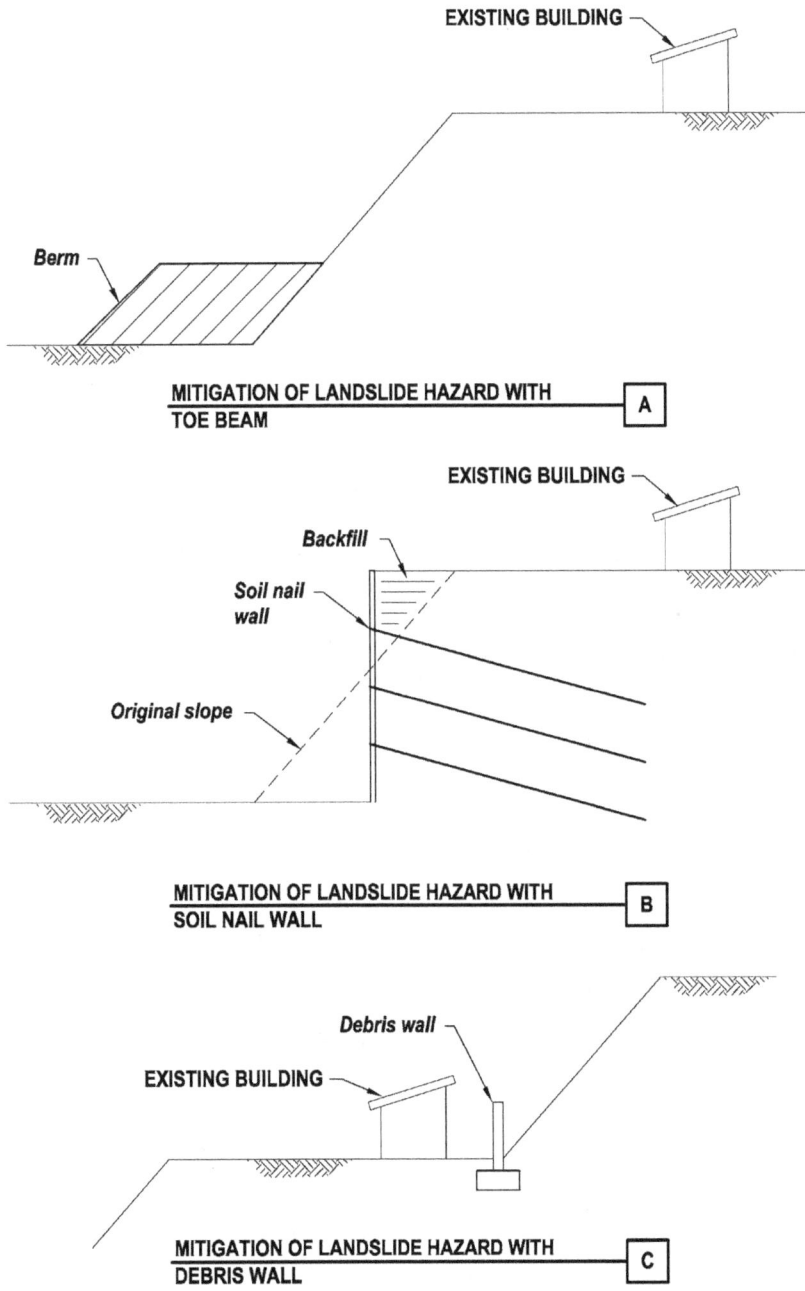

Figure 23.10.4-2: Mitigation of Landslide Hazards

23.11 References

ASCE, 2003, *Standard for the Seismic Evaluation of Buildings*, ASCE 31-03, Structural Engineering Institute of the American Society of Structural Engineers, Reston, VA.

ASTM D1143, 1994, *Standard Test Method for Piles Under Static Axial Compressive Load*, ASTM.

ASTM D1194, 1994, *Standard Test for Bearing Capacity of Soil for Static Load and Spread Footings*, ASTM.

ASTM D3689, 1995, *Standard Test Method for Piles Under Static Axial Tensile Load*, ASTM.

ATC, 1996, *The Seismic Evaluation and Retrofit of Concrete Buildings,* ATC-40, Applied Technology Council, Redwood City, CA.

Bowles, J.E., 1996, *Foundation Analysis and Design*, McGraw-Hill, New York, NY.

FEMA, 1997, *NEHRP Commentary on the Guidelines for Seismic Rehabilitation of Buildings*, FEMA 274, Federal Emergency Management Agency, Washington, D.C., November.

FEMA, 2000, *Prestandard and Commentary for the Seismic Rehabilitation of* Buildings, FEMA 356, Federal Emergency Management Agency, Washington, D.C., November.

FEMA, 2005, *Improvement of Nonlinear Static Seismic Procedures*, FEMA 440, Federal Emergency Management Agency, Washington, D.C.

FHWA, 2000, *Micropile Design and Construction Guidelines Implementation Manual*, Report No. FHWA-SA-97-070., June.

PTI, 1996, *Post-Tensioning Institute Recommendations for Prestressed Rock and Soil Anchors*, Post-Tensioning Institute (PTI), Phoenix, AR.

Warner, J., 2004, *Practical Handbook of Grouting*, John Wiley & Sons, Hoboken, NJ.

Chapter 24 - Reducing Seismic Demand

24.1   Overview

Most seismic rehabilitation projects utilize rehabilitation strategies involving adding strength, stiffness, ductility, and/or improvement load path details. Another approach, less commonly employed, is to reduce the seismic demand on the structure. This chapter covers three methods of reducing seismic demand on the structure: reducing the effective seismic weight, seismic isolation, and passive damping.

24.2   Reduction of Seismic Weight

Reduction of seismic weight may reduce the seismic demand on an existing structure in certain cases; however, the engineer must carefully evaluate the dynamic effects of such an approach before adopting it as part of a retrofit scheme. Techniques may include replacing heavy cladding with a curtain wall system, removing high permanent live loads, or removing upper stories. Since removing upper stories results in a loss of usable floor area, this approach is usually considered after an owner has built a new adjacent building that provides replacement space.

While the reduction of seismic weight may potentially improve performance by changing a structure's yielding sequence, reducing story drifts, reducing global overturning, or reducing base shear, these reductions in demand, particularly base shear, may not be directly proportional to the decrease in seismic weight. For example, the removal of a building's upper stories will typically shorten the structure's fundamental period of vibration, often leading to an increased spectral acceleration. If this increase in spectral acceleration is greater than the corresponding decrease in seismic weight, the demand base shear on the structure will increase. Buildings with periods in the velocity-sensitive region of the response spectrum should be evaluated for this effect early in the development of a rehabilitation strategy. In particular, tall buildings in which much of the seismic weight is concentrated in the lower stories are likely to have very limited benefits in base shear reduction associated with the removal of upper stories.

The following examples in Figures 24.2-1, 24.2-2 and 24.2-3 illustrate how the removal of a building's upper stories may not significantly decrease the calculated base shear demand. All three examples assume a concrete moment frame system and employ ASCE 7-05 (ASCE, 2005) base shear equations. The first examines a generic model building with uniform story heights and masses, using the ASCE 7-05 approximate period calculation. The second examines a similar model building, in which additional weight is concentrated in the lower stories. The third examines a real building, using fundamental periods calculated from computer analysis. In all three examples, the removal of upper stories decreases global overturning demands but does not significantly decrease base shear demands.

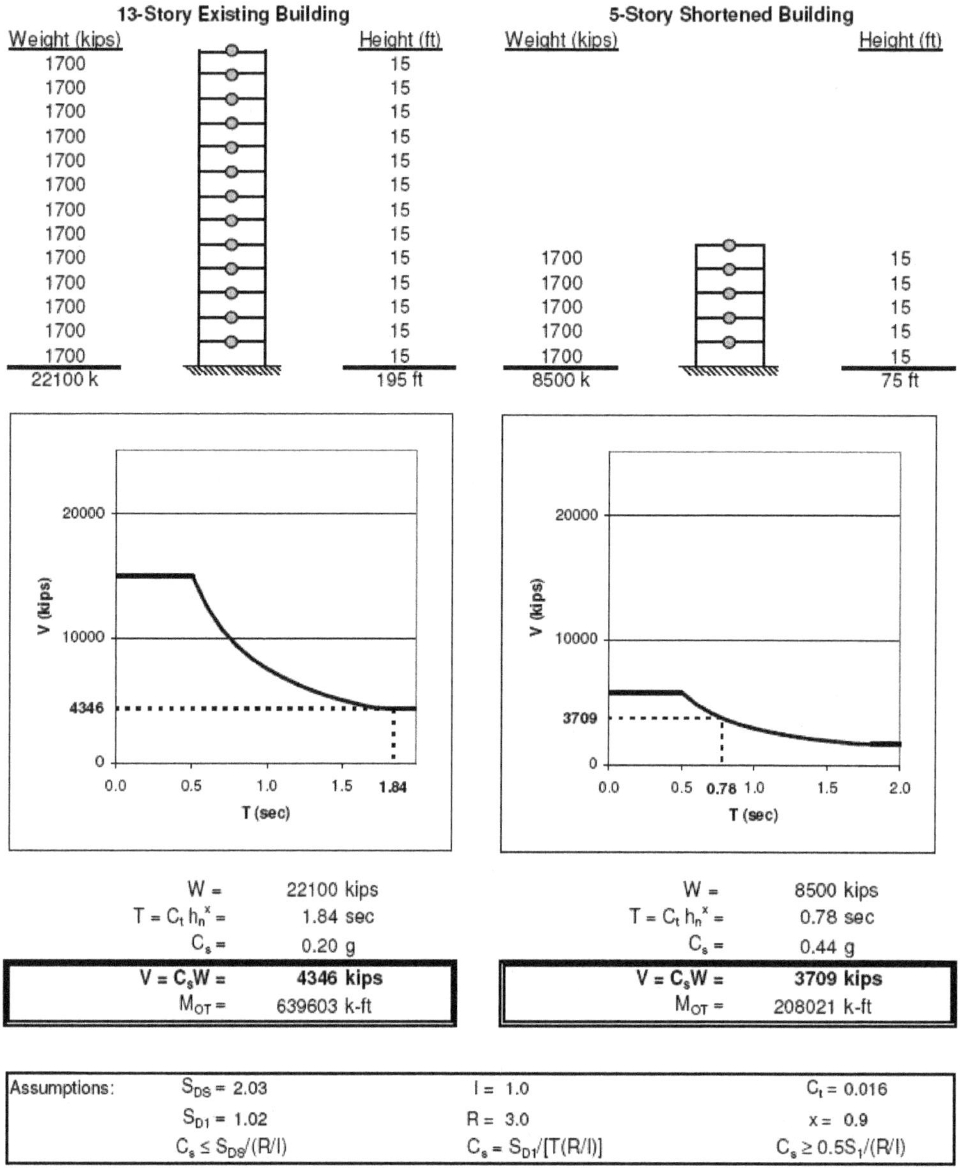

13-Story Existing Building

Weight (kips)		Height (ft)
1700		15
1700		15
1700		15
1700		15
1700		15
1700		15
1700		15
1700		15
1700		15
1700		15
1700		15
1700		15
1700		15
22100 k		195 ft

5-Story Shortened Building

Weight (kips)		Height (ft)
1700		15
1700		15
1700		15
1700		15
1700		15
8500 k		75 ft

13-Story Existing Building:

$$W = \quad 22100 \text{ kips}$$
$$T = C_t h_n^x = \quad 1.84 \text{ sec}$$
$$C_s = \quad 0.20 \text{ g}$$
$$V = C_s W = \quad \mathbf{4346 \text{ kips}}$$
$$M_{OT} = \quad 639603 \text{ k-ft}$$

5-Story Shortened Building:

$$W = \quad 8500 \text{ kips}$$
$$T = C_t h_n^x = \quad 0.78 \text{ sec}$$
$$C_s = \quad 0.44 \text{ g}$$
$$V = C_s W = \quad \mathbf{3709 \text{ kips}}$$
$$M_{OT} = \quad 208021 \text{ k-ft}$$

Assumptions:	$S_{DS} = 2.03$	$I = 1.0$	$C_t = 0.016$
	$S_{D1} = 1.02$	$R = 3.0$	$x = 0.9$
	$C_s \leq S_{DS}/(R/I)$	$C_s = S_{D1}/[T(R/I)]$	$C_s \geq 0.5S_1/(R/I)$

Figure 24.2-1: Generic Building Example of Decreasing Base Shear with Decreasing Height

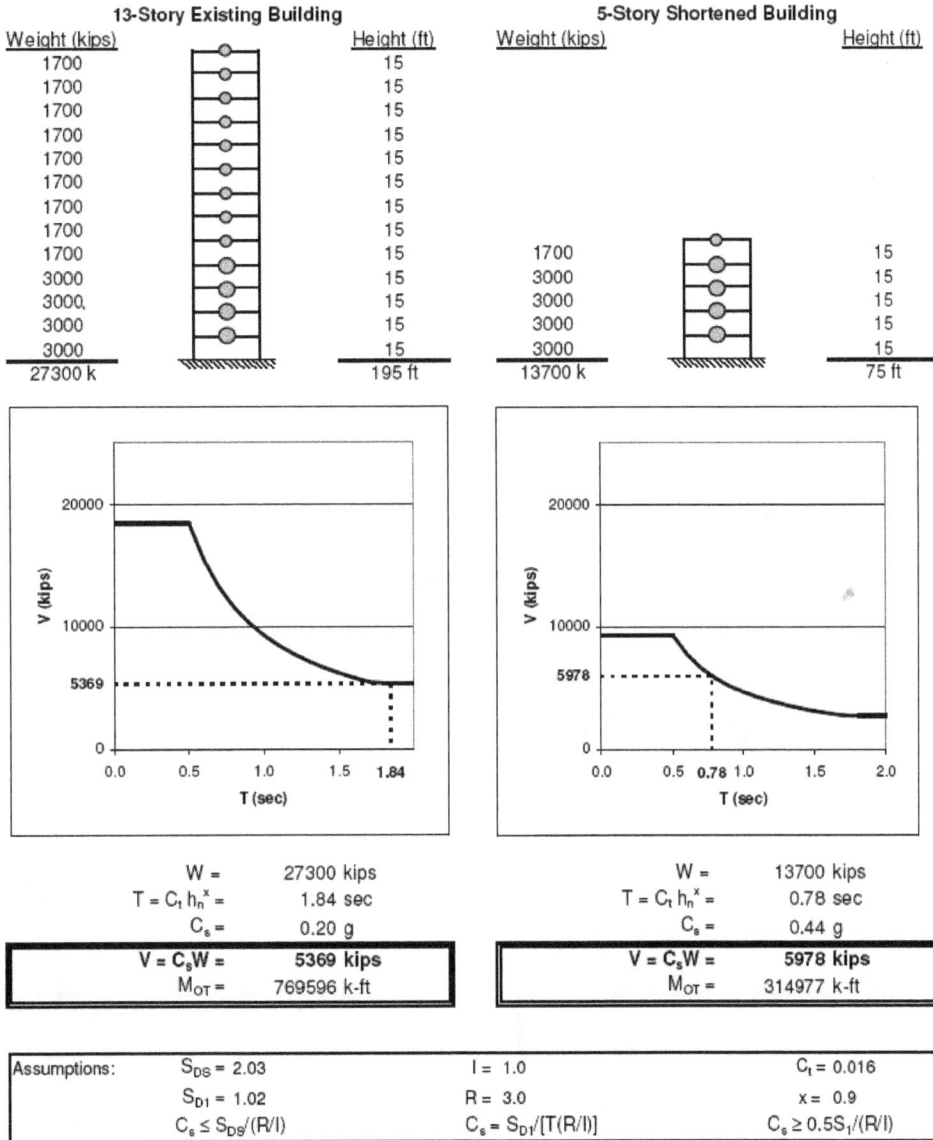

Figure 24.2-2: Generic Building Example of Increasing Base Shear with Decreasing Height

13-Story Existing Building

Weight (kips)	Height (ft)
1621	17
1779	13.33
1685	13.33
1685	13.33
1688	13.33
1688	13.33
1695	13.33
1695	13.33
1709	13.33
2480	26.5
3436	15.5
2551	15.5
3236	19.75
26948 k	**201 ft**

5-Story Shortened Building

Weight (kips)	Height (ft)
1709	13.33
2480	26.5
3436	15.5
2551	15.5
3236	19.75
13412 k	**91 ft**

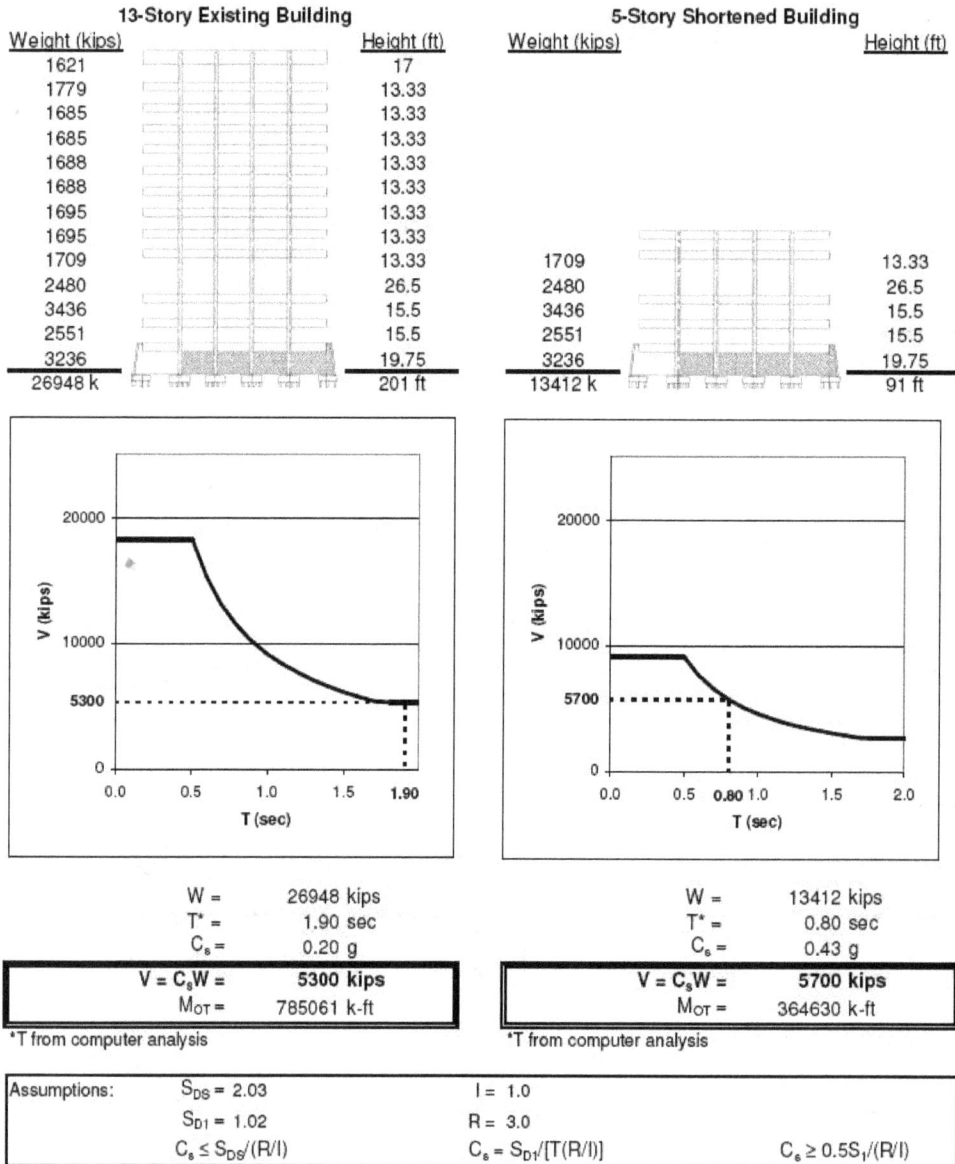

W =	26948 kips
T^* =	1.90 sec
C_s =	0.20 g
$V = C_s W$ =	**5300 kips**
M_{OT} =	785061 k-ft

*T from computer analysis

W =	13412 kips
T^* =	0.80 sec
C_s =	0.43 g
$V = C_s W$ =	**5700 kips**
M_{OT} =	364630 k-ft

*T from computer analysis

Assumptions:	S_{DS} = 2.03	I = 1.0	
	S_{D1} = 1.02	R = 3.0	
	$C_s \leq S_{DS}/(R/I)$	$C_s = S_{D1}/[T(R/I)]$	$C_s \geq 0.5S_1/(R/I)$

Figure 24.2-3: Real Building Example of Increasing Base Shear with Decreasing Height

24.3   Seismic Isolation

Seismic or base isolation involves lengthening a building's fundamental period of vibration to reduce the seismic demand transmitted from the ground to the building. It has been more commonly used in new building design, but it has also been employed in the United States for several high profile existing buildings as the key strategy in the rehabilitation design.

Types of isolation components: Isolation components include elastomeric bearings and sliding bearings. Elastomeric bearings include high damped rubber, low damped rubber, and low damped rubber with lead cores. Sliding bearings include the friction pendulum system. Dampers are often part of the isolation system to limit displacements. See Section 24.4 for some discussion on dampers.

Applicable buildings: The period range for isolated buildings is from about 2 seconds to 4 seconds. As such, buildings on very soft soils and very tall, flexible buildings may not achieve much benefit from isolation. Seismic isolation is usually a very expensive rehabilitation strategy and has been primarily applied in the United States to important historic structures, usually as a way of minimizing the amount of superstructure strengthening an impact on the historic fabric of the building. Isolation displacements are highly site specific, but in high seismic zones the Maximum Capable Earthquake (MCE) displacements are often on the order of 30" or more. Sufficient clearance from adjacent structures is necessary to avoid pounding during seismic response

System elements: In addition to the bearings and dampers, a complete isolation system will require a number of other special elements, including a moat around the building to accommodate the displacements. The moat has to go down past the plane of isolation. There is usually a complete or partial moat cover at the top of the moat for aesthetic or security considerations. Elevators are typically hung from the superstructure, as they cannot cross the isolation plane without special detailing. Utilities entering the building need to be able to accommodate the isolation displacements; this often triggers special vaults outside the building or areas under the building for joint details. A foundation is needed below the isolators to take the forces they impart, and a structural system is needed above the isolators as well to deliver forces to the isolators and resist the moments that are induced. All of these elements add to the cost of isolating the building.

Analysis and design requires special expertise: The analytical and design effort for an isolated building is typically much more extensive than in fixed base rehabilitations. Time history analysis, where at least the isolators are modeled nonlinearly, is standard practice. Material properties must be achievable with components in the marketplace and must account for material variability from manufacturing, loading, temperature, velocity, wear, aging, and other effects. Experience with this type of work and the properties of the various vendor's components and associated issues is quite useful.

Determining the plane of isolation: Selecting the plane of isolation is a critical design choice. It is usually near the base of the building, though there are examples of isolation elements placed at the top of columns under heavy roofs to limit the forces in the columns. Isolation at the base is either done either in the basement level, leading to some loss of use of the basement or under the basement, leading to additional excavation to place the isolators, the foundation below them and the superstructure assembly above them. The elements directly above and below the isolators are designed to take the Design Basis Earthquake (DBE) motions, without force reductions, which is a comparatively severe demand. The different types of isolation components have different sizes and means of transferring moments. With rubber bearings, P-delta moments are assumed to be split with half of the moment going up and half going down. With the traditional friction pendulum system, all of the P-delta moment goes up or down depending on which way the dish is oriented. How moments are resisted can lead to the selection of specific types of isolators.

Reducing tension: Rubber bearings are much less stiff and have much less strength to resist tension forces. Lead cores in low damped rubber bearings also have limited capacity for resisting tension. Many engineers have concerns about sliding bearings under tension, though sliding bearings that resist some tension have recently come into the market. As a result, isolation layout and superstructure design is often aimed at minimizing tension in bearings.

Transferring load from the existing building foundations to the new bearings: When a new isolated building is built, the columns and remaining elements of the superstructure can be erected directly on top of the isolation bearings. In an existing building, the superstructure is already in place. A key issue in design is developing details that delineate and facilitate the load transfer process of shoring the existing building, cutting the base of columns free, installing new foundations and new horizontal structure above the isolator, installing the isolators, and transferring load to the isolators without damaging movements of the superstructure.

Proprietary/bidding considerations: Detailing for rubber and sliding bearings is quite different. If multiple vendors are necessary, vendors of different types of systems are usually considered. Often they are procured in an early package, due to the long lead time. This also permits the design engineer to move into final design knowing which type of system will be used.

24.4   Energy Dissipation

Adding damping to an existing structure, like seismic isolation, is a relatively unusual seismic rehabilitation strategy. The added damping reduces overall building displacement and acceleration response, and local interstory drifts; but it can impart additional localized forces that must be addressed.

Types of damping components: FEMA 356 provides guidance for displacement-dependent devices or velocity-dependent devices. Displacement-dependent devices include devices that exhibit rigid-plastic (friction devices), bilinear (metallic yielding devices) or trilinear hysteresis. Velocity-dependent devices include solid and fluid viscoelastic devices and fluid viscous devices. There are other devices as well, including shape-memory alloys, friction-spring assemblies with recentering capability, and fluid restoring force-damping devices.

Applicable buildings: Most engineers believe that adding damping is most relevant in flexible buildings, such as steel or concrete moment frames. Damping is also a common element in the seismic isolation system, but there it must accommodate very large displacements. See Section 24.3.

System elements: Figures 24.4-1 and 24.4-2 show examples of adding damping devices in an existing steel moment frame building to minimize drifts and demands on the beam-column joints. Other dampers, such as wall dampers, are possible but not shown. The damper must be connected to the existing structure and potentially the foundation. Installing dampers is similar to installing braced frames. See Chapter 8 for detailed discussion on adding braced frames to a steel building and Chapter 12 for adding a braced frame to a concrete building. Some damper devices and orientations require out-of-plane bracing for stability, such as those shown in Figure 24.4-2.

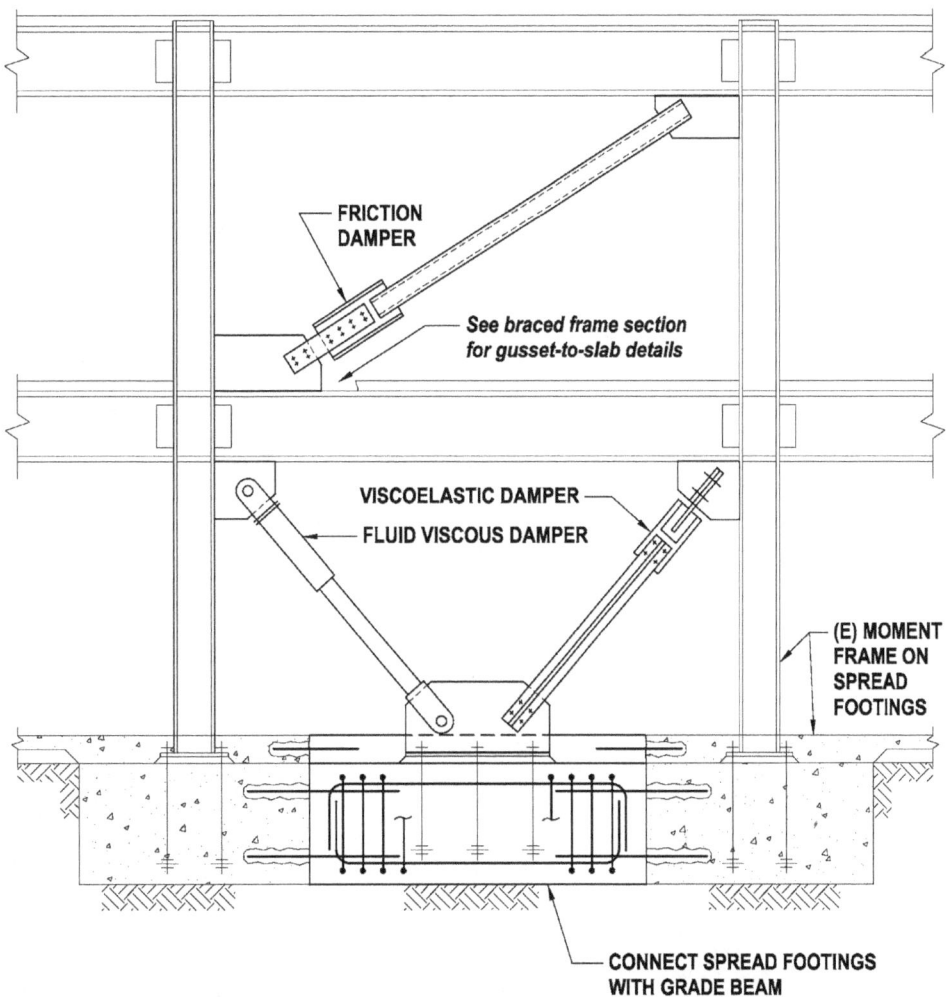

Figure 24.4-1: Damper Alternatives for Rehabilitating an Existing Moment Frame

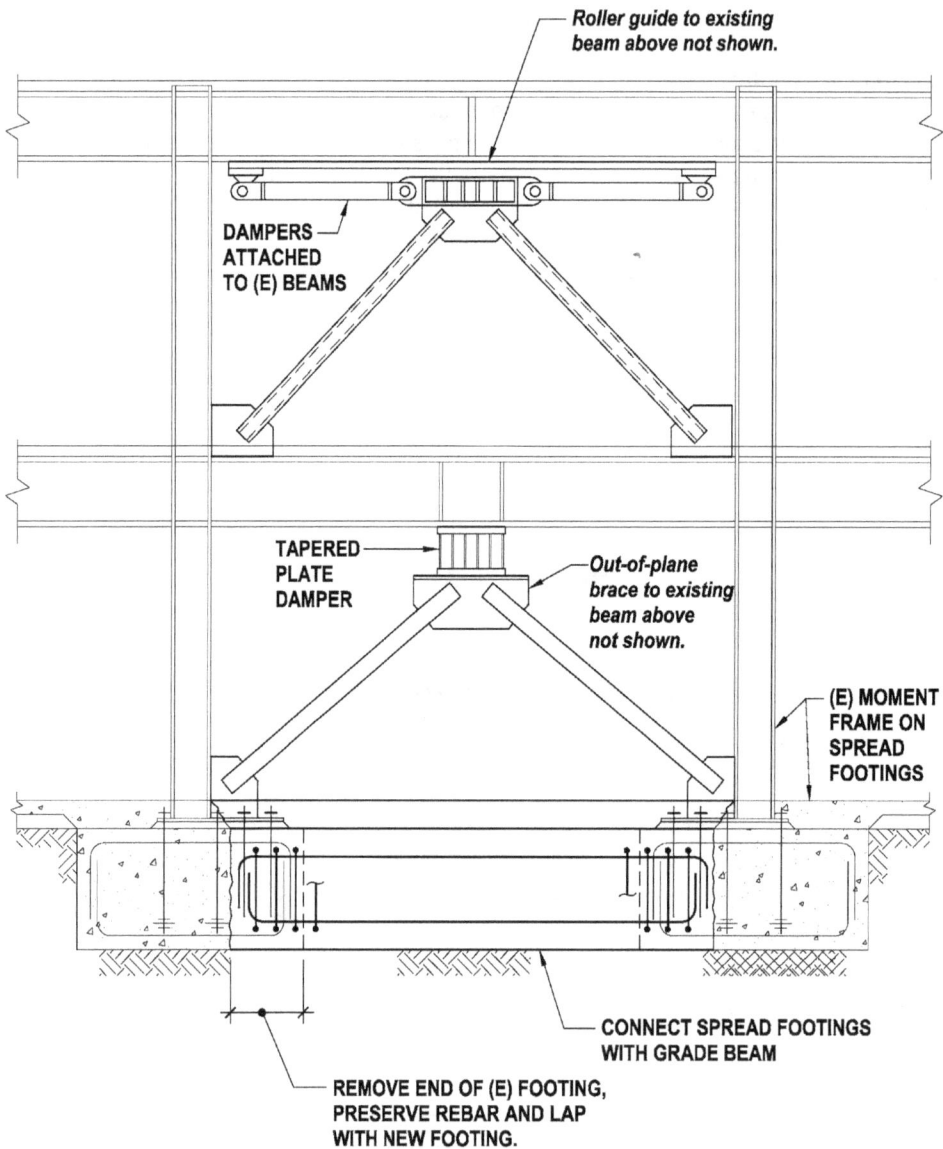

**Figure 24.4-2: Additional Damper Alternatives for Rehabilitating an Existing
Moment Frame**

Analysis can require special expertise: The analytical and design effort for a rehabilitation
design involving damping can be more extensive than in fixed base rehabilitations. Time history
analysis, where at least the dampers are modeled nonlinearly, is common. Material properties
must be achievable with components in the marketplace and must account for material variability
from manufacturing, temperature, velocity, wear, aging, and other effects. Experience with this

type of work and the properties of the various vendor's components and associated issues is useful. Hanson and Soong (2001) is a comprehensive monograph that covers analysis of buildings with supplemental energy dissipation devices, and it includes several design examples of seismic rehabilitation using damping devices.

Aesthetic impact: Adding dampers looks very similar to adding a braced frame, with the resulting visual and programmatic impacts. Some dampers or their connections can be particularly visually obtrusive.

Checking the existing structure: When dampers are added to the structure, the loads they impart locally must be considered in the design.

Proprietary issues: Most dampers available on the market are proprietary. Material properties, testing histories, limitations and detailing considerations are obtained from the manufacturer. Like seismic isolation components, the particular category of damper such as a fluid viscous damper or a friction damper is usually selected early in the design because the analysis and detailing can be significantly different between categories. There is also a patent regarding certain techniques for connecting bracing and dampers to beams when sliding is employed.

24.5 References

ASCE, 2005, *Minimum Design Loads for Buildings and Other Structures*, ASCE 7-05, American Society of Civil Engineers, Reston, VA.

Hanson, R.D. and T.T. Soong, 2001, *Seismic Design with Supplemental Energy Dissipation Devices*, Monograph MNO-8, Earthquake Engineering Research Institute, Oakland, CA.

Glossary

BLOCKED DIAPHRAGM: A diaphragm in which all sheathing edges not occurring on framing members are supported on and connected to blocking.

BOUNDARY ELEMENT: An element at the edge of an opening or at the perimeter of a shear wall or diaphragm.

BOUNDARY NAILING: Nailing at the perimeter edge of a wood diaphragm to framing members and blocking below.

BRACED FRAME: An essentially vertical truss, or its equivalent, of the concentric or eccentric type that is provided in a building frame or dual system to resist lateral forces.

CHEVRON BRACING: Bracing where a pair of braces, located either both above or both below a beam, terminates at a single point within the clear beam span.

CHORD: See DIAPHRAGM CHORD.

COLLECTOR: A member or element provided to transfer lateral forces from a portion of a structure to vertical elements of the lateral-force-resisting system (also called a drag strut).

CONCENTRICALLY BRACED FRAME (CBF): A braced frame in which the members are subjected primarily to axial forces.

CONTINUITY PLATES: Steel column stiffeners at the top and bottom of the panel zone. They are also known as transverse stiffeners.

CONTINUITY TIES: Structural members and connections that provide a load path between diaphragm chords to distribute out-of-plane wall loads.

COUPLING BEAM: A structural element connecting adjacent shear walls.

DAMPING: The internal energy absorption characteristic of a structural system that acts to attenuate induced free vibration.

DEMAND: The prescribed design forces required to be resisted by a structural element, subsystem, or system.

DIAPHRAGM: A horizontal, or nearly horizontal, system designed to transmit lateral forces to the vertical elements of the lateral-force-resisting system. The term "diaphragm" includes horizontal bracing systems.

DIAPHRAGM CHORD: The boundary element of a diaphragm or shear wall that is assumed to take axial tension or compression.

DIAPHRAGM STRUT: The element of a diaphragm parallel to the applied load that collects and transfers diaphragm shear to vertical-resisting elements or distributes loads within the diaphragm. Such members may take axial tension or compression. Also refers to drag strut, tie, or collector.

DOUBLER PLATE: A steel plate added to a panel zone to increase panel zone strength.

DRAG STRUT: See COLLECTOR.

DRIFT: See STORY DRIFT.

DUCTILITY: The ability of a structure or element to dissipate energy inelastically when displaced beyond its elastic limit without a significant loss in load-carrying capacity.

ECCENTRICALLY BRACED FRAME (EBF): A diagonal braced frame in which at least one end of each brace frames into a beam a short distance from a beam-column joint or from another diagonal brace.

EDGE NAILING: Nailing at the perimeter edge of a wood structural panel in a shear wall or diaphragm to framing members and blocking.

FIELD NAILING: Nailing within the interior of a wood structural panel in a shear wall or diaphragm to framing members.

FUNDAMENTAL PERIOD OF VIBRATION: The time it takes the predominant mode of a structure to move back and forth when vibrating freely.

HORIZONTAL BRACING SYSTEM: A horizontal truss system that serves the same function as a diaphragm.

K-BRACING: Bracing where a pair of braces located on one side of a column terminates at a single point within the clear column height.

LAMINATED VENEER LUMBER: An engineered wood product created by layering dried and graded wood veneers with waterproof adhesive into blocks of material. It is also known as structural composite lumber.

LATERAL FORCE-RESISTING SYSTEM: That part of the structural system assigned to resist lateral forces.

LINK BEAM: That part or segment of a beam in an eccentrically braced frame that is designed to yield in shear and/or bending so that buckling or tension failure of the diagonal brace is prevented.

MOMENT RESISTING SPACE FRAME: A structural system with an essentially complete space frame providing support for vertical loads.

NOTCH TOUGHNESS: A measure of material ductility related to the ability to resist fracture. It is typically measured with Charpy V-notch (CVN) test standards.

PANEL ZONE: Area of the beam-to-column connection delineated by beam and column flanges.

REDUNDANCY: A measure of the number of alternate load paths that exist for primary structural elements and/or connections such that if one element or connection fails, the capacity of alternate elements or connections are available to satisfactorily resist the demand loads.

RE-ENTRANT CORNER: A corner on the exterior of a building that is directed inward such as the inside corner of an L-shaped building.

SHEAR WALL: A wall, bearing or nonbearing, designed to resist lateral forces acting in the plane of the wall.

SHOTCRETE: Concrete that is pneumatically placed on vertical or near vertical surfaces typically with a minimum use of forms.

SOFT STORY: A story in which the lateral stiffness is less than 70 percent of the stiffness of the story above.

SOIL-STRUCTURE RESONANCE: The coincidence of the natural period of a structure with a dominant frequency in the ground motion.

STORY DRIFT: The displacement of one level relative to the level above or below.

STRUCTURE: An assemblage of framing members designed to support gravity loads and resist lateral forces. Structures may be categorized as building structures or nonbuilding structures.

SUBDIAPHRAGM: A portion of a larger wood diaphragm designed to anchor and transfer local forces to primary diaphragm struts and the main diaphagm.

SUBSYSTEMS: One of the following three principle lateral-force-resisting systems in a building: vertical resisting elements, diaphragms, and foundations.

SUPPLEMENTAL ELEMENT: A new member added to an existing lateral-force-resisting subsystem that shares in resisting lateral loads with existing members of that subsystem.

TIE-DOWN: A prefabricated steel element consisting of a tension rod, end brackets and bolts or lags used to transfer tension across wood connections. It is also known as a hold-down.

V-BRACING: Chevron bracing that intersects a beam from above. Inverted V-bracing is that form of chevron bracing that intersects a beam from below.

VERTICAL-RESISTING ELEMENTS: That part of the structural system located in a vertical or near vertical plane that resists lateral loads (typically a moment frame, shear wall, or braced frame).

WEAK STORY: A story in which the lateral strength is less than 80 percent of that in the story above.

WOOD STRUCTURAL PANEL: A wood-based panel product that satisfies the requirements of Voluntary National Product Standard PS-1 or PS-2 and is bonded with waterproof adhesive. Included under this designation are plywood, oriented strand board (OSB) and composite panels.

X-BRACING: Bracing where a pair of diagonal braces crosses near mid-length of the bracing members.

Abbreviations

The following abbreviations are commonly used by structural engineers and have been used in figures and/or text throughout the document.

B.N.   Boundary nailing

BRBF   Buckling-restrained braced frame

CIP   Cast-in-place

C.J.   Construction joint

CJP   Complete joint penetration weld

CL   Centerline

CP   Complete penetration weld

(E)   Existing

EA.   Each

E.N.   Edge nailing

F.N.   Field nailing

FTG.   Footing

MC   Moisture content

M/E/P   Mechanical/electrical/plumbing

(N)   New

PL   Plate

PP   Partial penetration weld

RBS   Reduced beam section (in a beam-to-column moment frame connection)

SPSW   Steel plate shear wall

TYP.   Typical

WSMF   Welded steel moment frame

Workshop Participants

A workshop was held on September 20-21, 2005 in Oakland, California to solicit comment on a draft of the final document. Workshop participants included the following individuals.

ICSSC Members

Krishna Banga
Department of Veteran Affairs
Washington, D.C.

Richard Kahler
U.S. Navy
Norfolk, VA

Cathleen Carlisle
FEMA
Washington, D.C.

H.S. Lew
NIST
Gaithersburg, MD

James A. Caulder
Air Force Civil Engineer Support Agency
Tyndall AFB, FL

Dai H. Oh
U.S. Department of State
McLean, VA

James Farasatpour
Federal Aviation Administration
Inglewood, CA

Raymond F. Schuler
NASA Ames Research Center
Moffett Field, CA

Jack Hayes
U.S. Army
Champaign, IL

Academics and Industry

Daniel Abrams
University of Illinois
Urbana, IL

Andrew W. Taylor
KPFF
Seattle, WA

Melvyn Green
Melvyn Green & Associates
Torrance, CA Richard Howe
Memphis, TN

James Parker
Simpson, Gumpertz & Heger
Waltham, MA

Barry H. Welliver
Draper, UT

Lawrence Reaveley
University of Utah
Salt Lake City, UT

Project Review Panel

Daniel Dolan   Jim Harris
Terry Dooley   Bela I. Palfalvi
Kurt Gustafson   Daniel Shapiro
Robert D. Hanson
Neil Hawkins

Technical Update Team and Contributing Staff

Kelly Cobeen   Bret Lizundia
William T. Holmes   James Malley
Jack Hsueh   Karl Telleen

www.ingramcontent.com/pod-product-compliance
Lightning Source LLC
Chambersburg PA
CBHW060946210326
41598CB00031B/4739